Chrysler 300 Dodge Charger, Magnum & Challenger

Automotive Repair Manual

by Joe L. Hamilton and John H Haynes

Member of the Guild of Motoring Writers

Models covered:

Chrysler 300 (2005 through 2018)
Dodge Charger (2006 through 2018)
Dodge Magnum (2005 through 2008)
Dodge Challenger (2008 through 2018)

Does not include information specific to diesel engine, all-wheel drive models or Hellcat/Demon models

Haynes Group Limited *(25027-10X9)*
Sparkford Nr Yeovil
Somerset BA22 7JJ England

ABCDE
FGHIJ
KLMNO
PQRST
2

Haynes North America, Inc.
2801 Townsgate Road, Suite 340
Thousand Oaks, CA 91361 USA

www.haynes.com

Acknowledgements

Technical writers who contributed to this project include John Wegmann, Rob Maddox, Mike Stubblefield, Jeff Killingsworth, Demian Hurst and Scott "Gonzo" Weaver. Wiring diagrams provided exclusively for Haynes North America, Inc by Solution Builders and Haynes Pro BV.

A book in the Haynes Automotive Repair Manual Series

Printed in Malaysia

ISBN-13: 978-1-62092-335-1
ISBN-10: 1-62092-335-1

Library of Congress Control Number: 2018962332

While every attempt is made to ensure that the information in this manual is correct, no liability can be accepted by the authors or publishers for loss, damage or injury caused by any errors in, or omissions from, the information given.

Contents

Haynes mechanic and photographer with a Dodge Challenger

About this manual

Its purpose

The purpose of this manual is to help you get the best value from your vehicle. It can do so in several ways. It can help you decide what work must be done, even if you choose to have it done by a dealer service department or a repair shop; it provides information and procedures for routine maintenance and servicing; and it offers diagnostic and repair procedures to follow when trouble occurs.

We hope you use the manual to tackle the work yourself. For many simpler jobs, doing it yourself may be quicker than arranging an appointment to get the vehicle into a shop and making the trips to leave it and pick it up. More importantly, a lot of money can be saved by avoiding the expense the shop must pass on to you to cover its labor and overhead costs. An added benefit is the sense of satisfaction and accomplishment that you feel after doing the job yourself.

Using the manual

The manual is divided into Chapters. Each Chapter is divided into numbered Sections, which are headed in bold type between horizontal lines. Each Section consists of consecutively numbered paragraphs.

The reference numbers used in illustration captions pinpoint the pertinent Section and the Step within that Section. That is, illustration 3.2 means the illustration refers to Section 3 and Step (or paragraph) 2 within that Section.

Procedures, once described in the text, are not normally repeated. When it's necessary to refer to another Chapter, the reference will be given as Chapter and Section number. Cross references given without use of the word "Chapter" apply to Sections and/or paragraphs in the same Chapter. For example, "see Section 8" means in the same Chapter.

References to the left or right side of the vehicle assume you are sitting in the driver's seat, facing forward.

Even though we have prepared this manual with extreme care, neither the publisher nor the author can accept responsibility for any errors in, or omissions from, the information given.

NOTE

A **Note** provides information necessary to properly complete a procedure or information which will make the procedure easier to understand.

CAUTION

A **Caution** provides a special procedure or special steps which must be taken while completing the procedure where the Caution is found. Not heeding a Caution can result in damage to the assembly being worked on.

WARNING

A **Warning** provides a special procedure or special steps which must be taken while completing the procedure where the Warning is found. Not heeding a Warning can result in personal injury.

Introduction

Chrysler 300 and Dodge Charger models are available in a four-door sedan body style, Dodge Challenger models are available in a two-door coupe body style, while Dodge Magnum models are available in a four-door wagon body style.

Available engines across the model line include a 2.7L DOHC V6, 3.5L SOHC V6, 3.6L DOHC V6 or 5.7L V8, 6.1L V8 or a 6.4L V8.

Power from the engine is transferred through a four-, five- or eight-speed automatic transmission or a six-speed manual transmission, then through a driveshaft and a rear differential which drives the rear wheels through a pair of driveaxles.

Suspension is a fully independent, long-arm/short arm front suspension system with upper and lower control arms, tension links and shock absorber/coil spring assemblies. A stabilizer bar controls body roll. Each steering knuckle is positioned by a series of balljoints in the ends of the upper and lower control arms and the tension links. The rear suspension consists of an independent multi-link suspension with shock absorbers, coil springs and several specialized links. Some models incorporate a stabilizer bar to control body roll.

The steering system consists of a rack-and-pinion steering gear and two adjustable tie-rods. Power assist is standard.

The brakes are disc at the front and disc at the rear, with power assist standard.

Vehicle identification numbers

Modifications are a continuing and unpublicized process in vehicle manufacturing. Since spare parts lists and manuals are compiled on a numerical basis, the individual vehicle numbers are necessary to correctly identify the component required.

Vehicle Identification Number (VIN)

This very important identification number is stamped on a plate attached to the dashboard inside the windshield on the driver's side of the vehicle (see illustration). The VIN also appears on the Vehicle Certificate of Title and Registration. It contains information such as where and when the vehicle was manufactured, the model year and the body style.

VIN engine year code

Two particularly important pieces of information found in the VIN are the engine code and the model year code. Counting from the left, the engine code letter designation is the 8th digit and the model year code letter designation is the 10th digit.

On the models covered by this manual the engine codes are:

R	2.7L DOHC V6 - 2005 through 2008
T	2.7L DOHC V6 (Magnum) - 2005 through 2008
D	2.7L DOHC V6 - 2009 and 2010

G	3.5L SOHC V6 - 2005 through 2008
V	3.5L SOHC V6 (Magnum) - 2006 and 2010
G	3.6L V6 DOHC VVT V6 - 2011 through 2018
H	5.7L Hemi V8 - 2005 through 2008
2	5.7L Hemi V8 (Magnum) - 2006 through 2008
T	5.7L Hemi V8 - 2009 through 2018
W	6.1L Hemi V8 - 2005 through 2010
J	6.4L Hemi V8 - 2012 through 2018

VIN model year code

On the models covered by this manual the engine codes are:

5	2005
6	2006
7	2007
8	2008
9	2009
A	2010
B	2011
C	2012
D	2013
E	2014
F	2015
G	2016
H	2017
J	2018

Vehicle Certification Label

The Vehicle Certification Label is attached to the driver's side door (see illustration). Information on this label includes the name of the manufacturer, the month and year of production, as well as information on the options with which it is equipped. This label is especially useful for matching the color and type of paint for repair work.

Engine identification number

Labels containing the engine code, engine number and build date can be found on the valve cover. The engine number is also stamped onto a machined pad on the external surface of the engine block.

Automatic transmission identification number

The automatic transmission ID number is affixed to a label on the left side of the case (see illustration).

Vehicle Emissions Control Information label

This label is found in the engine compartment. See Chapter 6 for more information on this label.

The VIN number is visible through the windshield on the driver's side

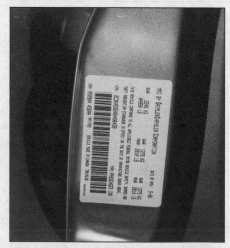

The vehicle certification label is affixed to the driver's side door

The automatic transmission identification tag is located on the left side of the transmission

Recall information

Vehicle recalls are carried out by the manufacturer in the rare event of a possible safety-related defect. The vehicle's registered owner is contacted at the address on file at the Department of Motor Vehicles and given the details of the recall. Remedial work is carried out free of charge at a dealer service department.

If you are the new owner of a used vehicle which was subject to a recall and you want to be sure that the work has been carried out, it's best to contact a dealer service department and ask about your individual vehicle - you'll need to furnish them your Vehicle Identification Number (VIN).

The table below is based on information provided by the National Highway Traffic Safety Administration (NHTSA), the body which oversees vehicle recalls in the United States. The recall database is updated constantly. For the latest information on vehicle recalls, check the NHTSA website at www.nhtsa.gov, or call the NHTSA hotline at 1-888-327-4236.

Recall date	Recall campaign number	Model(s) affected	Concern
Jul 02, 2004	04V333000	2005 Chrysler 300, Dodge Magnum	On certain passenger vehicles, a spot welding operation for the rear floorpan reinforcement may not have been performed. This could cause the rear seat belt anchors and/or the child seat anchors to separate in certain crash conditions, which can increase the risk of injury to rear seat passengers.
Jul 02, 2004	04V334000	2005 Chrysler 300, Dodge Magnum	On certain passenger vehicles, the battery cable fasteners at the bulkhead stud may not have been properly tightened. A loose fastener can cause an instrument panel fire.
Jul 02, 2004	04V335000	2005 Chrysler 300, Dodge Magnum	On certain passenger vehicles, the front shoulder belt adjustable turning loop (ATL) bolts may not have been properly tightened. This could cause shoulder belt separation in certain crash conditions, which can increase the risk of injury to front seat occupants.
Oct 04, 2005	05V460000	2006 300/Charger/Magnum	On some vehicles with the 42RLE automatic transmission, the plug that retains the park pawl anchor shaft might be incorrectly installed. If the shaft moves out of position, the transmission might not be able to go into the Park position. If this occurs and the the parking brake is not applied, the vehicle could roll and cause a crash
May 03, 2006	06V149000	2006 300/Charger/Magnum	On some vehicles with a 2.7L engine and an MK25E brake system, there is insufficient clearance between the rear brake lines and the Exhaust Gas Recirculation (EGR) line. Over an extended period, contact between the brake lines and EGR line might cause the brake lines to wear through, which could seriously impair braking force and potentially cause an engine fire.
Sep 06, 2006	06V341000	2007 300/Charger/Magnum	On some vehicles with automatic transmissions, the software programmed into the Powertrain Control Module (PCM) can cause a momentary lock-up of the drive wheels if the vehicle is traveling over 40 mph and the driver shifts from Drive to Neutral, then back to Drive. If the wheels lock up, loss of vehicle control could occur, which could cause a crash.
Nov 05, 2006	08V583000	2006 through 2009 Charger/Magnum	On some vehicles equipped with the police package and a column-mounted shift lever, the shift control cable might detach from the steering column mounting bracket and cause an incorrect transmission gear position display, which could allow the vehicle to move in an unexpected direction and cause a crash without warning.
Dec 22, 2006	06V493000	2007 300/Charger/Magnum	On some vehicles, the Antilock Brake System (ABS) control module software might cause the rear brakes to lock up during certain braking conditions, which could result in loss of vehicle control, which could cause a crash.
Jul 01, 2008	08V295000	2008 300/Charger/Magnum	The torque retention crimp feature was missed during the manufacture of one lot of rear rear axle hub nuts, which could cause the nut to loosen and allow the driveaxle assembly to separate from the hub. If this happens, it could cause the vehicle to lose power and could result in a crash.
Sep 10, 2008	08V458000	2008 and 2009 Challenger	Chrysler is recalling certain vehicles equipped with automatic transmissions and "Keyless Go" option. If the stop/start button is pressed and held and the engine turns off, the electronic key code is removed from the vehicle. Since this can occur in a transmission or gear selector position other than "park," and the transmission or gear selector does not become locked in "park" as a direct result of key removal, it was determined that a noncompliance may exist. This standard specifies vehicle performance requirements intended to reduce the incidence of crashes resulting from theft and accidental rollaway of motor vehicles

Recall information

Recall date	Recall campaign number	Model(s) affected	Concern
Dec 04, 2008	08V642000	2009 300/Charger	On some vehicles, the Tire Pressure Monitor (TPM) sensors might not transmit the correct tire pressure, which could prevent the driver from knowing that the tire pressure is low, which could damage the tires and might cause a crash.
Oct 29, 2009	09V420000	2009 300	On some rear-drive vehicles, no spindle nut was installed, which could result in the vehicle losing a front wheel and tire while driving, which could result in a crash.
Sep 29, 2011	11V487000	2012 Charger	Chrysler is recalling certain vehicles that are equipped with 3.6L engines. These engines may experience connecting rod bearing failure due to debris inside the engine block. Connecting rod failure may lead to engine seizure which may increase the risk of a crash
Jan 06, 2012	12V004000	2012 Charger	Chrysler is recalling certain vehicles that fail to comply with the requirements of Federal Motor Vehicle Safety Standard No. 138, "Tire pressure monitoring system" (TPMS). As built, the TPMS does not warn the driver when a tire is 25 percent below the recommended pressure. The recalled vehicles notify the customers when the tire pressure reaches 22 psi rather than the required 24 psi. Underinflated tires can result in tire overloading and overheating, which would lead to a blowout and possible crash.
May 02, 2012	12V197000	2011 and 2012 300/Charger	Chrysler is recalling certain vehicles that may lose ABS/ESC system function due to an overheated power distribution center. This could lead to loss of vehicle control, increasing the risk of a crash.
Mar 21, 2013	13V103000	2013 Challenger	Chrysler is recalling certain vehicles equipped with a V6 engine. The battery positive cable at the starter motor may experience an electrical short circuit to ground. A short circuit could lead to a vehicle fire.
Apr 03, 2013	13V118000	2011 and 2012 300/Charger/ Challenger	Chrysler is recalling certain vehicles that may have improperly sized terminal crimps on the seat side-airbag wiring harness which may cause the seat side-airbags to malfunction and illuminate the airbag warning light. In the event of a crash necessitating airbag deployment the airbags may not operate as designed, increasing the risk of injury.
Mar 04, 2014	14V101000	2011 and 2012 Charger	Chrysler is recalling certain vehicles that may have a sub-harness for the headlights that may overheat and cause the low beam headlights to go out. Loss of headlights reduces the driver's visibility, increasing the risk of a crash.
Oct 08, 2014	14V634000	2011 through 2014 300/Charger/ Challenger	Chrysler is recalling certain vehicles equipped with a 3.6L engine and a 160 amp alternator. In the affected vehicles, the alternator may suddenly fail. If the alternator fails, the vehicle may stall without warning, increasing the risk of a crash.
Nov 24, 2014	14V749000	2015 Challenger	Chrysler is recalling certain vehicles in which the instrument cluster may become inoperative. This could cause the gauges to oscillate at zero and the Vehicle Theft Alarm could remain illuminated. Thus, the affected vehicles fail to conform to Federal Motor Vehicle Safety Standard (FMVSS) No. 101, "Controls and displays." If the instrument cluster becomes inoperative, it can increase the risk of a vehicle crash.
May 13, 2015	15V282000	2015 Challenger	Chrysler is recalling certain vehicles in which the driver's side air bag inflated curtain (SABIC) may have a loose or missing rear mounting bolt. A loose or missing (SABIC) rear mounting bolt may alter the air bag inflated curtain deployment, increasing the risk of occupant injury in the event of a side impact or roll-over crash.

Recall date	Recall campaign number	Model(s) affected	Concern
May 27, 2015	15V313000	2005 through 2010 300, 2006 through 2010 Charger	Chrysler is recalling certain vehicles equipped with a dual-stage driver frontal air bag that may be susceptible to moisture intrusion which, over time, could cause the inflator to rupture. In the event of a crash necessitating deployment of the driver's frontal air bag, the inflator could rupture with metal fragments striking the driver or other occupants resulting in serious injury or death.
Jul 10, 2015	15V444000	2008, 2009, 2010 Challenger	Chrysler is recalling certain vehicles equipped a dual-stage driver frontal air bag that may be susceptible to moisture intrusion which, over time, could cause the inflator to rupture. In the event of a crash necessitating deployment of the driver's frontal air bag, the inflator could rupture with metal fragments striking the driver or other occupants resulting in serious injury or death
Jul 23, 2015	15V461000	2015 300/Charger/ Challenger	Chrysler is recalling certain vehicles equipped with radios that have software vulnerabilities that can allow third-party access to certain networked vehicle control systems. Exploitation of the software vulnerability may result in unauthorized remote modification and control of certain vehicle systems, increasing the risk of a crash.
Jul 27, 2015	15V467000	2011 through 2014 Charger	Chrysler is recalling certain vehicles where the side impact sensor calibrations may be overly sensitive, and as a result, the side air bag inflatable curtains and seat air bags may unexpectedly deploy and the seat belt pre-tensioners may activate. Air bags that unexpectedly deploy increase the risk of a crash or injury.
Jan 28, 2016	16V043000	2011 through 2016 Charger	Chrysler is recalling certain vehicles that when the vehicle lifted using the supplied tire jack without chocking the wheels, the vehicle's body-side sill may give, causing the tire jack to fail and the car to fall. A car that falls unexpectedly increases the risk of injury.
Apr 25, 2016	16V240000	2012 through 2014 300/Charger	Chrysler is recalling certain vehicles equipped with an eight-speed automatic transmission and a monostable gear selector, may not adequately warn the driver when driver's door is opened and the vehicle is not in PARK, allowing them to exit the vehicle while the vehicle is still in gear. Drivers thinking that their vehicle's transmission is in the PARK position may be struck by the vehicle and injured if they attempt to get out of the vehicle while the engine is running and the parking brake is not engaged.
May 24, 2016	16V352000	2005 through 2012 300/Magnum, 2006 through 2012 Charger, 2008 through 2012 Challenger	Chrysler is recalling certain vehicles equipped with frontal air bags inflator assembled as part of the passenger frontal air bag modules, used as original equipemnt or replacement equipment. These inflators may rupture due to propellant degradation occurring after long-term exposure to absolute humidity and temperature cycling. In the event of a crash necessitating deployment of the driver's frontal air bag, the inflator could rupture with metal fragments striking the driver or other occupants resulting in serious injury or death.
Jul 7, 2017	17V431000	2017 Challenger	Chrysler is recalling certain vehicles equipped with 5.7L V8 engines and eight-speed automatic transmissions. The transmission may not remain in the PARK position if that gear is selected, the engine is still running, and the driver exits the vehicle exited. If the vehicle is exited without the transmission in the PARK position and without the parking brake set, the vehicle may roll, increasing the risk of a crash.
Jul 10, 2017	17V435000	2011 through 2014 300/Charger/ Challenger	Chrysler is recalling certain vehicles with electro-hydraulic power steering (EHPS) and are equipped with a 5.7L or a 3.6L engine and a 160, 180 or 220 amp alternator. In the affected vehicles, the alternator may suddenly fail. If the alternator fails, the vehicle may stall without warning, increasing the risk of a crash. There is also the possibility that the alternator may short circuit, increasing the risk of a fire.

Recall date	Recall campaign number	Model(s) affected	Concern
Aug 8, 2017	17V496000	2017 Charger/ Challenger	Chrysler is recalling certain vehicles with engine oil coolers, the hoses may fail resulting in a rapid loss of engine oil. If a hose fails, oil may spray onto the windshield or contact a hot surface, impairing driver visibility and increasing the risk of a crash or fire.
Sep 17, 2017	14V567000	2008 300/Charger/ Magnum	Chrysler is recalling vehicles where the the ignition key (FOBIK) may not return to the "ON" position after being rotated to the "START" position, instead sticking between the two positions. Potentially, while the vehicle is being driven, the ignition may inadvertently move through the "RUN" position to the "ACCESSORY" or "OFF" ignition position, turning off the engine. If the ignition stays between "START" and "ON" positions it could cause loss of certain electrical functions. If the ignition key is inadvertently moves into the "OFF" or "ACCESSORY" position, the engine will turn off, which will then depower various key safety systems including but not limited to the air bags, power steering, and power braking. Loss of functionality of these systems may increase the risk of a crash and/or increase the risk of injury in the event of a crash.
Nov 21, 2017	17V741000	2014 300/Charger/ Challenger	Chrysler is recalling certain vehicle equipped with a 3.6L engine and a 160 amp alternator. In the affected vehicles, the alternator may suddenly fail. If the alternator fails, the vehicle may stall without warning, increasing the risk of a crash
Dec 21, 2017	17V824000	2017 and 2018 300/Charger/ Challenger	Chrysler is recalling certain vehicles equipped with Kidde Plastic-Handle or Push Button 'Pindicator' Fire Extinguishers. These extinguishers may become clogged, preventing the extinguisher from discharging as expected or requiring excessive force to activate the extinguisher. Additionally, in certain models, the nozzle may detach from the valve assembly with enough force that it could cause injury and also render the product inoperable. If the fire extinguisher does not function properly, it can increase the risk of injury in the event of a fire.
Jan 09, 2018	18V021000	2009 through 2013 300/Charger/ Challenger	Chrysler is recalling certain vehicles that in the event of a crash necessitating deployment of the passenger frontal air bag, the passenger airbag inflator may explode due to propellant degradation occurring after long-term exposure to absolute humidity and temperature cycling. An inflator explosion may result in sharp metal fragments striking the driver or other occupants resulting in serious injury or death.
May 1, 2018	18V280000	2018 Charger/ Challenger/300	Chrysler is recalling certain vehicles that may have an incorrect transmission park lock rod installed in the transmission. If the incorrect park lock rod is installed, the transmission may not shift into "PARK" and keep the vehicle from moving, increasing the risk of unintended vehicle movement and the risk of a crash.
May 17, 2018	18V332000	2014 through 2018 300/Charger, 2015 through 2018 Challenger	Chrysler is recalling certain vehicle to address a defect that could prevent the cruise control system from disengaging. If, when using cruise control, there is a short circuit within the vehicle's wiring, the driver may not be able to shut off the cruise control either by depressing the brake pedal or manually turning the system off once it has been engaged, resulting in either the vehicle maintaining its current speed or possibly accelerating. If the vehicle maintains its speed or accelerates despite attempts to deactivate the cruise control, there would be an increased risk of a crash.

Buying parts

Replacement parts are available from many sources, which generally fall into one of two categories - authorized dealer parts departments and independent retail auto parts stores. Our advice concerning these parts is as follows:

Retail auto parts stores: Good auto parts stores will stock frequently needed components which wear out relatively fast, such as clutch components, exhaust systems, brake parts, tune-up parts, etc. These stores often supply new or reconditioned parts on an exchange basis, which can save a considerable amount of money. Discount auto parts stores are often very good places to buy materials and parts needed for general vehicle maintenance such as oil, grease, filters, spark plugs, belts, touch-up paint, bulbs, etc. They also usually sell tools and general accessories, have convenient hours, charge lower prices and can often be found not far from home.

Authorized dealer parts department: This is the best source for parts which are unique to the vehicle and not generally available elsewhere (such as major engine parts, transmission parts, trim pieces, etc.).

Warranty information: If the vehicle is still covered under warranty, be sure that any replacement parts purchased - regardless of the source - do not invalidate the warranty!

To be sure of obtaining the correct parts, have engine and chassis numbers available and, if possible, take the old parts along for positive identification.

Maintenance techniques, tools and working facilities

Maintenance techniques

There are a number of techniques involved in maintenance and repair that will be referred to throughout this manual. Application of these techniques will enable the home mechanic to be more efficient, better organized and capable of performing the various tasks properly, which will ensure that the repair job is thorough and complete.

Fasteners

Fasteners are nuts, bolts, studs and screws used to hold two or more parts together. There are a few things to keep in mind when working with fasteners. Almost all of them use a locking device of some type, either a lockwasher, locknut, locking tab or thread adhesive. All threaded fasteners should be clean and straight, with undamaged threads and undamaged corners on the hex head where the wrench fits. Develop the habit of replacing all damaged nuts and bolts with new ones. Special locknuts with nylon or fiber inserts can only be used once. If they are removed, they lose their locking ability and must be replaced with new ones.

Rusted nuts and bolts should be treated with a penetrating fluid to ease removal and prevent breakage. Some mechanics use turpentine in a spout-type oil can, which works quite well. After applying the rust penetrant, let it work for a few minutes before trying to loosen the nut or bolt. Badly rusted fasteners may have to be chiseled or sawed off or removed with a special nut breaker, available at tool stores.

If a bolt or stud breaks off in an assembly, it can be drilled and removed with a special tool commonly available for this purpose. Most automotive machine shops can perform this task, as well as other repair procedures, such as the repair of threaded holes that have been stripped out.

Flat washers and lockwashers, when removed from an assembly, should always be replaced exactly as removed. Replace any damaged washers with new ones. Never use a lockwasher on any soft metal surface (such as aluminum), thin sheet metal or plastic.

Fastener sizes

For a number of reasons, automobile manufacturers are making wider and wider use of metric fasteners. Therefore, it is important to be able to tell the difference between standard (sometimes called U.S. or SAE) and metric hardware, since they cannot be interchanged.

All bolts, whether standard or metric, are sized according to diameter, thread pitch and length. For example, a standard 1/2 - 13 x 1 bolt is 1/2 inch in diameter, has 13 threads per inch and is 1 inch long. An M12 - 1.75 x 25 metric bolt is 12 mm in diameter, has a thread pitch of 1.75 mm (the distance between threads) and is 25 mm long. The two bolts are nearly identical, and easily confused, but they are not interchangeable.

In addition to the differences in diameter, thread pitch and length, metric and standard bolts can also be distinguished by examining the bolt heads. To begin with, the distance across the flats on a standard bolt head is measured in inches, while the same dimension on a metric bolt is sized in millimeters (the same is true for nuts). As a result, a standard wrench should not be used on a metric bolt and a metric wrench should not be used on a standard bolt. Also, most standard bolts have slashes radiating out from the center of the head to denote the grade or strength of the bolt, which is an indication of the amount of torque that can be applied to it. The greater the number of slashes, the greater the strength of the bolt. Grades 0 through 5 are commonly used on automobiles. Metric bolts have a property class (grade) number, rather than a slash, molded into their heads to indicate bolt strength. In this case, the higher the number, the stronger the bolt. Property class numbers 8.8, 9.8 and 10.9 are commonly used on automobiles.

Strength markings can also be used to distinguish standard hex nuts from metric hex nuts. Many standard nuts have dots stamped into one side, while metric nuts are marked with a number. The greater the number of dots, or the higher the number, the greater the strength of the nut.

Metric studs are also marked on their ends according to property class (grade). Larger studs are numbered (the same as metric bolts), while smaller studs carry a geometric code to denote grade.

It should be noted that many fasteners, especially Grades 0 through 2, have no distinguishing marks on them. When such is the case, the only way to determine whether it is standard or metric is to measure the thread pitch or compare it to a known fastener of the same size.

Standard fasteners are often referred to as SAE, as opposed to metric. However, it should be noted that SAE technically refers to a non-metric fine thread fastener only. Coarse thread non-metric fasteners are referred to as USS sizes.

Since fasteners of the same size (both standard and metric) may have different strength ratings, be sure to reinstall any bolts, studs or nuts removed from your vehicle in their original locations. Also, when replacing a fastener with a new one, make sure that the new one has a strength rating equal to or greater than the original.

Tightening sequences and procedures

Most threaded fasteners should be tightened to a specific torque value (torque is the twisting force applied to a threaded component such as a nut or bolt). Overtightening the fastener can weaken it and cause it to break, while undertightening can cause it to eventually come loose. Bolts, screws and studs, depending on the material they are made of and their thread diameters, have specific torque values, many of which are noted in the Specifications at the beginning of each Chapter. Be sure to follow the torque recommen-

Grade 1 or 2   Grade 5   Grade 8

Bolt strength marking (standard/SAE/USS; bottom - metric)

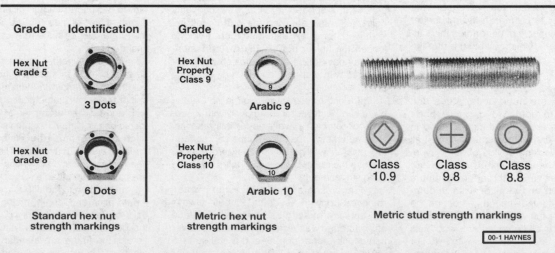

Grade	Identification
Hex Nut Grade 5	3 Dots
Hex Nut Grade 8	6 Dots

Standard hex nut strength markings

Grade	Identification
Hex Nut Property Class 9	Arabic 9
Hex Nut Property Class 10	Arabic 10

Metric hex nut strength markings

Class 10.9   Class 9.8   Class 8.8

Metric stud strength markings

00-1 HAYNES

dations closely. For fasteners not assigned a specific torque, a general torque value chart is presented here as a guide. These torque values are for dry (unlubricated) fasteners threaded into steel or cast iron (not aluminum). As was previously mentioned, the size and grade of a fastener determine the amount of torque that can safely be applied to it. The figures listed here are approximate for Grade 2 and Grade 3 fasteners. Higher grades can tolerate higher torque values.

Fasteners laid out in a pattern, such as cylinder head bolts, oil pan bolts, differential cover bolts, etc., must be loosened or tightened in sequence to avoid warping the component. This sequence will normally be shown in the appropriate Chapter. If a specific pattern is not given, the following procedures can be used to prevent warping.

Initially, the bolts or nuts should be assembled finger-tight only. Next, they should be tightened one full turn each, in a crisscross or diagonal pattern. After each one has been tightened one full turn, return to the first one and tighten them all one-half turn, following the same pattern. Finally, tighten each of them one-quarter turn at a time until each fastener has been tightened to the proper torque. To loosen and remove the fasteners, the procedure would be reversed.

Metric thread sizes	Ft-lbs	Nm
M-6	6 to 9	9 to 12
M-8	14 to 21	19 to 28
M-10	28 to 40	38 to 54
M-12	50 to 71	68 to 96
M-14	80 to 140	109 to 154

Pipe thread sizes		
1/8	5 to 8	7 to 10
1/4	12 to 18	17 to 24
3/8	22 to 33	30 to 44
1/2	25 to 35	34 to 47

U.S. thread sizes		
1/4 - 20	6 to 9	9 to 12
5/16 - 18	12 to 18	17 to 24
5/16 - 24	14 to 20	19 to 27
3/8 - 16	22 to 32	30 to 43
3/8 - 24	27 to 38	37 to 51
7/16 - 14	40 to 55	55 to 74
7/16 - 20	40 to 60	55 to 81
1/2 - 13	55 to 80	75 to 108

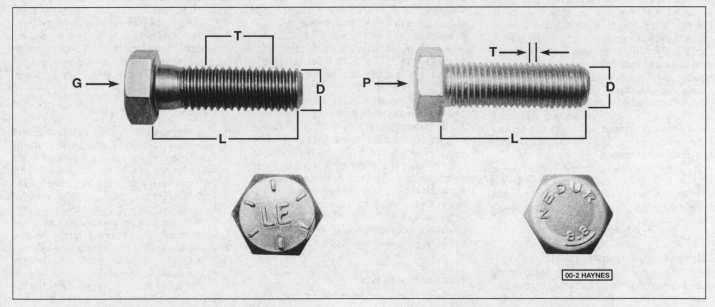

Standard (SAE and USS) bolt dimensions/grade marks

- G   Grade marks (bolt strength)
- L   Length (in inches)
- T   Thread pitch (number of threads per inch)
- D   Nominal diameter (in inches)

Metric bolt dimensions/grade marks

- P   Property class (bolt strength)
- L   Length (in millimeters)
- T   Thread pitch (distance between threads in millimeters)
- D   Diameter

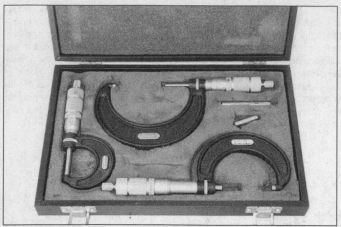

Micrometer set

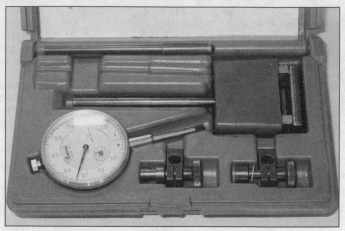

Dial indicator set

Component disassembly

Component disassembly should be done with care and purpose to help ensure that the parts go back together properly. Always keep track of the sequence in which parts are removed. Make note of special characteristics or marks on parts that can be installed more than one way, such as a grooved thrust washer on a shaft. It is a good idea to lay the disassembled parts out on a clean surface in the order that they were removed. It may also be helpful to make sketches or take instant photos of components before removal.

When removing fasteners from a component, keep track of their locations. Sometimes threading a bolt back in a part, or putting the washers and nut back on a stud, can prevent mix-ups later. If nuts and bolts cannot be returned to their original locations, they should be kept in a compartmented box or a series of small boxes. A cupcake or muffin tin is ideal for this purpose, since each cavity can hold the bolts and nuts from a particular area (i.e. oil pan bolts, valve cover bolts, engine mount bolts, etc.). A pan of this type is especially helpful when working on assemblies with very small parts, such as the carburetor, alternator, valve train or interior dash and trim pieces. The cavities can be marked with paint or tape to identify the contents.

Whenever wiring looms, harnesses or connectors are separated, it is a good idea to identify the two halves with numbered pieces of masking tape so they can be easily reconnected.

Gasket sealing surfaces

Throughout any vehicle, gaskets are used to seal the mating surfaces between two parts and keep lubricants, fluids, vacuum or pressure contained in an assembly.

Many times these gaskets are coated with a liquid or paste-type gasket sealing compound before assembly. Age, heat and pressure can sometimes cause the two parts to stick together so tightly that they are very difficult to separate. Often, the assembly can

be loosened by striking it with a soft-face hammer near the mating surfaces. A regular hammer can be used if a block of wood is placed between the hammer and the part. Do not hammer on cast parts or parts that could be easily damaged. With any particularly stubborn part, always recheck to make sure that every fastener has been removed.

Avoid using a screwdriver or bar to pry apart an assembly, as they can easily mar the gasket sealing surfaces of the parts, which must remain smooth. If prying is absolutely necessary, use an old broom handle, but keep in mind that extra clean up will be necessary if the wood splinters.

After the parts are separated, the old gasket must be carefully scraped off and the gasket surfaces cleaned. Stubborn gasket material can be soaked with rust penetrant or treated with a special chemical to soften it so it can be easily scraped off. **Caution:** *Never use gasket removal solutions or caustic chemicals on plastic or other composite components.* A scraper can be fashioned from a piece of copper tubing by flattening and sharpening one end. Copper is recommended because it is usually softer than the surfaces to be scraped, which reduces the chance of gouging the part. Some gaskets can be removed with a wire brush, but regardless of the method used, the mating surfaces must be left clean and smooth. If for some reason the gasket surface is gouged, then a gasket sealer thick enough to fill scratches will have to be used during reassembly of the components. For most applications, a non-drying (or semi-drying) gasket sealer should be used.

Hose removal tips

Warning: *If the vehicle is equipped with air conditioning, do not disconnect any of the A/C hoses without first having the system depressurized by a dealer service department or a service station.*

Hose removal precautions closely parallel gasket removal precautions. Avoid scratching or gouging the surface that the

hose mates against or the connection may leak. This is especially true for radiator hoses. Because of various chemical reactions, the rubber in hoses can bond itself to the metal spigot that the hose fits over. To remove a hose, first loosen the hose clamps that secure it to the spigot. Then, with slip-joint pliers, grab the hose at the clamp and rotate it around the spigot. Work it back and forth until it is completely free, then pull it off. Silicone or other lubricants will ease removal if they can be applied between the hose and the outside of the spigot. Apply the same lubricant to the inside of the hose and the outside of the spigot to simplify installation.

As a last resort (and if the hose is to be replaced with a new one anyway), the rubber can be slit with a knife and the hose peeled from the spigot. If this must be done, be careful that the metal connection is not damaged.

If a hose clamp is broken or damaged, do not reuse it. Wire-type clamps usually weaken with age, so it is a good idea to replace them with screw-type clamps whenever a hose is removed.

Tools

A selection of good tools is a basic requirement for anyone who plans to maintain and repair his or her own vehicle. For the owner who has few tools, the initial investment might seem high, but when compared to the spiraling costs of professional auto maintenance and repair, it is a wise one.

To help the owner decide which tools are needed to perform the tasks detailed in this manual, the following tool lists are offered: *Maintenance and minor repair*, *Repair/overhaul* and *Special*.

The newcomer to practical mechanics should start off with the *maintenance and minor repair* tool kit, which is adequate for the simpler jobs performed on a vehicle. Then, as confidence and experience grow, the owner can tackle more difficult tasks, buying additional tools as they are needed. Eventually the basic kit will be expanded into the *repair and overhaul* tool set. Over a period of time, the

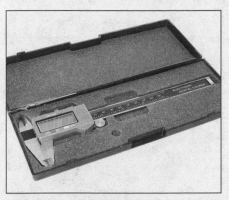

Dial caliper

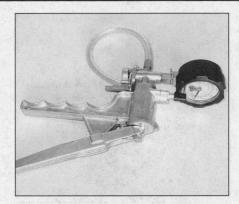

Hand-operated vacuum pump

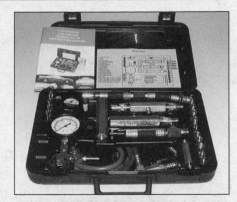

Fuel pressure gauge set

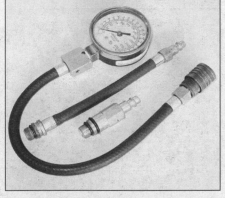

Compression gauge with spark plug hole adapter

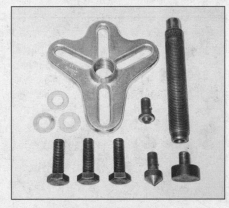

Damper/steering wheel puller

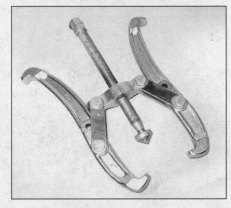

General purpose puller

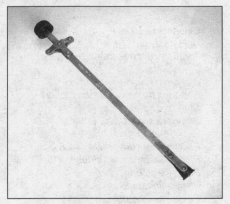

Hydraulic lifter removal tool

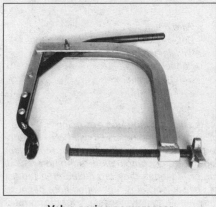

Valve spring compressor

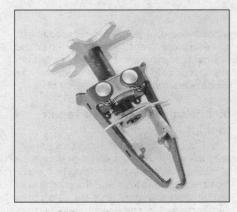

Valve spring compressor

Ridge reamer

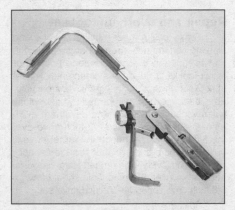

Piston ring groove cleaning tool

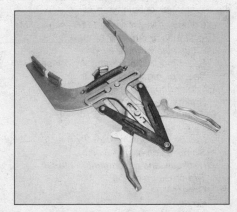

Ring removal/installation tool

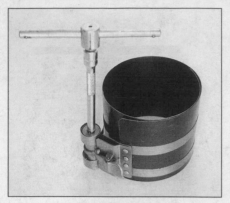

Ring compressor

Cylinder hone

Brake hold-down spring tool

Torque angle gauge

Clutch plate alignment tool

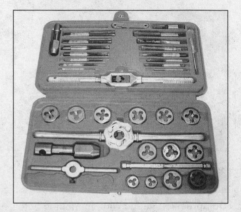

Tap and die set

experienced do-it-yourselfer will assemble a tool set complete enough for most repair and overhaul procedures and will add tools from the special category when it is felt that the expense is justified by the frequency of use.

Maintenance and minor repair tool kit

The tools in this list should be considered the minimum required for performance of routine maintenance, servicing and minor repair work. We recommend the purchase of combination wrenches (box-end and open-end combined in one wrench). While more expensive than open end wrenches, they offer the advantages of both types of wrench.

Combination wrench set (1/4-inch to 1 inch or 6 mm to 19 mm)
Adjustable wrench, 8 inch
Spark plug wrench with rubber insert
Spark plug gap adjusting tool
Feeler gauge set
Brake bleeder wrench
Standard screwdriver (5/16-inch x 6 inch)
Phillips screwdriver (No. 2 x 6 inch)
Combination pliers - 6 inch
Hacksaw and assortment of blades
Tire pressure gauge
Grease gun
Oil can
Fine emery cloth

Wire brush
Battery post and cable cleaning tool
Oil filter wrench
Funnel (medium size)
Safety goggles
Jackstands (2)
Drain pan

Note: *If basic tune-ups are going to be part of routine maintenance, it will be necessary to purchase a good quality stroboscopic timing light and combination tachometer/dwell meter. Although they are included in the list of special tools, it is mentioned here because they are absolutely necessary for tuning most vehicles properly.*

Repair and overhaul tool set

These tools are essential for anyone who plans to perform major repairs and are in addition to those in the maintenance and minor repair tool kit. Included is a comprehensive set of sockets which, though expensive, are invaluable because of their versatility, especially when various extensions and drives are available. We recommend the 1/2-inch drive over the 3/8-inch drive. Although the larger drive is bulky and more expensive, it has the capacity of accepting a very wide range of large sockets. Ideally, however, the mechanic should have a 3/8-inch drive set and a 1/2-inch drive set.

Socket set(s)
Reversible ratchet

Extension - 10 inch
Universal joint
Torque wrench (same size drive as sockets)
Ball peen hammer - 8 ounce
Soft-face hammer (plastic/rubber)
Standard screwdriver (1/4-inch x 6 inch)
Standard screwdriver (stubby - 5/16-inch)
Phillips screwdriver (No. 3 x 8 inch)
Phillips screwdriver (stubby - No. 2)
Pliers - vise grip
Pliers - lineman's
Pliers - needle nose
Pliers - snap-ring (internal and external)
Cold chisel - 1/2-inch
Scribe
Scraper (made from flattened copper tubing)
Centerpunch
Pin punches (1/16, 1/8, 3/16-inch)
Steel rule/straightedge - 12 inch
Allen wrench set (1/8 to 3/8-inch or 4 mm to 10 mm)
A selection of files
Wire brush (large)
Jackstands (second set)
Jack (scissor or hydraulic type)

Note: *Another tool which is often useful is an electric drill with a chuck capacity of 3/8-inch and a set of good quality drill bits.*

Special tools

The tools in this list include those which are not used regularly, are expensive to buy, or which need to be used in accordance with their manufacturer's instructions. Unless these tools will be used frequently, it is not very economical to purchase many of them. A consideration would be to split the cost and use between yourself and a friend or friends. In addition, most of these tools can be obtained from a tool rental shop on a temporary basis.

This list primarily contains only those tools and instruments widely available to the public, and not those special tools produced by the vehicle manufacturer for distribution to dealer service departments. Occasionally, references to the manufacturer's special tools are included in the text of this manual. Generally, an alternative method of doing the job without the special tool is offered. However, sometimes there is no alternative to their use. Where this is the case, and the tool cannot be purchased or borrowed, the work should be turned over to the dealer service department or an automotive repair shop.

Valve spring compressor
Piston ring groove cleaning tool
Piston ring compressor
Piston ring installation tool
Cylinder compression gauge
Cylinder ridge reamer
Cylinder surfacing hone
Cylinder bore gauge
Micrometers and/or dial calipers
Hydraulic lifter removal tool
Balljoint separator
Universal-type puller
Impact screwdriver
Dial indicator set
Stroboscopic timing light (inductive pick-up)
Hand operated vacuum/pressure pump
Tachometer/dwell meter
Universal electrical multimeter
Cable hoist
Brake spring removal and installation tools
Floor jack

Buying tools

For the do-it-yourselfer who is just starting to get involved in vehicle maintenance and repair, there are a number of options available when purchasing tools. If maintenance and minor repair is the extent of the work to be done, the purchase of individual tools is satisfactory. If, on the other hand, extensive work is planned, it would be a good idea to purchase a modest tool set from one of the large retail chain stores. A set can usually be bought at a substantial savings over the individual tool prices, and they often come with a tool box. As additional tools are needed, add-on sets, individual tools and a larger tool box can be purchased to expand the tool selection. Building a tool set gradually allows the cost of the tools to be spread over a longer period of time and gives the mechanic the freedom to choose only those tools that will actually be used.

Tool stores will often be the only source of some of the special tools that are needed, but regardless of where tools are bought, try to avoid cheap ones, especially when buying screwdrivers and sockets, because they won't last very long. The expense involved in replacing cheap tools will eventually be greater than the initial cost of quality tools.

Care and maintenance of tools

Good tools are expensive, so it makes sense to treat them with respect. Keep them clean and in usable condition and store them properly when not in use. Always wipe off any dirt, grease or metal chips before putting them away. Never leave tools lying around in the work area. Upon completion of a job, always check closely under the hood for tools that may have been left there so they won't get lost during a test drive.

Some tools, such as screwdrivers, pliers, wrenches and sockets, can be hung on a panel mounted on the garage or workshop wall, while others should be kept in a tool box or tray. Measuring instruments, gauges, meters, etc. must be carefully stored where they cannot be damaged by weather or impact from other tools.

When tools are used with care and stored properly, they will last a very long time. Even with the best of care, though, tools will wear out if used frequently. When a tool is damaged or worn out, replace it. Subsequent jobs will be safer and more enjoyable if you do.

How to repair damaged threads

Sometimes, the internal threads of a nut or bolt hole can become stripped, usually from overtightening. Stripping threads is an all-too-common occurrence, especially when working with aluminum parts, because aluminum is so soft that it easily strips out.

Usually, external or internal threads are only partially stripped. After they've been cleaned up with a tap or die, they'll still work. Sometimes, however, threads are badly damaged. When this happens, you've got three choices:

1) *Drill and tap the hole to the next suitable oversize and install a larger diameter bolt, screw or stud.*

2) *Drill and tap the hole to accept a threaded plug, then drill and tap the plug to the original screw size. You can also buy a plug already threaded to the original size. Then you simply drill a hole to the specified size, then run the threaded plug into the hole with a bolt and jam nut. Once the plug is fully seated, remove the jam nut and bolt.*

3) *The third method uses a patented thread repair kit like Heli-Coil or Slimsert. These easy-to-use kits are designed to repair damaged threads in straight-through holes and blind holes. Both are available as kits which can handle a variety of sizes and thread patterns. Drill the hole, then tap it with the special included tap. Install the Heli-Coil and the hole is back to its original diameter and thread pitch.*

Regardless of which method you use, be sure to proceed calmly and carefully. A little impatience or carelessness during one of these relatively simple procedures can ruin your whole day's work and cost you a bundle if you wreck an expensive part.

Working facilities

Not to be overlooked when discussing tools is the workshop. If anything more than routine maintenance is to be carried out, some sort of suitable work area is essential.

It is understood, and appreciated, that many home mechanics do not have a good workshop or garage available, and end up removing an engine or doing major repairs outside. It is recommended, however, that the overhaul or repair be completed under the cover of a roof.

A clean, flat workbench or table of comfortable working height is an absolute necessity. The workbench should be equipped with a vise that has a jaw opening of at least four inches.

As mentioned previously, some clean, dry storage space is also required for tools, as well as the lubricants, fluids, cleaning solvents, etc. which soon become necessary.

Sometimes waste oil and fluids, drained from the engine or cooling system during normal maintenance or repairs, present a disposal problem. To avoid pouring them on the ground or into a sewage system, pour the used fluids into large containers, seal them with caps and take them to an authorized disposal site or recycling center. Plastic jugs, such as old antifreeze containers, are ideal for this purpose.

Always keep a supply of old newspapers and clean rags available. Old towels are excellent for mopping up spills. Many mechanics use rolls of paper towels for most work because they are readily available and disposable. To help keep the area under the vehicle clean, a large cardboard box can be cut open and flattened to protect the garage or shop floor.

Whenever working over a painted surface, such as when leaning over a fender to service something under the hood, always cover it with an old blanket or bedspread to protect the finish. Vinyl covered pads, made especially for this purpose, are available at auto parts stores.

Jacking and towing

Jacking

Warning: *The jack supplied with the vehicle should only be used for changing a tire or placing jackstands under the frame. Never work under the vehicle or start the engine while this jack is being used as the only means of support.*

The vehicle should abe on level ground. Place the shift lever in Park, and block the wheel diagonally opposite the wheel being changed. Set the parking brake.

Remove the spare tire and jack from stowage. Remove the wheel cover and trim ring (if so equipped) with the tapered end of the lug nut wrench by inserting and twisting the handle and then prying against the back of the wheel cover. Loosen the wheel lug nuts about 1/4-to-1/2 turn each.

Place the scissors-type jack under the side of the vehicle and adjust the jack height until it fits in the notch in the vertical rocker panel flange nearest the wheel to be changed. There is a front and rear jacking point on each side of the vehicle (see illustration).

Turn the jack handle clockwise until the tire clears the ground. Remove the lug nuts and pull the wheel off. Replace it with the spare.

Install the lug nuts with the beveled edges facing in. Tighten them snugly. Don't attempt to tighten them completely until the vehicle is lowered or it could slip off the jack. Turn the jack handle counterclockwise to lower the vehicle. Remove the jack and tighten the lug nuts in a diagonal pattern.

Install the cover (and trim ring, if used) and be sure it's snapped into place all the way around.

Stow the tire, jack and wrench. Unblock the wheels.

Towing

As a general rule, the vehicle should be towed with the rear wheels off the ground. If they can't be raised, either place them on a dolly or disconnect the driveshaft from the differential. When a vehicle is towed with the rear wheels raised, the steering wheel must be clamped in the straight ahead position with a special device designed for use during towing. The ignition key must be in the OFF position, since the steering lock mechanism isn't strong enough to hold the front wheels straight while towing.

Vehicles can be towed from the front only with all four wheels on the ground, provided that speeds don't exceed 30 mph and the distance is not over 30 miles. Release the parking brake, put the transmission in Neutral and place the ignition key in the OFF position.

Equipment specifically designed for towing should be used. It should be attached to the main structural members of the vehicle, not the bumpers or brackets.

Safety is a major consideration when towing and all applicable state and local laws must be obeyed. A safety chain system must be used at all times. Remember that power steering and power brakes will not work with the engine off.

Place the jack so it engages the notch in the rocker panel nearest the wheel to be raised

Booster battery (jump) starting

Observe the following precautions when using a booster battery to start a vehicle:

a) Before connecting the booster battery, make sure the ignition switch is in the Off position.

b) Turn off the lights, heater and other electrical loads.

c) Your eyes should be shielded. Safety goggles are a good idea.

d) Make sure the booster battery is the same voltage as the dead one in the vehicle.

e) The two vehicles MUST NOT TOUCH each other!

f) Make sure the transmission is in Neutral (manual) or Park (automatic).

g) If the booster battery is not a maintenance-free type, remove the vent caps and lay a cloth over the vent holes.

Connect one jumper lead between the positive (+) terminals of the two batteries (see illustration). Note: These vehicles are equipped with a remote positive terminal located on the side of the underhood fuse/relay box, to make jumper cable connection easier (see illustration). Connect the other jumper lead first to the negative (-) terminal of the booster battery, then to a good engine ground on the vehicle to be started. Note: These vehicles are also equipped with a remote grounding point, located on the right shock tower (near the remote positive terminal). Attach the lead at least 18 inches from the battery, if possible. Make sure the that the jumper leads will not contact the fan, drivebelt or other moving parts of the engine.

Start the engine using the booster battery, then, with the engine running at idle speed, disconnect the jumper cables in the reverse order of connection.

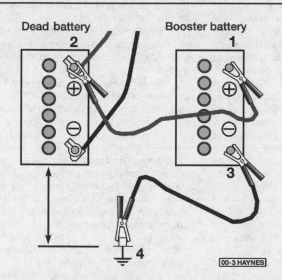

Make the booster battery cable connections in the numerical order shown (note that the negative cable of the booster battery is NOT attached to the negative terminal of the dead battery)

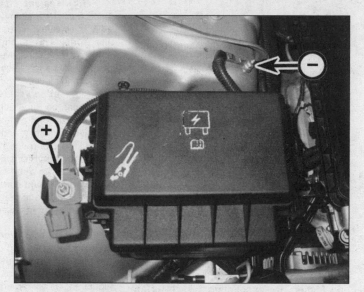

Remote jump starting terminals

Automotive chemicals and lubricants

A number of automotive chemicals and lubricants are available for use during vehicle maintenance and repair. They include a wide variety of products ranging from cleaning solvents and degreasers to lubricants and protective sprays for rubber, plastic and vinyl.

Cleaners

Carburetor cleaner and choke cleaner is a strong solvent for gum, varnish and carbon. Most carburetor cleaners leave a dry-type lubricant film which will not harden or gum up. Because of this film it is not recommended for use on electrical components.

Brake system cleaner is used to remove brake dust, grease and brake fluid from the brake system, where clean surfaces are absolutely necessary. It leaves no residue and often eliminates brake squeal caused by contaminants.

Electrical cleaner removes oxidation, corrosion and carbon deposits from electrical contacts, restoring full current flow. It can also be used to clean spark plugs, carburetor jets, voltage regulators and other parts where an oil-free surface is desired.

Demoisturants remove water and moisture from electrical components such as alternators, voltage regulators, electrical connectors and fuse blocks. They are non-conductive and non-corrosive.

Degreasers are heavy-duty solvents used to remove grease from the outside of the engine and from chassis components. They can be sprayed or brushed on and, depending on the type, are rinsed off either with water or solvent.

Lubricants

Motor oil is the lubricant formulated for use in engines. It normally contains a wide variety of additives to prevent corrosion and reduce foaming and wear. Motor oil comes in various weights (viscosity ratings) from 0 to 50. The recommended weight of the oil depends on the season, temperature and the demands on the engine. Light oil is used in cold climates and under light load conditions. Heavy oil is used in hot climates and where high loads are encountered. Multi-viscosity oils are designed to have characteristics of both light and heavy oils and are available in a number of weights from 0W-20 to 20W-50.

Gear oil is designed to be used in differentials, manual transmissions and other areas where high-temperature lubrication is required.

Chassis and wheel bearing grease is a heavy grease used where increased loads and friction are encountered, such as for wheel bearings, balljoints, tie-rod ends and universal joints.

High-temperature wheel bearing grease is designed to withstand the extreme temperatures encountered by wheel bearings in disc brake equipped vehicles. It usually contains molybdenum disulfide (moly), which is a dry-type lubricant.

White grease is a heavy grease for metal-to-metal applications where water is a problem. White grease stays soft under both low and high temperatures (usually from -100 to +190-degrees F), and will not wash off or dilute in the presence of water.

Assembly lube is a special extreme pressure lubricant, usually containing moly, used to lubricate high-load parts (such as main and rod bearings and cam lobes) for initial start-up of a new engine. The assembly lube lubricates the parts without being squeezed out or washed away until the engine oiling system begins to function.

Silicone lubricants are used to protect rubber, plastic, vinyl and nylon parts.

Graphite lubricants are used where oils cannot be used due to contamination problems, such as in locks. The dry graphite will lubricate metal parts while remaining uncontaminated by dirt, water, oil or acids. It is electrically conductive and will not foul electrical contacts in locks such as the ignition switch.

Moly penetrants loosen and lubricate frozen, rusted and corroded fasteners and prevent future rusting or freezing.

Heat-sink grease is a special electrically non-conductive grease that is used for mounting electronic ignition modules where it is essential that heat is transferred away from the module.

Sealants

RTV sealant is one of the most widely used gasket compounds. Made from silicone, RTV is air curing, it seals, bonds, waterproofs, fills surface irregularities, remains flexible, doesn't shrink, is relatively easy to remove, and is used as a supplementary sealer with almost all low and medium temperature gaskets.

Anaerobic sealant is much like RTV in that it can be used either to seal gaskets or to form gaskets by itself. It remains flexible, is solvent resistant and fills surface imperfections. The difference between an anaerobic sealant and an RTV-type sealant is in the curing. RTV cures when exposed to air, while an anaerobic sealant cures only in the absence of air. This means that an anaerobic sealant cures only after the assembly of parts, sealing them together.

Thread and pipe sealant is used for sealing hydraulic and pneumatic fittings and vacuum lines. It is usually made from a Teflon compound, and comes in a spray, a paint-on liquid and as a wrap-around tape.

Chemicals

Anti-seize compound prevents seizing, galling, cold welding, rust and corrosion in fasteners. High-temperature anti-seize, usually made with copper and graphite lubricants, is used for exhaust system and exhaust manifold bolts.

Anaerobic locking compounds are used to keep fasteners from vibrating or working loose and cure only after installation, in the absence of air. Medium strength locking compound is used for small nuts, bolts and screws that may be removed later. High-strength locking compound is for large nuts, bolts and studs which aren't removed on a regular basis.

Oil additives range from viscosity index improvers to chemical treatments that claim to reduce internal engine friction. It should be noted that most oil manufacturers caution against using additives with their oils.

Gas additives perform several functions, depending on their chemical makeup. They usually contain solvents that help dissolve gum and varnish that build up on carburetor, fuel injection and intake parts. They also serve to break down carbon deposits that form on the inside surfaces of the combustion chambers. Some additives contain upper cylinder lubricants for valves and piston rings, and others contain chemicals to remove condensation from the gas tank.

Miscellaneous

Brake fluid is specially formulated hydraulic fluid that can withstand the heat and pressure encountered in brake systems. Care must be taken so this fluid does not come in contact with painted surfaces or plastics. An opened container should always be resealed to prevent contamination by water or dirt.

Weatherstrip adhesive is used to bond weatherstripping around doors, windows and trunk lids. It is sometimes used to attach trim pieces.

Undercoating is a petroleum-based, tar-like substance that is designed to protect metal surfaces on the underside of the vehicle from corrosion. It also acts as a sound-deadening agent by insulating the bottom of the vehicle.

Waxes and polishes are used to help protect painted and plated surfaces from the weather. Different types of paint may require the use of different types of wax and polish. Some polishes utilize a chemical or abrasive cleaner to help remove the top layer of oxidized (dull) paint on older vehicles. In recent years many non-wax polishes that contain a wide variety of chemicals such as polymers and silicones have been introduced. These non-wax polishes are usually easier to apply and last longer than conventional waxes and polishes.

Conversion factors

Length (distance)
Inches (in)	X	25.4	= Millimeters (mm)	X	0.0394	= Inches (in)
Feet (ft)	X	0.305	= Meters (m)	X	3.281	= Feet (ft)
Miles	X	1.609	= Kilometers (km)	X	0.621	= Miles

Volume (capacity)
Cubic inches (cu in; in³)	X	16.387	= Cubic centimeters (cc; cm³)	X	0.061	= Cubic inches (cu in; in³)
Imperial pints (Imp pt)	X	0.568	= Liters (l)	X	1.76	= Imperial pints (Imp pt)
Imperial quarts (Imp qt)	X	1.137	= Liters (l)	X	0.88	= Imperial quarts (Imp qt)
Imperial quarts (Imp qt)	X	1.201	= US quarts (US qt)	X	0.833	= Imperial quarts (Imp qt)
US quarts (US qt)	X	0.946	= Liters (l)	X	1.057	= US quarts (US qt)
Imperial gallons (Imp gal)	X	4.546	= Liters (l)	X	0.22	= Imperial gallons (Imp gal)
Imperial gallons (Imp gal)	X	1.201	= US gallons (US gal)	X	0.833	= Imperial gallons (Imp gal)
US gallons (US gal)	X	3.785	= Liters (l)	X	0.264	= US gallons (US gal)

Mass (weight)
Ounces (oz)	X	28.35	= Grams (g)	X	0.035	= Ounces (oz)
Pounds (lb)	X	0.454	= Kilograms (kg)	X	2.205	= Pounds (lb)

Force
Ounces-force (ozf; oz)	X	0.278	= Newtons (N)	X	3.6	= Ounces-force (ozf; oz)
Pounds-force (lbf; lb)	X	4.448	= Newtons (N)	X	0.225	= Pounds-force (lbf; lb)
Newtons (N)	X	0.1	= Kilograms-force (kgf; kg)	X	9.81	= Newtons (N)

Pressure
Pounds-force per square inch (psi; lbf/in²; lb/in²)	X	0.070	= Kilograms-force per square centimeter (kgf/cm²; kg/cm²)	X	14.223	= Pounds-force per square inch (psi; lbf/in²; lb/in²)
Pounds-force per square inch (psi; lbf/in²; lb/in²)	X	0.068	= Atmospheres (atm)	X	14.696	= Pounds-force per square inch (psi; lbf/in²; lb/in²)
Pounds-force per square inch (psi; lbf/in²; lb/in²)	X	0.069	= Bars	X	14.5	= Pounds-force per square inch (psi; lbf/in²; lb/in²)
Pounds-force per square inch (psi; lbf/in²; lb/in²)	X	6.895	= Kilopascals (kPa)	X	0.145	= Pounds-force per square inch (psi; lbf/in²; lb/in²)
Kilopascals (kPa)	X	0.01	= Kilograms-force per square centimeter (kgf/cm²; kg/cm²)	X	98.1	= Kilopascals (kPa)

Torque (moment of force)
Pounds-force inches (lbf in; lb in)	X	1.152	= Kilograms-force centimeter (kgf cm; kg cm)	X	0.868	= Pounds-force inches (lbf in; lb in)
Pounds-force inches (lbf in; lb in)	X	0.113	= Newton meters (Nm)	X	8.85	= Pounds-force inches (lbf in; lb in)
Pounds-force inches (lbf in; lb in)	X	0.083	= Pounds-force feet (lbf ft; lb ft)	X	12	= Pounds-force inches (lbf in; lb in)
Pounds-force feet (lbf ft; lb ft)	X	0.138	= Kilograms-force meters (kgf m; kg m)	X	7.233	= Pounds-force feet (lbf ft; lb ft)
Pounds-force feet (lbf ft; lb ft)	X	1.356	= Newton meters (Nm)	X	0.738	= Pounds-force feet (lbf ft; lb ft)
Newton meters (Nm)	X	0.102	= Kilograms-force meters (kgf m; kg m)	X	9.804	= Newton meters (Nm)

Vacuum
Inches mercury (in. Hg)	X	3.377	= Kilopascals (kPa)	X	0.2961	= Inches mercury
Inches mercury (in. Hg)	X	25.4	= Millimeters mercury (mm Hg)	X	0.0394	= Inches mercury

Power
Horsepower (hp)	X	745.7	= Watts (W)	X	0.0013	= Horsepower (hp)

Velocity (speed)
Miles per hour (miles/hr; mph)	X	1.609	= Kilometers per hour (km/hr; kph)	X	0.621	= Miles per hour (miles/hr; mph)

Fuel consumption*
Miles per gallon, Imperial (mpg)	X	0.354	= Kilometers per liter (km/l)	X	2.825	= Miles per gallon, Imperial (mpg)
Miles per gallon, US (mpg)	X	0.425	= Kilometers per liter (km/l)	X	2.352	= Miles per gallon, US (mpg)

Temperature
Degrees Fahrenheit = (°C x 1.8) + 32   Degrees Celsius (Degrees Centigrade; °C) = (°F - 32) x 0.56

*It is common practice to convert from miles per gallon (mpg) to liters/100 kilometers (l/100km),
where mpg (Imperial) x l/100 km = 282 and mpg (US) x l/100 km = 235

DECIMALS to MILLIMETERS

Decimal	mm	Decimal	mm
0.001	0.0254	0.500	12.7000
0.002	0.0508	0.510	12.9540
0.003	0.0762	0.520	13.2080
0.004	0.1016	0.530	13.4620
0.005	0.1270	0.540	13.7160
0.006	0.1524	0.550	13.9700
0.007	0.1778	0.560	14.2240
0.008	0.2032	0.570	14.4780
0.009	0.2286	0.580	14.7320
		0.590	14.9860
0.010	0.2540		
0.020	0.5080		
0.030	0.7620		
0.040	1.0160	0.600	15.2400
0.050	1.2700	0.610	15.4940
0.060	1.5240	0.620	15.7480
0.070	1.7780	0.630	16.0020
0.080	2.0320	0.640	16.2560
0.090	2.2860	0.650	16.5100
		0.660	16.7640
0.100	2.5400	0.670	17.0180
0.110	2.7940	0.680	17.2720
0.120	3.0480	0.690	17.5260
0.130	3.3020		
0.140	3.5560		
0.150	3.8100		
0.160	4.0640	0.700	17.7800
0.170	4.3180	0.710	18.0340
0.180	4.5720	0.720	18.2880
0.190	4.8260	0.730	18.5420
		0.740	18.7960
0.200	5.0800	0.750	19.0500
0.210	5.3340	0.760	19.3040
0.220	5.5880	0.770	19.5580
0.230	5.8420	0.780	19.8120
0.240	6.0960	0.790	20.0660
0.250	6.3500		
0.260	6.6040		
0.270	6.8580	0.800	20.3200
0.280	7.1120	0.810	20.5740
0.290	7.3660	0.820	20.8280
		0.830	21.0820
0.300	7.6200	0.840	21.3360
0.310	7.8740	0.850	21.5900
0.320	8.1280	0.860	21.8440
0.330	8.3820	0.870	22.0980
0.340	8.6360	0.880	22.3520
0.350	8.8900	0.890	22.6060
0.360	9.1440		
0.370	9.3980		
0.380	9.6520		
0.390	9.9060	0.900	22.8600
0.400	10.1600	0.910	23.1140
0.410	10.4140	0.920	23.3680
0.420	10.6680	0.930	23.6220
0.430	10.9220	0.940	23.8760
0.440	11.1760	0.950	24.1300
0.450	11.4300	0.960	24.3840
0.460	11.6840	0.970	24.6380
0.470	11.9380	0.980	24.8920
0.480	12.1920	0.990	25.1460
0.490	12.4460	1.000	25.4000

FRACTIONS to DECIMALS to MILLIMETERS

Fraction	Decimal	mm	Fraction	Decimal	mm
1/64	0.0156	0.3969	33/64	0.5156	13.0969
1/32	0.0312	0.7938	17/32	0.5312	13.4938
3/64	0.0469	1.1906	35/64	0.5469	13.8906
1/16	0.0625	1.5875	9/16	0.5625	14.2875
5/64	0.0781	1.9844	37/64	0.5781	14.6844
3/32	0.0938	2.3812	19/32	0.5938	15.0812
7/64	0.1094	2.7781	39/64	0.6094	15.4781
1/8	0.1250	3.1750	5/8	0.6250	15.8750
9/64	0.1406	3.5719	41/64	0.6406	16.2719
5/32	0.1562	3.9688	21/32	0.6562	16.6688
11/64	0.1719	4.3656	43/64	0.6719	17.0656
3/16	0.1875	4.7625	11/16	0.6875	17.4625
13/64	0.2031	5.1594	45/64	0.7031	17.8594
7/32	0.2188	5.5562	23/32	0.7188	18.2562
15/64	0.2344	5.9531	47/64	0.7344	18.6531
1/4	0.2500	6.3500	3/4	0.7500	19.0500
17/64	0.2656	6.7469	49/64	0.7656	19.4469
9/32	0.2812	7.1438	25/32	0.7812	19.8438
19/64	0.2969	7.5406	51/64	0.7969	20.2406
5/16	0.3125	7.9375	13/16	0.8125	20.6375
21/64	0.3281	8.3344	53/64	0.8281	21.0344
11/32	0.3438	8.7312	27/32	0.8438	21.4312
23/64	0.3594	9.1281	55/64	0.8594	21.8281
3/8	0.3750	9.5250	7/8	0.8750	22.2250
25/64	0.3906	9.9219	57/64	0.8906	22.6219
13/32	0.4062	10.3188	29/32	0.9062	23.0188
27/64	0.4219	10.7156	59/64	0.9219	23.4156
7/16	0.4375	11.1125	15/16	0.9375	23.8125
29/64	0.4531	11.5094	61/64	0.9531	24.2094
15/32	0.4688	11.9062	31/32	0.9688	24.6062
31/64	0.4844	12.3031	63/64	0.9844	25.0031
1/2	0.5000	12.7000	1	1.0000	25.4000

Safety first!

Regardless of how enthusiastic you may be about getting on with the job at hand, take the time to ensure that your safety is not jeopardized. A moment's lack of attention can result in an accident, as can failure to observe certain simple safety precautions. The possibility of an accident will always exist, and the following points should not be considered a comprehensive list of all dangers. Rather, they are intended to make you aware of the risks and to encourage a safety conscious approach to all work you carry out on your vehicle.

Essential DOs and DON'Ts

DON'T rely on a jack when working under the vehicle. Always use approved jackstands to support the weight of the vehicle and place them under the recommended lift or support points.

DON'T attempt to loosen extremely tight fasteners (i.e. wheel lug nuts) while the vehicle is on a jack - it may fall.

DON'T start the engine without first making sure that the transmission is in Neutral (or Park where applicable) and the parking brake is set.

DON'T remove the radiator cap from a hot cooling system - let it cool or cover it with a cloth and release the pressure gradually.

DON'T attempt to drain the engine oil until you are sure it has cooled to the point that it will not burn you.

DON'T touch any part of the engine or exhaust system until it has cooled sufficiently to avoid burns.

DON'T siphon toxic liquids such as gasoline, antifreeze and brake fluid by mouth, or allow them to remain on your skin.

DON'T inhale brake lining dust - it is potentially hazardous (see *Asbestos* below).

DON'T allow spilled oil or grease to remain on the floor - wipe it up before someone slips on it.

DON'T use loose fitting wrenches or other tools which may slip and cause injury.

DON'T push on wrenches when loosening or tightening nuts or bolts. Always try to pull the wrench toward you. If the situation calls for pushing the wrench away, push with an open hand to avoid scraped knuckles if the wrench should slip.

DON'T attempt to lift a heavy component alone - get someone to help you.

DON'T rush or take unsafe shortcuts to finish a job.

DON'T allow children or animals in or around the vehicle while you are working on it.

DO wear eye protection when using power tools such as a drill, sander, bench grinder, etc. and when working under a vehicle.

DO keep loose clothing and long hair well out of the way of moving parts.

DO make sure that any hoist used has a safe working load rating adequate for the job.

DO get someone to check on you periodically when working alone on a vehicle.

DO carry out work in a logical sequence and make sure that everything is correctly assembled and tightened.

DO keep chemicals and fluids tightly capped and out of the reach of children and pets.

DO remember that your vehicle's safety affects that of yourself and others. If in doubt on any point, get professional advice.

Steering, suspension and brakes

These systems are essential to driving safety, so make sure you have a qualified shop or individual check your work. Also, compressed suspension springs can cause injury if released suddenly - be sure to use a spring compressor.

Airbags

Airbags are explosive devices that can **CAUSE** injury if they deploy while you're working on the vehicle. Follow the manufacturer's instructions to disable the airbag whenever you're working in the vicinity of airbag components.

Asbestos

Certain friction, insulating, sealing, and other products - such as brake linings, brake bands, clutch linings, torque converters, gaskets, etc. - may contain asbestos or other hazardous friction material. Extreme care must be taken to avoid inhalation of dust from such products, since it is hazardous to health. If in doubt, assume that they do contain asbestos.

Fire

Remember at all times that gasoline is highly flammable. Never smoke or have any kind of open flame around when working on a vehicle. But the risk does not end there. A spark caused by an electrical short circuit, by two metal surfaces contacting each other, or even by static electricity built up in your body under certain conditions, can ignite gasoline vapors, which in a confined space are highly explosive. Do not, under any circumstances, use gasoline for cleaning parts. Use an approved safety solvent.

Always disconnect the battery ground (-) cable at the battery before working on any part of the fuel system or electrical system. Never risk spilling fuel on a hot engine or exhaust component. It is strongly recommended that a fire extinguisher suitable for use on fuel and electrical fires be kept handy in the garage or workshop at all times. Never try to extinguish a fuel or electrical fire with water.

Fumes

Certain fumes are highly toxic and can quickly cause unconsciousness and even death if inhaled to any extent. Gasoline vapor falls into this category, as do the vapors from some cleaning solvents. Any draining or pouring of such volatile fluids should be done in a well ventilated area.

When using cleaning fluids and solvents, read the instructions on the container carefully. Never use materials from unmarked containers.

Never run the engine in an enclosed space, such as a garage. Exhaust fumes contain carbon monoxide, which is extremely poisonous. If you need to run the engine, always do so in the open air, or at least have the rear of the vehicle outside the work area.

The battery

Never create a spark or allow a bare light bulb near a battery. They normally give off a certain amount of hydrogen gas, which is highly explosive.

Always disconnect the battery ground (-) cable at the battery before working on the fuel or electrical systems.

If possible, loosen the filler caps or cover when charging the battery from an external source (this does not apply to sealed or maintenance-free batteries). Do not charge at an excessive rate or the battery may burst.

Take care when adding water to a non maintenance-free battery and when carrying a battery. The electrolyte, even when diluted, is very corrosive and should not be allowed to contact clothing or skin.

Always wear eye protection when cleaning the battery to prevent the caustic deposits from entering your eyes.

Household current

When using an electric power tool, inspection light, etc., which operates on household current, always make sure that the tool is correctly connected to its plug and that, where necessary, it is properly grounded. Do not use such items in damp conditions and, again, do not create a spark or apply excessive heat in the vicinity of fuel or fuel vapor.

Secondary ignition system voltage

A severe electric shock can result from touching certain parts of the ignition system (such as the spark plug wires) when the engine is running or being cranked, particularly if components are damp or the insulation is defective. In the case of an electronic ignition system, the secondary system voltage is much higher and could prove fatal.

Hydrofluoric acid

This extremely corrosive acid is formed when certain types of synthetic rubber, found in some O-rings, oil seals, fuel hoses, etc. are exposed to temperatures above 750-degrees F (400-degrees C). The rubber changes into a charred or sticky substance containing the acid. *Once formed, the acid remains dangerous for years. If it gets onto the skin, it may be necessary to amputate the limb concerned.*

When dealing with a vehicle which has suffered a fire, or with components salvaged from such a vehicle, wear protective gloves and discard them after use.

Troubleshooting

Contents

Engine

1 Engine will not rotate when attempting to start

1   Battery terminal connections loose or corroded. Check the cable terminals at the battery; tighten cable clamp and/or clean off corrosion as necessary (see Chapter 1).
2   Battery discharged or faulty. If the cable ends are clean and tight on the battery posts, turn the key to the On position and switch on the headlights or windshield wipers. If they won't run, the battery is discharged.
3   Automatic transmission not engaged in park (P) or Neutral (N).
4   Broken, loose or disconnected wires in the starting circuit. Inspect all wires and connectors at the battery, starter solenoid and ignition switch (on steering column).
5   Starter motor pinion jammed in flywheel/driveplate ring gear. Remove the starter (Chapter 5) and inspect the pinion and ring gear (Chapter 2A).
6   Starter solenoid faulty (Chapter 5).
7   Starter motor faulty (Chapter 5).
8   Ignition switch faulty (Chapter 12).
9   Engine seized. Try to turn the crankshaft with a large socket and breaker bar on the pulley bolt.
10 Starter relay faulty (Chapter 4).
11 Transmission Range (TR) sensor out of adjustment or defective (Chapter 6).

2 Engine rotates but will not start

1   Fuel tank empty.
2   Battery discharged (engine rotates slowly).
3   Battery terminal connections loose or corroded.
4   Fuel not reaching fuel injectors. Check for clogged fuel filter or lines and defective fuel pump. Also make sure the tank vent lines aren't clogged (Chapter 4).
5   Low cylinder compression. Check as described in Chapter 2A.
6   Water in fuel. Drain tank and fill with new fuel.
7   Defective ignition coil(s) (Chapter 5).
8   Dirty or clogged fuel injector(s) (Chapter 4).
9   Wet or damaged ignition components (Chapters 1 and 5).
10 Worn, faulty or incorrectly gapped spark plugs (Chapter 1).
11 Broken, loose or disconnected wires in the starting circuit (see previous Section).
12 Timing belt or chain failure or wear affecting valve timing (Chapter 2A).
13 Fuel injection or engine control systems failure (Chapters 4 and 6).
14 Defective MAP sensor (Chapter 6)

3 Starter motor operates without turning engine

1   Starter pinion sticking. Remove the starter (Chapter 5) and inspect.
2   Starter pinion or flywheel/driveplate teeth worn or broken. Remove the inspection cover and inspect.

4 Engine hard to start when cold

1   Battery discharged or low. Check as described in Chapter 1.
2   Fuel not reaching the fuel injectors. Check the fuel filter, lines and fuel pump (Chapters 1 and 4).
3   Defective spark plugs (Chapter 1).
4   Defective engine coolant temperature sensor (Chapter 6).
5   Fuel injection or engine control systems malfunction (Chapters 4 and 6).

5 Engine hard to start when hot

1   Air filter dirty (Chapter 1).
2   Fuel not reaching the fuel injectors (see Chapter 4, Section 4). Check for a vapor lock situation, brought about by clogged fuel tank vent lines.
3   Bad engine ground connection.
4   Fuel injection or engine control systems malfunction (Chapters 4 and 6).

6 Starter motor noisy or engages roughly

1   Pinion or driveplate teeth worn or broken. Remove the inspection cover on the left side of the engine and inspect.
2   Starter motor mounting bolts loose or missing.

7 Engine starts but stops immediately

1   Loose or damaged wire harness connections at coil or alternator.
2   Intake manifold vacuum leaks. Make sure all mounting bolts/nuts are tight and all vacuum hoses connected to the manifold are attached properly and in good condition.
3   Insufficient fuel pressure (see Chapter 4).
4   Fuel injection or engine control systems malfunction (Chapters 4 and 6).

8 Engine 'lopes' while idling or idles erratically

1   Vacuum leaks. Check mounting bolts at the intake manifold for tightness. Make sure that all vacuum hoses are connected and in good condition. Use a stethoscope or a length of fuel hose held against your ear to listen for vacuum leaks while the engine is running. A hissing sound will be heard. A soapy water solution will also detect leaks. Check the intake manifold gasket surfaces.
2   Leaking EGR valve or plugged PCV valve (see Chapters 1 and 6).
3   Air filter clogged (Chapter 1).
4   Fuel pump not delivering sufficient fuel (Chapter 4).
5   Leaking head gasket. Perform a cylinder compression check (Chapter 2A).
6   Timing chain(s) worn (Chapter 2A).
7   Camshaft lobes worn (Chapter 2A).
8   Valves burned or otherwise leaking (Chapter 2A).
9   Ignition system not operating properly (Chapters 1 and 5).
10 Fuel injection or engine control systems malfunction (Chapters 4 and 6).

9 Engine misses at idle speed

1   Spark plugs faulty or not gapped properly (Chapter 1).
2   Faulty spark plug wires (Chapter 1).
3   Short circuits in ignition, coil or spark plug wires.
4   Sticking or faulty emissions systems (see Chapter 6).
5   Clogged fuel filter and/or foreign matter in fuel. Remove the fuel filter (Chapter 1) and inspect.
6   Vacuum leaks at intake manifold or hose connections. Check as described in Chapter 2E, Section 4.
7   Low or uneven cylinder compression. Check as described in Chapter 2A.
8   Fuel injection or engine control systems malfunction (Chapters 4 and 6).

10 Excessively high idle speed

1   Sticking throttle linkage (Chapter 4).
2   Vacuum leaks at intake manifold or hose connections.
3   Fuel injection or engine control systems malfunction (Chapters 4 and 6).

11 Battery will not hold a charge

1   Alternator drivebelt defective or not adjusted properly (Chapter 1).
2   Battery cables loose or corroded (Chapter 1).
3   Alternator not charging properly (Chapter 5).
4   Loose, broken or faulty wires in the charging circuit (Chapter 5).
5   Short circuit causing a continuous drain on the battery.
6   Battery defective internally.

12 Alternator light stays on

1   Fault in alternator or charging circuit (Chapter 5).
2   Alternator drivebelt defective or not properly adjusted (Chapter 1).

13 Alternator light fails to come on when key is turned on

1   Faulty bulb (Chapter 12).
2   Defective alternator (Chapter 5).
3   Fault in the printed circuit, dash wiring or bulb holder (Chapter 12).

14 Engine misses throughout driving speed range

1   Fuel filter clogged and/or impurities in the fuel system. Check fuel filter (Chapter 1) or clean system (Chapter 4).
2   Faulty or incorrectly gapped spark plugs (Chapter 1).
3   Emissions system components faulty (Chapter 6).
4   Low or uneven cylinder compression pressures. Check as described in Chapter 2A.
5   Weak or faulty ignition coil(s) (Chapter 5).
6   Weak or faulty ignition system (Chapter 5).
7   Vacuum leaks at intake manifold or vacuum hoses (see Chapter 2E, Section 4).
8   Dirty or clogged fuel injector(s) (Chapter 4).
9   Leaky EGR valve (Chapter 6).
10   Fuel injection or engine control systems malfunction (Chapters 4 and 6).

15 Hesitation or stumble during acceleration

1   Ignition system not operating properly (Chapter 5).
2   Dirty or clogged fuel injector(s) (Chapter 4).
3   Low fuel pressure. Check for proper operation of the fuel pump and for restrictions in the fuel filter and lines (Chapter 4).
4   Fuel injection or engine control systems malfunction (Chapters 4 and 6).

16 Engine stalls

1   Fuel filter clogged and/or water and impurities in the fuel system (Chapter 4).
2   Emissions system components faulty (Chapter 6).
3   Faulty or incorrectly gapped spark plugs (Chapter 1).
4   Vacuum leak at the intake manifold or vacuum hoses.

5   Fuel injection or engine control systems malfunction (Chapters 4 and 6).

17 Engine lacks power

1   Faulty or incorrectly gapped spark plugs (Chapter 1).
2   Air filter dirty (Chapter 1).
3   Faulty ignition coil(s) (Chapter 5).
4   Brakes binding (Chapters 1 and 9).
5   Automatic transmission fluid level incorrect, causing slippage (Chapter 1).
6   Fuel filter clogged and/or impurities in the fuel system (Chapter 4).
7   EGR system not functioning properly (Chapter 6).
8   Use of sub-standard fuel. Fill tank with proper octane fuel.
9   Low or uneven cylinder compression pressures. Check as described in Chapter 2A.
10   Vacuum leak at intake manifold or vacuum hoses (check as described in Chapter 2E, Section 4).
11   Dirty or clogged fuel injector(s) (Chapters 1 and 4).
12   Fuel injection or engine control systems malfunction (Chapters 4 and 6).
13 Restricted exhaust system (Chapter 4).

18 Engine backfires

1   EGR system not functioning properly (Chapter 6).
2   Vacuum leak (refer to Chapter 2E, Section 4).
3   Damaged valve springs or sticking valves.
4   Vacuum leak at the intake manifold or vacuum hoses (see Section 8).

19 Engine surges while holding accelerator steady

1   Vacuum leak at the intake manifold or vacuum hoses (see Section 8).
2   Restricted air filter (Chapter 1).
3   Fuel pump or pressure regulator defective (Chapter 4).
4   Fuel injection or engine control systems malfunction (Chapters 4 and 6).

20 Pinging or knocking engine sounds when engine is under load

1   Incorrect grade of fuel. Fill tank with fuel of the proper octane rating.
2   Carbon build-up in combustion chambers. Remove cylinder head(s) and clean combustion chambers (Chapter 2A).
3   Incorrect spark plugs (Chapter 1).
4   Fuel injection or engine control systems malfunction (Chapters 4 and 6).
5   Restricted exhaust system (Chapter 4).

21 Engine diesels (continues to run) after being turned off

1   Incorrect spark plug heat range (Chapter 1).
2   Vacuum leak at the intake manifold or vacuum hoses.
3   Carbon build-up in combustion chambers. Remove the cylinder head(s) and clean the combustion chambers (Chapter 2A).
4   Valves sticking (Chapter 2A).
5   EGR system not operating properly (Chapter 6).
6   Fuel injection or engine control systems malfunction (Chapters 4 and 6).
7   Check for causes of overheating (see Section 27).

22 Low oil pressure

1   Improper grade of oil.
2   Oil pump worn or damaged (Chapter 2A).
3   Engine overheating (refer to Section 27).
4   Clogged oil filter (Chapter 1).
5   Clogged oil strainer (Chapter 2A).
6   Oil pressure gauge not working properly (Chapter 2A).

23 Excessive oil consumption

1   Loose oil drain plug.
2   Loose bolts or damaged oil pan gasket (Chapter 2A).
3   Loose bolts or damaged front cover gasket (Chapter 2A).
4   Front or rear crankshaft oil seal leaking (Chapter 2A).
5   Loose bolts or damaged valve cover gasket (Chapter 2A).
6   Loose oil filter (Chapter 1).
7   Loose or damaged oil pressure switch (Chapter 2A).
8   Pistons and cylinders excessively worn (Chapter 2A).
9   Piston rings not installed correctly on pistons (Chapter 2A).
10   Worn or damaged piston rings (Chapter 2A).
11   Intake and/or exhaust valve oil seals worn or damaged.
12   Worn valve stems or guides.
13   Worn or damaged valves/guides.
14   Faulty or incorrect PCV valve allowing too much crankcase airflow.

24 Excessive fuel consumption

1   Dirty or clogged air filter element (Chapter 1).
2   Low tire pressure or incorrect tire size (Chapter 10).
3   Inspect for binding brakes.
4   Fuel leakage. Check all connections,

lines and components in the fuel system (Chapter 4).

5 Dirty or clogged fuel injectors (Chapter 4).

6 Fuel injection or engine control systems malfunction (Chapter 4 and Chapter 6).

7 Thermostat stuck open or not installed.

8 Improperly operating transmission.

25 Fuel odor

1 Fuel leakage. Check all connections, lines and components in the fuel system (Chapter 4).

2 Fuel tank overfilled. Fill only to automatic shut-off.

3 Charcoal canister in Evaporative Emissions Control system clogged (Chapter 1).

4 Vapor leaks from Evaporative Emissions Control system lines (Chapter 6).

26 Miscellaneous engine noises

1 A strong dull noise that becomes more rapid as the engine accelerates indicates worn or damaged crankshaft bearings or an unevenly worn crankshaft. To pinpoint the trouble spot, disconnect the electrical connector from one coil at a time and crank the engine over. If the noise stops, the cylinder with the disconnected coil indicates the problem area. Replace the bearing and/or service or replace the crankshaft (Chapter 2A).

2 A similar (yet slightly higher pitched) noise to the crankshaft knocking described in the previous paragraph, that becomes more rapid as the engine accelerates, indicates worn or damaged connecting rod bearings (Chapter 2A). The procedure for locating the problem cylinder is the same as described in Paragraph 1.

3 An overlapping metallic noise that increases in intensity as the engine speed increases, yet diminishes as the engine warms up indicates abnormal piston and cylinder wear (Chapter 2A). To locate the problem cylinder, use the procedure described in Paragraph 1.

4 A rapid clicking noise that becomes faster as the engine accelerates indicates a worn piston pin or piston pin hole. This sound will happen each time the piston hits the highest and lowest points in the stroke (Chapter 2A). The procedure for locating the problem piston is described in Paragraph 1.

5 A metallic clicking noise coming from the water pump indicates worn or damaged water pump bearings or pump. Replace the water pump with a new one (Chapter 3).

6 A rapid tapping sound or clicking sound that becomes faster as the engine speed increases indicates "valve tapping." This can be identified by holding one end of a section of hose to your ear and placing the other end at different spots along the valve cover. The point where the sound is loudest indicates the problem valve. If the pushrod and rocker arm components are in good shape, you likely have a collapsed valve lifter. Changing the engine oil and adding a high viscosity oil treatment will sometimes cure a stuck lifter problem. If the problem persists, the lifters, pushrods and rocker arms must be removed for inspection (see Chapter 2A).

7 A steady metallic rattling or rapping sound coming from the area of the timing chain cover indicates a worn, damaged or out-of-adjustment timing chain. Service or replace the chain and related components (Chapter 2A).

Cooling system

27 Overheating

1 Insufficient coolant in system (Chapter 1).

2 Drivebelt defective or not adjusted properly (Chapter 1).

3 Radiator core blocked or dirty and restricted (Chapter 3).

4 Thermostat faulty (Chapter 3).

5 Cooling fan not functioning properly (Chapter 3).

6 Expansion tank cap not maintaining proper pressure. Have cap pressure tested by gas station or repair shop.

7 Defective water pump (Chapter 3).

8 Improper grade of engine oil.

9 Inaccurate temperature gauge (Chapter 12).

28 Overcooling

1 Thermostat faulty (Chapter 3).

2 Inaccurate temperature gauge (Chapter 12).

29 External coolant leakage

1 Deteriorated or damaged hoses. Loose clamps at hose connections (Chapter 1).

2 Water pump seals defective. If this is the case, water will drip from the weep hole in the water pump body (Chapter 3).

3 Leakage from radiator core or header tank. This will require the radiator to be replaced or professionally repaired (see Chapter 3 for removal procedures).

4 Leakage from the expansion tank or cap.

5 Engine drain plugs or water jacket freeze plugs leaking.

6 Leak from coolant temperature switch (Chapter 3).

7 Leak from damaged gaskets or small cracks (Chapter 2A).

30 Internal coolant leakage

Note: *Internal coolant leaks can usually be detected by examining the oil. Check the dipstick and the underside of the engine oil filler cap for water deposits and an oil consistency like that of a milkshake.*

1 Leaking cylinder head gasket. Have the system pressure tested or remove the cylinder head (Chapter 2A) and inspect.

2 Cracked cylinder bore or cylinder head. Dismantle engine and inspect (Chapter 2A).

31 Abnormal coolant loss

1 Overfilled cooling system (Chapter 1).

2 Coolant boiling away due to overheating (see causes in Section 27).

3 Internal or external leakage (see Sections 29 and 30).

4 Faulty expansion tank cap. Have the cap pressure tested.

5 Cooling system being pressurized by engine compression. This could be due to a cracked head or block or leaking head gasket(s). Have the system tested for the presence of combustion gas in the coolant at a shop. (Combustion leak detectors are also available at some auto parts stores.)

32 Poor coolant circulation

1 Inoperative water pump (Chapter 3).

2 Restriction in cooling system. Drain, flush and refill the system (Chapter 1). If necessary, remove the radiator (Chapter 3) and have it reverse flushed or professionally cleaned.

3 Defective drivebelt tensioner (Chapter 1).

4 Thermostat sticking (Chapter 3).

5 Insufficient coolant (Chapter 1).

33 Corrosion

1 Excessive impurities in the water. Soft, clean water is recommended. Distilled or rainwater is satisfactory.

2 Insufficient antifreeze solution (refer to Chapter 1 for the proper ratio of water to antifreeze).

3 Infrequent flushing and draining of system. Regular flushing of the cooling system should be carried out at the specified intervals as described in (Chapter 1).

Clutch

34 Fails to release (pedal pressed to the floor - shift lever does not move freely in and out of Reverse)

Note: *All clutch service information is located in Chapter 8, unless otherwise noted.*
1 Clutch plate warped, distorted or otherwise damaged.
2 Diaphragm spring fatigued. Remove clutch cover/pressure plate assembly and inspect.
3 Insufficient pedal stroke. Check and adjust as necessary.
4 Lack of grease on pilot bushing.

35 Clutch slips (engine speed increases with no increase in vehicle speed)

1 Worn or oil soaked clutch plate.
2 Clutch plate not broken in. It may take 30 or 40 normal starts for a new clutch to seat.
3 Diaphragm spring weak or damaged. Remove clutch cover/pressure plate assembly and inspect.
4 Flywheel warped (Chapter 2C or Chapter 2D).

36 Grabbing (chattering) as clutch is engaged

1 Oil on clutch plate. Remove and inspect. Repair any leaks.
2 Worn or loose engine or transmission mounts. They may move slightly when clutch is released. Inspect mounts and bolts.
3 Worn splines on transmission input shaft. Remove clutch components and inspect.
4 Warped pressure plate or flywheel. Remove clutch components and inspect.
5 Diaphragm spring fatigued. Remove clutch cover/pressure plate assembly and inspect.
6 Clutch linings hardened or warped.
7 Clutch lining rivets loose.

37 Squeal or rumble with clutch engaged (pedal released)

1 Improper pedal adjustment. Adjust pedal free play.
2 Release bearing binding on transmission shaft. Remove clutch components and check bearing. Remove any burrs or nicks, clean and relubricate before reinstallation.
3 Pilot bushing worn or damaged.
4 Clutch rivets loose.
5 Clutch plate cracked.
6 Fatigued clutch plate torsion springs. Replace clutch plate.

38 Squeal or rumble with clutch disengaged (pedal depressed)

1 Worn or damaged release bearing.
2 Worn or broken pressure plate diaphragm fingers.

39 Clutch pedal stays on floor when disengaged

Sticking cable or release bearing. Inspect cable or remove clutch components as necessary.

Manual transmission

40 Noisy in Neutral with engine running

Note: *All manual transmission service information is located in Chapter 7A, unless otherwise noted.*
1 Input shaft bearing worn.
2 Damaged main drive gear bearing.
3 Insufficient transmission oil (Chapter 1).
4 Transmission oil in poor condition. Drain and fill with proper grade oil. Check old oil for water and debris (Chapter 1).
5 Noise can be caused by variations in engine torque. Change the idle speed and see if noise disappears.

41 Noisy in all gears

1 Any of the above causes, and/or:
2 Worn or damaged output gear bearings or shaft.

42 Noisy in one particular gear

1 Worn, damaged or chipped gear teeth.
2 Worn or damaged synchronizer.

43 Slips out of gear

1 Transmission loose on clutch housing.
2 Stiff shift lever seal.
3 Shift linkage binding.
4 Broken or loose input gear bearing retainer.
5 Dirt between clutch lever and engine housing.
6 Worn linkage.
7 Damaged or worn check balls, fork rod ball grooves or check springs.
8 Worn mainshaft or countershaft bearings.
9 Loose engine mounts (Chapter 2C or Chapter 2D).
10 Excessive gear end play.
11 Worn synchronizers.

44 Oil leaks

1 Excessive amount of lubricant in transmission (see Chapter 1 for correct checking procedures). Drain lubricant as required.
2 Rear oil seal or speedometer oil seal damaged.
3 To pinpoint a leak, first remove all built-up dirt and grime from the transmission. Degreasing agents and/or steam cleaning will achieve this. With the underside clean, drive the vehicle at low speeds so the air flow will not blow the leak far from its source. Raise the vehicle and determine where the leak is located.

45 Difficulty engaging gears

1 Clutch not releasing completely.
2 Loose or damaged shift linkage. Make a thorough inspection, replacing parts as necessary.
3 Insufficient transmission oil (Chapter 1).
4 Transmission oil in poor condition. Drain and fill with proper grade oil. Check oil for water and debris (Chapter 1).
5 Sticking or jamming gears.

46 Noise occurs while shifting gears

1 Check for proper operation of the clutch (Chapter 8).
2 Faulty synchronizer assemblies. Measure baulk ring-to-gear clearance. Also, check for wear or damage to baulk rings or any parts of the synchromesh assemblies.

Automatic transmission

47 Fluid leakage

1 Automatic transmission fluid leaks should not be confused with engine oil, which can easily be blown by airflow to the transmission.
2 To pinpoint a leak, first remove all built-up dirt and grime from the transmission. Degreasing agents and/or steam cleaning will achieve this. With the underside clean, drive the vehicle at low speeds so the air flow will not blow the leak far from its source. Raise the vehicle and determine where the leak is located. Common areas of leakage are:
a) *Fluid pan: tighten mounting bolts and/or replace pan gasket as necessary (Chapter 1).*
b) *Rear extension: tighten bolts and/or replace oil seal as necessary.*
c) *Filler pipe: replace the rubber oil seal where pipe enters transmission case.*
d) *Transmission oil lines: tighten fittings where lines enter transmission case and/or replace lines.*

e) *Vent pipe: transmission overfilled and/or water in fluid (see checking procedures, Chapter 1).*

f) *Vehicle speed sensor: replace the O-ring where speed sensor enters transmission case.*

48 General shift mechanism problems

1   Chapter 7B deals with checking and adjusting the shift cable on automatic transmissions. Common problems which may be caused by out of adjustment cable are:

a) *Engine starting in gears other than P (Park) or N (Neutral).*

b) *Indicator pointing to a gear other than the one actually engaged.*

c) *Vehicle moves with transmission in P (Park) position.*

49 Transmission will not downshift with the accelerator pedal pressed to the floor

Since these transmissions are electronically controlled, check for any diagnostic trouble codes stored in the PCM. The actual repair will most likely have to be performed by a qualified repair shop with the proper equipment.

50 Engine will start in gears other than Park or Neutral

1   Shift cable out of adjustment (Chapter 7B).
2   Defective Transmission Range (TR) sensor (42RLE transmission) or Park/Neutral Contact (NAG1 transmission) inside transmission. Check for stored trouble codes (see Chapter 6). Further diagnosis and repair should be entrusted to a professional.

51 Transmission slips, shifts rough, is noisy or has no drive in forward or Reverse gears

1   There are many probable causes for the above problems, but the home mechanic should concern himself only with one possibility: fluid level.
2   Before taking the vehicle to a shop, check the fluid level and condition as described in Chapter 1. Add fluid, if necessary, or change the fluid and filter if needed. If problems persist, have a professional diagnose the transmission.

Driveshaft

52 Leaks at front of driveshaft

Defective transmission extension housing oil seal. See Chapter 7B for replacement procedure. As this is done, check the output shaft flange for burrs or roughness that could damage the new seal. Remove burrs with a fine file or whetstone.

53 Knock or clunk when transmission is under initial load (just after transmission is put into gear)

1   Loose or disconnected rear suspension components.
2   Loose driveshaft bolts. Inspect all bolts and nuts and tighten them securely.

54 Metallic grating sound consistent with vehicle speed

Pronounced wear in the driveshaft center support bearing. Remove the driveshaft and have the center support bearing replaced by a driveline specialist (special tools are required).

55 Vibration

Note: *Before blaming the driveshaft, make sure the tires are perfectly balanced and perform the following test.*
1   Install a tachometer inside the vehicle to monitor engine speed as the vehicle is driven. Drive the vehicle and note the engine speed at which the vibration (roughness) is most pronounced. Now shift the transmission to a different gear and bring the engine speed to the same point.
2   If the vibration occurs at the same engine speed (rpm) regardless of which gear the transmission is in, the driveshaft is NOT at fault since the driveshaft speed varies.
3   If the vibration decreases or is eliminated when the transmission is in a different gear at the same engine speed, refer to the following probable causes:

a) *Bent or dented driveshaft. Inspect and replace as necessary.*

b) *Undercoating or built-up dirt, etc., on the driveshaft. Clean the shaft thoroughly.*

c) *Driveshaft and/or companion flange out of balance. Check for missing weights on the shaft. Remove driveshaft and reinstall 180-degrees from original position, then recheck. Have the driveshaft balanced if problem persists.*

d) *Loose driveshaft mounting bolts/nuts.*

e) *Worn transmission rear bushing (Chapter 7B).*

56 Scraping noise

Make sure there is nothing, such as an exhaust heat shield, rubbing on the driveshaft.

Axle(s) and differential

57 Noise - same when in drive as when vehicle is coasting

1   Road noise. No corrective action available.
2   Tire noise. Inspect tires and check tire pressures (Chapter 1).
3   Front wheel bearings worn or damaged (Chapter 10).
4   Insufficient differential oil (Chapter 1).
5   Defective differential.

58 Knocking sound when starting or shifting gears

Defective or incorrectly adjusted differential.

59 Noise when turning

Defective differential.

60 Vibration

See probable causes under Driveshaft. Proceed under the guidelines listed for the driveshaft. If the problem persists, check the rear wheel bearings by raising the rear of the vehicle and spinning the wheels by hand. Listen for evidence of rough (noisy) bearings. Remove and inspect (Chapter 8).

61 Oil leaks

1   Pinion oil seal damaged (Chapter 8).
2   Axleshaft oil seals damaged (Chapter 8).
3   Differential cover leaking. Tighten mounting bolts or replace the gasket as required.
4   Loose filler plug on differential (Chapter 1).
5   Clogged or damaged breather on differential.

Brakes

62 Vehicle pulls to one side during braking

1   Defective, damaged or contaminated brake pads on one side. Inspect as described in Chapter 1. Refer to Chapter 9 if replacement is required.
2   Excessive wear of brake pad material or disc on one side. Inspect and repair as necessary.
3   Loose or disconnected front suspension components.
4   Defective front brake caliper assembly. Remove caliper and inspect for stuck piston or damage.
5   Scored or out-of-round disc.
6   Loose brake caliper mounting bolts.

63 Noise (high-pitched squeal or scraping sound)

1   Brake pads worn out. Replace pads with new ones immediately!
2   Glazed or contaminated pads.
3   Dirty or scored disc.
4   Bent support plate.

64 Excessive brake pedal travel

1   Partial brake system failure. Inspect entire system (Chapter 1) and correct as required.
2   Insufficient fluid in master cylinder. Check (Chapter 1) and add fluid - bleed system if necessary.
3   Air in system. Bleed system.
4   Defective master cylinder.

65 Brake pedal feels spongy when depressed

1   Air in brake lines. Bleed the brake system.
2   Deteriorated rubber brake hoses. Inspect all system hoses and lines. Replace parts as necessary.
3   Master cylinder mounting nuts loose. Inspect master cylinder bolts (nuts) and tighten them securely.
4   Master cylinder faulty.
5   Incorrect brake pad clearance.
6   Clogged reservoir cap vent hole.
7   Deformed rubber brake lines.
8   Soft or swollen caliper seals.
9   Poor quality brake fluid. Bleed entire system and fill with new approved fluid.

66 Excessive effort required to stop vehicle

1   Power brake booster not operating properly.
2   Excessively worn brake pads. Check and replace if necessary.
3   One or more caliper pistons seized or sticking. Inspect and rebuild as required.
4   Brake pads contaminated with oil or grease. Inspect and replace as required.
5   Worn or damaged master cylinder or caliper assemblies. Check particularly for frozen pistons.

67 Pedal travels to the floor with little resistance

Little or no fluid in the master cylinder reservoir caused by leaking caliper piston(s) or loose, damaged or disconnected brake lines. Inspect entire system and repair as necessary.

68 Brake pedal pulsates during brake application

1   Wheel bearings damaged, worn or out of adjustment.
2   Caliper not sliding properly due to improper installation or obstructions. Remove and inspect.
3   Disc not within specifications. Check for excessive lateral runout and parallelism. Have the discs resurfaced or replace them with new ones. Also make sure that all discs are the same thickness.

69 Brakes drag (indicated by sluggish engine performance or wheels being very hot after driving)

1   Master cylinder piston seized in bore. Replace master cylinder.
2   Caliper piston seized in bore.
3   Parking brake assembly will not release.
4   Clogged or internally split brake lines.

70 Rear brakes lock up under light brake application

1   Tire pressures too high.
2   Tires excessively worn (Chapter 1).
3   Defective proportioning valve.

71 Rear brakes lock up under heavy brake application

1   Tire pressures too high.
2   Tires excessively worn (Chapter 1).
3   Front brake pads contaminated with oil, mud or water. Clean or replace the pads.
4   Front brake pads excessively worn.
5   Defective proportioning valve (models without ABS).

Suspension and steering

72 Vehicle pulls to one side

1   Tire pressures uneven (Chapter 1).
2   Defective tire (Chapter 1).
3   Excessive wear in suspension or steering components (Chapter 1).
4   Front end alignment incorrect.
5   Front brakes dragging. Inspect as described in Chapter 9, Section 2.
6   Wheel bearings worn (Chapter 10).
7   Wheel lug nuts loose.

73 Shimmy, shake or vibration

1   Tire or wheel out of balance or out of round.
2   Worn wheel bearings (Chapter 10).
3   Shock absorbers and/or suspension components worn or damaged (see Chapter 10).

74 Excessive pitching and/or rolling around corners or during braking

1   Defective shock absorbers. Replace as a set.
2   Sagging springs.
3   Worn or damaged stabilizer bar or bushings.

75 Wandering or general instability

1   Improper tire pressures.
2   Incorrect front end alignment.
3   Worn or damaged steering gear or suspension components.
4   Improperly adjusted steering gear.
5   Out-of-balance wheels.
6   Loose wheel lug nuts.
7   Worn rear shock absorbers.

76 Excessively stiff steering

1   Lack of fluid in the power steering fluid reservoir, where appropriate (Chapter 1).
2   Incorrect tire pressures (Chapter 1).
3   Front end out of alignment.
4   Steering gear out of adjustment or lacking lubrication.
5   Worn or damaged steering gear.
6   Low tire pressures.
7   Worn or damaged balljoints.
8   Worn or damaged tie-rod ends.

77 Excessive play in steering

1   Worn wheel bearings (Chapter 1).
2   Excessive wear in suspension bushings (Chapter 1).
3   Steering gear worn.
4   Incorrect front end alignment.
5   Steering gear mounting bolts loose.
6   Worn or damaged tie-rod ends.

78 Lack of power assistance

1   Steering pump drivebelt faulty or tensioner defective (Chapter 1).
2   Fluid level low (Chapter 1).
3   Hoses or pipes restricting the flow. Inspect and replace parts as necessary.
4   Air in power steering system. Bleed system.
5   Defective power steering pump.

79 Steering wheel fails to return to straight-ahead position

1   Incorrect front end alignment.
2   Tire pressures low.
3   Worn or damaged balljoint.

4   Worn or damaged tie-rod end.
5   Lack of fluid in power steering pump.

80 Steering effort not the same in both directions

1   Leaks in steering gear.
2   Clogged fluid passage in steering gear.

81 Noisy power steering pump

1   Insufficient fluid in pump.
2   Clogged hoses or oil filter in pump.
3   Drivebelt faulty or tensioner defective (Chapter 1).
4   Defective pump.

82 Miscellaneous noises

1   Improper tire pressures.
2   Defective balljoint or tie-rod end.
3   Loose or worn steering gear or suspension components.
4   Defective shock absorber.
5   Defective wheel bearing.
6   Worn or damaged suspension bushings.
7   Loose wheel lug nuts.
8   Worn or damaged shock absorber mounting bushing.
9   Worn stabilizer bar bushings.
10   Incorrect rear axle endplay.
11   See also causes of noises at the rear axle and driveshaft.

83 Excessive tire wear (not specific to one area)

1   Incorrect tire pressures.
2   Tires out of balance.

3   Wheels damaged. Inspect and replace as necessary.
4   Suspension or steering components worn (Chapter 1).
5   Front end alignment incorrect.
6   Lack of proper tire rotation routine. See Routine Maintenance Schedule, Chapter 1.

84 Excessive tire wear on outside edge

1   Incorrect tire pressure.
2   Excessive speed in turns.
3   Front end alignment incorrect.

85 Excessive tire wear on inside edge

1   Incorrect tire pressure.
2   Front end alignment incorrect.

86 Tire tread worn in one place

1   Tires out of balance.
2   Damaged wheel. Inspect and replace if necessary.
3   Defective tire.

Notes

Chapter 1
Tune-up and routine maintenance

Contents

Specifications

Recommended lubricants and fluids

Note: *Listed here are manufacturer recommendations at the time this manual was written. Manufacturers occasionally upgrade their fluid and lubricant specifications, so check with your local auto parts store for current recommendations.*

Engine oil type	API grade "certified for gasoline engines"
Oil viscosity	
2.7L and 3.6L V6 engines	5W-20
3.5L V6 engine	10W-30
5.7L Hemi engine	5W-20
6.1L Hemi engine	5W-40 full synthetic
6.4L Hemi engine	0W-40 full synthetic
Automatic transmission fluid type	
All models except 8-speed transmissions	Mopar® type ATF+4 or equivalent
8-speed automatic transmission	Mopar® type ZF 8&9-speed ATF
Coolant	
2012 and earlier models	50/50 mixture of Mopar® 5 year/100,000 mile Formula antifreeze/coolant with HOAT (Hybrid Organic Additive Technology) and water
2013 and later models	50/50 mixture of Mopar® 10 year/150,000 mile Formula antifreeze/coolant with OAT (Organic Additive Technology) and water

Note: *Models are filled with a 50/50 mixture of Mopar® 5 year/100,000 mile (HOAT) coolant or 10 year/150,000 mile (OAT) coolant shouldn't be mixed with other coolants. Always refill with the correct coolant.*

Recommended lubricants and fluids (continued)

Note: *Listed here are manufacturer recommendations at the time this manual was written. Manufacturers occasionally upgrade their fluid and lubricant specifications, so check with your local auto parts store for current recommendations.*

Manual transmission... Mopar® type ATF+4 or equivalent
Differential lubricant type
    2010 and earlier models.. SAE 75W-140 GL-5 synthetic gear lubricant
    2011 through 2014 models
        300 and Charger
            V6 and 5.7L V8 models .... SAE 75W-140 GL-5 synthetic gear lubricant
            6.4L V8 models .... SAE 75W-90 GL-5 synthetic gear lubricant*
        Challenger
            V6 models .... SAE 75W-140 GL-5 synthetic gear lubricant
            V8 models .... SAE 75W-90 GL-5 synthetic gear lubricant*
    2015 and later models
        195, 220 and 230 rear axles .... SAE 75W-85 GL-5 MOPAR OD synthetic gear lubricant or equivalent
        230 limited-slip rear axle .... SAE 75W-85 GL-5 MOPAR LSD synthetic gear lubricant or equivalent
Brake fluid type... DOT 3 brake fluid
Clutch fluid type.. DOT 3 brake fluid
Power steering fluid .... MOPAR® type ATF+4 or equivalent

Note: * *On limited-slip rear axles, add 4 ounces of MOPAR Limited Slip Additive.*

Capacities*

Cooling system
    2.7L V6 engine
        2006 and earlier models.. 9.7 quarts
        2007 and later models.. 9.9 quarts
    3.5L V6 engine
        2006 and earlier models.. 10.6 quarts
        2007 and later models.. 11.1 quarts
    3.6L V6 engine
        Charger and 300 models .... 10 quarts
        Challenger models
            2012 and earlier models.................................... 10.4 quarts
            2013 and later models...................................... 11.1 quarts
    5.7L V8 engine .... 14.5 to 14.7 quarts
    6.1L and 6.4L V8 engines .... 15 to 15.2 quarts
Engine oil (with filter change)
    2.7L V6 engine .... 6 quarts
    3.5L V6 engine .... 6 quarts
    3.6L V6 engine .... 6 quarts
    V8 engines .... 7 quarts
Automatic transmission (drain and refill)
    42RLE .... 1-1/2 quarts**
    845RE and 8HP70 .... 9.5 quarts (dry fill)**
    NAG1 7.4 quarts**
Manual transmmission
    TR6060 .... N/A

Note: *The best way to determine the amount of fluid to add during a routine fluid change is to measure the amount drained. It is important not to overfill the transmission.*

Rear axle
    195RIA .... 1.0 quarts
    198RII .... 1.5 quarts
    210RII .... 1.7 quarts
    215RII .... 1.7 quarts
    220 and 230RIA .... 1.16 quarts

All capacities approximate. Add as necessary to bring the appropriate levels.

Ignition system

Spark plug type
    2.7L V6 engine RE10PMC5
    3.5L V6 engine .... NGK ZFR5LP-13G or Mopar equivalent
    3.6L V6 engine
        2015 and earlier models.. SP149125AD, Champion - RER8ZWYCB4 or Mopar equivalent
        2016 and later models.. SZFR5LP-13G or Mopar equivalent
    5.7L V8 engine .... NGK LZFR5C-11G or Champion - RE14MCC4
    6.1L V8 engine .... NGK PLZTR5A-13 or equivalent
    6.4L engine .... NGK LZTR6AP11-EG or equivalent

Ignition system (continued)

Spark plug gap
    2.7L V6 engine .... 0.048 inch to 0.058 inch
    3.5L V6 engine .... 0.048 inch to 0.058 inch
    3.6L V6 engine .... 0.043 inch
    V8 engines .... 0.045 inch
Firing order
    V6 engines 1-2-3-4-5-6
    V8 engine 1-8-4-3-6-5-7-2

Brakes

Disc brake pad lining thickness (minimum) .... 1/8 inch
Parking brake shoe lining thickness (minimum) .... 1/16 inch

Front

Cylinder locations - V6 engines

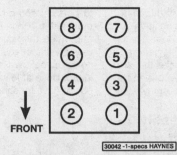

FRONT

Cylinder locations - V8 engines

Torque specifications   Ft-lbs (unless otherwise indicated)

Note: *One foot-pound (ft-lb) of torque is equivalent to 12 inch-pounds (in-lbs) of torque. Torque values below approximately 15 ft-lbs are expressed in inch-pounds, because most foot-pound torque wrenches are not accurate at these smaller values.*

Automatic transmission pan bolts
    42LE .... 174 in-lbs
    845RE and 8HP70 .... 89 in-lbs
    NAG1 .... 70 in-lbs
845RE automatic transmission drain plug .... 89 in-lbs
8HP70 automatic transmission drain plug .... 80 in-lbs
845RE and 8HP70 automatic transmission fill plug 26
Manual transmission drain and fill plug 20
Drivebelt tensioner mounting bolt
    2.7 V6 and 3.5L V6 models 40
    3.6L V6 models 37
    V8 models
        2010 and earlier models 40
        2011 and later models 30
Engine oil drain plug 20
Engine oil filter cap (3.6L engine) 18
Differential drain/fill plug
    195RIA 26
    198RII 44
    210RII 37
    215RII 37
    220 and 230RIA 26
Spark plugs
    2.7L V6 engine .... 150 in-lbs
    3.5L V6 engine 20
    3.6L V6 engine .... 156 in-lbs
    5.7L V8 engine
        2010 and earlier models .... 156 in-lbs
        2011 and later models .... 18.5 to 22
    6.1L V8 engines .... 156 in-lbs
    6.4L V8 engine (Critical taper design, DO NOT exceed 15 ft-lbs)...... 156 in-lbs
Wheel lug nuts
    2013 and earlier models... 110
    2014 and later models
        Base models 130
        SRT.. 111

1 Chrysler 300, Dodge Charger, Magnum and Challenger maintenance schedule

1   The maintenance intervals in this manual are provided with the assumption that you, not the dealer, will be doing the work. These are the minimum maintenance intervals recommended by the factory for vehicles that are driven daily. If you wish to keep your vehicle in peak condition at all times, you may wish to perform some of these procedures even more often. Because frequent maintenance enhances the efficiency, performance and resale value of your car, we encourage you to do so. If you drive in dusty areas, tow a trailer, idle or drive at low speeds for extended periods or drive for short distances (less than four miles) in below freezing temperatures, shorter intervals are also recommended.

2   When your vehicle is new, it should be serviced by a factory authorized dealer service department to protect the factory warranty. In many cases, the initial maintenance check is done at no cost to the owner.

Every 250 miles or weekly, whichever comes first

Check the engine oil level (see Section 4)
Check the engine coolant level (see Section 4)
Check the brake fluid level (see Section 4)
Check the power steering fluid level (see Section 4)
Check the windshield washer fluid level (see Section 4)
Check the tires and tire pressures (see Section 5)
Check the operation of all lights
Check the horn operation

Every 3,000 miles or 3 months, whichever comes first

All items listed above, plus:
Change the engine oil and filter (see Section 6)
Check and replace, if necessary, the air filter element (Section 7)

Every 6,000 miles or 6 months, whichever comes first

All items listed above, plus:
Check the wiper blade condition (see Section 8)
Check and clean the battery and terminals (see Section 9)
Rotate the tires (see Section 10)
Check the seatbelts (see Section 11)
Inspect underhood hoses (see Section 12)
Check the cooling system hoses and connections for leaks and damage (see Section 13)
Check the exhaust pipes and hangers (see Section 14)

Every 15,000 miles or 12 months, whichever comes first

All items listed above, plus:
Check the differential lubricant (Section 4)
Check the manual transmission lubricant (Section 4)

Replace the cabin air filter (see Section 15)
Check the brake system (see Section 16)*
Check the suspension, steering components and driveaxle boots (see Section 17)
Check the fuel system hoses and connections for leaks and damage (see Section 18)
Check the drivebelts and replace if necessary (see Section 19)

Every 30,000 miles or 24 months, whichever comes first

All items listed above, plus:
Replace the air filter element (see Section 7)
Change the brake fluid (see Section 20)
Replace the spark plugs (V8 engine) (see Section 21)
Check the ignition coil(s) (V8 engine) (see Section 22)
Change the differential lubricant (see Section 23)
Change the manual transmission lubricant (see Section 24)

Every 60,000 miles or 48 months, whichever comes first

All items listed above, plus:
Check and replace, if necessary, the PCV valve (see Section 25)

Every 90,000 miles or 72 months, whichever comes first

Replace the spark plugs (V6 engines) (see Section 21)

Every 60 months (regardless of mileage)

Service the cooling system (drain, flush and refill) (Section 26)

Every 100,000 miles

Replace the timing belt (see Chapter 2A)
Change the automatic transmission fluid and filter (Section 27)**

This item is affected by "severe" operating conditions as described below. If your vehicle is operated under "severe" conditions, perform all maintenance indicated with an asterisk (*) at 3000 mile/3 month intervals (unless otherwise specified in the schedule). Severe conditions are indicated if you mainly operate your vehicle under one or more of the following conditions:

    Operating in dusty areas
    Towing a trailer

    Idling for extended periods and/or low speed operation
    Operating in extended temperatures below freezing (32-degrees F/0-degrees C)
    If more the half of your driving is at high speeds in temperatures above 90-degrees F (32-degrees C)

**If operated under one or more of the following conditions, change the or automatic transmission fluid lubricant every 30,000 miles:*

    In heavy city traffic where the outside temperature regularly reaches 90-degrees F (32-degrees C) or higher
    In hilly or mountainous terrain
    Frequent towing of a trailer

Engine compartment layout (3.5L V6 300 model shown)

1	Automatic transmission fluid filler tube	5	Air filter housing
2	Brake fluid reservoir access cover	6	Engine oil filler cap
3	Coolant expansion tank pressure cap	7	Upper radiator hose
4	Power steering fluid reservoir	8	Engine oil dipstick

9	Windshield washer fluid reservoir
10	Remote jump start positive battery terminal
11	Underhood fuse/relay block

Engine compartment layout (3.6L V6 Challenger model shown)

1	Brake fluid reservoir access cover	5	Engine oil filter cap/filter	9 Remote jump start positive
2	Coolant expansion tank (pressure cap)	6	Windshield washer fluid reservoir	battery terminal
3	Engine oil filler cap	7	Underhood fuse/relay block	10 Remote jump start negative
4	Air filter housing	8	Engine oil dipstick	battery terminal

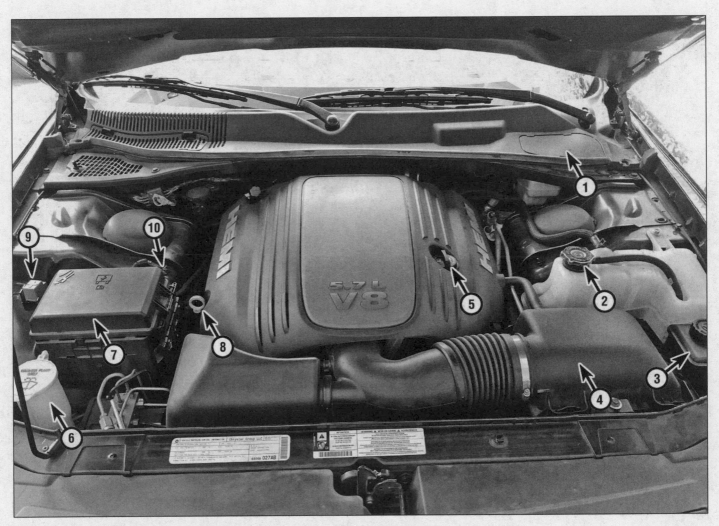

Engine compartment layout (5.7L Hemi V8 Challenger model shown)

1	Brake fluid reservoir access cover	5	Engine oil filler cap	9	Remote jump start positive battery terminal
2	Coolant expansion tank pressure cap	6	Windshield washer fluid reservoir	10	Remote jump start negative battery terminal
3	Power steering fluid reservoir	7	Underhood fuse/relay block		
4	Air filter housing	8	Engine oil dipstick		

Typical underside components (3.5L V6 300 shown)

1	Engine oil filter	3	Steering gear boot	5	Automatic transmission fluid pan
2	Engine oil drain plug	4	Front disc brake caliper	6	Exhaust pipe

Typical underside components (3.6L V6 Challenger shown)

1	Cooling system drain hose	3	Steering gear boot	5	Exhaust pipes
2	Engine oil filter	4	Front disc brake caliper	6	Automatic transmission fluid pan

Typical rear underside components

1	Driveaxle boot	3	Fuel tank filler neck hose
2	Rear differential	4	Rear disc brake caliper

5	Muffler
6	Fuel tank

2   Introduction

1   This Chapter is designed to help the home mechanic maintain the Chrysler 300, Dodge Charger, Magnum and Challenger with the goals of maximum performance, economy, safety and reliability in mind.

2   Included is a master maintenance schedule, followed by procedures dealing specifically with each item on the schedule. Visual checks, adjustments, component replacement and other helpful items are included. Refer to the accompanying illustrations of the engine compartment and the underside of the vehicle for the locations of various components.

3   Servicing your vehicle in accordance with the mileage/time maintenance schedule and the step-by-step procedures will result in a planned maintenance program that should produce a long and reliable service life. Keep in mind that it's a comprehensive plan, so maintaining some items but not others at the specified intervals will not produce the same results.

4   As you service your vehicle, you will discover that many of the procedures can - and should - be grouped together because of the nature of the particular procedure you're performing or because of the close proximity of two otherwise unrelated components to one another.

5   For example, if the vehicle is raised for chassis lubrication, you should inspect the exhaust, suspension, steering and fuel systems while you're under the vehicle. When you're rotating the tires, it makes good sense to check the brakes since the wheels are already removed. Finally, let's suppose you have to borrow or rent a torque wrench. Even if you only need it to tighten the spark plugs, you might as well check the torque of as many critical fasteners as time allows.

6   The first step in this maintenance program is to prepare yourself before the actual work begins. Read through all the procedures you're planning to do, then gather up all the parts and tools needed. If it looks like you might run into problems during a particular job, seek advice from a mechanic or an experienced do-it-yourselfer.

Owner's manual

7   Your vehicle owner's manual was written for your year and model and contains very specific information on component locations, specifications, fuse ratings, part numbers, etc. The owner's manual is an important resource for the do-it-yourselfer to have; if one was not supplied with your vehicle, it can generally be ordered from a dealer parts department.

8   Among other important information, the Vehicle Emissions Control Information (VECI) label contains specifications and procedures for applicable tune-up adjustments and, in some instances, spark plugs. The information on this label is the exact maintenance data recommended by the manufacturer. This data often varies by intended operating altitude,

4.2 Engine oil dipstick

local emissions regulations, month of manufacture, etc.

9   This Chapter contains procedural details, safety information and more ambitious maintenance intervals than you might find in manufacturer's literature. However, you may also find procedures or specifications in your owner's manual or VECI label that differ with what's printed here. In these cases, the owner's manual or VECI label can be considered correct, since it is specific to your particular vehicle.

3   Tune-up general information

1   The term tune-up is used in this manual to represent a combination of individual operations rather than one specific procedure.

2   If, from the time the vehicle is new, the routine maintenance schedule is followed closely and frequent checks are made of fluid levels and high wear items, as suggested throughout this manual, the engine will be kept in relatively good running condition and the need for additional work will be minimized.

3   More likely than not, however, there will be times when the engine is running poorly due to lack of regular maintenance. This is even more likely if a used vehicle, which has not received regular and frequent maintenance checks, is purchased. In such cases, an engine tune-up will be needed outside of the regular routine maintenance intervals.

4   The first step in any tune-up or diagnostic procedure to help correct a poor running engine is a cylinder compression check. A compression check (see Chapter 2E) will help determine the condition of internal engine components and should be used as a guide for tune-up and repair procedures. If, for instance, a compression check indicates serious internal engine wear, a conventional tune-up will not improve the performance of the engine and would be a waste of time and money. Because of its importance, the compression check should be done by someone with the right equipment and the knowledge to

use it properly.

5   The following procedures are those most often needed to bring a generally poor running engine back into a proper state of tune.

Minor tune-up

Check all engine-related fluids (Section 4)
Clean, inspect and test the battery (Section 9)
Check all underhood hoses (Section 12)
Check the cooling system (Section 13)

Major tune-up

All items listed under Minor tune-up, plus...
Replace the air filter (Section 7)
Replace the spark plugs (Section 21)
Check the drivebelt (Section 19)
Replace the PCV valve (Section 25)
Check the charging system (Chapter 5)

4   Fluid level checks (every 250 miles or weekly)

1   Fluids are an essential part of the lubrication, cooling, brake and windshield washer systems. Because the fluids gradually become depleted and/or contaminated during normal operation of the vehicle, they must be periodically replenished. See Recommended lubricants and fluids at the beginning of this Chapter before adding fluid to any of the following components.

Note: *The vehicle must be on level ground when fluid levels are checked.*

Engine oil

2   The oil level is checked with a dipstick, which is located on the side of the engine **(see illustration)**. The dipstick extends through a metal tube down into the oil pan.

3   The oil level should be checked before the vehicle has been driven, or about 5 minutes after the engine has been shut off. If the oil is checked immediately after driving the vehicle, some of the oil will remain in the upper part of the engine, resulting in an inaccurate reading on the dipstick.

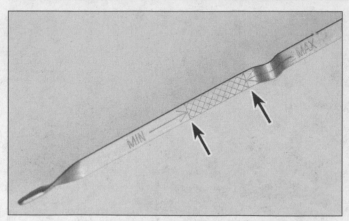

4.4 At its highest point, the level should be between the MIN and MAX marks on the dipstick

4.6 Engine oil filler cap

4   Pull the dipstick out of the tube and wipe all the oil from the end with a clean rag or paper towel. Insert the clean dipstick all the way back into the tube and pull it out again. Note the oil at the end of the dipstick. At its highest point, the level should be between the MIN and MAX marks on the dipstick **(see illustration)**.

5   It takes one quart of oil to raise the level from the MIN mark to the MAX mark on the dipstick. Do not allow the level to drop below the MIN mark or oil starvation may cause engine damage. Conversely, overfilling the engine (adding oil above the MAX mark) may cause oil fouled spark plugs, oil leaks or oil seal failures.

6   To add oil, remove the filler cap from the valve cover **(see illustration)**. After adding oil, wait a few minutes to allow the level to stabilize, then pull out the dipstick and check the level again. Add more oil if required. Install the filler cap and tighten it by hand only.

7   Checking the oil level is an important preventive maintenance step. A consistently low oil level indicates oil leakage through damaged seals, defective gaskets or past worn rings or valve guides. If the oil looks milky in color or has water droplets in it, the cylinder head gasket(s) may be blown or the head(s) or block may be cracked. The engine should be checked immediately. The condition of the oil should also be checked. Whenever you check the oil level, slide your thumb and index finger up the dipstick before wiping off the oil. If you see small dirt or metal particles clinging to the dipstick, the oil should be changed (see Section 6).

Engine coolant

Warning: *Do not allow antifreeze to come in contact with your skin or painted surfaces of the vehicle. Flush contaminated areas immediately with plenty of water. Don't store new coolant or leave old coolant lying around where it's accessible to children or pets - they're attracted by its sweet smell. Ingestion of even a small amount of coolant can be fatal! Wipe up garage floor and drip pan spills immediately. Keep antifreeze containers covered and repair cooling system leaks as soon as they're noticed.*

8   All vehicles covered by this manual are equipped with a pressurized coolant recovery system. An expansion tank is located in the left side of the engine compartment.

9   The coolant level in the tank should be checked regularly. The level in the tank varies with the temperature of the engine. When the engine is cold, the coolant level should be at the COLD FILL MAX mark on the expansion tank. If it isn't, remove the cap from the tank and add the appropriate coolant mixture as listed in this Chapter's Specifications **(see illustration)**.

Warning: *Do not remove the expansion tank cap to check the coolant level when the engine is warm!*

10   Drive the vehicle and recheck the coolant level. If only a small amount of coolant is required to bring the system up to the proper level, water can be used. However, repeated additions of water will dilute the antifreeze and water solution. In order to maintain the proper ratio of antifreeze and water, always top up the coolant level with the correct mixture. Don't use rust inhibitors or additives. An empty plastic milk jug or bleach bottle makes an excellent container for mixing coolant.

11   If the coolant level drops consistently, there may be a leak in the system. Inspect the radiator, hoses, filler cap, drain plugs and water pump (see Section 13). If no leaks are noted, have the expansion tank cap pressure tested by a service station.

12   If you have to remove the expansion tank cap, wait until the engine has cooled completely, then wrap a thick cloth around the cap and turn it to the first stop. If coolant or steam escapes, or if you hear a hissing noise, let the engine cool down longer, then remove the cap.

13   Check the condition of the coolant as well. It should be relatively clear. If it's brown or rust colored, the system should be drained, flushed and refilled. Even if the coolant appears to be normal, the corrosion inhibitors wear out, so it must be replaced at the specified intervals.

Brake fluid

14   The brake master cylinder is located in the driver's side of the engine compartment, near the firewall **(see illustration)**.

15   To check the fluid level of the brake master cylinder, remove the access cover from the

4.9 The coolant level should be at the COLD FILL MAX mark on the expansion tank

4.14 Remove the cover to access the brake fluid reservoir

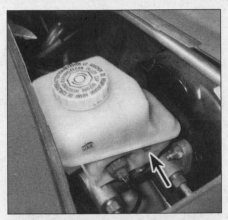

4.15 Never let the brake fluid level drop below the MIN mark

4.23 Power steering reservoir

24   For the check, the front wheels should be pointed straight ahead and the engine should be off.

25   Use a clean rag to wipe off the reservoir cap and the area around the cap. This will help prevent any foreign matter from entering the reservoir during the check.

26   Twist off the cap and check the temperature of the fluid at the end of the dipstick with your finger.

27   Wipe off the fluid with a clean rag, reinsert the dipstick, then withdraw it and read the fluid level. The fluid should be at the proper level, depending on whether it was checked hot or cold **(see illustration)**. Never allow the fluid level to drop below the lower mark on the dipstick.

28   If additional fluid is required, pour the specified type directly into the reservoir, using a funnel to prevent spills.

29   If the reservoir requires frequent fluid additions, all power steering hoses, hose connections, steering gear and the power steering pump should be carefully checked for leaks.

left cowl cover, then look at the MAX and MIN marks on the reservoir **(see illustration)**. The level should be within the specified distance from the maximum fill line.

16   If the level is low, wipe the top of the reservoir cover with a clean rag to prevent contamination of the brake system before lifting the cover.

17   Add only the specified brake fluid to the brake reservoir (refer to *Recommended lubricants and fluids* at the front of this Chapter or to your owner's manual). Mixing different types of brake fluid can damage the system. Fill the brake master cylinder reservoir only to the MAX line.

Warning: *Use caution when filling the reservoir - brake fluid can harm your eyes and damage painted surfaces. Do not use brake fluid that is more than one year old or has been left open. Brake fluid absorbs moisture from the air. Excess moisture can cause a dangerous loss of braking.*

18   While the reservoir cap is removed, inspect the master cylinder reservoir for contamination. If deposits, dirt particles or water droplets are present, the system should be drained and refilled.

19   After filling the reservoir to the proper level, make sure the lid is properly seated to

prevent fluid leakage and/or system pressure loss.

20   The fluid in the brake master cylinder will drop slightly as the brake pads at each wheel wear down during normal operation. If the master cylinder requires repeated replenishing to keep it at the proper level, this is an indication of leakage in the brake system, which should be corrected immediately. If the brake system shows an indication of leakage, check all brake lines and connections, along with the calipers and master cylinder (see Section 16 for more information).

21   If, upon checking the brake master cylinder fluid level, you discover the reservoir empty or nearly empty, the system should be bled (see Chapter 9).

Power steering fluid

22   Check the power steering fluid level periodically to avoid steering system problems, such as damage to the pump.

Caution: *DO NOT hold the steering wheel against either stop (extreme left or right turn) for more than five seconds. If you do, the power steering pump could be damaged.*

23   The power steering reservoir is located at the left side of the engine compartment **(see illustration)**.

Windshield washer fluid

30   Fluid for the windshield washer system is stored in a plastic reservoir located at the right front of the engine compartment **(see illustration)**.

31   In milder climates, plain water can be used in the reservoir, but it should be kept no more than 2/3 full to allow for expansion if the water freezes. In colder climates, use windshield washer system antifreeze, available at any auto parts store, to lower the freezing point of the fluid. Mix the antifreeze with water in accordance with the manufacturer's directions on the container.

Caution: *Do not use cooling system antifreeze - it will damage the vehicle's paint.*

Differential fluid

Note: *It isn't necessary to check this lubricant weekly, every 15,000 miles or 12 months will be adequate.*

32   To check the fluid level, raise the vehicle and support it securely on jackstands. On the

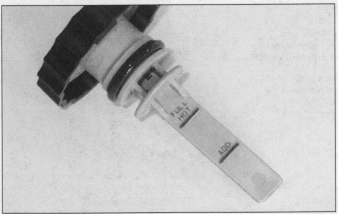

4.27 The fluid should be at the proper level, depending on whether it was checked hot or cold

4.30 Windshield washer fluid reservoir

axle housing, remove the check/fill plug **(see illustration)**. If the lubricant level is correct, it should be up to the lower edge of the hole.

33 If the differential needs more lubricant (if the level is not up to the hole), use a syringe or a gear oil pump to add more. Stop filling the differential when the lubricant begins to run out the hole.

34 Install the plug and tighten it securely. Drive the vehicle a short distance, then check for leaks.

Manual transmission

Note: *It isn't necessary to check this lubricant weekly, every 15,000 miles or 12 months will be adequate.*

35 The manual transmission does not have a dipstick. To check the fluid level, raise the vehicle and support it securely on jackstands. Remove the center splash shield (belly pan) mounting bolts and remove the center splash shield.

36 On the left side of the transmission housing toward the back you will see a fill plug about half-way up on the transmission case. Remove the plug; if the lubricant level is correct, it should be up to the lower edge of the hole.

37 If the transmission needs more lubricant (if the level is not up to the hole), use a syringe or a gear oil pump to add more. Stop filling the transmission when the lubricant begins to run out of the hole.

38 Install the plug and tighten it to the torque listed in this Chapter's Specifications.

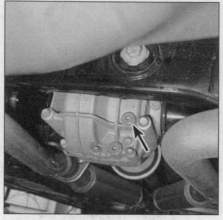

4.32 Differential check/fill plug

39 Drive the vehicle for a short distance, then check the fill plug for leakage. If no leaks are present install the center splash shield and tighten the mounting bolts securely.

5 Tire and tire pressure checks (every 250 miles or weekly)

1 Periodic inspection of the tires may spare you the inconvenience of being stranded with a flat tire. It can also provide you with vital information regarding possible problems in

5.2 A tire tread depth indicator should be used to monitor tire wear - they are available at auto parts stores and service stations and cost very little

the steering and suspension systems before major damage occurs.

2 The original tires on this vehicle are equipped with 1/2-inch wide bands that will appear when tread depth reaches 1/16-inch, at which point they can be considered worn out. Tread wear can be monitored with a simple, inexpensive device known as a tread depth indicator **(see illustration)**.

3 Note any abnormal tread wear **(see illustration)**. Tread pattern irregularities such as cupping, flat spots and more wear on one

UNDERINFLATION

CUPPING

Cupping may be caused by:
● Underinflation and/or mechanical irregularities such as out-of-balance condition of wheel and/or tire, and bent or damaged wheel.
● Loose or worn steering tie-rod or steering idler arm.
● Loose, damaged or worn front suspension parts.

OVERINFLATION

INCORRECT TOE-IN OR EXTREME CAMBER

FEATHERING DUE TO MISALIGNMENT

5.3 This chart will help you determine the condition of your tires, the probable cause(s) of abnormal wear and the corrective action necessary

5.4a If a tire loses air on a steady basis, check the valve core first to make sure it's snug (special inexpensive wrenches are commonly available at auto parts stores)

5.4b If the valve core is tight, raise the corner of the vehicle with the low tire and spray a soapy water solution onto the tread as the tire is turned slowly - slow leaks will cause small bubbles to appear

5.8 To extend the life of your tires, check the air pressure at least once a week with an accurate gauge (don't forget the spare!)

side than the other are indications of front end alignment and/or balance problems. If any of these conditions are noted, take the vehicle to a tire shop or service station to correct the problem.

4   Look closely for cuts, punctures and embedded nails or tacks. Sometimes a tire will hold air pressure for a short time or leak down very slowly after a nail has embedded itself in the tread. If a slow leak persists, check the valve stem core to make sure it is tight **(see illustration)**. Examine the tread for an object that may have embedded itself in the tire or for a plug that may have begun to leak (radial tire punctures are repaired with a plug that is installed in a puncture). If a puncture is suspected, it can be easily verified by spraying a solution of soapy water onto the puncture area **(see illustration)**. The soapy solution will bubble if there is a leak. Unless the puncture is unusually large, a tire shop or service station can usually repair the tire.

5   Carefully inspect the inner sidewall of each tire for evidence of brake fluid leakage. If you see any, inspect the brakes immediately.

6   Correct air pressure adds miles to the life span of the tires, improves mileage and enhances overall ride quality. Tire pressure cannot be accurately estimated by looking at a tire, especially if it's a radial. A tire pressure gauge is essential. Keep an accurate gauge in the glove compartment. The pressure gauges attached to the nozzles of air hoses at gas stations are often inaccurate.

7   Always check tire pressure when the tires are cold. Cold, in this case, means the vehicle has not been driven over a mile in the three hours preceding a tire pressure check. A pressure rise of four to eight pounds is not uncommon once the tires are warm.

8   Unscrew the valve cap protruding from the wheel or hubcap and push the gauge firmly onto the valve stem **(see illustration)**. Note the reading on the gauge and compare the figure to the recommended tire pressure

shown on the tire placard on the driver's side door. Be sure to reinstall the valve cap to keep dirt and moisture out of the valve stem mechanism. Check all four tires and, if necessary, add enough air to bring them up to the recommended pressure.

9   Don't forget to keep the spare tire inflated to the specified pressure (refer to the pressure molded into the tire sidewall).

6   Engine oil and filter change

1   Frequent oil changes are the most important preventive maintenance procedures that can be done by the home mechanic. As engine oil ages, it becomes diluted and contaminated, which leads to premature engine wear.

2   Although some sources recommend oil filter changes every other oil change, we feel that the minimal cost of an oil filter and the relative ease with which it is installed dictate that a new filter be installed every time the oil is changed.

3   Gather together all necessary tools and materials before beginning this procedure **(see illustration)**.

4   You should have plenty of clean rags and newspapers handy to mop up any spills. Access to the underside of the vehicle is greatly improved if the vehicle can be lifted on a hoist, driven onto ramps or supported by jackstands.

Warning: *Do not work under a vehicle which is supported only by a bumper, hydraulic or scissors-type jack.*

5   If this is your first oil change, get under the vehicle and familiarize yourself with the locations of the oil drain plug and the oil filter. The engine and exhaust components will be warm during the actual work, so note how they are situated to avoid touching them when working under the vehicle.

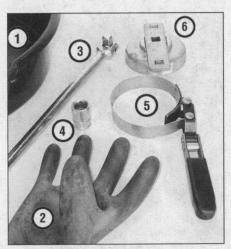

6.3 These tools are required when changing the engine oil and filter

1   *Drain pan* - It should be fairly shallow in depth, but wide in order to prevent spills

2   *Rubber gloves* - When removing the drain plug and filter, it is inevitable that you will get oil on your hands (the gloves will prevent burns)

3   *Breaker bar* - Sometimes the oil drain plug is pretty tight and a long breaker bar is needed to loosen it

4   *Socket* - To be used with the breaker bar or a ratchet (must be the correct size to fit the drain plug)

5   *Filter wrench* - This is a metal band-type wrench, which requires clearance around the filter to be effective

6   *Filter wrench* - This type fits on the bottom of the filter and can be turned with a ratchet or beaker bar (different size wrenches are available for different types of filters)

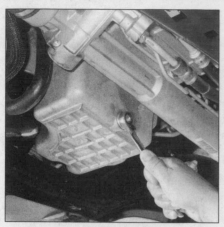

6.9 Use the correct size box-end wrench or six-point socket to avoid rounding off the drain plug when removing or installing it

6.14 The oil filter is on very tight and will require a special wrench for removal - DO NOT use the wrench to tighten the new filter

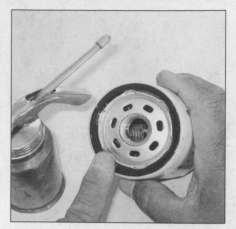

6.18 Lubricate the oil filter gasket with clean engine oil before installing the filter on the engine

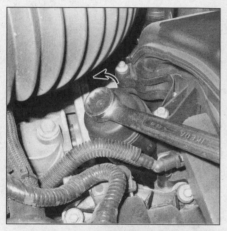

6.21 Using a box end wrench or socket, turn the filter cap counterclockwise to remove it (3.6L V6 engine)

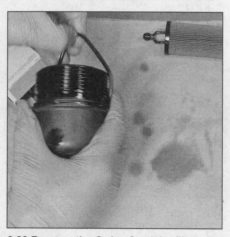

6.23 Remove the O-ring from the filter cap, and replace it with a new one

6   Warm the engine to normal operating temperature. If the new oil or any tools are needed, use this warm-up time to gather everything necessary for the job. The correct type of oil for your application can be found in *Recommended lubricants and fluids* at the beginning of this Chapter.

7   With the engine oil warm (warm engine oil will drain better and more built-up sludge will be removed with it), raise and support the vehicle. Make sure it's safely supported!

8   Move all necessary tools, rags and newspapers under the vehicle. Set the drain pan under the drain plug. Keep in mind that the oil will initially flow from the pan with some force; position the pan accordingly.

9   Being careful not to touch any of the hot exhaust components, use a wrench to remove the drain plug near the bottom of the oil pan **(see illustration)**. Depending on how hot the oil is, you may want to wear gloves while unscrewing the plug the final few turns.
Caution: *When performing an engine oil change on the 3.6L V6, the oil filter cap must*

be removed. Removing the oil filter cap releases oil held within the oil filter cavity and allows it to drain into the oil pan. If the cap is not removed prior to installation of the oil drain plug, some old oil will not completely drain from the engine.

10   Allow the old oil to drain into the pan. It may be necessary to move the pan as the oil flow slows to a trickle.

11   After all the oil has drained, wipe off the drain plug with a clean rag. Small metal particles may cling to the plug and would immediately contaminate the new oil.

12   Clean the area around the drain plug opening and reinstall the plug. Tighten the plug securely with the wrench. If a torque wrench is available, use it to tighten the plug.

13   Move the drain pan into position under the oil filter.

All except 3.6L V6 engines

14   Use the filter wrench to loosen the oil filter **(see illustration)**. Chain or metal band

filter wrenches may distort the filter canister, but it doesn't matter since the filter will be discarded anyway.

15   Completely unscrew the old filter. Be careful; it's full of oil. Empty the oil inside the filter into the drain pan.

16   Compare the old filter with the new one to make sure they're the same type.

17   Use a clean rag to remove all oil, dirt and sludge from the area where the oil filter mounts to the engine. Check the old filter to make sure the rubber gasket isn't stuck to the engine. If the gasket is stuck to the engine (use a flashlight if necessary), remove it.

18   Apply a light coat of clean oil to the rubber gasket on the new oil filter **(see illustration)**.

19   Attach the new filter to the engine, following the tightening directions printed on the filter canister or packing box. Most filter manufacturers recommend against using a filter wrench due to the possibility of overtightening and damage to the seal.

3.6L V6 engines

20   Lift up on the outer edges of the engine cover to disengage the rubber mounts from the ballstuds, and remove the engine cover.

21   Place a rag at the base of the filter housing, then loosen the filter cap by turning it counterclockwise **(see illustration)**.

22   Remove the cap and filter from the engine, then pull the filter out of the cap.

23   Remove and discard the O-ring from the filter cap **(see illustration)**.

24   Insert a new filter into the cap, making sure the filter clips lock into the cap **(see illustration)**.

25   Install a new O-ring onto the cap and apply a light amount of engine oil to the O-ring.

26   Place the filter assembly into the filter housing and carefully thread the cap into the housing. Tighten the cap to the torque listed in this Chapter's Specifications.

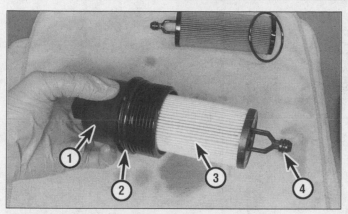

6.24 Oil filter details (3.6L engine)

1	Filter cap	3	Filter element
2	O-ring	4	Filter clip/O-ring

7.1a Release the tabs that secure the two halves of the air cleaner housing . . .

All models

27 Remove all tools, rags, etc., from under the vehicle, being careful not to spill the oil in the drain pan, then lower the vehicle.

28 Move to the engine compartment and locate the oil filler cap.

29 Start by pouring five quarts of fresh oil into the engine. Wait a few minutes to allow the oil to drain into the pan, then check the level on the oil dipstick (see Section 4 if necessary). If the oil level is above the ADD mark, start the engine and allow the new oil to circulate.

30 Run the engine for only about a minute checking the pressure gauge or indicator light to make sure normal oil pressure is achieved. Shut off the engine. Immediately look under the vehicle and check for leaks at the oil pan drain plug and around the oil filter. If either is leaking, tighten with a bit more force.

31 With the new oil circulated and the filter now completely full, recheck the level on the dipstick and add more oil as necessary.

32 During the first few trips after an oil change, make it a point to check frequently for leaks and proper oil level.

33 The old oil drained from the engine cannot be reused in its present state and should be disposed of. Oil reclamation centers, and some recycling centers will accept the oil, which can be refined and used again. After

the oil has cooled it can be drained into a suitable container (capped plastic jugs, topped bottles, milk cartons, etc.) for transport to one of these disposal sites.

Oil change indicator resetting

Note: *It is possible that, driving under the best possible conditions, the oil life monitoring system may not indicate the oil needs to be changed. The manufacturer states that the oil and filter must be changed every 8,000 miles (at the most) or six months and the oil life monitor reset.*

Note: *If the "Oil Change Required" message comes on when the vehicle is immediately restarted, the oil life monitor was not reset and the reset procedure must be done again.*

Note: *If the message is not reset, it will continue to show up each time you turn the ignition switch On or start the vehicle. It is possible to temporarily turn off the message by pressing and releasing the "Menu" button.*

Vehicles not equipped with keyless Enter-N-Go

34 Turn the ignition key to the On position but DO NOT start the engine.

35 Slowly depress the accelerator pedal all the way to the floor, three times within 10 seconds.

36 Turn the ignition key to the Off or Lock position, then start the vehicle.

Vehicles equipped with keyless Enter-N-Go

37 Without pressing the brake pedal, push the Engine Start/Stop button and cycle the ignition to the On/Run position but DO NOT start the engine.

38 Slowly depress the accelerator pedal all the way to the floor, three times within 10 seconds.

39 Without pressing the brake pedal, push the Engine Start/Stop button once to return the ignition to the Off/Lock position.

---

7 Air filter check and replacement (every 3000 miles or 3 months)

---

1 The air filter is located inside a housing at the left (driver's) side of the engine compartment. To remove the air filter, release the tabs or loosen the screws that secure the two halves of the air filter housing together, then separate the cover halves and remove the air filter element **(see illustrations)**.

2 Inspect the outer surface of the filter element. If it is dirty, replace it. If it is only mod-

7.1b . . . or, on later models, loosen the screws . . .

7.1c . . . then lift the air filter housing cover and remove the air filter element

7.5 These tangs must fit into the slots in the filter housing base

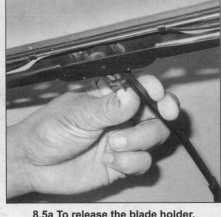

8.5a To release the blade holder, push the release pin . . .

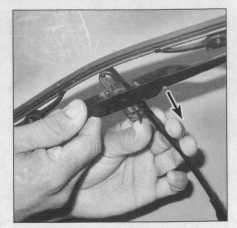

8.5b . . . and pull the wiper blade in the direction of the arrow to separate it from the arm

9.1 Tools and materials required for battery maintenance

1   *Face shield/safety goggles - When removing corrosion with a brush, the acidic particles can easily fly up into your eyes*
2   *Baking soda - A solution of baking soda and water can be used to neutralize corrosion*
3   *Petroleum jelly - A layer of this on the battery posts will help prevent corrosion*
4   *Battery post/cable cleaner - This wire brush cleaning tool will remove all traces of corrosion from the battery posts and cable clamps*
5   *Treated felt washers - Placing one of these on each post, directly under the cable clamps, will help prevent corrosion*
6   *Puller - Sometimes the cable clamps are very difficult to pull off the posts, even after the nut/bolt has been completely loosened. This tool pulls the clamp straight up and off the post without damage*
7   *Battery post/cable cleaner - Here is another cleaning tool which is a slightly different version of number 4 above, but it does the same thing*
8   *Rubber gloves - Another safety item to consider when servicing the battery; remember that's acid inside the battery!*

erately dusty, it can be reused by blowing it clean from the back to the front surface with compressed air. Because it is a pleated paper type filter, it cannot be washed or oiled. If it cannot be cleaned satisfactorily with compressed air, discard and replace it. While the cover is off, be careful not to drop anything down into the housing.
Caution: *Never drive the vehicle with the air filter element removed. Excessive engine wear could result.*
3   Wipe out the inside of the air cleaner housing.
4   Place the new filter into the air cleaner housing, making sure it seats properly.
5   Installation of the housing is the reverse of removal. Make sure the tangs on the filter housing lid engage properly with the lower half of the housing **(see illustration)**.

---

8   Windshield wiper blade inspection and replacement (every 6000 miles or 6 months)

---

1   The windshield wiper and blade assembly should be inspected periodically for damage, loose components and cracked or worn blade elements.
2   Road film can build up on the wiper blades and affect their efficiency, so they should be washed regularly with a mild detergent solution.
3   The action of the wiping mechanism can loosen bolts, nuts and fasteners, so they should be checked and tightened, as necessary, at the same time the wiper blades are checked.
4   If the wiper blade elements are cracked, worn or warped, or no longer clean adequately, they should be replaced with new ones.
5   Lift the arm assembly away from the glass for clearance, press the release lever, then slide the wiper blade assembly out of the hook at the end of the arm **(see illustrations)**.

6   Attach the new wiper to the arm. Connection can be confirmed by an audible click.

---

9   Battery check, maintenance and charging (every 6000 miles or 6 months)

Warning: *Certain precautions must be followed when checking and servicing the battery. Hydrogen gas, which is highly flammable, is always present in the battery cells, so keep lighted tobacco and all other open flames and sparks away from the battery. The electrolyte inside the battery is actually diluted sulfuric acid, which will cause injury if splashed on your skin or in your eyes. It will also ruin clothes and painted surfaces. When removing the battery cables, always detach the negative cable first and hook it up last!*
Note: *The battery is located in the trunk, to the right of the spare tire.*
1   A routine preventive maintenance program for the battery in your vehicle is the only way to ensure quick and reliable starts. But before performing any battery maintenance, make sure that you have the proper equipment necessary to work safely around the battery **(see illustration)**.
2   There are also several precautions that should be taken whenever battery maintenance is performed. Before servicing the battery, always turn the engine and all accessories off and disconnect the cable from the negative terminal of the battery (see Chapter 5, Section 1).
Caution: *On 2015 and later models, if equipped with an Intelligent Battery Sensor (IBS), disconnect the IBS connector first before disconnecting the negative battery cable.*
3   The battery produces hydrogen gas, which is both flammable and explosive. Never create a spark, smoke or light a match around the battery. Always charge the battery in a ventilated area.
4   Electrolyte contains poisonous and corrosive sulfuric acid. Do not allow it to get in

9.6a Battery terminal corrosion usually appears as light, fluffy powder

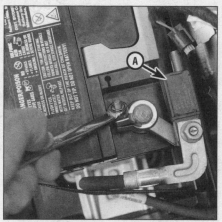

9.6b Removing a cable from the battery post with a wrench - sometimes a pair of special battery pliers are required for this procedure if corrosion has caused deterioration of the nut hex. If equipped with an Intelligent Battery Sensor (A), be sure to disconnect the sensor connector before detaching the cable

9.7a When cleaning the cable clamps, all corrosion must be removed

9.7b Regardless of the type of tool used to clean the battery posts, a clean, shiny surface should be the result

your eyes, on your skin on your clothes. Never ingest it. Wear protective safety glasses when working near the battery. Keep children away from the battery.

5 Note the external condition of the battery. If the positive terminal and cable clamp on your vehicle's battery is equipped with a rubber protector, make sure that it's not torn or damaged. It should completely cover the terminal. Look for any corroded or loose connections, cracks in the case or cover or loose hold-down clamps. Also check the entire length of each cable for cracks and frayed conductors.

6 If corrosion, which looks like white, fluffy deposits **(see illustration)** is evident, particularly around the terminals, the battery should be removed for cleaning. Loosen the cable buts with a wrench, being careful to remove the ground cable first, and slide them off the terminals **(see illustration)**. Then disconnect the hold-down clamp bolt and nut, remove the clamp and lift the battery from the engine compartment.

Warning: *Always remove the ground (-) cable first and hook it up last!*

7 Clean the cable clamps thoroughly with a battery brush or a terminal cleaner and a solution of warm water and baking soda **(see illustration)**. Wash the terminals and the top of the battery case with the same solution but make sure that the solution doesn't get into the battery. When cleaning the cables, terminals and battery top, wear safety goggles and rubber gloves to prevent any solution from coming in contact with your eyes or hands. Wear old clothes too - even diluted, sulfuric acid splashed onto clothes will burn holes in them. If the terminals have been extensively corroded, clean them up with a terminal cleaner **(see illustration)**. Thoroughly wash all cleaned areas with plain water.

8 Make sure that the battery tray is in good condition and the hold-down clamp fasteners are tight. If the battery is removed from the tray, make sure no parts remain in the bot-

tom of the tray when the battery is reinstalled. When reinstalling the hold-down clamp bolts, do not overtighten them.

9 Information on removing and installing the battery can be found in Chapter 5. If you disconnected the cable(s) from the negative and/or positive battery terminals, see Chapter 5, Section 1. Information on jump starting can be found at the front of this manual. For more detailed battery checking procedures, refer to the *Haynes Automotive Electrical Manual.*

Cleaning

10 Corrosion on the hold-down components, battery case and surrounding areas can be removed with a solution of water and baking soda. Thoroughly rinse all cleaned areas with plain water.

11 Any metal parts of the vehicle damaged by corrosion should be covered with a zinc-based primer, then painted.

Charging

Warning: *When batteries are being charged, hydrogen gas, which is very explosive and flammable, is produced. Do not smoke or allow open flames near a charging or a recently charged battery. Wear eye protection when near the battery during charging. Also, make sure the charger is unplugged before connecting or disconnecting the battery from the charger.*

12 Slow-rate charging is the best way to restore a battery that's discharged to the point where it will not start the engine. It's also a good way to maintain the battery charge in a vehicle that's only driven a few miles between starts. Maintaining the battery charge is particularly important in the winter when the bat-

tery must work harder to start the engine and electrical accessories that drain the battery are in greater use.

13 It's best to use a one or two-amp battery charger (sometimes called a "trickle" charger). They are the safest and put the least strain on the battery. They are also the least expensive. For a faster charge, you can use a higher amperage charger, but don't use one rated more than 1/10th the amp/hour rating of the battery. Rapid boost charges that claim to restore the power of the battery in one to two hours are hardest on the battery and can damage batteries not in good condition. This type of charging should only be used in emergency situations.

14 The average time necessary to charge a battery should be listed in the instructions that come with the charger. As a general rule, a trickle charger will charge a battery in 12 to 16 hours.

10   Tire rotation (every 6000 miles or 6 months)

1   The tires should be rotated at the specified intervals and whenever uneven wear is noticed.

2   Refer to the accompanying **illustrations** for the preferred tire rotation pattern.

3   Refer to the information in *Jacking and towing* at the front of this manual for the proper procedures to follow when raising the vehicle and changing a tire. If the brakes are to be checked, don't apply the parking brake as stated. Make sure the tires are blocked to prevent the vehicle from rolling as it's raised.

4   Preferably, the entire vehicle should be raised at the same time. This can be done on a hoist or by jacking up each corner and then lowering the vehicle onto jackstands placed under the frame rails. Always use four jackstands and make sure the vehicle is safely supported.

5   After rotation, check and adjust the tire pressures as necessary. Tighten the lug nuts to the torque listed in this Chapter's Specifications.

11   Seat belt check (every 6000 miles or 6 months)

1   Check seat belts, buckles, latch plates and guide loops for obvious damage and signs of wear.

2   Where the seat belt receptacle bolts to the floor of the vehicle, check that the bolts are secure.

3   See if the seat belt reminder light comes on when the key is turned to the Run or Start position.

12   Underhood hose check and replacement (every 6000 miles or 6 months)

General

Caution: *Replacement of air conditioning hoses must be left to a dealer service department or air conditioning shop that has the equipment to depressurize the system safely and recover the refrigerant. Never remove air conditioning components or hoses until the system has been depressurized.*

1   High temperatures in the engine compartment can cause the deterioration of the rubber and plastic hoses used for engine, accessory and emission systems operation. Periodic inspection should be made for cracks, loose clamps, material hardening and leaks. Information specific to the cooling system hoses can be found in Section 13.

2   Some, but not all, hoses are secured to their fittings with clamps. Where clamps are used, check to be sure they haven't lost their tension, allowing the hose to leak. If clamps aren't used, make sure the hose has not

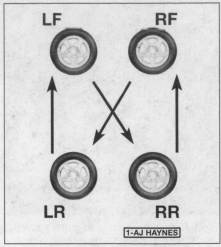

10.2a The recommended four-tire rotation pattern for non-directional tires

expanded and/or hardened where it slips over the fitting, allowing it to leak.

Vacuum hoses

3   It's quite common for vacuum hoses, especially those in the emissions system, to be color-coded or identified by colored stripes molded into them. Various systems require hoses with different wall thickness, collapse resistance and temperature resistance. When replacing hoses, be sure the new ones are made of the same material.

4   Often the only effective way to check a hose is to remove it completely from the vehicle. If more than one hose is removed, be sure to label the hoses and fittings to ensure correct installation.

5   When checking vacuum hoses, be sure to include any plastic T-fittings in the check. Inspect the fittings for cracks and the hose where it fits over the fitting for distortion, which could cause leakage.

6   A small piece of vacuum hose (1/4-inch inside diameter) can be used as a stethoscope to detect vacuum leaks. Hold one end of the hose to your ear and probe around vacuum hoses and fittings, listening for the "hissing" sound characteristic of a vacuum leak. **Warning:** *When probing with the vacuum hose stethoscope, be very careful not to come into contact with moving engine components such as the drivebelt, cooling fan, etc.*

Fuel hose

Warning: *There are certain precautions that must be taken when inspecting or servicing fuel system components. Work in a well-ventilated area and do not allow open flames (cigarettes, appliances, etc.) or bare light bulbs near the work area. Mop up any spills immediately and do not store fuel soaked rags where they could ignite. The fuel system is under high pressure, so if any fuel lines are to be disconnected, the pressure in the system must be relieved first (see Chapter 4 for more information).*

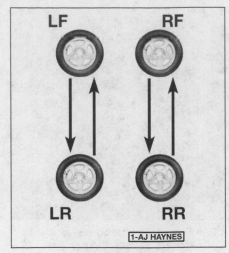

10.2b The recommended four-tire rotation pattern for directional tires

7   Check all rubber fuel lines for deterioration and chafing. Check especially for cracks in areas where the hose bends and just before fittings, such as where a hose attaches to the fuel filter.

8   High quality fuel line, made specifically for high-pressure fuel injection systems, must be used for fuel line replacement. Never, under any circumstances, use unreinforced vacuum line, clear plastic tubing or water hose for fuel lines.

9   Spring-type clamps are commonly used on fuel lines. These clamps often lose their tension over a period of time, and can be "sprung" during removal. Replace all spring-type clamps with screw clamps whenever a hose is replaced.

Metal lines

10   Sections of metal line are routed along the frame, between the fuel tank and the engine. Check carefully to be sure the line has not been bent or crimped and that cracks have not started in the line.

11   If a section of metal fuel line must be replaced, only seamless steel tubing should be used, since copper and aluminum tubing don't have the strength necessary to withstand normal engine vibration.

12   Check the metal brake lines where they enter the master cylinder and brake proportioning unit for cracks in the lines or loose fittings. Any sign of brake fluid leakage calls for an immediate and thorough inspection of the brake system.

13   Cooling system check (every 6000 miles or 6 months)

1   Many major engine failures can be attributed to a faulty cooling system. If the vehicle is equipped with an automatic transmission, the cooling system also cools the transmission fluid and thus plays an important role in prolonging transmission life.

Check for a chafed area that could fail prematurely.

Check for a soft area indicating the hose has deteriorated inside.

Overtightening the clamp on a hardened hose will damage the hose and cause a leak.

Check each hose for swelling and oil-soaked ends. Cracks and breaks can be located by squeezing the hose.

13.4 Hoses, like drivebelts, have a habit of failing at the worst possible time - to prevent the inconvenience of a blown radiator or heater hose, inspect them carefully as shown here

14.2a Inspect the muffler (A) for signs of deterioration, and all hangers (B)

14.2b Inspect all flanged joints (arrow indicates pipe-to-manifold joint) for signs of exhaust gas leakage

2   The cooling system should be checked with the engine cold. Do this before the vehicle is driven for the day or after it has been shut off for at least three hours.

3   Remove the cooling system pressure cap and thoroughly clean the cap, inside and out, with clean water. Also clean the filler neck on the radiator. All traces of corrosion should be removed. The coolant inside the radiator should be relatively transparent. If it is rust-colored, the system should be drained, flushed and refilled (see Section 26). If the coolant level is not up to the top, add additional anti-freeze/coolant mixture (see Section 4).

4   Carefully check the large upper and lower radiator hoses along with the smaller diameter heater hoses that run from the engine to the firewall. Inspect each hose along its entire length, replacing any hose that is cracked, swollen or shows signs of deteriora-

tion. Cracks may become more apparent if the hose is squeezed (see illustration). Regardless of condition, it's a good idea to replace hoses with new ones every two years.

5   Make sure all hose connections are tight. A leak in the cooling system will usually show up as white or rust-colored deposits on the areas adjoining the leak. If wire-type clamps are used at the ends of the hoses, it may be a good idea to replace them with more secure screw-type clamps.

6   Use compressed air or a soft brush to remove bugs, leaves, etc., from the front of the radiator or air conditioning condenser. Be careful not to damage the delicate cooling fins or cut yourself on them.

7   Every other inspection, or at the first indication of cooling system problems, have the cap and system pressure tested. If you don't have a pressure tester, most repair shops will do this for a minimal charge.

14   Exhaust system check (every 6000 miles or 6 months)

1   With the engine cold (at least three hours after the vehicle has been driven),

check the complete exhaust system from the manifold to the end of the tailpipe. Be careful around the catalytic converter, which may be hot even after three hours. The inspection should be done with the vehicle on a hoist to permit unrestricted access. If a hoist isn't available, raise the vehicle and support it securely on jackstands.

2   Check the exhaust pipes and connections for signs of leakage and/or corrosion indicating a potential failure. Make sure that all brackets and hangers are in good condition and tight (see illustrations).

3   Inspect the underside of the body for holes, corrosion, open seams, etc., which may allow exhaust gasses to enter the passenger compartment. Seal all body openings with silicone sealant or body putty.

4   Rattles and other noises can often be traced to the exhaust system, especially the hangers, mounts and heat shields. Try to move the pipes, mufflers and catalytic converter. If the components can come in contact with the body or suspension parts, secure the exhaust system with new brackets and hangers.

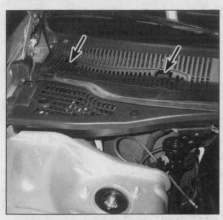

15.2 Release the two tabs to open the air inlet grille

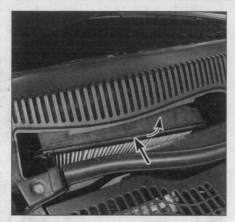

15.3a Swing the top of the filter carrier open (later models) . . .

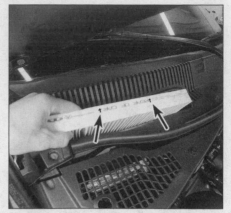

15.3b . . . then remove the filter, noting the direction of the arrows on the filter (indicates airflow direction)

16.7a With the wheel off, check the thickness of the inner pad through the inspection hole (front disc shown, rear disc caliper similar)

16.7b The outer pad is more easily checked at the edge of the caliper

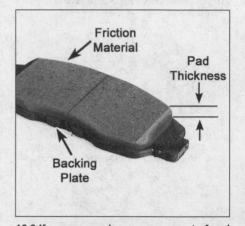

16.9 If a more precise measurement of pad thickness is necessary, remove the pads and measure the remaining friction material

15   Cabin air filter replacement (every 15,000 miles or 12 months)

1   Some models are equipped with an air filtering element in the air conditioning system, located in a housing under the cowl cover.
2   Remove the air inlet grille from the cowl cover **(see illustration)**.
3   Open the filter door by sliding it forward and down (early models) or by lifting the forward edge of the filter carrier upwards (later models), then remove the filter. Note the direction of the arrow(s) on the filter, indicating airflow direction **(see illustrations)**.
4   Installation is the reverse of the removal procedure.

16   Brake system check (every 15,000 miles or 12 months)

Warning: *The dust created by the brake system is harmful to your health. Never blow it out with compressed air and don't inhale any of it. An approved filtering mask should be worn when working on the brakes. Do not, under*

any circumstances, use petroleum-based solvents to clean brake parts. Use brake system cleaner only!
Note: *For detailed photographs of the brake system, refer to Chapter 9.*
1   In addition to the specified intervals, the brakes should be inspected every time the wheels are removed or whenever a defect is suspected.
2   Any of the following symptoms could indicate a potential brake system defect: The vehicle pulls to one side when the brake pedal is depressed; the brakes make squealing or dragging noises when applied; brake pedal travel is excessive; the pedal pulsates; or brake fluid leaks, usually onto the inside of the tire or wheel. A brake pedal that sinks slowly to the floor, with no apparent external fluid leakage, indicates a faulty master cylinder.
Note: *A faulty master cylinder can leak fluid into the power brake booster.*
3   Loosen the wheel lug nuts.
4   Raise the vehicle and place it securely on jackstands.
5   Remove the wheels (see *Jacking and towing* at the front of this book, or your owner's manual, if necessary).

Disc brakes

6   There are two pads (an outer and an inner) in each caliper. The pads are visible with the wheels removed.
7   Check the pad thickness by looking at each end of the caliper and through the inspection window in the caliper body **(see illustrations)**. If the lining material is less than the thickness listed in this Chapter's Specifications, replace the pads.
Note: *Keep in mind that the lining material is riveted or bonded to a metal backing plate and the metal portion is not included in this measurement.*
8   If it is difficult to determine the exact thickness of the remaining pad material by the above method, or if you are at all concerned about the condition of the pads, remove the caliper(s), then remove the pads from the calipers for further inspection (refer to Chapter 9).
9   Once the pads are removed from the calipers, clean them with brake cleaner and re-measure them with a ruler or a vernier caliper **(see illustration)**.
10   Measure the disc thickness with a micrometer to make sure that it still has service life remaining. If any disc is thinner than

17.6 Check the shocks for leakage at the indicated area

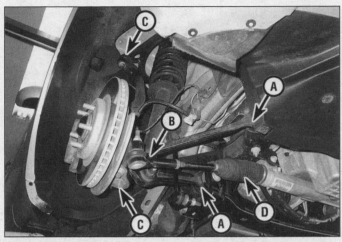

17.9 Examine the mounting points for the upper and lower control arms on the front suspension (A), lower shown, upper not visible), the tie-rod ends (B), the balljoints (C), and the steering gear boots (D)

the specified minimum thickness, replace it (see Chapter 9). Even if the disc has service life remaining, check its condition. Look for scoring, gouging and burned spots. If these conditions exist, remove the disc and have it resurfaced (see Chapter 9).

11   Before installing the wheels, check all brake lines and hoses for damage, wear, deformation, cracks, corrosion, leakage, bends and twists, particularly in the vicinity of the rubber hoses at the calipers. Check the clamps for tightness and the connections for leakage. Make sure that all hoses and lines are clear of sharp edges, moving parts and the exhaust system. If any of the above conditions are noted, repair, reroute or replace the lines and/ or fittings as necessary (see Chapter 9).

Brake booster check

12   Sit in the driver's seat and perform the following sequence of tests.

13   With the brake fully depressed, start the engine - the pedal should move down a little when the engine starts.

14   With the engine running, depress the brake pedal several times - the travel distance should not change.

15   Depress the brake, stop the engine and hold the pedal in for about 30 seconds - the pedal should neither sink nor rise.

16   Restart the engine, run it for about a minute and turn it off. Then firmly depress the brake several times - the pedal travel should decrease with each application.

17   If your brakes do not operate as described, the brake booster has failed. Refer to Chapter 9 for the replacement procedure.

Parking brake

18   One method of checking the parking brake is to park the vehicle on a steep hill with the parking brake set and the transmission in Neutral (be sure to stay in the vehicle for this check!). If the parking brake cannot prevent the vehicle from rolling, check the parking brake (see Chapter 9).

---

17 Suspension, steering and driveaxle boot check (every 15,000 miles or 12 months)

Note: *The steering linkage and suspension components should be checked periodically. Worn or damaged suspension and steering linkage components can result in excessive and abnormal tire wear, poor ride quality and vehicle handling and reduced fuel economy. For detailed illustrations of the steering and suspension components, refer to Chapter 10.*

Shock absorber check

1   Park the vehicle on level ground, turn the engine off and set the parking brake. Check the tire pressures.

2   Push down at one corner of the vehicle, then release it while noting the movement of the body. It should stop moving and come to rest in a level position within one or two bounces.

3   If the vehicle continues to move up-and-down or if it fails to return to its original position, a worn or weak shock absorber is probably the reason.

4   Repeat the above check at each of the three remaining corners of the vehicle.

5   Raise the vehicle and support it securely on jackstands.

6   Check the shock absorbers for evidence of fluid leakage **(see illustration)**. A light film of fluid is no cause for concern. Make sure that any fluid noted is from the shocks and not from some other source. If leakage is noted, replace the shocks as a set.

7   Check the shocks to be sure that they are securely mounted and undamaged. Check the upper mounts for damage and wear. If damage or wear is noted, replace the shocks as a set (front or rear).

8   If the shocks must be replaced, refer to Chapter 10 for the procedure.

Steering and suspension check

9   Visually inspect the steering and suspension components (front and rear) for damage and distortion. Look for damaged seals, boots and bushings and leaks of any kind. Examine the bushings where the control arms meet the chassis **(see illustration)**.

10   Clean the lower end of the steering knuckle. Have an assistant grasp the lower edge of the tire and move the wheel in-and-out while you look for movement at the steering knuckle-to-control arm balljoint. If there is any movement the suspension balljoint(s) must be replaced.

11   Grasp each front tire at the front and rear edges, push in at the front, pull out at the rear and feel for play in the steering system components. If any freeplay is noted, check the idler arm and the tie-rod ends for looseness **(see illustration)**.

17.11 With the steering wheel in the locked position and the vehicle raised, grasp the front tire as shown and try to move it back-and-forth - if any play is noted, check the steering gear mounts and tie-rod ends for looseness

17.14 Inspect the inner and outer driveaxle boots for loose clamps, cracks or signs of leaking lubricant

12   Additional steering and suspension system information and illustrations can be found in Chapter 10.

Driveaxle boot check

13   The driveaxle boots are very important because they prevent dirt, water and foreign material from entering and damaging the constant velocity (CV) joints. Oil and grease can cause the boot material to deteriorate prematurely, so it's a good idea to wash the boots with soap and water. Because it constantly pivots back and forth following the steering action of the front hub, the outer CV boot wears out sooner and should be inspected regularly.

14   Inspect the boots for tears and cracks as well as loose clamps (**see illustration**). If there is any evidence of cracks or leaking lubricant, it is recommended that the driveaxle be replaced with a rebuilt unit as described in Chapter 8.

---

18   Fuel system check
(every 15,000 miles or 12 months)

Warning: *Gasoline is flammable, so take extra precautions when you work on any part of the fuel system. Don't smoke or allow open flames or bare light bulbs near the work area, and don't work in a garage where a gas-type appliance (such as a water heater or clothes dryer) is present. Since fuel is carcinogenic, wear fuel-resistant gloves when there's a possibility of being exposed to fuel, and, if you spill any fuel on your skin, rinse it off immediately with soap and water. Mop up any spills immediately and do not store fuel-soaked rags where they could ignite. When you perform any kind of work on the fuel system, wear safety glasses and have a Class B type fire extinguisher on hand. The fuel system is under constant pressure, so, before any lines are disconnected, the fuel system pressure must be relieved (see Chapter 4).*

1   If you smell fuel while driving or after the

vehicle has been sitting in the sun, inspect the fuel system immediately.

2   Remove the fuel filler cap and inspect it for damage and corrosion. The gasket should have an unbroken sealing imprint. If the gasket is damaged or corroded, install a new cap.

3   Inspect the fuel feed line for cracks. Make sure that the connections between the fuel lines and the fuel injection system and between the fuel lines and the in-line fuel filter are tight.

Warning: *Your vehicle is fuel injected, so you must relieve the fuel system pressure before servicing fuel system components. The fuel system pressure relief procedure is outlined in Chapter 4.*

4   Since some components of the fuel system - the fuel tank and part of the fuel feed and return lines, for example - are underneath the vehicle, they can be inspected more easily with the vehicle raised on a hoist. If that's not possible, raise the vehicle and support it on jackstands.

5   With the vehicle raised and safely supported, inspect the fuel tank and filler neck for punctures, cracks and other damage. The connection between the filler neck and the tank is particularly critical. Sometimes a rubber filler neck will leak because of loose clamps or deteriorated rubber. Inspect all fuel tank mounting brackets and straps to be sure that the tank is securely attached to the vehicle.

Warning: *Do not, under any circumstances, try to repair a fuel tank (except rubber components).*

6   Carefully check all rubber hoses and metal lines leading away from the fuel tank. Check for loose connections, deteriorated hoses, crimped lines and other damage. Repair or replace damaged sections as necessary.

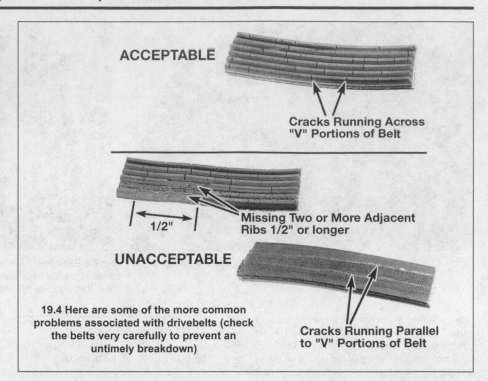

ACCEPTABLE

Cracks Running Across "V" Portions of Belt

1/2"

Missing Two or More Adjacent Ribs 1/2" or longer

UNACCEPTABLE

Cracks Running Parallel to "V" Portions of Belt

19.4 Here are some of the more common problems associated with drivebelts (check the belts very carefully to prevent an untimely breakdown)

---

19   Drivebelt check and replacement
(every 15,000 miles or 12 months)

1   The drivebelt is located at the front of the engine and plays an important role in the overall operation of the vehicle and its components. Due to its function and material make-up, the drivebelt is prone to failure after a period of time and should be inspected and adjusted periodically to prevent major engine damage.

2   The vehicles covered by this manual are equipped with a single self-adjusting serpentine drivebelt, which is used to drive all of the accessory components such as the alternator, power steering pump, water pump and air conditioning compressor.

Inspection

3   With the engine off, open the hood and locate the drivebelt at the front of the engine. Using your fingers (and a flashlight, if necessary), move along the belts checking for cracks and separation of the belt plies. Also check for fraying and glazing, which gives the belt a shiny appearance. Both sides of each belt should be inspected, which means you will have to twist the belt to check the underside.

4   Check the ribs on the underside of the belt. They should all be the same depth, with none of the surface uneven (**see illustration**).

5   The tension of the belt is automatically adjusted by the belt tensioner and does not require any adjustments. Drivebelt wear can be checked visually by inspecting the wear indicator marks located on the side of the tensioner body. Locate the belt tensioner at

19.5 Belt wear indicator marks are located on the tensioner body - when the belt reaches the maximum wear mark it must be replaced (3.5L V6 engine shown)

1   *Normal range*
2   *Maximum wear mark*

19.6 Rotate the tensioner arm to relieve belt tension

19.11 The drivebelt tensioner is secured by a single bolt

the front of the engine, then find the tensioner operating marks **(see illustration)**. If the indicator mark is outside the operating range, the belt should be replaced.

Replacement

6   To replace the belt, remove the air inlet tube (see Chapter 4) then rotate the tensioner to relieve the tension on the belt **(see illustration)**. All models have a square hole in the tensioner arm that will accept a breaker bar or ratchet.
7   Remove the belt from the auxiliary components and carefully release the tensioner.
8   Route the new belt over the various pulleys, again rotating the tensioner to allow the belt to be installed, then release the belt tensioner. Make sure the belt fits properly into the pulley grooves - it must be completely engaged.
Note: *Most models have a drivebelt routing decal on the upper radiator panel to help during drivebelt installation.*
9   Reconnect the battery.

Tensioner replacement

10   Remove the drivebelt.
11   Remove the bolt that secures the drivebelt to the engine, then remove the tensioner **(see illustration)**.
12   Installation is the reverse of removal. Tighten the mounting bolt to the torque listed in this Chapter's Specifications.

20   Brake fluid change (every 30,000 miles or 24 months)

Warning: *Brake fluid can harm your eyes and damage painted surfaces, so use extreme caution when handling or pouring it. Do not use brake fluid that has been standing open or is more than one year old. Brake fluid ab-*

sorbs moisture from the air. Excess moisture can cause a dangerous loss of braking effectiveness.

1   At the specified intervals, the brake fluid should be drained and replaced. Since the brake fluid may drip or splash when pouring it, place plenty of rags around the master cylinder to protect any surrounding painted surfaces.
2   Before beginning work, purchase the specified brake fluid (see *Recommended lubricants and fluids* at the beginning of this Section).
3   Remove the cap from the master cylinder reservoir.
4   Using a hand suction pump or similar device, withdraw the fluid from the master cylinder reservoir.
5   Add new fluid to the master cylinder until it rises to the base of the filler neck.
6   Bleed the brake system as described in Chapter 9 at all four brakes until new and uncontaminated fluid is expelled from the bleeder screw. Be sure to maintain the fluid level in the master cylinder as you perform the bleeding process. If you allow the master cylinder to run dry, air will enter the system.
7   Refill the master cylinder with fluid and check the operation of the brakes. The pedal should feel solid when depressed, with no sponginess.
Warning: *Do not operate the vehicle if you are in doubt about the effectiveness of the brake system.*

21   Spark plug check and replacement (see Maintenance schedule for intervals)

1   The spark plugs are located in the cylinder head(s).
2   In most cases the tools necessary for spark plug replacement include a spark plug socket which fits onto a ratchet (this special socket is padded inside to protect the porce-

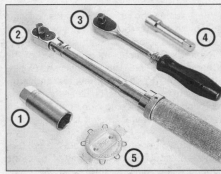

21.2 Tools required for changing spark plugs

1   *Spark plug socket - This will have special padding inside to protect the spark plug porcelain insulator*
2   *Torque wrench - Although not mandatory, use of this tool is the best way to ensure that the plugs are tightened properly*
3   *Ratchet - Standard hand tool to fit the plug socket*
4   *Extension - Depending on model and accessories, you may need special extensions and universal joints to reach one or more of the plugs*
5   *Spark plug gap gauge - This gauge for checking the gap comes in a variety of styles. Make sure the gap for your engine is included*

lain insulators on the new plugs and hold them in place), various extensions and a feeler gauge to check and adjust the spark plug gap **(see illustration)**. Since these engines are equipped with an aluminum cylinder head, a torque wrench should be used when tightening the spark plugs.
3   The best approach when replacing the spark plugs is to purchase the new spark plugs beforehand, adjust them to the proper gap and then replace each plug one at a time.

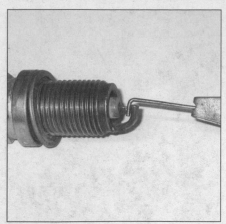

21.5 Spark plug manufacturers recommend using a wire-type gauge when checking the gap - the wire should slide between the electrodes with a slight drag

21.8 Remove the mounting bolt and the individual coil(s) to access the spark plug(s)

21.10 Use a socket and extension to unscrew the spark plugs

When buying the new spark plugs, be sure to obtain the correct plug for your specific engine. This information can be found in this Chapter's Specifications at the front of this Chapter, or on the Vehicle Emissions Control Information (VECI) label located under the hood. If differences exist between the sources, purchase the spark plug type specified on the VECI label as it was printed for your specific engine.

4   Allow the engine to cool completely before attempting to remove any of the plugs. During this cooling-off time, each of the new spark plugs can be inspected for defects and the gaps can be checked.

5   The gap is checked by inserting the proper thickness gauge between the electrodes at the tip of the plug **(see illustration)**. The gap between the electrodes should be as listed in this Chapter's Specifications or in your owner's manual. The wire should touch each of the electrodes. Also, at this time check for cracks in the spark plug body (if any are found, the plug must not be used).

6   Cover the fender to prevent damage to the paint, then remove the engine cover if equipped.

Note: *Fender covers are available from auto parts stores but an old blanket will work just fine.*

7   If you're working on a V6 model, remove the upper intake manifold (see Chapter 2A, Chapter 2B or Chapter 2C).

8   All models are equipped with individual ignition coils which must be removed first to access the spark plugs **(see illustration).**

9   If compressed air is available, use it to blow any dirt or foreign material away from the spark plug area. This will eliminate the possibility of material falling into the cylinder through the spark plug hole as the spark plug is removed.

Warning: *Wear eye protection when using compressed air!*

10   Place the spark plug socket over the plug and remove it from the engine by turning it in a counterclockwise direction **(see illustration).**

11   Compare the spark plug with the chart on the inside back cover of this manual to get an indication of the overall running condition of the engine.

12   It's a good idea to lightly coat the threads of the spark plugs with an anti-seize compound **(see illustration)** to insure that the spark plugs do not seize in the aluminum cylinder head.

13   It's often difficult to insert spark plugs into their holes without cross-threading them. To avoid this possibility, fit a piece of rubber hose over the end of the spark plug **(see illustration).** The flexible hose acts as a universal joint to help align the plug with the plug hole. Should the plug begin to cross-thread, the hose will slip on the spark plug, preventing thread damage. Install the spark plug and tighten it to the torque listed in this Chapter's Specifications.

14   Before pushing the ignition coil onto the end of the plug, inspect the ignition coil following the procedures outlined in Section 22.

15   Repeat the procedure for the remaining spark plugs.

22   Ignition coil check (every 30,000 miles or 24 months)

1   Clean the coils with a dampened cloth and dry them thoroughly.

2   Inspect each coil for cracks, damage and carbon tracking. Make sure the coil fits securely onto the spark plug(s). If damage exists, replace the coil (see Chapter 5).

23   Differential lubricant change (every 30,000 miles or 24 months)

1   This procedure should be performed after the vehicle has been driven so the lubricant will be warm and therefore will flow out of the differential more easily. Raise the vehicle and support it securely on jackstands.

2   Remove the check/fill plug, then remove

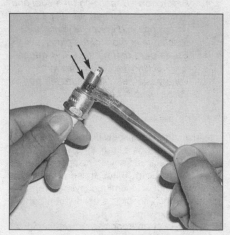

21.12 Apply a thin film of anti-seize compound to the spark plug threads to prevent damage to the cylinder head

21.13 A length of snug-fitting rubber hose will save time and prevent damaged threads when installing the spark plugs

23.2a Differential drain plug (early models)

23.2b Differential check/fill plug (A) and
drain plug (B) (later models)

25.2a On the 3.5L V6 engine, the PCV
valve is located at the rear of the
right valve cover

the drain plug and drain the lubricant **(see illustration)**.

3   Reinstall the drain plug and tighten it to the torque listed in this Chapter's Specifications.

4   Use a hand pump, syringe or funnel to fill differential housing with the specified lubricant until the lubricant level is up to the bottom of the hole.

Note: *If friction modifier is required, add it to the differential first.*

5   Install the filler plug and tighten it to the torque listed in this Chapter's Specifications.

24   Manual transmission lubricant change (every 30,000 miles or 24 months)

1   This procedure should be performed after the vehicle has been driven so the lubricant will be warm and therefore flow out of the transmission more easily.

2   Raise the vehicle and support it securely

on jackstands. Remove the center splash shield (belly pan) mounting bolts and remove the center splash shield.

3   Position a drain pan under the transmission. Remove the transmission fill plug on the right side of the transmission tail housing and allow the lubricant to drain into the pan.

4   After the lubricant has drained completely, reinstall the drain plug and tighten it to the torque listed in this Chapter's Specifications.

5   Using a hand pump, syringe or funnel, fill the transmission with the specified lubricant until it is level with the lower edge of the filler hole. Reinstall the fill plug and tighten it to the torque listed in this Chapter's Specifications.

6   Lower the vehicle.

7   Drive the vehicle for a short distance, then check the drain and fill plugs for leakage. If no leaks are present install the center splash shield and tighten the mounting bolts securely.

8   The old lubricant drained from the transmission cannot be reused in its present state and should be disposed of. Check with your

local auto parts store, disposal facility or environmental agency to see if they will accept the lubricant for recycling. After the lubricant has cooled it can be drained into a container (capped plastic jugs, topped bottles, milk cartons, etc.) for transport to one of these disposal sites. Don't dispose of the lubricant by pouring it on the ground or down a drain!

25   Positive Crankcase Ventilation (PCV) valve replacement (every 60,000 miles or 48 months)

1   Remove the engine cover by pulling it straight up.

V6 engines

2   Detach the hose from the valve, then unscrew the valve from the upper intake manifold (2.7L engine) or the right valve cover (3.5L engine) **(see illustration)**. On 3.6L engines, the valve is secured to the rear of the right valve cover with screws **(see illustration)**.

Note: *On some 3.6L models, the valve is retained by two screws; on others, three screws are used. Also, be sure to replace the PCV valve seal in the valve cover.*

V8 engines

3   To replace the valve, twist the valve and pull it out from the intake manifold, noting its installed position **(see illustration)**.

All models

4   When purchasing a replacement PCV valve, make sure it's for your particular vehicle and engine size. Compare the old valve with the new one to make sure they're the same.

5   Inspect the rubber O-rings for damage and hardening. Replace it with a new one if necessary.

6   Install the PCV valve securely into position.

25.2b On the 3.6L V6, the valve is also
located at the rear of the right valve cover,
but it's retained by screws
(one not visible here)

25.3 On V8 engines, the PCV valve is
located on the top of the intake manifold

26.4 The radiator drain fitting is located at the lower corner of the radiator

26.5 The block drain plugs are generally located about one to two inches above the oil pan - there is one on each side of the engine block

26.15a Coolant bleeder valve location on the 3.5L V6 engine

26.15b On the 3.6L V6 engine, the coolant bleeder screw is located on the thermostat housing

26   Cooling system servicing (draining, flushing and refilling) (every 60 months)

Warning: *Do not allow antifreeze to come in contact with your skin or painted surfaces of the vehicle. Rinse off spills immediately with plenty of water. Antifreeze is highly toxic if ingested. Never leave antifreeze lying around in an open container or in puddles on the floor; children and pets are attracted by its sweet smell and may drink it. Check with local authorities about disposing of used antifreeze. Many communities have collection centers which will see that antifreeze is disposed of safely.*

Warning: *The engine must be completely cool before beginning this procedure.*

1   Periodically, the cooling system should be drained, flushed and refilled to replenish the antifreeze mixture and prevent formation of rust and corrosion, which can impair the performance of the cooling system and

cause engine damage. When the cooling system is serviced, all hoses and the expansion tank cap should be checked and replaced if necessary.

Draining

2   Apply the parking brake and block the wheels. If the vehicle has just been driven, wait several hours to allow the engine to cool down before beginning this procedure.

3   Once the engine is completely cool, remove the expansion tank cap.

4   Move a large container under the radiator drain to catch the coolant. Attach a length of hose to the drain fitting to direct the coolant into the container, then open the drain fitting (a pair of pliers may be required to turn it) **(see illustration)**.

Note: *Some models are equipped with a drain hose as standard equipment.*

5   After the coolant stops flowing out of the radiator, move the container under the engine block drain plugs and allow the coolant in the block to drain **(see illustration)**.

6   While the coolant is draining, check the condition of the radiator hoses, heater hoses and clamps (refer to Section 13 if necessary). Replace any damaged clamps or hoses.

7   Apply thread sealant to the block drain plugs. Reinstall the drain plugs and tighten them securely. Close the radiator drain plug.

Flushing

8   Close the radiator and engine block drain plugs. Fill the cooling system with clean water, following the *Refilling* procedure (see Step 14).

9   Start the engine and allow it to reach normal operating temperature, then rev up the engine a few times.

10   Turn the engine off and allow it to cool completely, then drain the system as described earlier.

11   Repeat Steps 8 through 10 until the water being drained is free of contaminants.

12   Severe cases of radiator contamination

or clogging will require removing the radiator (see Chapter 3) and reverse flushing it. This involves inserting a hose in the bottom radiator outlet to allow the clean water to run against the normal flow, draining out through the top. A radiator repair shop should be consulted if further cleaning or repair is necessary.

13   When the coolant is regularly drained and the system refilled with the correct coolant mixture, there should be no need to employ chemical cleaners or descalers.

Refilling

14   Place the heater temperature control in the maximum heat position.

15   On models with a coolant bleeder valve, connect a hose to the bleeder valve and direct the other end of the hose into a container. On models with a bleeder screw or plug, this will not be possible. Be prepared for coolant spillage and cover any rubber parts that are in the way.

   a) *On 2.7L V6 engines the bleeder valve is located at the front of the engine, on the water outlet.*

   b) *On 3.5L V6 engines the bleeder valve is located at the front of the intake manifold, just to the rear of where the upper radiator hose connects to the water outlet* **(see illustration)**.

   c) *On 3.6L V6 models, the bleeder screw is located on the thermostat housing* **(see illustration)**.

   d) *2010 and earlier V8 engines are equipped with a plug in place of a bleeder valve* **(see illustration)**, *at the front of the engine on the water outlet housing; you'll have to obtain a bleeder valve (1/4-18 npt thread) at a dealer parts department or an auto parts store and install it in place of the plug.*

16   Open the bleeder valve (be sure to do this before adding coolant).

17   Make sure to use the proper coolant listed in this Chapter's Specifications. Slowly

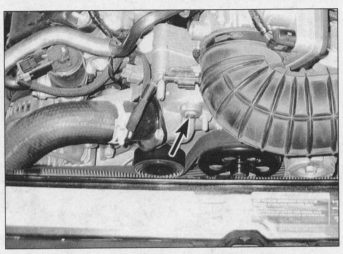

26.15c Coolant bleeder plug location on 2010
and earlier V8 engines

27.18 845RE and 8HP70 transmission pan drain plug location

fill the cooling system with the recommended mixture of antifreeze and water until a bubble-free stream flows from the bleeder valve, screw or plug, then tighten the bleeder valve, screw or plug.

18   Continue adding coolant to the expansion tank until the level is up to the MIN mark. Wait five minutes and recheck the coolant level, adding if necessary.

19   Install the expansion tank cap and run the engine in a well-ventilated area until the thermostat opens (coolant will begin flowing through the radiator and the upper radiator hose will become hot).

20   Turn the engine off and let it cool. Add more coolant mixture to bring the level between the MIN and MAX marks on the expansion tank.

21   Squeeze the upper radiator hose to expel air, then add more coolant mixture if necessary. Install the expansion tank cap.

22   Start the engine, allow it to reach normal operating temperature and check for leaks. Also, set the heater and blower controls to the maximum setting and check to see that the heater output from the air ducts is warm. This is a good indication that all air has been purged from the cooling system.

---

27   Automatic transmission fluid and filter change/fluid level check (every 100,000 miles)

Note: *On 845RE and 8HP70 transmissions, the filter is integrated into the oil pan and is not serviceable.*

Note: *On 42RLE and NAG1 transmissions, before beginning this procedure, read the Note preceding Step 24.*

1   At the specified time intervals, the transmission fluid should be drained and replaced. Since the fluid will remain hot long after driving, perform this procedure only after everything has cooled down completely.

2   Before beginning work, purchase the specified transmission fluid (see *Recommended lubricants and fluids* at the front of this Chapter) and a new filter.

3   Other tools necessary for this job include jackstands to support the vehicle in a raised position, a drain pan capable of holding several quarts, newspapers and clean rags.

4   Raise and support the vehicle on jackstands.

42RLE and NAG1 transmissions

5   With a drain pan in place, remove the front and side transmission pan mounting bolts.

6   Loosen the rear pan bolts one turn.

7   Carefully pry the transmission pan loose with a screwdriver, allowing the fluid to drain.

8   Remove the remaining bolts, pan and gasket. Carefully clean the gasket surface of the transmission to remove all traces of the old gasket and sealant.

9   Drain the fluid from the transmission pan, clean the pan with solvent and dry it with compressed air. Be careful not to lose the magnet.

10   Remove the filter and pry out the seal.

11   Push a new filter seal fully into its bore, then install the new filter.

12   Make sure the gasket surface on the transmission pan is clean.

13   If you're working on a NAG1 transmission, install a new gasket. Put the pan in place against the transmission and install the bolts.

14   If you're working on a 42LE transmission, clean the pan and transmission surfaces thoroughly with brake system cleaner and apply a continuous bead of ATF-resistant RTV sealant to the pan, then bolt it in place and tighten the bolts to the torque listed in this Chapter's Specifications within five minutes.

15   Working around the pan, tighten each bolt a little at a time until the final torque fig-

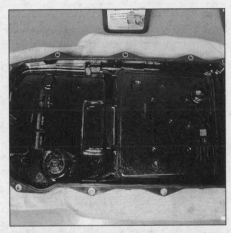

27.20 Make sure the sealing surface of the pan is clean and the new gasket is aligned properly on the pan.

ure is reached.

16   Lower the vehicle and add the specified amount of automatic transmission fluid through the filler tube and check the fluid level (see Steps 24 through 28).

17   Check under the vehicle for leaks during the first few trips.

845RE and 8HP70 transmissions

18   Remove the transmission oil pan drain plug **(see illustration)** and allow the transmission fluid to drain. Measure and record the amount of fluid drained to use for the filling procedure.

19   Remove the transmission pan mounting bolts in the reverse order of the tightening sequence **(see illustration 27.21)** and lower the oil pan.

20   Wipe out the oil pan surface with a clean rag then install a new gasket, aligning the gasket with the bolt holes in the pan **(see illustration)**.

27.21 Transmission fluid pan bolt tightening sequence

27.22 Transmission filler plug location

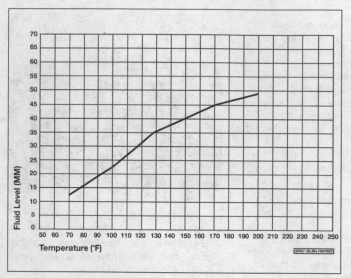

27.27a Automatic transmission fluid level vs. temperature chart - 42RLE transmission

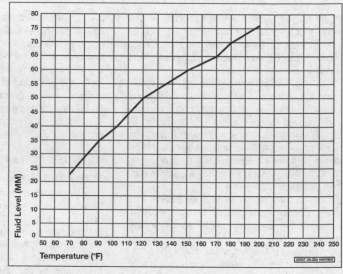

27.27b Automatic transmission fluid level vs. temperature chart - NAG1 transmission

21   Install the transmission pan and tighten the fasteners in sequence **(see illustration)** and to the torque listed in this Chapter's Specifications.

22   Remove the filler plug **(see illustration)** and add the measured amount of automatic transmission fluid through the filler plug hole and check the fluid level (see Steps 30 through 36).

Fluid level check

23   The manufacturer states that routine checks of the automatic transmission fluid are not necessary; this procedure should only be used when refilling the transmission after the fluid has been drained, unless an obvious leak has been detected. Low fluid level can lead to slipping or loss of drive, while overfilling can cause foaming and loss of fluid.

42RLE and NAG1 transmissions

Note: *To perform this procedure you'll have to obtain a dipstick (part number 9336), which you'll have to obtain through a dealer parts department (unlike most vehicles, a dipstick is not present in the transmission filler tube). A scan tool capable of reading the temperature of the automatic transmission fluid will also be required.*

24   With the parking brake set, start the engine, then move the shift lever through all the gear ranges, ending in Neutral. Wait at least two minutes.

Note: *The fluid level must be checked with the vehicle level and the engine running at idle.*

25   With the transmission at normal operating temperature, insert the dipstick into the filler tube and pull out again, and note the fluid level.

26   Using a scan tool capable of reading the temperature of the automatic transmission fluid, check the temperature of the automatic transmission fluid.

27   Determine the correct fluid level by using the temperature and the height of the fluid on the dipstick with the accompanying chart(s) **(see illustrations)**.

28   If additional fluid is required, add it directly into the tube using a funnel. Add the fluid a little at a time and keep checking the level until it's correct. Once the oil level is correct, install the dipstick tube cap.

845RE and 8HP70 transmissions

29   If not already done, raise the vehicle in the front and the rear and support the vehicle securely on jackstands.

Note: *The vehicle must be in a level position for proper fluid level readings.*

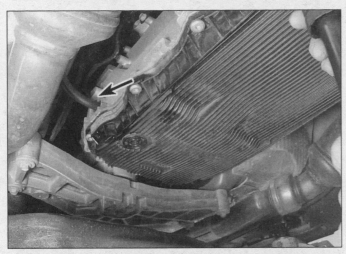

27.30 Insert the hose from a hand pump into the fill plug opening and fill the transmission

27.35 The transmission fluid must be between 86°F (30°C) and 122°F (50°C)

30   Remove the transmission fill plug **(see illustration 27.22)** then insert the hose from a hand pump into the opening **(see illustration)** and fill the transmission using the fluid level measurement from Step 18.

31   Install the fill plug and tighten it securely at this time.

32   With the tires at least eight inches off the the ground, apply the brakes and start the vehicle. Then shift the transmission into reverse and hold for five seconds.

33   With the brakes still applied, shift the transmission into drive and hold for 5 seconds then release the brakes, slowly accelerate to 2nd gear and hold for 5 seconds.

34   Apply the brakes and shift the transmission in Neutral and raise the engine speed to 2000 RPM for five seconds

35   With the brakes applied, return the engine to idle and shift the transmission into Park. The fluid should now be at the correct temperature as shown on the vehicle instrument display (select "Vehicle Info/Trans Temp") **(see illustration)**.

36   With the temperature at the proper range, remove the transmission fill plug **(see illustration 27.30)**. Fluid should trickle out if it is full. If the fluid does not trickle out, add fluid as needed (see Step 30) until it drains out from the hole. If there is too much fluid, allow the excess to drain until there is just a slight trickle of fluid coming from the hole.

37   Install the fill plug and tighten the plug to the torque listed in this Chapter's Specifications.

38   Lower the vehicle and check under the vehicle for leaks during the first few trips.

Notes

Chapter 2 Part A
2.7L V6 engine

Contents

Specifications

General

Firing order	1-2-3-4-5-6
Bore and stroke	3.386 x 3.091 inches
Displacement	167 cubic inches
Cylinder numbers (front to rear)	
Left bank	2-4-6
Right bank	1-3-5
Compression pressure	See Chapter 2E

Front

Cylinder identification diagram

Cylinder head

Warpage limit	0.008 inch

Camshaft

Endplay	0.0051 to 0.0110 inch
Lobe wear limit	
Standard	0.001 inch
Service limit	0.010 inch
Camshaft bearing oil clearance	
Standard	0.0020 to 0.0035 inch
Service limit	0.0040 inch
Camshaft journal diameter	0.9441 to 0.9449 inch
Camshaft bore diameter	0.9469 to 0.9476 inch

Oil pump

Cover warpage limit	0.001 inch
Inner and outer rotor thickness	
2005	0.3731 to 0.3741 inch
2006 and later	0.3919 to 0.3929 inch
Outer rotor diameter (minimum)	3.5109 inches
Outer rotor-to-housing clearance (maximum)	0.015 inch
Inner rotor-to-outer rotor lobe clearance	0.008 inch
Oil pump housing-to-rotor side clearance	0.003 inch

Torque specifications

Ft-lbs (unless otherwise indicated)

Note: *One foot-pound (ft-lb) of torque is equivalent to 12 inch-pounds (in-lbs) of torque. Torque values below approximately 15 foot-pounds are expressed in inch-pounds, because most foot-pound torque wrenches are not accurate at these smaller values.*

Camshaft sprocket bolts	21
Camshaft bearing cap bolts **(see illustration 7.20)**	105 in-lbs
Camshaft timing chain tensioner bolts (secondary)	105 in-lbs
Crankshaft pulley bolt	125
Cylinder head bolts (in sequence - **see illustration 10.19**)	
Step 1 (tighten bolts 1 through 8)	35
Step 2 (tighten bolts 1 through 8)	55
Step 3 (tighten bolts 1 through 8)	55
Step 4 (tighten bolts 1 through 8)	Tighten an additional 90 degrees
Step 5 (tighten bolts 9 through 11)	21
Drivebelt tensioner	40
Driveplate bolts	70
Exhaust manifold bolts	200 in-lbs
Exhaust manifold heat shield bolts	105 in-lbs
Engine mount heat shield bolts	97 in-lbs
Engine mount to bracket bolts	55
Engine mount retaining nuts	55
Engine rear mount to transmission bolts	35
Idler pulley bolt	21
Intake manifold bolts	
Upper intake manifold bolts **(see illustration 8.8)**	105 in-lbs
Lower intake manifold bolts **(see illustration 8.16)**	105 in-lbs
Oil pan bolts	
6-mm bolts	105 in-lbs
6-mm nuts	105 in-lbs
8-mm bolts	21
Oil pan drain plug	20
Oil pan support brace **(see illustration 13.6)**	
Step 1	
Support brace-to-oil pan bolts	10 in-lbs
Support brace-to-transmission bolts	40
Step 2	
Support brace-to-oil pan bolts	40
Oil pick-up tube mounting bolts	21
Oil pump mounting bolts	21
Oil pump cover screws	105 in-lbs
Rear main oil seal retainer bolts	105 in-lbs
Timing chain cover bolts	
6 mm	105 in-lbs
10 mm	40
Timing chain guide bolts	21
Timing chain guide access plugs	15
Timing chain tensioner arm pivot bolt	21
Timing chain tensioner (primary)	105 in-lbs
Timing chain tensioner (secondary)	105 in-lbs
Transmission crossmember bolts	50
Valve cover bolts	105 in-lbs
Water outlet housing mounting bolts	105 in-lbs
Water outlet tube-to-housing bolts	30 in-lbs

1 General information

1   This Part of Chapter 2 is devoted to in-vehicle repair procedures for the 2.7L V6 engine. These engines utilize an aluminum block with six cylinders arranged in a "V" shape at a 60-degree angle between the two banks. The overhead camshaft aluminum cylinder heads are equipped with replaceable valve guides and seats. Stamped steel rocker arms with an integral roller bearing actuate the valves.

2   Information concerning engine removal and installation and engine overhaul can be found in Chapter 2E.

3   The following repair procedures are based on the assumption that the engine is installed in the vehicle. If the engine has been removed from the vehicle and mounted on a stand, many of the steps outlined in this Part of Chapter 2 will not apply.

2 Repair operations possible with the engine in the vehicle

1   Many major repair operations can be accomplished without removing the engine from the vehicle.

2   Clean the engine compartment and the exterior of the engine with some type of degreaser before any work is done. It will make the job easier and help keep dirt out of the internal areas of the engine.

3   Depending on the components involved, it may be helpful to remove the hood to improve access to the engine as repairs are performed (refer to Chapter 11, if necessary). Cover the fenders to prevent damage to the paint. Special pads are available, but an old bedspread or blanket will also work.

4   If vacuum, exhaust, oil or coolant leaks develop, indicating a need for gasket or seal replacement, the repairs can generally be made with the engine in the vehicle. The intake and exhaust manifold gaskets, oil pan gasket, crankshaft oil seals and cylinder head gaskets are all accessible with the engine in place.

5   Exterior engine components, such as the intake and exhaust manifolds, the oil pan, the oil pump, the water pump (see Chapter 3), the starter motor, the alternator and the fuel system components (see Chapter 4) can be removed for repair with the engine in place.

6   Since the cylinder heads can be removed without pulling the engine, valve component servicing can also be accomplished with the engine in the vehicle. Replacement of the camshafts, timing chains and sprockets are also possible with the engine in the vehicle.

7   In extreme cases caused by a lack of necessary equipment, repair or replacement of piston rings, pistons, connecting rods and rod bearings is possible with the engine in the

3.5 A compression gauge can be used in the number one plug hole to assist in finding TDC

vehicle. However, this practice is not recommended because of the cleaning and preparation work that must be done to the components involved.

3 Top Dead Center (TDC) for number one piston - locating

1   Top Dead Center (TDC) is the highest point in the cylinder that each piston reaches as it travels up the cylinder bore. Each piston reaches TDC on the compression stroke and again on the exhaust stroke, but TDC generally refers to piston position on the compression stroke.

2   Positioning the piston(s) at TDC is an essential part of many procedures such as valve timing, camshaft and timing chain/belt sprocket removal.

3   Before beginning this procedure, be sure to place the transmission in Neutral and apply the parking brake or block the rear wheels. Also, disconnect the cable from the negative terminal of the battery, then remove the ignition coils (see Chapter 5) and the spark plugs (see Chapter 1).

4   The preferred method is to turn the crankshaft with a socket and ratchet or breaker bar attached to the bolt threaded into the front of the crankshaft. When looking at the front of the engine, normal crankshaft rotation is clockwise. Turn the bolt in a clockwise direction only.

5   Install a compression pressure gauge in the number one spark plug hole (refer to Chapter 2E). It should be a gauge with a screw-in fitting and a hose at least six inches long **(see illustration)**.

6   Rotate the crankshaft using the method described above while observing for pressure on the compression gauge. The moment the gauge shows pressure, indicates that the number one cylinder is on the compression stroke.

7   Once the compression stroke has begun, TDC for the compression stroke of the number one cylinder is reached by bringing the piston to the top of the cylinder.

8   Insert a long dowel into the number one spark plug hole until it rests on top of the piston crown. Continue rotating the crankshaft slowly until the dowel levels off (piston reaches top of travel). This will be approximate TDC for number 1 piston.

9   These engines are not equipped with components (vibration damper, flywheel, timing hole, etc.) that are marked to identify the position of number 1 TDC. Therefore the only method to double-check the exact location of TDC number 1 on V8 engines is to remove the timing chain cover to access timing chain sprockets (see Section 6), or with the use of a degree wheel on the crankshaft vibration damper and a positive stop threaded into the spark plug hole, as you would use in the process of degreeing a camshaft (this procedure is described in detail in the *Haynes Chrysler Engine Overhaul Manual*).

10   After the No. 1 piston has been positioned at TDC on the compression stroke, TDC for any of the remaining pistons can be located by turning the crankshaft and following the firing order. Divide the crankshaft pulley into three equal sections with chalk marks at each point, each indicating 120-degrees of crankshaft rotation. Rotating the engine past TDC no. 1 to the next mark will place the engine at TDC for cylinder no. 2.

4 Valve cover - removal and installation

Removal

1   Disconnect the cable from the negative terminal of the battery (see Chapter 5, Section 1).

2   Remove the upper intake manifold (see Section 8) and cover the lower intake manifold with rags to keep out dirt.

3   Remove the ignition coils (see Chapter 5).

4   Remove the PCV hose from the grommet on the right valve cover (see Chapter 6). Remove the intake hose from the left valve cover.

4.5 Valve cover mounting bolts

5.5 Using a shaft-type valve spring compressor, depress the valve spring just enough to remove the rocker arm

5   Pull the wire harness up from the valve cover studs and remove the valve cover studs/nuts and bolts **(see illustration)**.
6   Detach the valve cover.
Note: *If the cover sticks to the cylinder head, use a block of wood and a hammer to dislodge it. If the cover still won't come loose, pry on it carefully, but don't distort the sealing flange.*

Installation

7   The mating surfaces of each cylinder head and valve cover must be perfectly clean when the covers are installed. Inspect the rubber gaskets; if they're in good condition and there were no oil leaks, they can be re-used.
8   Inspect the spark plug tube seals. Replace them if they're cracked or flattened, or if the rubber has hardened. Make sure the spark plug tube seals are in position before installing the valve cover.
9   Clean the mounting bolt threads with a die if necessary to remove any corrosion and restore damaged threads. Use a tap to clean the threaded holes in the heads.
10   Place the valve cover in position, then install the bolts. Tighten the bolts evenly, in several steps, to the torque listed in this Chapter's Specifications.
11   Complete the installation by reversing the removal procedure. Start the engine and check carefully for oil leaks.

---

5   Rocker arms and hydraulic lash adjusters - removal, inspection and installation

Note: *A universal shaft-type valve spring compressor (available from most aftermarket specialty tool manufacturers) will be required for this procedure. The only other alternative to accomplishing this task without the use of this special tool is to remove the timing chains and the camshafts which requires major dis-assembly of the engine and surrounding components.*

1   Before beginning this procedure, be sure to place the transmission in Neutral and apply the parking brake or block the rear wheels. Also, disable the ignition system by disconnecting the primary electrical connectors at the ignition coils and remove the spark plugs (see Chapter 1).
2   Remove the upper intake manifold (see Section 8) and the valve cover(s) (see Section 4).
3   Rotate the engine with a socket and ratchet attached to the crankshaft pulley bolt until the cam lobe for the rocker arm to be removed is located on its base circle. Turn the crankshaft in a clockwise direction.
4   Before the rocker arms and lash adjusters are removed, arrange to label and store them, so they can be kept separate and reinstalled on the same valve they were removed from.
5   Mount the valve spring compressor on the cylinder head. Depress the valve spring just enough to release tension on the rocker arm to be removed. Once tension on the rocker arm is relieved, the rocker arm can be removed by simply pulling it out **(see illustration)**.
6   If you're replacing or removing all of the rocker arms or lash adjusters, begin with cylinder number one and work on the rocker arms for one cylinder at a time. Move from cylinder-to-cylinder following the firing order sequence (see this Chapter's Specifications). Remember to keep the rocker arm and lash adjuster for each valve together so they can be reinstalled in the same locations
7   Once the rocker arms are removed, the lash adjusters can be pulled out of the cylinder head and stored with the corresponding rocker arm **(see illustration)**.
8   Inspect each rocker arm for wear, cracks and other damage. Make sure the rollers turn freely and show no signs of wear, also check the pivot area for wear, cracks and galling **(see illustration)**.

5.7 Pull the lash adjuster up and out of its bore to remove it from the cylinder head

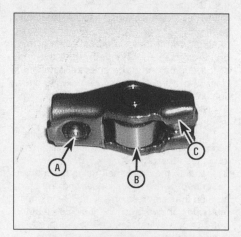

5.8 Inspect the rocker arms at the following locations

A   *Lash adjuster pocket*
B   *Roller*
C   *Valve stem seat*

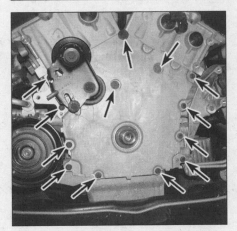

6.11 Timing chain cover retaining bolts

6.12a Make sure the mark on the crankshaft sprocket (A) aligns with the colored link on the primary timing chain and the mark on the oil pump housing (B)

6.12b Also verify that the mark on the left camshaft sprocket is flanked by the two colored links . . .

9   Inspect the lash adjuster contact surfaces for wear or damage. Make sure the lash adjusters move up and down freely in their bores on the cylinder head without excessive side to side play.

10   Installation is the reverse of removal with the following exceptions: Always install the lash adjuster first and make sure they're at least partially full of oil before installation. This is indicated by little or no lash adjuster plunger travel.

6   Timing chain and sprockets - removal, inspection and installation

Note: *Special tools are necessary to complete this procedure. Read through the entire procedure and obtain the special tools before beginning work.*

Note: *Because of work necessary to get at the timing chain and replace it, and because the water pump is in this area, it is recommended that the water pump be thoroughly inspected and replaced if necessary during this procedure (see Chapter 3).*

Note: *The 2.7L engine utilizes three timing chains to produce proper valve timing. The primary timing chain runs around the crankshaft sprocket, the water pump and around two intake camshaft sprockets. This chain synchronizes the valve timing with the crankshaft and pistons, while two secondary timing chains run around separate intake and exhaust camshaft sprockets to synchronize the intake and exhaust camshaft events. Refer to Step 19 for primary timing chain inspection procedures prior to timing chain removal.*

Removal

1   Disconnect the cable from the negative terminal of the battery (see Chapter 5, Section 1).

2   Relieve the fuel system pressure (see Chapter 4) and drain the cooling system (see Chapter 1).

6.12c . . . and the marks on the right camshaft sprocket align with the colored link on the primary chain

3   Remove the cooling fan assembly, the coolant reservoir and the radiator (see Chapter 3).

4   Remove the accessory drivebelts (see Chapter 1).

5   Remove the crankshaft pulley (see Section 11).

6   Remove the upper intake manifold (see Section 8) and the valve covers (see Section 4).

7   Remove the spark plugs (see Chapter 1) and position the number one piston at TDC on the compression stroke (see Section 3).

8   Unbolt the air conditioning compressor and set it aside without disconnecting the lines (see Chapter 3).

9   Remove the water outlet housing from the engine block (see Chapter 3).

10   Remove the idler pulley.

11   Remove the timing chain cover **(see illustration)**. Note that various types and sizes of bolts are used. They must be reinstalled in their original locations. Mark each bolt or make a sketch to help remember where they go.

6.14a A 3/8-inch drive extension and ratchet inserted into the end of the camshaft hub is used as leverage while the sprocket bolts are loosened - make note of the vibration damper-to-camshaft sprocket alignment hole

12   Rotate the engine until the crankshaft sprocket timing mark is aligned with the mark on the oil pump housing and the colored links on the chain are aligned with the timing marks on the camshaft sprockets and the crankshaft sprocket **(see illustrations)**. This position is approximately 60-degrees after TDC.

13   Remove the primary timing chain tensioner from the right cylinder head.

14   Remove the retaining bolts from the primary camshaft sprockets **(see illustrations)**. **Warning:** *Pressure from the valve springs will make the camshafts rotate clockwise as the bolts are removed. Do not rotate the crankshaft or camshaft separately after the primary timing chain is loosened or removed as piston or valve damage may occur. The only exception to this rule is when the camshafts must be rotated counterclockwise slightly, to realign the primary camshaft sprockets with the camshafts during installation.*

6.14b Left camshaft sprocket retaining bolts

6.16a Primary timing chain guide access plugs

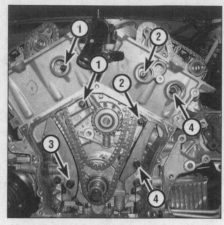

6.16b Primary timing chain guide mounting details

1   *Upper right timing chain guide mounting bolts*
2   *Upper left timing chain guide mounting bolts*
3   *Tensioner arm pivot bolt*
4   *Lower timing chain guide mounting bolts*

15   Pull the primary sprockets off the camshaft hubs one at a time, lower the sprocket(s) into the cylinder head opening until the chain can be displaced from around the sprocket, then remove the primary camshaft sprockets from the engine. Note that the left (primary) camshaft sprocket is identified by the camshaft position sensor ring which is mounted on the front of the sprocket by two nuts. It is not necessary to remove the sensor ring from the sprocket during this procedure unless damage to the sensor ring or sprocket has occurred. Also note that the right (primary) camshaft sprocket is identified by a vibration damper which is mounted in front of the sprocket. Make note of the alignment holes on the damper and the right (primary) sprocket as these two components will separate as the sprocket is removed from the camshaft. Always install the damper in the same position from which it was removed.
16   Remove the access plugs from the front of the cylinder heads and detach the primary timing chain guides and tensioner arm **(see illustrations)**.
17   Remove the primary timing chain.
Note: *The secondary timing chains are removed as an assembly with the camshafts and therefore covered in camshaft removal (see Section 7).*
18   If the crankshaft sprocket needs to be

replaced after thorough inspection or the sprocket simply needs to be removed for other procedures such as oil pump removal, proceed to Step 36.

Inspection

19   Inspect the camshaft, water pump and crankshaft sprockets for wear on the teeth and keyways. Inspect the chains for cracks or excessive wear of the rollers. Inspect the facing of the chain guides and tensioner arm for excessive wear. If any of the components show signs of excessive wear they must be replaced.

Installation

20   If removed, install the crankshaft sprocket as outlined in Step 38 and 39.
21   If you purchased a new timing chain, verify that you have the correct timing chain for your vehicle by counting the number of links the chain has and comparing the new chain with the old chain. Also compare the position of the colored links in the new chain with the position of the colored links in the old chain.
22   Verify the crankshaft sprocket is still aligned with the mark on the oil pump housing **(see illustration 6.12a)** and install the upper chain guides back into position on the engine **(see illustration 6.16b)**.

23   Prepare to install the primary timing chain by aligning the left (primary) camshaft sprocket mark between the light colored links on the chain **(see illustration 6.14b)**. Then lower the sprocket and chain assembly down through the opening in the left cylinder head and reposition the left camshaft sprocket over the camshaft hub.
24   Install the right (primary) camshaft sprocket, vibration damper and timing chain onto the engine by looping the chain around the crankshaft sprocket and aligning the light colored chain link with the mark on the crankshaft sprocket. Then place the chain around the water pump sprocket and install the right (primary) camshaft sprocket over the right camshaft hub making sure the colored links align with their respective marks on the sprockets **(see illustrations 6.12a, 6.12b and 6.12c)**.
25   Install the lower chain guide and tensioner arm **(see illustration 6.16b)**.
26   Use the special tool to purge the oil from the tensioner. Place the check ball end of the tensioner into the shallow end of the tool over the pin, apply hand pressure and slowly depress the tensioner until the oil has been purged from the tensioner **(see illustration)**.
27   After the oil has been purged, the tensioner must be reset before installation. Place the tensioner plunger into the deep end of the tool and depress the plunger until it locks into place. Once the tensioner is reset it will be about 1-1/2 inches shorter.
28   Install the reset tensioner and tighten the tensioner retaining plate bolts to the torque listed in this Chapter's Specifications. Always use a new O-ring or gasket on the tensioner housing or tensioner retaining plate.
Caution: *Be sure to align the tensioner retaining plate dowel pin with the hole in the cylinder head.*

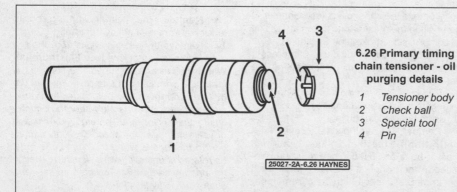

6.26 Primary timing chain tensioner - oil purging details

1   *Tensioner body*
2   *Check ball*
3   *Special tool*
4   *Pin*

25027-2A-6.26 HAYNES

6.32 With all slack removed from the left side of the chain and the colored links on the chain aligned with their respective marks on the sprockets, engage the tensioner by pushing the tensioner arm inward slightly, then release the tensioner arm to extend the tensioner

6.37 The crankshaft sprocket can be removed with a conventional three jaw puller

7.4 Secondary timing chain tensioner retaining bolts (right cylinder head shown, left cylinder head similar)

29   Install the camshaft sprocket retaining bolts by inserting a 3/8-inch drive extension and ratchet into the primary camshaft hub and slightly rotating the camshafts counter-clockwise until the camshaft sprocket bolt holes align with the camshaft hub bolt holes. Perform the procedure on the right (primary) camshaft sprocket first, then proceed to the left (primary) camshaft sprocket.

30   Leaving the 3/8-inch extension and ratchet inserted into the camshaft hub as leverage, tighten the camshaft sprocket bolts to the torque listed in this Chapter's Specifications.

31   Rotate the engine clockwise just enough to remove any slack from the left side of the timing chain.

32   Reconfirm that the timing marks on the camshaft and crankshaft sprockets are aligned with the colored links on the chain and release the primary timing chain tensioner from its reset position **(see illustration)**.

33   Remove all traces of old sealant from the timing chain cover and the cover bolts.

34   Apply a bead of RTV sealant to the timing chain cover gasket and sealing surfaces. Place the timing cover in position on the engine and install the bolts in their original locations and tighten the bolts to the torque listed in this Chapter's Specifications.

35   The remainder of the installation is the reverse of removal. Refill the cooling system (see Chapter 1).

Crankshaft sprocket

36   If the crankshaft sprocket is to be removed and installed, it will require the use of several special tools: A three jaw puller, a propane torch or crankshaft sprocket installation tool and a machinist ruler or dial caliper.

37   After the primary timing chain has been removed, the crankshaft sprocket can be removed with a three jaw puller **(see illustration)**.

38   To install the crankshaft sprocket it will be necessary to purchase a sprocket installation tool or a common household propane torch. If a sprocket installation tool is purchased, follow the installation procedures outlined in the tools instructions and install the sprocket to the proper depth (see Step 39). If a propane torch is used, simply place the sprocket in a vise and heat the sprocket hub until the sprocket has expanded enough to slide over the crankshaft.

Caution: *After heating the sprocket always handle the sprocket with a pair of pliers or other insulated tool to avoid serious injury and never heat an object when gasoline or other volatile chemicals are present.*

39   Install the sprocket until it bottoms on the shoulder on the crankshaft. Using a machinist ruler or a caliper, make sure it has been installed completely by measuring from the end of the crankshaft to the face of the crankshaft sprocket; it should measure 1.537 inches plus-or-minus 0.020 inch (39.05 +/- 0.5 mm).

7   Camshafts - removal, inspection and installation

Removal

1   Disconnect the cable from the negative terminal of the battery (see Chapter 5, Section 1).

2   Remove the valve covers (see Section 4).

3   Remove the primary timing chain and the primary camshaft sprockets (see Section 6).

4   Remove the retaining bolts from the secondary timing chain tensioner **(see illustration)**.

5   Verify the markings on the camshaft bearing caps. The caps should be marked from 1 to 5, and with an "I" or an "E", to indicate intake or exhaust. Also verify that there are arrow marks on the caps indicating the front of the engine **(see illustration)**. Loosen the camshaft bearing caps in two or three steps in the reverse order of the tightening sequence **(see illustration 7.19)**.

Caution: *Keep the caps in order. They must go back in the same location they were removed from.*

7.5 Verify that the camshaft bearing caps are marked to ensure correct reinstallation

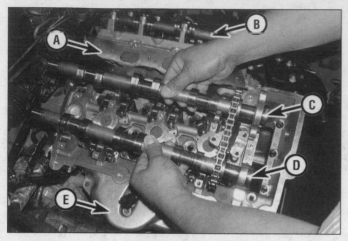

7.6 Camshaft installation details

A   Intake manifold location
B   Intake camshaft (left cylinder head)
C   Intake camshaft (right cylinder head)
D   Exhaust camshaft (right cylinder head)
E   Exhaust manifold

7.10 Check the diameter of each camshaft bearing journal to pinpoint excessive wear and out-of-round conditions

6   Detach the bearing caps, then remove the camshafts, secondary timing chain and the secondary tensioner as an assembly from the cylinder head. Make a note that the intake camshafts have a flanged hub at the front and the exhaust camshafts do not. Also note that the intake camshafts are installed in the cylinder head (toward the center of the engine) next to the intake manifold and the exhaust camshafts are installed (toward the sides of the engine) next to the exhaust manifolds **(see illustration)**.
7   Remove the secondary tensioner and the timing chain from the camshafts.
8   Inspect the camshaft secondary sprockets for wear on the teeth. Inspect the chains for cracks or excessive wear of the rollers. Inspect the facing of the secondary chain tensioners for excessive wear. If any of the components show signs of excessive wear they must be replaced.

Inspection

9   After the camshaft has been removed from the engine, cleaned with solvent and dried, inspect the bearing journals for uneven wear, pitting and evidence of seizure. If the camshaft journals are damaged, inspect the cylinder head and the camshaft bearing caps.
10   Measure the bearing journals with a micrometer to determine if they are excessively worn or out-of-round **(see illustration)**. Compare the measurements with the Specifications listed in this Chapter.
11   Measure the lobe height of each cam lobe on the intake camshaft and record your measurements **(see illustration)**. Compare the measurements for excessive variations. If the lobe heights vary more than 0.010 inch (0.254 mm), replace the camshaft. Compare the lobe height measurements on the exhaust camshaft and follow the same procedure. Do

not compare intake camshaft lobe heights with exhaust camshaft lobe heights, as they are different. Only compare intake lobes with intake lobes and exhaust lobes with other exhaust lobes.
12   Check the camshaft lobes for heat discoloration, score marks, chipped areas, pitting and uneven wear **(see illustration)**. If the lobes are in good condition and if the lobe lift variation measurements recorded earlier are within the limits, the camshaft can be reused.

Installation

13   Install the secondary timing chain(s) over the camshaft sprockets while aligning the colored links on the chain with the marks on the secondary camshaft sprockets in a 12 o'clock position.
Note: *The colored links on the chain must face outward toward the front of the engine.*
14   Using a paperclip, fabricate a U-shaped tool to use as a tensioner locking pin. Then place the secondary tensioner in a vise and compress the tensioner until the U-shaped tool can be inserted into the locking holes on the tensioner. This places the tensioner in the locked position so it can be reinstalled **(see illustration)**.
15   Once the tensioner is locked into place it can be installed back into the place between the camshafts and the secondary timing chain.
16   Apply moly-based engine assembly lubricant to the camshaft lobes and journals. Make sure the rocker arms are properly seated on their respective lash adjuster and the valve stem tip.
17   Install the camshafts, secondary tensioner and the timing chain as an assembly in their original position, with the marks on the sprockets and the colored links on the chain facing up (90-degrees from the valve cover mating surface) and inline with the cylinder

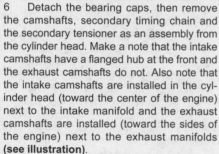

7.11 Measure the camshaft lobe height (greatest dimension) with a micrometer

7.12 Check the cam lobes for pitting, excessive wear and scoring. If scoring is excessive, as shown here, replace the camshaft

7.14 Compress the secondary tensioner in a vise until a paperclip can be inserted into the secondary tensioner locking holes

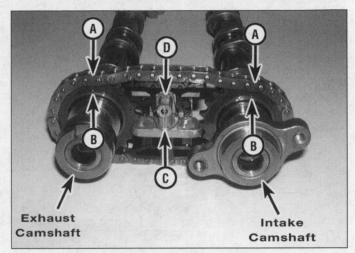

7.17 Camshaft installation details

A   Secondary timing chain colored links
B   Secondary camshaft sprocket marks
C   Tensioner locking pin
D   Secondary timing chain tensioner

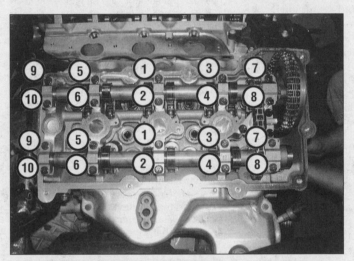

7.19 Camshaft bearing cap TIGHTENING sequence

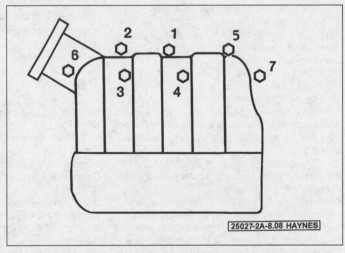

8.8 Upper intake manifold bolt TIGHTENING sequence

bank **(see illustration)**. **Note:***When installed correctly there should be 12 timing chain pins between the intake and exhaust camshaft marks.*

18   Install the bearing caps and bolts and tighten them hand-tight.

19   Tighten the bearing cap bolts in several steps, to the torque listed in this Chapter's Specifications, using the proper tightening sequence **(see illustration)**.

20   Tighten the tensioner mounting bolts to the torque listed in this Chapter's Specifications and remove the tensioner locking pin.

21   Install the primary timing chain and camshaft sprockets (see Section 6).

22   The remainder of installation is the reverse of removal.

8   Intake manifold - removal and installation

Upper intake manifold

1   Relieve the fuel pressure (see Chapter 4).

2   Disconnect the cable from the negative terminal of the battery (see Chapter 5, Section 1).

3   Remove the air intake duct and the air filter housing (see Chapter 4).

4   Label and disconnect the vacuum hoses and electrical connectors attached to the upper intake manifold and throttle body. Disconnect the electronic throttle control (ETC) system from the throttle body (see Chapter 4).

5   Remove the EGR pipe from the upper intake manifold (see Chapter 6).

6   Remove the upper intake manifold support brackets. Loosen the upper intake manifold bolts in the reverse order of the tightening sequence **(see illustration 8.8)**. Remove the upper intake manifold and the foam insulator pad.

7   Clean and inspect the upper intake manifold to lower intake manifold sealing surfaces. Inspect the gaskets for tears or cracks replacing them if necessary. The gaskets can be reused if not damaged.

8   Install the upper intake manifold onto the lower intake manifold and tighten the bolts in the recommended tightening sequence **(see illustration)** to the torque listed in this Chapter's Specifications. The remainder of installation is the reverse of removal.

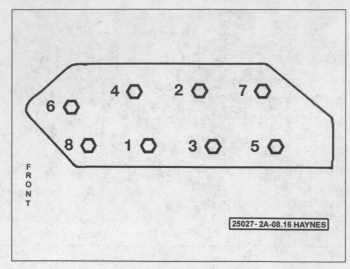

8.16 Lower intake manifold bolt TIGHTENING sequence

9.3 Front engine splash shield (1) and middle engine splash shield (2) fasteners

Lower intake manifold

9   Remove the upper intake manifold (see Steps 1 through 6).

10   Label and detach any remaining hoses which would interfere with the removal of the lower intake manifold.

11   Remove the fuel rail and injectors from the lower intake manifold (see Chapter 4).

12   Loosen the lower intake manifold mounting bolts/nuts in 1/4-turn increments, in the reverse order of the tightening sequence, until they can be removed by hand **(see illustration 8.16)**.

13   The manifold will probably be stuck to the cylinder heads and force may be required to break the gasket seal.

Caution: *Don't pry between the manifold and the heads or damage to the gasket sealing surfaces may occur, leading to vacuum leaks.*

14   Clean and inspect the lower intake manifold to cylinder head sealing surfaces. Inspect the gaskets for tears or cracks, replacing them if necessary. The gaskets can be reused if they aren't damaged.

15   Position the lower intake manifold onto the engine making sure the gaskets and manifold are aligned correctly over the cylinder heads. Install the fuel rail and injectors onto the cylinder heads and the lower manifold. Install the fuel rail retaining bolts into the manifold loosely to ensure correct gasket/manifold alignment, then install the remaining lower intake manifold bolts.

16   Following the recommended tightening sequence, tighten the bolts, in several steps, to the torque listed in this Chapter's Specifications **(see illustration)**.

17   The remainder of installation is the reverse of the removal procedure. Run the engine and check for fuel, vacuum and coolant leaks.

9   Exhaust manifold - removal and installation

Warning: *The engine must be completely cool before beginning this procedure.*

Removal

1   Disconnect the cable from the negative terminal of the battery (see Chapter 5, Section 1).

2   Block the rear wheels, set the parking brake, raise the front of the vehicle and support it securely on jackstands. Disconnect the downstream oxygen sensor connectors (see Chapter 6).

3   Remove the engine middle splash shield **(see illustration)**.

4   Unbolt the exhaust pipes at the exhaust manifolds.

5   Lower the vehicle. Working in the engine compartment, disconnect the oxygen sensor's electrical connector. If removing the right exhaust manifold, detach the EGR pipe from the top of the manifold. If removing the left manifold, remove the air filter housing (see Chapter 4) and the engine oil dipstick tube.

6   Remove the exhaust manifold upper heat shield mounting bolts and remove the heat shields.

7   Remove the mounting bolts and detach the manifold from the cylinder head. Be sure to spray penetrating lubricant onto the bolts and threads before attempting to remove them.

Installation

8   Clean the mating surfaces to remove all traces of old gasket material, then inspect the manifold for distortion and cracks. Warpage can be checked with a precision straightedge held against the mating flange. If a feeler gauge thicker than 0.030-inch can be inserted

between the straightedge and flange surface, take the manifold to an automotive machine shop for resurfacing.

9   Place the exhaust manifold in position with a new gasket and the lower heat shields and install the mounting bolts finger-tight.

Note: *Be sure to identify the exhaust manifold gaskets by the correct cylinder designation and the position of the exhaust ports on the gasket.*

10   Starting in the middle and working out toward the ends, tighten the mounting bolts in several increments, to the torque listed in this Chapter's Specifications.

11   Install the remaining components in the reverse order of removal.

12   Start the engine and check for exhaust leaks between the manifold and cylinder head and between the manifold and exhaust pipe.

10   Cylinder head(s) - removal and installation

Caution: *The engine must be completely cool before beginning this procedure.*

Removal

1   Refer to Section 6, Steps 1 through 18, and remove the primary timing chain and sprockets.

Caution: *Be careful not to disturb the crankshaft from its alignment marks during the remainder of this procedure.*

2   Remove the camshafts from the cylinder head (see Section 6).

3   Remove the rocker arms and hydraulic lash adjusters from the cylinder head (see Section 5). Before the rocker arms and lash adjusters are removed, arrange to label and store them, so they can be kept separate and reinstalled on the same valve from which they were removed.

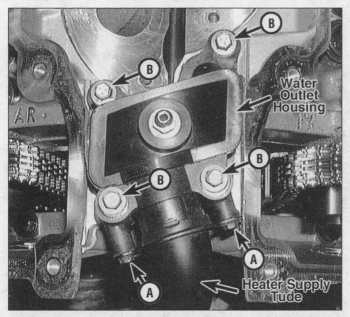

10.5 Remove the water outlet housing by detaching the heater supply tube retaining bolts (A), then slide the heater supply tube out of the water outlet housing and remove the water outlet housing retaining bolts (B)

10.19 Cylinder head bolt TIGHTENING sequence

4   Remove the lower intake manifold (see Section 8) and the exhaust manifold(s) (see Section 9).

5   Label and remove any remaining items attached to the cylinder head, such as coolant fittings, ground straps, cables, hoses, wires or brackets (see illustration).

6   Using a breaker bar and the appropriate sized socket, loosen the cylinder head bolts in 1/4-turn increments until they can be removed by hand. Loosen the bolts in the reverse order of the tightening sequence (see illustration 10.19) to avoid warping or cracking the head.

7   Lift the cylinder head off the engine block with the exhaust manifold attached. If it's stuck, very carefully pry up at the transmission end, beyond the gasket surface, at a casting protrusion.

8   Remove all external components from the cylinder head to allow for thorough cleaning and inspection before having the cylinder head serviced at a qualified automotive machine shop.

Installation

9   The mating surfaces of the cylinder head and block must be perfectly clean when the head is installed.

10   Use a gasket scraper to remove all traces of carbon and old gasket material from the cylinder head and engine block being careful not to gouge the aluminum, then clean the mating surfaces with lacquer thinner or acetone. If there's oil on the mating surfaces when the head is installed, the gasket may not seal correctly and leaks could develop. When working on the block, stuff the cylinders with clean shop rags to keep out debris. Use a vacuum cleaner to remove material that falls into the cylinders.

11   Check the block and head mating surfaces for nicks, deep scratches and other damage. If damage is slight, it can be removed with a file; if it's excessive, machining may be the only alternative.

12   Use a tap of the correct size to chase the threads in the head bolt holes, then clean the holes with compressed air - make sure that nothing remains in the holes. **Warning:** Wear eye protection when using compressed air!

13   Check each cylinder head bolt for stretching by laying a metal ruler or a straight-edge against the threads of the bolts. If the diameter of the bolt threads has necked down anywhere in the threaded area, the bolts have exceeded the maximum amount of stretch and will need to be replaced (all of the threads must touch the straightedge).

Note: It's a good idea to replace the head bolts with new ones.

14   Using a precision straightedge and feeler gauges, check the cylinder head for warpage. If warpage exceeds the amount listed in this Chapter's Specifications, have the heads machined at an automotive machine shop.

15   Install the components that were removed from the head.

16   Position the new cylinder head gasket over the dowel pins on the block, noting which direction on the gasket faces up.

17   Carefully set the head over the dowels on the block without disturbing the gasket.

18   Before installing the head bolts, apply a small amount of clean engine oil to the threads and hardened washers (if equipped). The chamfered side of the washers must face the bolt heads.

19   Install the bolts in their original locations and tighten them finger-tight. Then tighten all the bolts in several steps, following the proper sequence (see illustration), to the torque listed in this Chapter's Specifications.

20   Install the lash adjusters and the rocker arms in the cylinder head, then install the camshafts as described in Section 7,

21   Install the primary timing chain and sprockets as described in Section 7. The remaining installation steps are the reverse of removal.

22   Refill the cooling system and change the engine oil and filter (see Chapter 1).

23   Start the engine and check for oil and coolant leaks.

11   Crankshaft pulley - removal and installation

1   Disconnect the cable from the negative terminal of the battery (see Chapter 5, Section 1).

2   Remove the upper radiator crossmember, the cooling fan assembly and the radiator (see Chapter 3).

3   Remove the engine splash shield (see Section 9).

4   Remove the drivebelts (see Chapter 1) and position the belt tensioner away from the crankshaft pulley.

5   Use a strap wrench around the crankshaft pulley to hold it while using a breaker

11.5 Use a strap wrench to hold the crankshaft pulley while removing the center bolt (a chain-type wrench may be used if you wrap a section of old drivebelt or rag around the crankshaft pulley first)

11.6 The use of a three jaw puller will be necessary to remove the crankshaft pulley - always place the puller jaws around the pulley hub, not the outer ring

11.7 If the sealing surface of the pulley hub has a wear groove from contact with the seal, repair sleeves are available at most auto parts stores

bar and socket to remove the crankshaft pulley center bolt **(see illustration)**.

6   Pull the damper off the crankshaft with a puller **(see illustration)**.

Caution: *The jaws of the puller must only contact the hub of the pulley - not the outer ring.*

Caution: *A spacer should be placed against the end of the crankshaft nose (that fits in the pulley bore) or a long Allen-head bolt should be inserted into the crankshaft nose for the puller's tapered tip to push against to prevent damage to the crankshaft threads.*

7   Check the surface on the pulley hub that the oil seal rides on. If the surface has been grooved from long-time contact with the seal, a press-on sleeve may be available to renew the sealing surface **(see illustration)**. This sleeve is pressed into place with a hammer and a block of wood and is commonly available at auto parts stores for various applications.

8   Lubricate the pulley hub with clean engine oil and reinstall the crankshaft pulley.

Use a vibration damper installation tool to press the pulley onto the crankshaft.

9   Install the crankshaft pulley retaining bolt and tighten it to the torque listed in this Chapter's Specifications.

10   The remainder of installation is the reverse of the removal.

12   Crankshaft front oilseal - replacement

1   Remove the crankshaft pulley from the engine (see Section 11).

2   Carefully pry the seal out of the cover with a seal removal tool or a large screwdriver **(see illustration)**. **Caution:** *Be careful not to scratch, gouge or distort the area that the seal fits into or an oil leak will develop.*

3   Clean the bore to remove any old seal material and corrosion. Position the new seal in the bore with the seal lip (usually the side

with the spring) facing IN (toward the engine). A small amount of oil applied to the outer edge of the new seal will make installation easier - but don't overdo it!

4   Drive the seal into the bore with a seal driver or a large socket and hammer until it's completely seated **(see illustration)**. Select a socket that's the same outside diameter as the seal and make sure the new seal is pressed into place until it bottoms against the cover flange.

5   Lubricate the seal lips with engine oil and reinstall the crankshaft pulley.

6   The remainder of installation is the reverse of removal. Run the engine and check for oil leaks.

13   Oil pan - removal and installation

Removal

1   Disconnect the cable from the negative terminal of the battery (see Chapter 5, Section 1). Remove the engine oil dipstick.

2   Apply the parking brake and block the rear wheels. Raise the front of the vehicle and place it securely on jackstands. Remove the engine middle splash shield (see Section 9). Drain the engine oil and remove the oil filter (see Chapter 1).

3   Remove the intermediate shaft coupler at the steering gear (see Chapter 10) and separate the steering gear input shaft from the coupler.

4   Disconnect the power steering pressure switch and remove the power steering gear mounting bolts.

5   Remove the power steering fluid line bracket and position the power steering gear off to the side.

6   Remove the transmission-to-oil pan support brace at the rear of the pan **(see illustration)**. **Note:** *The support brace must be moved to the right of the engine block and to*

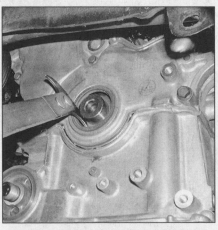

12.2 Pry the seal out very carefully with a seal removal tool or screwdriver, being careful not to nick or gouge the seal bore or the crankshaft

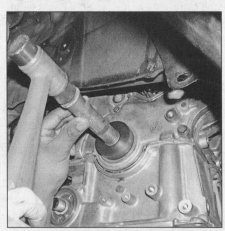

12.4 Use a seal driver or a large socket to drive the new seal into the cover

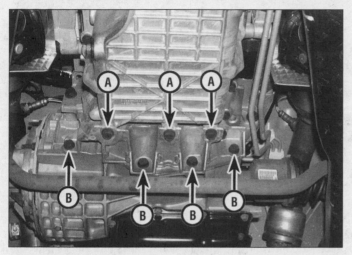

13.6 Transmission-to-oil pan support brace

A *Vertical bolts*   B *Horizontal bolts*

13.8 Insert a flathead screwdriver or small prybar on the casting protrusion at the front of the oil pan to break it loose

the rear to maneuver the support brace from the engine compartment.

7   Detach any clips securing the transmission oil cooler lines that would interfere with removal of the oil pan (see Chapter 3). Also remove the lower bolt securing the air conditioning compressor to the oil pan and the bolt securing the alternator bracket to the oil pan.

8   Remove the bolts and nuts, then carefully separate the oil pan from the block. Don't pry between the block and the pan or damage to the sealing surfaces could occur and oil leaks may develop. Instead, pry at the casting protrusion at the front of the pan **(see illustration)**.

Installation

9   Clean the pan with solvent and remove all old sealant and gasket material from the block and pan mating surfaces. Clean the mating surfaces with lacquer thinner or acetone and make sure the bolt holes in the block are clear. Check the oil pan flange for distortion, particularly around the bolt holes.

10   Apply a bead of RTV sealant to the oil pan rail parting lines at the front cover and at the rear main oil seal retainer. Install the gasket on the block.

11   Place the oil pan in position on the block and install the nuts/bolts.

12   After the fasteners are installed, tighten them to the torque listed in this Chapter's Specifications. Starting at the center, follow a criss-cross pattern and work up to the final torque in three steps.

13   Place the transmission-to-oil pan support brace in position and install the support brace-to-oil pan bolts. Torque the bolts to 10 in-lbs. Torque the center bolts first, followed by the outer bolts. Install the support brace-to-transmission bolts. Torque the bolts to 40 ft-lbs, then re-torque the support brace-to-oil pan bolts to 40 ft-lbs **(see illustration 13.6)**. Refer to the Specifications listed in this Chapter.

14   Install the power steering gear (see Chapter 10).

15   The remaining steps are the reverse of the removal procedure.

16   Refill the engine with oil (see Chapter 1), replace the filter, run it until normal operating temperature is reached and check for leaks.

14   Oil pump - removal, inspection and installation

Note: *The replacement oil pumps on early models may experience oil pressure problems because of a collapsing oil pressure relief valve and spring. These oil pumps are equipped with an expansion plug instead of the usual threaded plug to seal the oil pressure relief valve and spring inside the oil pump. Consult with a dealer parts department or other qualified automotive repair facility for the correct oil pump.*

14.3 Oil pump pick-up tube mounting bolts

Removal

1   Refer to Section 6 and remove the primary timing chain and the crankshaft sprocket.

Note: *Before removing the crankshaft sprocket, verify that the crankshaft sprocket marks align with the marks on the oil pump housing and do not rotate the crankshaft from this position at any time during this procedure. This position is approximately 60 degrees after TDC.*

2   Remove the oil pan (see Section 13).

3   Remove the oil pump pick-up tube **(see illustration)**.

4   Remove the oil pump-to-engine block bolts from the front of the engine **(see illustration)**.

5   Gently pry the oil pump housing outward enough to clear the dowel pins on the engine block and remove it from the engine.

Inspection

6   Remove the screws holding the front

14.4 Oil pump housing retaining bolts

14.6 Remove the screws and lift the cover off

14.9a Place a straightedge across the oil pump cover and check it for warpage with a feeler gauge

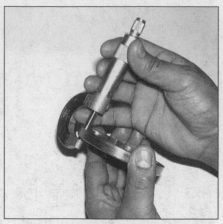

14.9b Use a micrometer or dial caliper to check the thickness and the diameter of the outer rotor

14.9c Use a micrometer or dial caliper to check the thickness of the inner rotor

14.9d Check the outer rotor-to-housing clearance

14.9e Check the clearance between the tips of the inner and outer rotors

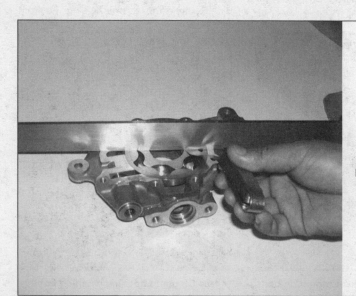

14.9f Using a straightedge and feeler gauge, check the side clearance between the surface of the oil pump and the inner and outer rotors

cover on the oil pump housing **(see illustration)**.

7   Clean all components with solvent, then inspect them for wear and damage.

8   Remove the oil pressure regulator cap, washer, spring and valve. Check the oil pressure relief valve sliding surface and valve spring. If either the spring or the valve is damaged, they must be replaced as a set. Small burrs can be removed with 400-grit wet sandpaper and oil.

9   Check the clearance of the following oil pump components with a feeler gauge and a micrometer or dial caliper **(see illustrations)** and compare the measurements to the clearance listed in this Chapter's Specifications :

    Cover flatness
    Outer rotor diameter and thickness
    Inner rotor thickness
    Outer rotor-to-body clearance
    Inner rotor to outer rotor tip clearance
    Cover-to-inner rotor side clearance
    Cover-to-outer rotor side clearance

10   If any clearance is excessive, replace the entire oil pump assembly.

11   Pack the pump with petroleum jelly to prime it. Assemble the oil pump and tighten all fasteners to the torque listed in this Chapter's Specifications. Install the oil pressure regulator valve, spring and washer, then tighten the oil pressure regulator valve cap.

Installation

12   To install the pump, turn the flats in the rotor so they align with the flats on the crankshaft.

13   Install the pump-to-block bolts and tighten them to the torque listed in this Chapter's Specifications.

14   The remainder of installation is the reverse of removal.

15   Driveplate - removal and installation

1   Raise the vehicle and support it securely on jackstands, then refer to Chapter 7B and remove the transmission.

Warning: *The engine must be supported from above with an engine hoist or an engine support fixture before working underneath the vehicle with the transmission removed.*

2   Now would be a good time to check and replace the transmission front pump seal.

3   Use paint or a center-punch to make alignment marks on the driveplate and crankshaft to ensure correct alignment during reinstallation.

4   Remove the bolts that secure the driveplate to the crankshaft. If the crankshaft turns, jam a large screwdriver or prybar through the driveplate to keep the crankshaft from turning, then remove the mounting bolts and bolt plate.

5   Pull straight back on the driveplate to detach it from the crankshaft.

6   Installation is the reverse of removal. Be sure to align the matching paint marks. Use thread locking compound on the bolt threads and tighten them to the specified torque in a criss-cross pattern.

16   Rear main oil seal - replacement

1   All models use a one-piece rear main oil seal which is installed in a bolt-on housing. Replacing this seal requires removal of the transmission and driveplate. The rear main seal and housing are serviced as one complete assembly. This procedure will require several special tools.

2   Remove the transmission (see Chapter 7B) and the oil pan (see Section 13).

3   Remove the driveplate (see Section 15).

4   Remove the rear main oil seal housing bolts and the housing.

5   Install the new oil seal housing onto the crankshaft using a special guide tool.

6   Once the oil seal housing is attached and bolts slightly tight, install special alignment tools onto the oil pan rails.

7   While pressing the oil seal housing evenly, tighten the oil seal housing retainer bolts to the torque listed in this Chapter's Specifications. The housing must be flush with the plane of the oil pan mating surface of the engine to prevent oil leaks.

8   The remainder of installation is the reverse of the removal procedure.

17   Engine mounts - check and replacement

1   There are three powertrain mounts; left and right engine mounts attached to the engine block and to the subframe and a rear mount attached to the transmission and the subframe. Refer to Chapter 7B for check and replacement procedures for the transmission mount.

Check

2   During the check, the engine must be raised slightly to remove the weight from the mounts.

3   Raise the vehicle and support it securely on jackstands. Remove the front wheels and tires. Position two jacks, one under the crankshaft pulley and the other under the transmission bellhousing. Place a block of wood between the jack head and the crankshaft pulley or bellhousing, then carefully raise the engine/transmission just enough to take the weight off the mounts.

Warning: *DO NOT place any part of your body under the engine when it's supported only by a jack!*

4   Check the mounts to see if the rubber is cracked, hardened or separated from the metal plates. Sometimes the rubber will split right down the center.

5   Check for relative movement between the mount plates and the engine or subframe (use a large screwdriver or prybar to attempt to move the mounts). If movement is noted, lower the engine and tighten the mount fasteners.

6   Rubber preservative should be applied to the mounts to slow deterioration.

Replacement

7   Disconnect the cable from the negative terminal of the battery (see Chapter 5, Section 1) then raise the vehicle and support it securely on jackstands.

8   Remove the middle engine splash shield (see Section 9).

9   Remove the nuts holding the engine mounts to the subframe.

10   Raise the engine with a jack and block of wood under the oil pan until the engine mount studs clear the engine bracket and the subframe.

11   Remove the engine mount heat shield.

12   Remove the bolts holding the engine mount to the frame.

13   Remove the engine mount and replace it with the new one. Tighten the upper bolts to the torque listed in this Chapter's Specifications.

14   Lower the engine and install the engine mount nuts and tighten them to the torque listed in this Chapter's Specifications.

Notes

Chapter 2 Part B
3.5L V6 engine

Contents

Specifications

General

Firing order	1-2-3-4-5-6
Bore and stroke	3.780 x 3.189 inches
Displacement	214 cubic inches
Cylinder numbers (front to rear)	
Left bank	2-4-6
Right bank	1-3-5
Compression pressure	See Chapter 2E

Front

Cylinder identification diagram

Camshaft

Endplay	0.001 to 0.014 inch
Lobe wear limit	
Standard	0.001 inch
Service limit	0.010 inch
Camshaft journal diameter	1.6905 to 1.6913 inches
Camshaft bore diameter (inside)	1.6944 to 1.6953 inches
Camshaft bearing oil clearance	
Standard	0.003 to 0.0047 inch
Service limit	0.0059 inch

Oil pump

Cover warpage limit	0.001 inch
Inner and outer rotor thickness (minimum)	0.563 inch
Outer rotor diameter (minimum)	3.149 inch
Outer rotor-to-housing clearance (maximum)	0.015 inch
Inner rotor-to-outer rotor lobe clearance	0.008 inch
Oil pump housing-to-rotor side clearance	0.003 inch

Torque specifications

Ft-lbs (unless otherwise indicated)

Note: *One foot-pound (ft-lb) of torque is equivalent to 12 inch-pounds (in-lbs) of torque. Torque values below approximately 15 ft-lbs are expressed in inch-pounds, because most foot-pound torque wrenches are not accurate at these smaller values.*

Camshaft sprocket bolts*	
Step 1	75
Step 2	Tighten an additional 90-degrees
Camshaft thrust plate bolts	21
Crankshaft pulley bolt	70
Cylinder head bolts (in sequence - **see illustration 10.21**)	
First step	45
Second step	65
Third step	65
Fourth step	Tighten an additional 90-degrees
Drivebelt tensioner bolt	40
Driveplate-to-crankshaft bolts	70
Engine mount bracket-to-block bolts	55
Engine mount nuts	55
Exhaust manifold-to-cylinder head bolts	17
Exhaust heat shield bolts	105 in-lbs
Exhaust pipe bolts	25
Heater supply tube mounting bolts	105 in-lbs
Intake manifold bolts	
Upper intake manifold bolts	105 in-lbs
Lower intake manifold bolts	21
Oil pan drain plug	20
Oil pan bolts	
Oil pan-to-engine block bolts	
M6 bolts	105 in-lbs
M8 bolts	21
Oil pan-to-transaxle bolts	40
Oil pump pick-up tube mounting bolt	21
Oil pump cover (plate) bolts (Torx no. 30)	105 in-lbs
Rocker arm shaft bolts	23
Timing belt cover (front and rear) bolts	
M6	105 in-lbs
M8	21
M10	40
Timing belt tensioner	21
Timing belt tensioner pulley bolt	45
Throttle body bracket-to-throttle body	105 in-lbs
Throttle body bracket-to-cylinder head	21
Rear main oil seal retainer bolts	105 in-lbs
Valve cover-to-cylinder head bolts	105 in-lbs

*Replace with new bolts

4.4 Pull up the wire harness tabs from the valve cover grommets using a panel tool and position the harness off to the side

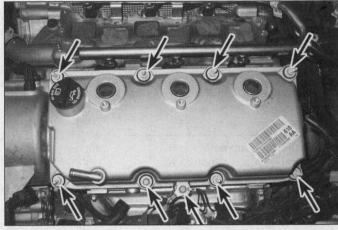

4.5 Location of the valve cover bolts and studs - left side shown, right side similar

1 General information

1   This Part of Chapter 2 is devoted to in-vehicle repair procedures for the 3.5L V6 engine. These engines utilize an aluminum block with six cylinders arranged in a "V" shape at a 60-degree angle between the two banks. The overhead camshaft aluminum cylinder heads are equipped with replaceable valve guides and seats. Aluminum roller rockers mounted on two shafts in each cylinder head actuate the valves.

2   Information concerning engine removal and installation and engine block overhaul can be found in Chapter 2E. The following repair procedures are based on the assumption the engine is installed in the vehicle. If the engine has been removed from the vehicle and mounted on a stand, many of the steps outlined in this Part of Chapter 2 will not apply.

2 Repair operations possible with the engine in the vehicle

1   Many major repair operations can be accomplished without removing the engine from the vehicle.

2   Clean the engine compartment and the exterior of the engine with some type of degreaser before any work is done. It'll make the job easier and help keep dirt out of the internal areas of the engine.

3   Depending on the components involved, it may be helpful to remove the hood to improve access to the engine as repairs are performed (see Chapter 11 if necessary). Cover the fenders to prevent damage to the paint. Special pads are available, but an old bedspread or blanket will also work.

4   If vacuum, exhaust, oil or coolant leaks develop, indicating a need for gasket or seal replacement, the repairs can generally be done with the engine in the vehicle. The intake and exhaust manifold gaskets, timing belt

cover gasket, oil pan gasket, crankshaft oil seals and cylinder head gaskets are all accessible with the engine in place.

5   Exterior engine components, such as the intake and exhaust manifolds, the oil pan, timing belt covers (and the oil pump), the water pump, the starter motor, the alternator and the fuel system components can be removed for repair with the engine in place.

6   Since the cylinder heads can be removed without pulling the engine, valve component servicing can also be accomplished with the engine in the vehicle. Replacement of the camshafts, timing belt and sprockets is also possible with the engine in the vehicle, although the cylinder head must be removed from the engine to remove the camshafts.

7   In extreme cases caused by a lack of necessary equipment, repair or replacement of piston rings, pistons, connecting rods and rod bearings is possible with the engine in the vehicle. However, this practice is not recommended because of the cleaning and preparation work that must be done to the components involved.

3 Top Dead Center (TDC) for number one piston - locating

1   This procedure is essentially the same as for the 2.7L V6 engine. Refer to Chapter 2A and follow the procedure outlined there.

4 Valve cover(s) - removal and installation

Removal

1   Disconnect the cable from the negative terminal of the battery (see Chapter 5, Section 1).

2   Remove the upper intake manifold (see Section 8) and cover the lower intake manifold with rags to keep out dirt.

3   Remove the ignition coils (see Chapter 5).

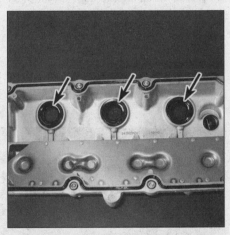

4.8 Install new spark plug tube seals before installing the valve cover

4   Pull up the wire harness tabs from the valve cover grommets and position the harness off to the side (see illustration).

5   Remove the valve cover bolts (see illustration). Note the locations of the stud bolts so they can be reinstalled in their proper locations.

6   Detach the valve cover.

Note: If the cover sticks to the cylinder head, use a block of wood and a hammer to dislodge it. If the cover still won't come loose, pry on it carefully, but don't distort the sealing flange.

Installation

7   The mating surfaces of each cylinder head and valve cover must be perfectly clean when the covers are installed. Inspect the rubber gaskets; if they're in good condition and there were no oil leaks, they can be re-used.

8   Inspect the spark plug tube seals. Replace them if they're cracked or flattened, or if the rubber has hardened. Make sure the spark plug tube seals are in position before installing the valve cover (see illustration).

9   Clean the mounting bolt threads with a die if necessary to remove any corrosion and

restore damaged threads. Use a tap to clean the threaded holes in the heads.

10　Place the valve cover and new gasket in position, then install the bolts. Tighten the bolts in several steps to the torque listed in this Chapter's Specifications.

11　Complete the installation by reversing the removal procedure. Start the engine and check carefully for oil leaks.

5　Rocker arm assembly and hydraulic lash adjusters - removal, inspection and installation

Caution: *To prevent air ingestion into the lash adjusters, do not turn the rocker arm assemblies over. Keep them in an upright position.*
Caution: *When the rocker arm assemblies are placed on the bench, do not allow them to rest on the lash adjusters as the plastic retainers and the lash adjusters may become damaged.*

Removal

1　Remove the valve cover(s) (see Section 4).
2　Mark the rocker arms and pedestals for identification before removing them. There are

four per cylinder, two intake and two exhaust.
3　Loosen each rocker arm shaft bolt a little at a time until they are all loose enough to be removed by hand. Remove the shaft and rockers as an assembly **(see illustration)**. **Caution:** *Never allow the rocker arm assembly to be placed upside down or to rest on the lash adjusters as damage may occur.*

Inspection

4　Inspect each rocker arm for wear, cracks and other damage. This engine uses aluminum rocker arms with steel roller tips and hydraulic lash adjusters. Make sure the rollers turn freely and show no signs of wear, also check the lash adjuster contact surfaces for wear or damage. Make sure the lash adjuster retaining caps hold all of the lash adjusters securely in place on the rocker arms.
Caution: *DO NOT at anytime remove the lash adjuster from the rocker arms or damage to the lash adjuster and the rocker arm will occur.*
5　Check each rocker arm pivot area and shaft for wear, cracks and galling. If the rocker arms or shafts are worn or damaged, replace them with new ones.
6　Make sure the axle for the roller tip is not sticking out one side more than the other, and make sure to check for wear, cracks and galling on the hydraulic lash adjuster to valve

stem tip contact surface.
7　To remove the rocker arms and pedestals from the shafts for inspection, the dowel pins from each pedestal must be pulled out. Use a 4 mm screw and nut, and a washer and spacer as a puller to extract the dowel pins from each pedestal **(see illustration)**. **Note:** *New dowel pins must be installed during installation.*

Installation

8　If the rockers arms or pedestals have been removed from the shafts for inspection, lubricate the shafts and install the rocker arms and pedestals in their marked order. Keep the intake rockers on the intake shaft, and exhaust rockers on the exhaust shaft.
Caution: *The notches in the rocker arm shafts must face UP when assembled or damage to the rocker arms will occur.*
9　Reinstall new dowels pins in each pedestal (they press in until they bottom out in the pedestals).
10　The rocker arm assemblies should be installed, and the bolts tightened, with the valvetrain in a "no load" situation. To accomplish this, the camshafts must be placed in a "neutral" position. Refer to Section 6 and remove the timing belt cover. Rotate the engine until the camshaft sprockets and the crankshaft sprocket are correctly aligned for TDC number 1 **(see illustrations 6.7a, 6.7b and 6.7c)**. Tighten the right side rocker arm assembly bolts in the proper sequence to the torque listed in this Chapter's Specifications.
11　Install the left rocker arm assembly and tighten the bolts in the proper sequence **(see illustration)** to the torque listed in this Chapter's Specifications. **Caution:** *The notches in the left rocker arm shafts must face UP and toward the front of the engine. The notches in the right rocker arm shafts must face UP and toward the rear of the engine (firewall side). Improper oiling of the rocker arms and shafts will result if this procedure is not followed!*
12　The remainder of installation is the reverse of the removal procedures. Start the engine and check for oil leaks.

5.3 Unscrew the rocker arm shaft bolts and lift the rocker arm assembly from the cylinder head - be sure to loosen them in the reverse order of the tightening sequence

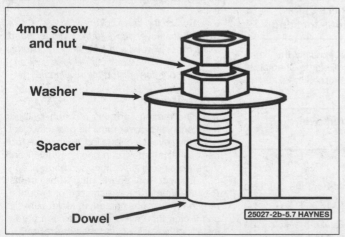

5.7 A 4 mm bolt, nut, washer and spacer can be used to extract the dowel pins from the pedestals

5.11 Rocker arm assembly TIGHTENING sequence

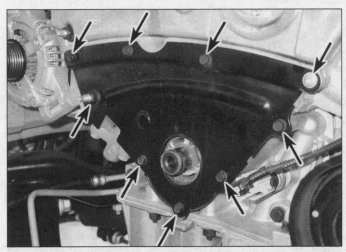

6.6a Location of the lower timing belt cover mounting fasteners

6.6b Location of the upper timing belt mounting fasteners

6 Timing belt - replacement

Caution: *Because this is an "interference" engine design, if the timing belt has broken there will be damage to the valves (and possibly the pistons) and will require removal of the cylinder heads.*

Note: *Because of work necessary to get at the timing belt and replace it, and because the water pump is driven by the timing belt, it is recommended that the water pump be replaced at the same time as the belt (see Chapter 3). The factory recommends replacing the timing belt at 100,000 miles.*

1 Disconnect the cable from the negative terminal of the battery (see Chapter 5, Section 1).

2 Remove the cooling fan assembly (see Chapter 3) and the accessory drivebelt (see Chapter 1).

3 Remove the valve covers (see Section 4) and loosen the rocker arm shaft bolts (see Section 5).

4 Raise the vehicle and support it securely on jackstands. Remove the crankshaft pulley (see Section 11) and the drivebelt tensioner (see Chapter 1).

5 Remove the power steering pump mounting bolts and position the power steering pump off to the side without disconnecting the power steering system fluid lines (see Chapter 10).

6 Remove the lower timing belt cover and the upper timing belt cover from the cylinder heads **(see illustrations)**. If the same belt is to be reinstalled, mark the direction of rotation on the belt using white paint.

7 Temporarily install the crankshaft pulley bolt and turn the crankshaft with the bolt to align the timing marks on the crankshaft and camshaft sprockets. The crankshaft sprocket arrow should line up with the TDC indicator on the oil pump cover and the camshaft sprocket marks should line up with the marks on the top of the rear covers **(see illustrations)**.

6.7a Rotate the crankshaft until the pointer on the crankshaft sprocket aligns with the TDC mark on the oil pump cover. . .

6.7b. . . the right camshaft sprocket timing mark aligns with the mark on the rear cover. . .

6.7c. . . and the left camshaft sprocket timing mark aligns with the mark on the rear cover

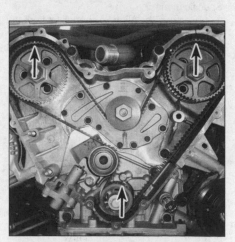

6.7d Overall view of the crankshaft and camshaft sprocket alignment marks on the 3.5L V6 engine

6.8 Remove the timing belt tensioner bolts, then remove the tensioner and the timing belt

6.9 Compress the plunger on the tensioner in a vise until a drill bit or Allen wrench can be inserted through the hole to retain the plunger - orient the pin in a way that won't interfere with installing the tensioner with the pin in place

7.4 Check the diameter of each camshaft bearing journal to pinpoint excessive wear and out-of-round conditions

8   Unbolt the timing belt tensioner and remove the belt **(see illustration)**.

9   Slowly compress the timing belt tensioner in a vise until a drill bit or Allen wrench can be inserted through the hole to lock it in **(see illustration)**. **Note:** *Total bleed down time of the tensioner should take approximately four to five minutes. DO NOT force tensioner to bleed down faster or damage to the tensioner may occur.*

10   Making sure that all timing marks are still aligned, start by installing the new belt at the crankshaft pulley, proceeding counterclockwise up to the left (driver's side) camshaft sprocket. After the belt is on the left camshaft sprocket, keep tension on the belt as it is fed under the water pump pulley, over the right camshaft sprocket, and past the tensioner. If the old timing belt is being reused, be sure the directional mark made in Step 6 is pointing in the proper direction.

11   Hold the tensioner pulley against the belt and install the tensioner. Tighten the tensioner bolts to the torque listed in this Chapter's Specifications.

12   Remove the Allen key or drill bit from the tensioner, allowing it to tension the pulley on the belt. Tighten the rocker arm shaft bolts, in the proper sequence **(see illustration 5.11)**, to the torque listed in this Chapter's Specifications.

13   Rotate the crankshaft pulley through two complete revolutions to check that the timing marks still remain aligned. If not, repeat the belt installation process.

14   The remainder of installation is the reverse of removal.

15   Start the engine and check for leaks.

Note: *Noise may be present upon initial start up, due to air entering the timing chain tensioner. If this situation occurs, raise the idle to 1,600 - 2,000 rpm for ten minutes to purge the air from the tensioner. This noise should last no longer than 15 minutes after initial start up.*

7   Camshafts - removal, inspection and installation

Removal

1   The cylinder head(s) must be removed to withdraw the camshafts from the rear of the heads. Refer to Section 10 for cylinder head removal.

2   Remove the camshaft thrust plate and O-ring seal and withdraw the camshaft from the cylinder head, being careful not to nick the cam bearings or journals. Also remove the camshaft oil seal.

Inspection

3   After the camshaft has been removed from the engine, cleaned with solvent and dried, inspect the bearing journals for uneven wear, pitting and evidence of seizure. If the camshaft journals are damaged, inspect the cylinder head also.

4   Measure the bearing journals with a micrometer to determine if they are excessively worn or out-of-round **(see illustration)**.

Compare the measurements with the Specifications listed in this Chapter. Use an inside micrometer to measure the journal diameters in the cylinder head. Compare the measurements with the Specifications listed in this Chapter.

5   Measure the lobe height of each cam lobe on the intake camshaft and record your measurements **(see illustration)**. Compare the measurements for excessive variations. Standard measurements should be 0.001 inch (0.025 mm). If the lobe heights vary more than 0.010 inch (0.254 mm), replace the camshaft. Compare the lobe height measurements on the exhaust camshaft and follow the same procedure. Do not compare intake camshaft lobe heights with exhaust camshaft lobe heights, as they are different. Only compare intake lobes with intake lobes and exhaust lobes with other exhaust lobes.

6   Check the camshaft lobes for heat discoloration, score marks, chipped areas, pitting and uneven wear **(see illustration)**. If the lobes are in good condition and if the lobe lift variation measurements recorded earlier are within the limits, the camshaft can be reused.

7.5 Measure the camshaft lobe height (greatest dimension) with a micrometer

7.6 Check the cam lobes for pitting, excessive wear and scoring. If scoring is excessive, as shown here, replace the camshaft

7.7 Coat both the cam lobes and the journals with camshaft installation lube before installing the camshaft

8.6a Location of the front upper intake manifold mounting bracket mounting studs/bolts

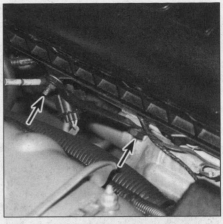

8.6b Location of the rear upper intake manifold mounting bracket mounting studs/bolts

Installation

7   Lubricate the camshaft bearing journals and cam lobes with camshaft installation lube **(see illustration)**.
8   Insert the camshaft carefully into the cylinder head, then install a new seal at the front of the head. Using a seal driver or a deep socket with an outside diameter slightly smaller than that of the seal, tap the new seal in.
9   Install the camshaft cover at the back of the head with a new O-ring.
10   The remainder of the installation is the reverse of removal and is covered in Sections 6 and 10.

---

8   Intake manifold - removal and installation

Upper intake manifold

1   Relieve the fuel system pressure (see Chapter 4).
2   Disconnect the cable from the negative terminal of the battery (see Chapter 5, Section 1).

3   Remove the air filter housing (see Chapter 4).
4   Label and disconnect the hoses and electrical connectors attached to the intake manifold and throttle body.
5   Remove the bolts securing the EGR pipe to the upper intake manifold (see Chapter 6).
6   Remove the brackets from the throttle body and the right side of the upper intake manifold **(see illustrations)**.
7   Loosen the upper intake manifold bolts starting with the outer bolts and working to the inside bolts and remove the upper intake manifold **(see illustration)**.
8   Clean the mounting surfaces of the lower intake manifold and the upper manifold. Inspect the rubber O-ring gasket for cracking, hardness or other signs of deterioration. If it's in good condition it can be reused.
9   Install the gasket into the groove in the upper intake manifold **(see illustration)**, then install the upper intake manifold onto the lower intake manifold. Tighten the bolts, starting with the center bolts and working to the outer bolts using a circular pattern, to the torque listed in this Chapter's Specifications. The remainder of the installation is the reverse of removal.

Lower intake manifold

Warning: *Wait until the engine is completely cool before beginning this procedure.*
10   Drain the cooling system (see Chapter 1).
11   Remove the upper intake manifold (see Steps 1 through 7).
12   Label and detach any remaining hoses which would interfere with the removal of the lower intake manifold
13   Remove the fuel rail and injectors from the lower intake manifold (see Chapter 4).
14   Remove the heater hoses and coolant hoses (see Chapter 3).
15   Loosen the remaining manifold mounting bolts/nuts in 1/4-turn increments until they can be removed by hand.
16   The manifold will probably be stuck to the cylinder heads and force may be required to break the gasket seal.
Caution: *Don't pry between the manifold and the heads or damage to the gasket sealing surfaces may occur, leading to vacuum leaks.*
17   Turn the manifold over, remove the heater supply tube and replace the O-ring with a new one.
18   Carefully use a scraper to remove all traces of old gasket material and sealant from

8.7 Lift the upper intake manifold from the engine

8.9 Install the upper intake manifold gasket, making sure it seats in the groove

8.19 Apply a bead of RTV sealant around the water passages at the ends of the heads

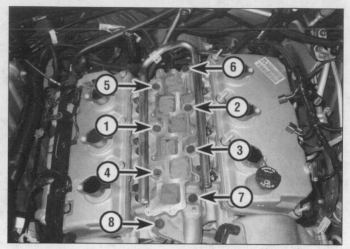

8.22 Lower intake manifold TIGHTENING sequence

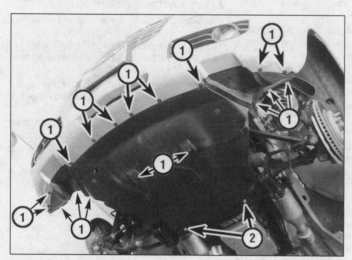

9.2 Front engine splash shield (1) and middle engine splash shield (2) fasteners

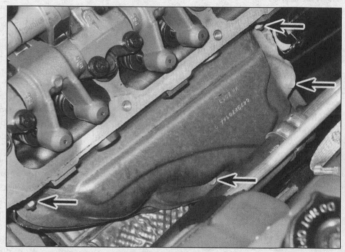

9.6 Location of the upper heat shield mounting bolts/nuts

the manifold and cylinder heads, then clean the mating surfaces with lacquer thinner or acetone.

19   Apply a bead of RTV sealant or equivalent around the front and rear cylinder head water passages before laying the new gaskets in place **(see illustration)**.

20   Install the intake manifold gaskets on the cylinder heads and apply RTV sealant to the manifold side of the gaskets, around the water passages.

21   Position the manifold on the engine, making sure the gaskets and manifold are aligned over the ports in the cylinder heads. Install the fuel rail and injectors onto the lower manifold, then install the manifold mounting bolts.

22   Following the recommended tightening sequence, tighten the nuts/bolts, in several steps, to the torque listed in this Chapter's Specifications **(see illustration)**.

23   The remainder of the installation is the reverse of the removal procedure. Refill the cooling system (see Chapter 1), then run the engine and check for fuel, vacuum and coolant leaks.

9   Exhaust manifold(s) - removal and installation

Warning: *The engine must be completely cool before beginning this procedure.*

Removal

1   Disconnect the cable from the negative terminal of the battery (see Chapter 5, Section 1).

2   Raise the front of the vehicle and support it securely on jackstands. Remove the engine splash shield **(see illustration)**.

3   Disconnect the oxygen sensors' electrical connectors and remove the sensors (see Chapter 6).

4   Detach the exhaust pipes from the exhaust manifolds (see Chapter 4, **illustration 14.4a**).

5   Lower the vehicle.

6   Remove the exhaust manifold heat shield mounting bolts/nuts **(see illustration)** and remove the heat shields.

7   Remove the mounting bolts and detach the manifold from the cylinder head **(see illustration)**. **Note:** *Be sure to spray penetrating oil on the bolts before attempting to remove them.*

Installation

8   Clean the mating surfaces to remove all traces of old gasket material, then inspect the manifold for distortion and cracks. Warpage can be checked with a precision straightedge held against the mating flange. If a feeler gauge thicker than 0.030-inch can be inserted between the straightedge and flange surface,

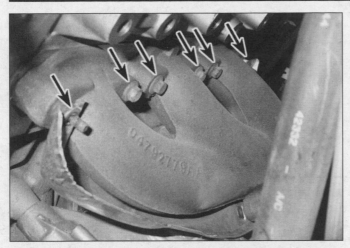

9.7 Location of the exhaust manifold mounting bolts/studs

10.9a With the engine raised (or the cooling fan assembly removed), hold the camshaft sprocket with a wrench while loosening the bolt with a socket and breaker bar - these bolts are very tight

take the manifold to an automotive machine shop for resurfacing.

9   Place the exhaust manifold in position with a new gasket and install the mounting bolts finger-tight.

Note: *Be sure to identify the exhaust manifold gaskets by the correct cylinder designation and the position of the exhaust ports on the gasket.*

10   Starting in the middle and working out toward the ends, tighten the mounting bolts in several increments, to the torque listed in this Chapter's Specifications.

11   Install the remaining components in the reverse order of removal.

12   Start the engine and check for exhaust leaks between the manifold and cylinder head and between the manifold and exhaust pipe.

10   Cylinder heads - removal and installation

Warning: *The engine must be completely cool before beginning this procedure.*

Note: *Special tools are necessary to complete this procedure. Read through the entire procedure and obtain the special tools before beginning work.*

Removal

1   Relieve the fuel system pressure (see Chapter 4). Disconnect the cable from the negative terminal of the battery (see Chapter 5, Section 1).

2   Drain the cooling system (see Chapter 1).

3   Remove the upper and lower intake manifolds (see Section 8).

4   Remove the ignition coils and the spark plugs (see Chapter 1 and Chapter 5, if necessary).

5   Detach the exhaust manifold from the cylinder head being removed (see Section 9).

6   Remove the valve cover(s) (see Section 4).

7   Remove the rocker arms and shafts (see Section 5).

8   Remove the timing belt covers, align the TDC marks and remove the timing belt (see Section 6).

Note: *If the belt is to be reused, mark its direction of rotation before removing it.*

9   Hold the camshaft sprocket hex with a wrench while using a socket and breaker bar to loosen the camshaft bolt, which is under considerable torque **(see illustrations)**.

Caution: *The two camshaft sprockets are not interchangeable, nor are the camshaft sprocket bolts. The left sprocket bolt is 10 inches while the right sprocket bolt is 8-3/8 inches. Always use NEW camshaft sprocket bolts.*

10   Remove the camshaft thrust plate from the rear of the head and slide the camshaft back approximately 3 - 1/2 inches. This will allow removal of the sprocket and bolt.

11   Remove the bolts securing the rear timing belt cover to the cylinder head. Don't dis-

turb any of the other timing belt cover bolts.

12   Using the new head gasket, outline the cylinders and bolt pattern on a piece of cardboard. Be sure to indicate the front of the engine for reference. Punch holes at the bolt locations. This cardboard holder will be used to accurately store each cylinder head bolt after they are removed from the cylinder head.

13   Loosen each of the cylinder head mounting bolts 1/4-turn at a time until they can be removed by hand - work from bolt-to-bolt in a pattern that's the reverse of the tightening sequence **(see illustration 10.21)**. Store the bolts in the cardboard holder as they're removed.

14   Lift the head(s) off the engine. If resistance is felt, don't pry between the head and block as damage to the mating surfaces will result. Recheck for head bolts that may have been overlooked, then use a hammer and block of wood to tap up on the head to break the gasket seal. Be careful because there are

10.9b The camshaft sprocket bolts are not interchangeable - The left sprocket bolt is 10 inches while the right sprocket bolt is 8-3/8 inches

10.9c The camshaft sprockets are also not interchangeable

10.16 Use a putty knife or gasket scraper to remove gasket material from the cylinder head and block

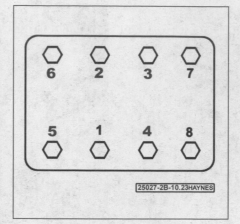

10.21 Cylinder head bolt tightening sequence

11.3 Insert a prybar or large screwdriver through the pulley to hold it while you loosen the crankshaft pulley bolt

11.4 Use a three jaw puller to remove the crankshaft pulley. A long Allen bolt must be inserted into the crankshaft for the tool to push against

locating dowels in the block which position each head. As a last resort, pry each head up at the rear corner only and be careful not to damage anything. After removal, place the head on blocks of wood to prevent damage to the gasket surfaces.

15   If the camshaft is to be removed for examination or replacement, it must be done with the cylinder head off the engine, as it comes out the back of the cylinder head (see Section 7).

Installation

16   The mating surfaces of each cylinder head and block must be perfectly clean when the head is installed. Use a gasket scraper to remove all traces of carbon and old gasket material **(see illustration)**, then clean the mating surfaces with lacquer thinner or acetone. If there's oil on the mating surfaces when the head is installed, the gasket may not seal correctly and leaks may develop. When working on the block, it's a good idea to

cover the valley with shop rags to keep debris out of the engine. Use a shop rag or vacuum cleaner to remove any debris that falls into the cylinders.

17   Check the block and head mating surfaces for nicks, deep scratches and other damage. If damage is slight, it can be removed with a file; if it's excessive, machining may be the only alternative.

18   Use a tap of the correct size to chase the threads in the head bolt holes. Dirt, corrosion, sealant and damaged threads will affect torque readings. Check the cleaned head bolts for stretch by holding them next to a steel ruler. If all the threads don't touch the ruler, the bolt should be replaced.

19   Position the new gasket over the dowel pins in the block. Some gaskets are marked TOP or FRONT to ensure correct installation.

20   Make sure the camshaft is installed in the cylinder head, then carefully position the head on the block without disturbing the gasket.

21   Tighten the bolts in the recommended sequence **(see illustration)** to the torque listed in this Chapter's Specifications.

22   Install the rear timing belt cover-to-cylinder head bolts. Tighten the bolts to the torque listed in this Chapter's Specifications.

23   Slide the camshaft back into the cylinder head and install the camshaft sprocket and new bolt, then move the camshaft back into place. Hold the hex on the camshaft sprocket with a wrench and tighten the camshaft bolt to the torque listed in this Chapter's Specifications **(see illustration 10.9a)**.

24   Install the camshaft thrust plate at the back of the head, using a new O-ring. Tighten the bolts to the torque listed in this Chapter's Specifications.

25   The remaining installation steps are the reverse of removal.

26   Change the oil and filter, refill the cooling system (see Chapter 1), run the engine and check for leaks.

11   Crankshaft pulley - removal and installation

1   Disconnect the cable from the negative terminal of the battery (see Chapter 5, Section 1). Remove the accessory drivebelts (see Chapter 1).

2   Remove the cooling fan assembly (see Chapter 3).

3   Position a large screwdriver through the crankshaft pulley to keep the crankshaft from turning and remove the pulley-to-crankshaft bolt **(see illustration)**.

4   Pull the crankshaft pulley off the crankshaft with a puller **(see illustration)**. **Caution:** *The jaws of the puller must only contact the hub of the pulley - not the outer ring.* **Note:** *A spacer that bears on the nose of the crankshaft or a long Allen-head bolt must be inserted into the crankshaft nose for the puller's tapered tip to push against. Don't allow the puller screw to bear against the crankshaft itself.*

5   To install the crankshaft pulley, slide the pulley onto the crankshaft as far as it will slide on, then use a vibration damper installation tool to press the pulley onto the crankshaft.

6   Install the crankshaft pulley retaining bolt and tighten it to the torque listed in this Chapter's Specifications.

7   The remainder of installation is the reverse of the removal.

12   Crankshaft front oil seal - replacement

1   Remove the crankshaft pulley (see Section 11).

2   Remove the timing belt (see Section 6).

3   Use a two-bolt puller to remove the crankshaft sprocket, with the bolts threaded into the holes in the sprocket.

4   With a hammer and punch, tap the dowel pin out of the crankshaft. Drive the pin into the

12.5 Using a hook-type seal remover, pry out the old seal

12.6 Drive the new seal in to the same depth as the old one, using a seal driver or a socket with an outside diameter the same as that of the seal

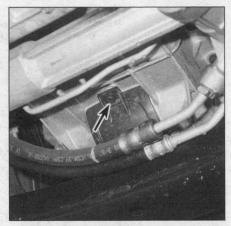

13.10a Remove this bolt and detach the torque converter access cover. . .

hollow, threaded area of the crankshaft snout, where it can be extracted with a small magnet.

5   Pry the old seal out with a hook-type seal tool, being very careful not to scratch the seal surface of the crankshaft **(see illustration)**. Note how the seal is installed - the new one must be installed to the same depth and facing the same way.

6   Lubricate the inner lip of the new seal with engine oil and drive it in with a seal driver or large deep socket and a hammer **(see illustration)**.

7   Installation is the reverse of removal. Tap the dowel pin back into the crankshaft, with 3/64-inch extending out.

13 Oil pan - removal and installation

Removal

1   Disconnect the cable from the negative terminal of the battery (see Chapter 5, Section 1).

2   Remove the engine oil dipstick.

3   Raise the front of the vehicle and support it securely on jackstands. Apply the parking brake and block the rear wheels to keep the vehicle from rolling off the stands.

4   Connect an engine hoist to the engine (see Chapter 2D).

5   Remove the engine splash shield (see Section 9).

6   Drain the engine oil and remove the oil filter (see Chapter 1).

7   Unbolt the engine oil cooler and set it aside (see Chapter 3).

8   Disconnect the steering intermediate shaft coupler at the steering gear (see Chapter 10).

9   Remove the steering gear mounting bolts at the subframe (see Chapter 10). Using rope or mechanics wire, tie the assembly off to the side without disconnecting the power steering fluid lines.

10   Remove the torque converter access

13.10b. . . then remove these two oil pan bolts

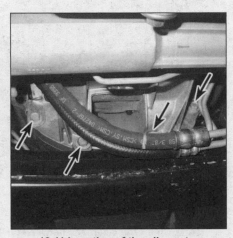

13.11 Location of the oil pan-to-transmission bolts

cover, then remove the two Allen-head bolts securing the rear of the oil pan to the engine block **(see illustrations)**.

11   Remove the oil pan-to-transmission bolts **(see illustration)**.

12   Remove the engine mount nuts at the subframe (see Section 17).

13   Remove the oil pan bolts, then carefully separate the oil pan from the block. Don't pry between the block and the pan or damage to the sealing surfaces could occur and oil leaks may develop. Instead, tap the pan with a soft-faced hammer to break the gasket seal.

14   Raise the engine with the hoist just enough to allow oil pan removal, then guide the pan out between the engine and the subframe.

Installation

15   Clean the pan with solvent and remove all old sealant and gasket material from the block and pan mating surfaces. Clean the mating surfaces with lacquer thinner or acetone and make sure the bolt holes in the block

are clear. Check the oil pan flange for distortion, particularly around the bolt holes.

16   Apply a bead of RTV sealant to the parting lines between the engine block and the oil pump housing and then to the engine block and rear main oil seal retainer. Install the gasket on the block.

17   Place the oil pan in position on the block and install the bolts, tightening them finger-tight.

18   Lower the engine and install the engine mount nuts, tightening them to the torque listed in this Chapter's Specifications.

19   Install the oil pan-to-transmission bolts and tighten them to approximately 12 in-lbs.

20   Tighten the oil pan fasteners to the torque listed in this Chapter's Specifications. Starting at the center, follow a criss-cross pattern and work up to the final torque in three steps. Don't forget the two small Allen bolts at the rear, accessible through the torque converter access opening.

21   Tighten the oil-pan-to-transmission bolts to the torque listed in this Chapter's Specifications.

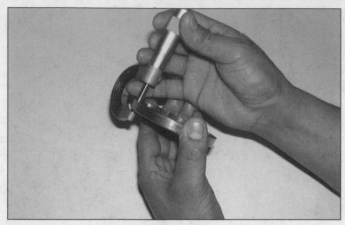

14.9 Use a micrometer to check the thickness of the outer rotor

14.11 Check the outer rotor-to-housing clearance

14.12 Check the clearance between the lobes of the inner and outer rotors

14.13 Using a straightedge and feeler gauge, check the side clearance between the surface of the oil pump housing and the rotors

22 Install the torque converter access cover.

23 The remaining steps are the reverse of the removal procedure.

24 Refill the engine with oil (see Chapter 1), replace the filter, run it until normal operating temperature is reached and check for leaks.

14 Oil pump - removal, inspection and installation

Caution: *Do not remove the oil pressure relief valve, spring and plunger from the oil pump. Replace the oil pump as one complete assembly if the oil pressure relief valve is damaged.*

Removal

1 Remove the timing belt and crankshaft sprocket (see Section 6 and Section 12).

2 Remove the oil pan (see Section 13).

3 Remove oil pump pick-up tube.

4 Remove the oil-pump-to-block bolts and pull the oil pump from the block.

Inspection

5 Unbolt the cover from the back of the oil pump housing.

6 Remove the inner and outer rotors from the oil pump body, noting their installed direction for reassembly.

7 Clean all parts thoroughly in solvent and carefully inspect the rotors, pump cover and oil pump housing for nicks, scratches or burrs. Replace the assembly if it is damaged.

8 Use a straightedge and measure the oil pump cover for warpage with a feeler gauge. If it's warped more than the limit listed in this Chapter's Specifications, the pump should be replaced.

9 Measure the thickness and the diameter of the outer rotor. If either measurement is less than the value listed in this Chapter's Specifications, the pump should be replaced **(see illustration)**.

10 Measure the thickness of the inner rotor. If the measurement is less than the value listed in this Chapter's Specifications, the pump should be replaced.

11 Insert the outer rotor into the oil pump housing and, while holding the rotor against one side of the housing with your finger, measure the clearance at the opposite side between the rotor and housing **(see illustration)**. If the measurement is more than the maximum allowable clearance listed in this Chapter's Specifications, the pump should be replaced.

12 Install the inner rotor in the oil pump assembly and measure the clearance between the lobes on the inner and outer rotors **(see illustration)**. If the clearance is more than the value listed in this Chapter's Specifications, the pump should be replaced. **Note:** *Install the inner rotor with the mark facing up.*

13 Position a straightedge across the face of the oil pump assembly **(see illustration)**. If the clearance between the pump surface and the rotors is greater than the limit listed in this Chapter's Specifications, the pump should be replaced.

Installation

14 Prime the oil pump by filling the housing with engine oil. Install the pump cover and tighten the bolts to the torque listed in this Chapter's Specifications.

17.1a Location of the left engine mount-to-subframe mounting nuts

17.1b Location of the right engine mount-to-subframe mounting nuts

15   To install the pump, place the gasket on the engine block mating surface, then turn the flats in the rotor so they align with the flats on the crankshaft and install the oil pump housing on to the engine block.

16   Install the pump-to-block bolts and tighten them to the torque listed in this Chapter's Specifications.

17   The remainder of installation is the reverse of removal.

15   Driveplate - removal and installation

1   This procedure is essentially the same as for the 2.7L V6 engine. Refer to Chapter 2A and follow the procedure outlined there, but use the bolt torque listed in this Chapter's Specifications.

16   Rear main oil seal - replacement

Note: *The rear main seal and housing are serviced as one complete assembly. This procedure will require several special tools. Read the procedure carefully before starting.*

1   This procedure is essentially the same as for the 2.7L V6 engine. Refer to Chapter 2A and follow the procedure outlined there, but use the bolt torque value listed in this Chapter's Specifications.

17   Engine mounts - check and replacement

1   This procedure is essentially the same as for the 2.7L V6 engine. Refer to Chapter 2A and follow the procedure and **illustrations** outlined there, but refer to the **illustrations** and use the torque values listed in this Chapter's Specifications.

Notes

Chapter 2 Part C
3.6L V6 engine

Contents

Specifications

General

Displacement	220 cubic inches
Bore	3.779 inches
Stroke	3.268 inches
Compression ratio	10.2:1
Cylinder numbers (front-to-rear)	
Left (driver's side) bank	2-4-6
Right bank	1-3-5
Firing order	1-2-3-4-5-6
Oil pressure	
At idle speed	5 psi (minimum)
At 1,200 to 3000 rpm	30 (warm) to 139 (cold) psi

Front

Cylinder locations

Camshaft

Bore diameter	
Cam tower 1	1.2606 to 1.2615 inches
Cam tower 2, 3, and 4	0.9457 to 0.9465 inch
Bearing journal diameter	
No. 1	1.2589 to 1.2596 inches
No. 2, 3, and 4	0.9440 to 0.9447 inch
Bearing clearance	
No. 1	0.0001 to 0.0026 inch
No. 2, 3, and 4	0.0009 to 0.0025 inch
End play	0.003 to 0.01 inch

Crankshaft main bearing

Main journal diameter	2.8345 ± 0.0035 inches
Clearance	0.0009 to 0.002 (limit) inch
End play	0.002 to 0.0114 (limit) inch

Cylinder head

Gasket thickness (compressed)	0.019 to 0.024 inch
Valve seat width	
Intake	0.04 to 0.05 inch
Exhaust	0.055 to 0.063 inch
Valve seat runout (maximum)	0.002 inch

Valves

Stem-to-guide clearance
  Intake
    Standard .... 0.0009 to 0.0024 inch
    Maximum .... 0.011 inch
  Exhaust
    Standard .... 0.0012 to 0.0027 inch
    Maximum .... 0.0146 inch

Torque specifications   Ft-lbs (unless otherwise indicated)

Note: *One foot-pound (ft-lb) of torque is equivalent to 12 inch-pounds (in-lbs) of torque. Torque values below approximately 15 ft-lbs are expressed in inch-pounds, since most foot-pound torque wrenches are not accurate at these smaller values.*

Crankshaft balancer bolt
  Step 1 30
  Step 2 .... Tighten an additional 105-degrees
Cylinder head bolts* (in sequence - **see illustrations 10.30a and 10.30b**)
  2013 and earlier models
    Step 1 22
    Step 2 33
    Step 3 .... Tighten an additional 75-degrees
    Step 4 .... Tighten an additional 50-degrees
    Step 5 .... Loosen all in reverse of tightening sequence
    Step 6 22
    Step 7 33
    Step 8 .... Tighten an additional 70-degrees
    Step 9 .... Tighten an additional 70-degrees
  2014 and later models**
    Step 1 22
    Step 2 33
    Step 3 33
    Step 4 .... Tighten an additional 125-degrees
    Step 5 .... Loosen all in reverse of tightening sequence
    Step 6 22
    Step 7 33
    Step 8 33
    Step 9 .... Tighten an additional 130-degrees
Drivebelt idler sprocket bolt 18
Driveplate-to-crankshaft bolts 70
Flywheel-to-crankshaft bolts* 55
Catalytic converter to cylinder head fasteners 27
Exhaust crossover bolts 21
Intake manifold bolts
  Upper .... 89 in-lbs
  Lower
    2013 and earlier models .... 106 in-lbs
    2014 and later models .... 71 in-lbs
Oil cooler
  Bolts .... 35 in-lbs
  Screws .... 106 in-lbs
Oil pan drain plug 20
Oil pan
  Lower pan-to-upper pan nut/bolts .... 97 in-lbs
  Upper pan-to-rear main seal housing (M6 bolts) .... 108 in-lbs
  Upper pan-to-cylinder block (M8 bolts) 18
  Upper pan-to-transaxle bolts 41
Oil pump pick-up tube mounting bolts .... 106 in-lbs
Oil pump cover (plate) screws .... 105 in-lbs
Oil pump-to-engine block fasteners .... 106 in-lbs
Rear main oil seal retainer bolts .... 105 in-lbs
Timing chain cover bolts
  M6 bolts .... 106 in-lbs
  M8 bolts 18
  M10 bolts 41
Timing gear splash shield bolts .... 35 in-lbs
Camshaft oil control valves 110
Oil pump timing chain sprocket (T45) 18
Valve cover-to-cylinder head bolts .... 106 in-lbs
Water pump bolts .... See Chapter 3

* *Use new bolts*
** *If using a new engine block, follow all nine Steps. If using the existing block, follow Steps 6 to 9.*

1 General information

1   This chapter is devoted to in-vehicle repair procedures for the 3.6L VVT V6 engine.

2   The 3.6 liter engine utilizes Variable Valve Timing (VVT), Dual Overhead Camshafts (DOHC), four timing chains, an aluminum cylinder block, steel cylinder sleeves or liners with six cylinders arranged in a "V"-shape, with 60-degrees between the two banks. The engine has a chain-driven oil pump with a multi-stage pressure regulator to increase fuel economy. The exhaust manifolds are integral with the cylinder heads to make the engine lighter. **Caution:** *This engine is not of a freewheeling design and severe engine damage will occur if the timing chain breaks.*

3   Information concerning engine removal and installation can be found in Chapter 2E. The following repair procedures are based on the assumption that the engine is installed in the vehicle. If the engine has been removed from the vehicle and mounted on a stand, many of the steps outlined in this Chapter do not apply.

2 Repair operations possible with the engine in the vehicle

1   Many major repair operations can be done without removing the engine from the vehicle.

2   Clean the engine compartment and the exterior of the engine with degreaser before any work is done. It'll make the job easier and help keep dirt out of internal parts of the engine.

3   It may be helpful to remove the hood to improve engine access when repairs are performed (see Chapter 11). Cover the fenders to prevent damage to the paint. Special pads are available, but an old bedspread or blanket will also work.

4   If vacuum, exhaust, oil, or coolant leaks develop, indicating a need for gasket or seal replacement, the repairs can generally be done with the engine in the vehicle. The intake and exhaust manifold gaskets, timing chain cover gasket, oil pan gasket, crankshaft oil seals, and cylinder head gaskets are all accessible with the engine in the vehicle.

5   Exterior engine components, such as the intake and exhaust manifolds, the oil pan, the oil pump, the timing chain cover, the water pump, the starter motor, the alternator, and fuel system components can be removed for repair with the engine in the vehicle.

6   Cylinder heads can be removed without pulling the engine. Valve component servicing can also be done with the engine in the vehicle. Replacement of the timing chain and sprockets is also possible with the engine in the vehicle, as is camshaft and valvetrain removal and installation.

7   Repair or replacement of piston rings, pistons, connecting rods, and rod bearings is possible with the engine in the vehicle, however, this practice is not recommended because of the cleaning and preparation work that must be done to the components.

3 Top Dead Center (TDC) for number one piston - locating

1   Top Dead Center (TDC) is the highest point in the cylinder that each piston reaches as it travels up the cylinder bore. Each piston reaches TDC on the compression stroke and again on the exhaust stroke, but TDC generally refers to piston position on the compression stroke.

2   Positioning the piston(s) at TDC is an essential part of certain procedures such as camshaft and timing chain/sprocket removal.

3   Before beginning this procedure, be sure to place the transmission in Neutral and apply the parking brake or block the rear wheels. Disconnect the cable from the negative terminal of the battery (see Chapter 5). Remove the ignition coils (see Chapter 5) and the spark plugs (see Chapter 1). **Caution:** *On 2015 and later models, if equipped with an Intelligent Battery Sensor (IBS), disconnect the IBS connector first before disconnecting the negative battery cable.*

4   Install a compression pressure gauge in the number one spark plug hole (refer to Chapter 2E). It should be a gauge with a screw-in fitting and a hose at least six inches long.

5   Rotate the crankshaft using a socket and breaker bar on the crankshaft pulley bolt while observing for pressure on the compression gauge. The moment the gauge shows pressure indicates that the number one cylinder has begun the compression stroke.

6   Once the compression stroke has begun, TDC for the compression stroke is reached by bringing the piston to the top of the cylinder.

7   These engines are not equipped with external components (crankshaft pulley, flywheel, timing hole, etc.) that are marked to identify the position of number 1 TDC. There-fore, the only method to double-check the location of TDC number 1 is to remove the valve cover to access the camshaft sprockets and alignment marks (see Section 4) and note the rocker arm position, or use a degree wheel and a positive stop timing device threaded into the spark plug hole for cylinder number 1.

8   After the number one piston has been positioned at TDC on the compression stroke, TDC for any of the remaining cylinders can be located by turning the crankshaft 120-degrees and following the firing order in (this Chapter's Specifications). For example, rotating the engine 120-degrees past TDC number 1 will put the engine at TDC compression for cylinder number 2.

4 Valve covers - removal and installation

Removal

1   Disconnect the cable from the negative terminal of the battery (see Chapter 5, Section 3). **Caution:** *On 2015 and later models, if equipped with an Intelligent Battery Sensor (IBS), disconnect the IBS connector first before disconnecting the negative battery cable.*

2   Remove the engine cover by pulling it straight up off of the ballstuds.

3   Remove the upper intake manifold (see Section 5). **Note:** *Cover the open ports on the lower intake manifold to prevent debris from entering the engine.* **Caution:** *Once the valve covers are removed, the magnetic timing wheels are exposed (see illustration 9.9). The magnetic timing wheels on the camshafts must not come in contact with any type of magnet or magnetic field. If contact is made, the timing wheels will need to be replaced.*

4   Remove insulator from the left valve cover, if equipped **(see illustration)**.

5   Before removing the variable valve timing solenoid connectors from the front of each valve cover, mark them appropriately so they can be reinstalled in their original locations.

4.4 Lift the insulator up and off of the retaining posts then remove it from the front valve cover

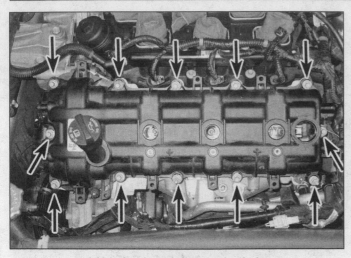

4.10 Left valve cover mounting bolts

5.4 Disconnect the ETC connector (A), the MAP sensor (B), the electrical harness (C) retainer and the PCV, brake booster and EVAP purge valve hoses (D)

6   Disconnect the wiring harness retainers from the valve cover and move the harnesses out of the way.

7   Remove the ignition coils (see Chapter 5).

8   Mark the Camshaft Position (CMP) sensors to each valve cover so they can be reinstalled in their original locations, then remove the sensor(s) (see Chapter 6).

9   Remove the PCV valve from the right-side cover (see Chapter 6).

10   Remove the valve cover fasteners **(see illustration)** and remove the cover(s).

Caution: *If the cover is stuck to the cylinder head, tap one end with a block of wood and a hammer to jar it loose. If that doesn't work, slip a flexible putty knife between the cylinder head and cover to break the gasket seal. Don't pry at the cover-to-cylinder head joint or damage to the sealing surfaces may occur (leading to future oil leaks).*

11   Remove the valve cover gasket, then remove the spark plug tube seals.

Note: *The cover gaskets can be reused if they are not damaged.*

Installation

12   The mating surfaces of each cylinder head and valve cover must be perfectly clean when the covers are installed. Use a gasket scraper to remove all traces of sealant and old gasket material, then clean the mating surfaces with brake system cleaner. If there's sealant or oil on the mating surfaces when the cover is installed, oil leaks may develop.

13   Inspect spark plug tube seals; if damaged, carefully remove the seals using an appropriate pry tool. Position the new seal with the part number facing the valve cover, then use a socket that contacts the outer edge to drive the seal in place.

14   Apply a dab of RTV sealant at the joints where the engine front cover meets the cylinder head.

15   Install the valve cover and bolts, then tighten the bolts to the torque listed in this Chapter's Specifications.

16   The remainder of installation is the reverse of removal.

5   Intake manifolds - removal and installation

Warning: *Wait until the engine is completely cool before beginning this procedure.*

Removal

1   If you will be removing the lower intake manifold, relieve the fuel system pressure (see Chapter 4, Section 3).

2   Disconnect the cable from the negative terminal of the battery (see Chapter 5, Section 3).

Caution: *On 2015 and later models, if equipped with an Intelligent Battery Sensor (IBS), disconnect the IBS connector first before disconnecting the negative battery cable.*

3   Remove the engine cover by pulling it straight up off the ballstuds.

Upper intake manifold

4   Disconnect the wiring harness from the MAP sensor and the throttle body **(see illustration)**.

5   Loosen the clamp and detach the intake duct from the throttle body.

6   Disconnect the PCV valve hose (see Chapter 6), EVAP purge hose and brake booster hoses.

7   Disconnect the wiring harness retainers from the upper intake support bracket and the retainer from the stud bolt **(see illustration)**.

8   Remove the nuts and stud bolt, then remove the upper intake manifold bracket **(see illustration)**.

9   Remove the nut from the bracket on the heater core return tube **(see illustration)**.

10   Remove the support bracket-to-upper manifold nuts **(see illustration)**.

11   Loosen, but do not remove, the bolts on the manifold, and remove the upper intake manifold **(see illustration)**.

12   Discard the six upper-to-lower intake manifold seals, and cover the open intake ports to prevent debris from entering the engine.

5.7 Pry the wiring harness retainer off of the bracket stud

5.8 Remove the stud bolt (A) and bracket nuts (B), and remove the bracket

5.9 Remove the nut from the bracket on the heater core return tube

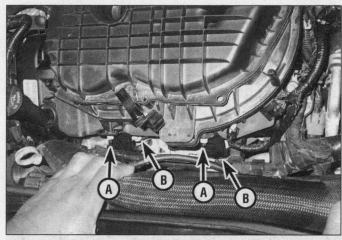

5.10 Remove the support bracket upper nuts (A), loosen the lower nuts (B) and remove the brackets from the upper intake manifold

13   If required, remove the insulator from the left valve cover **(see illustration 4.4)**.

Lower intake manifold

14   Remove the upper intake manifold (see Steps 4 through 13).

15   Remove the fuel injectors and fuel rail (see Chapter 4).
Note: *The lower intake manifold can be removed with the injectors and fuel rail in place. Be careful not to damage the fuel injectors once the manifold is removed.*

16   Pry the wiring harness retainer from the end of the manifold and move the harness out of the way.

17   Remove the lower intake manifold bolts **(see illustration)**, and remove the manifold from the cylinder heads.

18   Discard the six manifold-to-cylinder head seals.

Installation

Lower intake manifold

Note: *The mating surfaces of the cylinder*

heads, cylinder block, and the intake manifold *must be perfectly clean when the lower intake manifold is installed.*

19   Use a gasket scraper to remove all traces of sealant and old gasket material, then clean the mating surfaces with brake system cleaner. If there's old sealant or oil on the mating surfaces when the lower intake manifold is installed, oil or vacuum leaks may develop. Use a vacuum cleaner to remove any debris that falls into the intake ports or the valley between the cylinder banks.

20   Install the fuel injectors and the fuel rail (see Chapter 4).

21   Install new intake manifold seals to the manifold.
Note: *Remove any rags placed in the cylinder head ports.*

22   Carefully lower the lower intake manifold into place **(see illustration)** and install the mounting bolts finger-tight.

23   Tighten the mounting bolts in steps,

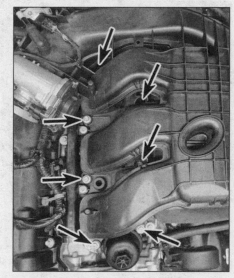

5.11 Upper intake manifold bolts

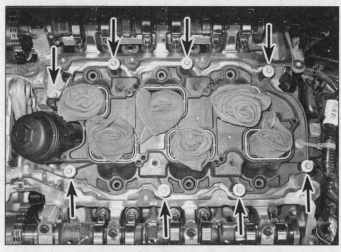

5.17 Lower intake manifold bolt locations

5.22 Install the manifold making sure the intake seals do not fall out of the manifold

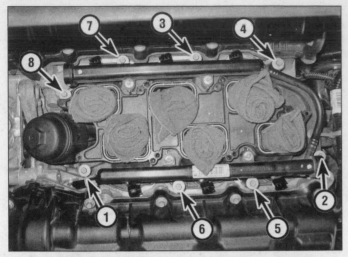

5.23 Lower intake manifold bolt tightening sequence

5.28 Upper intake manifold bolt tightening sequence

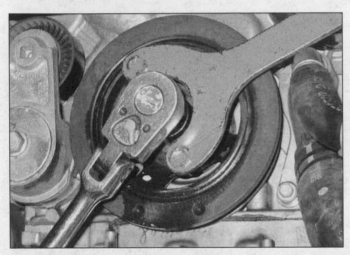

6.4 Using a special holding tool to prevent the crankshaft from turning, loosen, then remove the bolt

6.5 Slide the pulley from the end of the crankshaft; a puller shouldn't be required

following the tightening sequence (**see illustration**), to the torque listed in this Chapter's Specifications.

24   Install the upper intake manifold.

Upper intake manifold

25   Check the condition of the rubber seals that are installed into each intake runner on the upper intake manifold. If they are damaged, replace the seals in the upper intake manifold.

26   Place the insulator into position, if it was removed (**see illustration 4.4**).

27   Install the upper intake manifold onto the lower intake manifold while pulling the bolts up.

28   Tighten the mounting bolts in sequence (**see illustration**) to the torque listed in this Chapter's Specifications.

29   Installation of the remaining components is the reverse of removal.

30   Start the engine and check for leaks and proper operation.

6   Crankshaft balancer - removal and installation

Removal

1   Disconnect the cable from the negative terminal of the battery (see Chapter 5, Section 3).

Caution: *On 2015 and later models, if equipped with an Intelligent Battery Sensor (IBS), disconnect the IBS connector first before disconnecting the negative battery cable.*

2   Raise the front of the vehicle and support it securely on jackstands.

3   Remove the drivebelt (see Chapter 1).

4   The crankshaft balancer bolt is incredibly tight; using a breaker bar, socket and special tool #10198 or equivalent (**see illustration**), hold the balancer from turning while loosening the bolt.

5   Pull the crankshaft balancer off the crankshaft (**see illustration**).

Installation

6   Apply clean engine oil or multi-purpose grease to the seal contact surface of the balancer hub (if it isn't lubricated, the seal lip could be damaged and oil leakage would result).

7   Install the crankshaft balancer, aligning the keyway on the crankshaft with the slot in the balancer. Install the bolt and tighten it by hand.

8   Prevent the engine from rotating (see Step 5) then tighten the bolt to the torque listed in this Chapter's Specifications.

9   Installation of the remaining components is the reverse of removal.

7   Crankshaft front oil seal - replacement

1   Remove the crankshaft balancer (see Section 6).

2   Use a screwdriver or hook tool to care-

7.2 Use a hook tool and pry the seal from the timing cover

7.3 Another way of removing an old oil seal is to screw a self-tapping screw partially into the seal, then use pliers as a lever to pull it from the engine

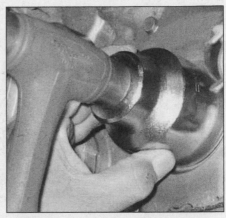

7.5 Drive the seal squarely into the cover using a socket and hammer

fully pry out the seal **(see illustration)**.
Note: *Be careful not to damage the oil pump cover bore where the seal is seated, or the nose and sealing surface of the crankshaft.*

3   Another procedure for removing the seal is to drill a small hole on each side of the seal and place a self-tapping screw in each hole **(see illustration)**. Use these screws as a means of pulling the seal out without having to pry on it.

4   If the seal is being replaced when the timing chain cover is removed, support the cover on top of two blocks of wood and drive the seal out from the backside with a hammer and punch.
Caution: *Be careful not to scratch, gouge or distort the area that the seal fits into or a leak will develop.*

5   Apply clean engine oil or multi-purpose grease to the outer edge of the new seal, then install it in the cover with the lip (spring side) facing IN. Drive the seal into place with a seal driver or a large socket and a hammer **(see illustration)**. Make sure the seal enters the bore squarely and stop when the front face is at the proper depth.
Note: *If a large socket isn't available, a piece of pipe will also work.*

6   Check the surface on the balancer hub that the oil seal rides on. If the surface has been grooved from long-time contact with the seal, the balancer will need to be replaced.

7   Lubricate the balancer hub with clean engine oil and install the crankshaft balancer (see Section 6).

8   The remainder of installation is the reverse of the removal.

---

8   Timing chain cover, chain and sprockets - removal, inspection and installation

Warning: *Wait until the engine is completely cool before beginning this procedure.*
Caution: *The timing system is complex, and severe engine damage will occur if you make*

any mistakes. Do not attempt this procedure unless you are highly experienced with this type of repair. If you are at all unsure of your abilities, be sure to consult an expert. Double-check all your work and be sure everything is correct before you attempt to start the engine.
Caution: *Do not rotate the crankshaft or cam-shafts separately during this procedure (with the timing chains removed), as damage to the valves may occur.*
Note: *Several special tools are required to complete these procedures, so read through the entire Section and obtain the special tools before beginning work.*

Removal

Timing chain cover

1   Disconnect the cable from the negative terminal of the battery (see Chapter 5, Section 3).
Caution: *On 2015 and later models, if equipped with an Intelligent Battery Sensor (IBS), disconnect the IBS connector first before disconnecting the negative battery cable.*

2   Raise the front of the vehicle and support it securely on jackstands. Drain the engine oil (see Chapter 1).

3   Drain the engine coolant (see Chapter 1).

4   Remove the drivebelt, drivebelt tensioner and idler pulley (see Chapter 1).

5   Remove the thermostat housing, upper radiator hose and disconnect the heater hose from the water pump (see Chapter 3).

6   Remove the heater core supply pipe fasteners from the right-side cylinder head and move the pipes out of the way.

7   Remove the power steering pump and tie it out of the way with wire (see Chapter 10).

8   Remove the crankshaft balancer (see Section 6).

9   Remove the valve covers (see Section 4).
Caution: *Once the valve covers are removed, the magnetic timing wheels at the front and rear of the camshafts are exposed (see illus-*

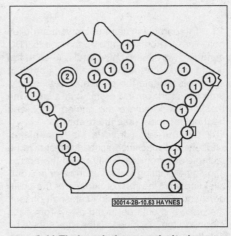

8.11 Timing chain cover bolt size and locations:

1   M6 size bolts locations
2   M8 size bolt location

tration 9.9). The magnetic timing wheels on the camshafts must not come in contact with any type of magnet or magnetic field. If contact is made, the timing wheels will need to be replaced.

10   Remove the lower and upper oil pans (see Section 11).

11   Remove the timing chain cover mounting bolts **(see illustration)**. There are seven indented prying points, one on top and three on each side; carefully pry the cover free of the engine block and cylinder heads. If it still sticks, slip a putty knife between the engine block and cover to break the bond (but be careful not to scratch the surfaces).

12   Once the cover is removed, discard the coolant housing and water pump gaskets from the back side of the timing chain cover.

Timing chain

Warning: *When the timing chains are removed, do not rotate the camshafts or crankshaft; the valves and pistons can be damaged if contact is made.*

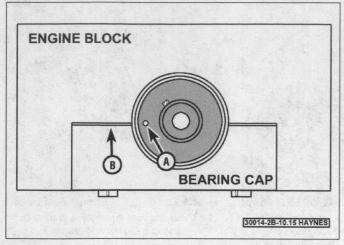

8.13 Align the dimple (A) on the crankshaft with the line (B) made where the engine block and bearing cap meet

8.15 Verity that there are 12 pins between the mark on each phaser

13   Temporarily install the crankshaft pulley bolt. Turn the crankshaft with the bolt to TDC number 1, on the exhaust stroke to align the timing marks on the crankshaft and camshaft sprockets. Rotate the engine clockwise only, until the mark on the crankshaft aligns with the line made where the engine block and bearing cap meet **(see illustration)**.

14   On the left (driver's side) camshaft phasers, the machined scribe lines should be facing away from each other, and the arrows should be pointing toward each other in a parallel line with the gasket surface of the cylinder head. On the right side camshaft phaser, the arrows should be facing away from each other and the machined scribe lines should be pointing toward each other in a parallel line with the gasket surface of the cylinder head **(see illustrations 9.10a and 9.10b)**. If, when you align the crankshaft mark with the bearing cap parting line, the camshaft marks are not in alignment as shown in **illustration 8.57**, rotate the engine one full revolution, realign the crankshaft mark, and verify that the camshaft marks are in proper alignment.

15   Verify the phaser marks are aligned with the plated links; if the plated links cannot be distinguished make sure there are 12 pins between the two marks **(see illustration)**.

Note: *Use paint or a permanent marker to mark the direction of rotation on all chains before removing them so they can be installed in the same direction.*

16   Starting with the right side chain tensioner, press the tensioner plunger in until special tool #8514 or a 3 mm Allen wrench can be inserted through both small holes in the top and bottom of the tensioner body, holding the plunger in the compressed position.

17   Working on the left side chain tensioner, locate the access hole on the side of the tensioner. Working through the hole, lift and hold the pawl off of the rack of the plunger in the tensioner. Press the plunger in until special tool #8514 or a 3 mm Allen wrench can be

inserted through both small holes in the top and bottom of the tensioner body, holding the plunger in the compressed position.

18   Remove the timing gear splash shield fasteners, then remove the shield from the oil pump housing.

19   Remove the oil pump tensioner and sprocket (see Section 12), then remove the oil pump chain from the crankshaft gear.

Note: *The oil pump chain and sprocket do not have to be timed, but the chain should be marked to make sure it is installed in the same direction of rotation.*

20   Starting with the right side chain, slide camshaft phaser lock tool # 10202-1 from the front, between the two camshaft phasers, toward the chain (with the tool number facing up).

Note: *It may be necessary to rotate the intake camshaft a few degrees using a wrench on the camshaft flat when installing the phaser lock tool.*

21   Using a large wrench on the camshaft flats and a socket and ratchet on the oil control valves, loosen, but do not remove, the oil control valves.

22   Remove the right side camshaft phaser lock tool, then unscrew the intake camshaft oil control valve from the center of the phaser.

23   Slide the intake camshaft phaser off of the end of the camshaft, then remove the right side timing chain.

Note: *If necessary, remove the exhaust camshaft oil control valve from the center of the phaser and remove the phaser.*

24   Working on the left side chain, slide camshaft phaser lock tool # 10202-2 from the front, between the two camshaft phasers, toward the chain (with the tool number facing up).

Note: *It may be necessary to rotate the intake camshaft a few degrees using a wrench on the camshaft flat when installing the phaser lock tool.*

25   Using a large wrench on the camshaft flats and a socket and ratchet on oil control

valves, loosen, but do not remove, the oil control valves.

26   Remove the left side camshaft phaser lock tool, then unscrew the exhaust camshaft oil control valve from the center of the phaser.

27   Slide the exhaust camshaft phaser off of the end of the camshaft, then remove the left side timing chain.

Note: *If necessary, remove the intake camshaft oil control valve from the center of the phaser and remove the phaser.*

28   Locate the primary chain tensioner to the side of the crankshaft chain and press the tensioner plunger in until special tool #8514 or a 3 mm Allen wrench can be inserted through the small hole in the side of the tensioner body, holding the plunger in the compressed position.

29   With the tensioner in the compressed position, remove the TORX (T30) mounting fasteners and the tensioner.

30   Remove the primary chain guide TORX (T30) mounting fasteners and the guide.

31   Remove the idler sprocket TORX (T45) mounting fastener and washer, then remove the idler sprocket, primary chain and crankshaft sprocket.

Note: *The chain should be marked to make sure it is installed in the same direction of rotation.*

32   If necessary, remove the chain tensioner (T30) fasteners and remove the tensioner(s), keeping the tensioners in the compressed position.

33   If necessary, remove the chain guide fasteners and guides for both chains.

Inspection

34   Inspect the timing chain dampener (guide) for cracks and wear and replace it, if necessary.

35   Clean the timing chain and sprockets with solvent and dry them with compressed air (if available).

Warning: *Wear eye protection when using compressed air.*

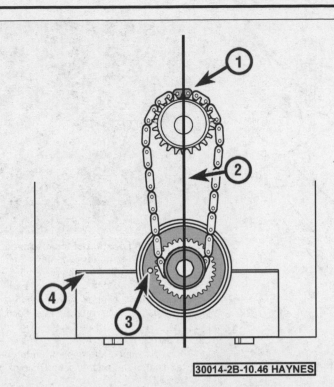

8.44 Primary chain alignment details;

1   *Primary chain plated link*
2   *12 o'clock position*
3   *Crankshaft dimple*
4   *Line formed where engine block and bearing cap meet*

30014-2B-10.46 HAYNES

36   Inspect the components for wear and damage. Look for teeth that are deformed, chipped, pitted, and cracked.

37   The timing chain and sprockets should be replaced with new ones if the engine has high mileage, the chain has visible damage, or total freeplay midway between the sprockets exceeds one inch. Failure to replace a worn timing chain and sprockets may result in erratic engine performance, loss of power, and decreased fuel mileage. Loose chains can jump timing. In the worst case, chain jumping or breakage will result in severe engine damage.

Installation

Caution: *Before starting the engine, carefully rotate the crankshaft by hand through at least two full revolutions (use a socket and breaker bar on the crankshaft pulley center bolt). If you feel any resistance, STOP! There is something wrong - most likely, valves are contacting the pistons. You must find the problem before proceeding.*

38   Use a plastic gasket scraper to remove all traces of old gasket material and sealant from the cover, engine block and cylinder heads. The cover is made of aluminum, so be careful not to nick or gouge it. Only clean the gasket sealing surfaces with rubbing alcohol (isopropyl) or brake system cleaner - do not use any oil based fluids.

39   If removed, install the chain guides and tensioners (still in the compressed position).

40   Make sure the keyway is installed on the crankshaft and the dimple on the crankshaft is aligned with the line made where the engine block and bearing cap meet **(see illustration 8.13)**.

41   Verify the camshafts are at TDC, with the alignment holes pointing up **(see illustration 9.29)**.

42   Place the primary chain on the crankshaft sprocket, with the plated link of the primary chain aligned with the arrow on the bottom of the sprocket. Insert the idler sprocket into the chain, aligning the other plated link with the machined mark of the idler sprocket.

43   Using clean engine oil, coat the sprockets and chain. Install the assembly while keeping the marks aligned, then install the idler sprocket mounting fastener finger-tight.

44   Check the alignment of the marks; the plated link on the idler sprocket should be on top (12 o'clock) and the machined mark on the crankshaft should be aligned with the line made where the engine block and bearing cap meet **(see illustration)**. If the marks are all aligned, tighten the idler sprocket fastener to the torque listed in this Chapter's Specifications.

45   Install the primary chain guide and tensioner, then tighten the fasteners to the torque listed in this Chapter's Specifications. Remove the special tool from the tensioner plunger.

46   Starting with the left side chain, install the intake camshaft phaser and oil control valve, then tighten the valve finger-tight, if removed.

47   Place the left side chain over the intake phaser and around the inside cogs of the idler sprocket so that the plate link of the chain is aligned with the machined arrow on the sprocket.

48   With the chain aligned at the idler sprocket, install the exhaust camshaft phaser so that the arrows are pointing toward each other and in a parallel line with the cylinder

head gasket surface **(see illustration 9.10a)**, then install the oil control valve finger-tight.

49   Slide camshaft phaser lock tool #10202-2 from the front, between the two camshaft phasers toward the chain with the tool number facing up.

50   Using a large wrench on the camshaft flats and a socket and ratchet on the oil control valves, tighten both valves to the torque listed in this Chapter's Specifications.

51   Working on the right (passenger's side) chain, install the exhaust camshaft phaser and oil control valve, tightening the valve finger-tight, if removed.

52   Place the right side chain over the intake phaser and around the outside cogs of the idler sprocket so that the plate link of the chain is aligned with the machined circle on the sprocket.

53   With the chain aligned at the idler sprocket, install the intake camshaft phaser so that the machined lines are pointing toward each other and in a parallel line with the gasket surface of the cylinder head **(see illustration 9.10b)**, then install the oil control valve finger-tight.

54   Slide camshaft phaser lock tool #10202-1 from the front, between the two camshaft phasers, toward the chain (with the tool number facing up).

55   Using a large wrench on the camshaft flats and a socket and ratchet on the oil control valves, tighten both valves to the torque listed in this Chapter's Specifications.

56   Install the oil pump timing chain, tensioner, sprocket and splash shield (see Section 12).

Note: *There are no timing or timing marks on the oil pump chain or sprocket.*

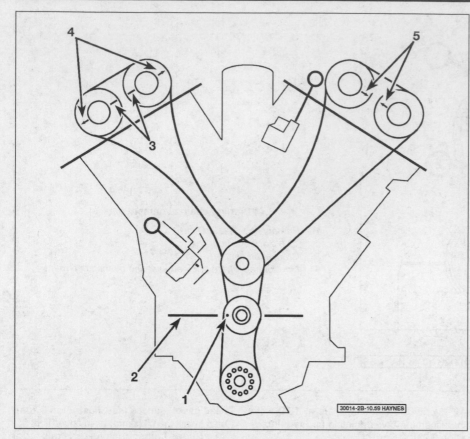

8.57 Timing mark alignment details

1 *Dimple on crankshaft*
2 *Junction of main bearing cap and cylinder block*
3 *Lines on right side cam phasers - must be pointing toward each other and parallel with cylinder head*
4 *Arrows on right side cam phasers - must be pointing away from each other*
5 *Arrows on left side cam phasers - must be pointing toward each other*

57 Verify all the marks are aligned **(see illustration),** then remove the special tool or Allen wrenches from the primary and secondary tensioners. Also remove the camshaft phaser lock tools.
58 Rotate the engine two complete turns using the machined mark on the crankshaft with the line made where the engine block and bearing caps meet as the reference. Verify all the marks are aligned and there are 12 pins between the phaser marks **(see illustration 8.15)**; if the marks are off, rotate the engine two more complete turns and check again.
59 Once the timing marks are correct, install the new coolant housing and water pump housing gaskets into the grooves on the back side of the timing cover.
60 Apply a 1/8-inch wide by 1/16-inch high, bead of RTV sealant to the sealing surface of the cover, then install the cover on the alignment dowels.
61 Install the cover bolts **(see illustration 8.11)** and tighten them in a criss-cross pattern, in three steps, to the torque listed in this Chapter's Specifications.
62 Installation of the remaining components is the reverse of removal.

63 Add oil and coolant (see Chapter 1), start the engine and check for leaks.

9 Camshaft(s) – removal, inspection and installation

Caution: *The timing system is complex, and severe engine damage will occur if you make any mistakes. Do not attempt this procedure unless you are highly experienced with this type of repair. If you are at all unsure of your abilities, be sure to consult an expert. Double-check all your work and be sure everything is correct before you attempt to start the engine.*
Caution: *Once the valve covers are removed, the magnetic timing wheels are exposed. The magnetic timing wheels on the camshafts must not come in contact with any type of magnet or magnetic field. If contact is made, the timing wheels will need to be replaced.*
Note: *The timing chain for each camshaft can be removed from the camshafts individually, without removing all the timing chains, using the tools outlined in this Section. If the tools are not available, the timing chain cover and*

9.9 Location of the magnetic timing wheels

all chains will need to be removed before the camshafts can be removed (see Section 8).

Removal

1 Disconnect the cable from the negative terminal of the battery (see Chapter 5, Section 3).
Caution: *On 2015 and later models, if equipped with an Intelligent Battery Sensor (IBS), disconnect the IBS connector first before disconnecting the negative battery cable.*
2 Raise the front of the vehicle and support it securely on jackstands, then drain the engine oil and coolant (see Chapter 1).
3 Remove the drivebelt (see Chapter 1).
4 Remove the air filter housing (see Chapter 4) and resonator.
5 Remove the intake manifolds (see Section 5) and unbolt the exhaust from the cylinder head(s).
6 Disconnect all wires and vacuum hoses from the cylinder heads. Label them to simplify reinstallation.
7 Disconnect the ignition coils and remove the spark plugs (see Chapter 1). Label the ignition coils to simplify installation.
8 Remove the valve covers (see Section 4).
9 Once the valve covers are removed, the magnetic timing wheels are exposed **(see illustration)**. The magnetic timing wheels on the camshafts must not come in contact with any type of magnet or magnetic field. If contact is made, the timing wheels will need to be replaced.
10 Rotate the crankshaft clockwise and place the #1 piston at TDC on the exhaust stroke. On the left (driver's side) camshaft phaser, the machined scribe lines should be facing away from each other, and the arrows should be pointing toward each other in a parallel line with the gasket surface of the cylinder head. On the right side camshaft phaser, the arrows should be facing away from each other, and the machined scribe lines should be pointing toward each other in a parallel line with the gasket's surface of the cylinder head **(see illustrations)**.

9.10a With the engine at TDC #1, the left (driver's side) camshaft phaser scribe marks (A) should be pointing away from each other, the arrow marks (B) should be pointing toward each other in a straight line and that line should be parallel with the cylinder head surface (C) . . .

9.10b . . . and the right-side phaser scribe marks should be pointing toward each other in a straight line (and that line should be parallel with the cylinder head surface)

11   Using a permanent marker or paint, mark the camshaft phasers to the timing chains for reinstallation.

12   Working from the top of the timing chain cover, insert special tool #10200-3 down the side of the tensioner to the access hole on the side of the tensioner. Working through the small hole in the side of the tensioner, lift and hold the pawl off of the rack of the plunger in the tensioner. Slide the chain holding tool #10200-1 between the cylinder head and the back side of the chain against the chain guide forcing the rack and plunger back into the tensioner body.

Caution: *The chain holding tool must remain in place while the phasers are removed or the timing chain will fall off into the timing cover.*

13   Slide camshaft phaser lock tool #10202-1 (right side) or 10202-2 (left side), from the front, between the two camshaft phasers, toward the chain.

Note: *It may be necessary to rotate the intake camshaft a few degrees using a wrench on the camshaft flat when installing the phaser lock tool.*

14   Using a large wrench on the camshaft flats and a socket and ratchet on the oil control valves, loosen, then remove each of the oil control valves from the phaser end of the camshaft.

15   At the same time, carefully slide both the intake and exhaust phaser (with the phaser lock securely between them) forward until they are off the end of the camshafts.

Caution: *Do not remove the phaser lock or try to disassemble the phasers.*

16   Using the alignment holes in the camshaft as a reference point, slowly rotate both camshafts counterclockwise approximately 30-degrees Before Top-Dead-Center (BTDC). In this position the camshafts are in a neutral or no load position.

Note: *The camshaft bearing caps are marked with a number and letter code; "1I" is for the*

9.22 Use a micrometer to measure cam lobe height

number one Intake camshaft bearing cap. The notch on the caps should always be installed toward the front.

17   Loosen the camshaft bearing cap bolts in the reverse order of the tightening sequence **(see illustration 9.29)**.

18   Remove the camshaft bearing caps and carefully lift the camshafts from the cylinder head.

19   With the camshafts removed, mark the rocker arms so they can be installed in the same locations; then remove the rocker arms.

20   Mark the hydraulic lash adjusters so they can be installed in the same locations, then remove them from the cylinder head.

Inspection

21   Check the camshaft bearing surfaces for pitting, score marks, galling, and abnormal wear. If the bearing surfaces are damaged, the cylinder head will have to be replaced.

22   Compare the camshaft lobe height by measuring each lobe with a micrometer **(see illustration)**. Measure each of the intake lobes

and record the measurements and relative positions. Then measure each of the exhaust lobes and record the measurements and relative positions. This will let you compare all of the intake lobes to one another and all of the exhaust lobes to one another. If the difference between the lobes exceeds 0.005 inch, the camshaft should be replaced. Do not compare intake lobe heights to exhaust lobe heights as lobe lift may be different. Only compare intake lobes to intake lobes and exhaust lobes to exhaust lobes for this comparison.

23   Check the rocker arms and shafts for abnormal wear, pits, galling, score marks, and rough spots. Don't attempt to restore rocker arms by grinding the pad surfaces. Replace defective parts.

Installation

Caution: *Before starting the engine, carefully rotate the crankshaft by hand through at least two full revolutions (use a socket and breaker bar on the crankshaft pulley center bolt). If you feel any resistance, STOP! There is some-*

9.28 Camshaft bearing cap tightening sequence 3.6L engines –
left (front) side shown, right side is identical

9.29 Locate the alignment holes on the camshafts and make sure
they are in the neutral position (pointing straight up) –
right side shown, left side is identical

thing wrong - *most likely, valves are contacting the pistons. You must find the problem before proceeding. Check your work and see if any updated repair information is available.*

24   Dip the hydraulic lash adjusters in clean engine oil and install them into their original locations.

25   Apply moly-base grease or engine assembly lube to the rocker arm contact points and rollers and install them into their original locations.

26   Lubricate the camshaft bearing journals and lobes with moly-base grease or engine assembly lube, then install them carefully in the cylinder head about 30-degrees before (counterclockwise of) TDC. Don't scratch the bearing surfaces with the cam lobes!

Caution: *Do not rotate the camshafts more than a few degrees to prevent the valves from contacting the pistons.*

27   Install the camshaft bearing caps, then install the mounting bolts and finger-tighten them.

28   Tighten the bearing caps in sequence **(see illustration)** to the torque listed in this Chapter's Specifications.

29   Rotate the camshafts clockwise 30-degrees, verify the alignment holes in the camshafts are at 12 o'clock (pointing straight up) or neutral position **(see illustration)**.

30   Carefully slide both the intake and exhaust phaser (with the phaser lock tool securely between them) onto the camshafts and verify the marks are aligned.

31   Install the oil control valves onto the camshaft phasers and install the bolts, then tighten the bolts to the torque listed in this Chapter's Specifications. Remove the chain holding tool and release the tensioner plunger.

Caution: *Make sure to prevent the camshafts from turning by holding the camshaft with a large wrench on the camshaft flats.*

32   Slowly rotate the engine two complete turns (360-degrees) and verify the alignment marks are correct **(see illustrations 9.10a and 9.10b)**.

33   The remainder of installation is the reverse of removal.

---

10   Cylinder heads - removal and installation

Warning: *Wait until the engine is completely cool before beginning this procedure.*

Removal

1   Disconnect the cable from the negative terminal of the battery (see Chapter 5, Section 3).

Caution: *On 2015 and later models, if equipped with an Intelligent Battery Sensor (IBS), disconnect the IBS connector first before disconnecting the negative battery cable.*

2   Raise the front of the vehicle and support it securely on jackstands. Drain the engine oil and coolant (see Chapter 1).

3   Remove the drivebelt (see Chapter 1).

4   Remove the air filter housing and resonator (see Chapter 4).

5   Disconnect all wires and vacuum hoses from the intake manifolds and cylinder heads. Label them to simplify reinstallation.

6   Remove the intake manifolds (see Section 5).

7   Disconnect the ignition coils and remove the spark plugs (see Chapter 1).

8   Remove the catalytic converter(s) (see Chapter 6).

9   Remove the valve covers (see Section 4).

10   Remove the crankshaft balancer (see Section 6).

11   Remove the oil pans (see Section 11).

12   Remove the alternator (see Chapter 5).

13   If you're working on the left cylinder head:

a) *Remove the air conditioning compressor (see Chapter 3), without disconnecting the refrigerant lines.*

b) *Disconnect the main engine harness connectors at the rear of the cylinder and move the harness and retainers out of the way.*

c) *Remove upper intake support bracket nuts and bracket.*

14   If you're working on the right cylinder head:

a) *Remove the power steering pump (see Chapter 10).*

b) *Remove the heater core tube fasteners and move the tube away from the cylinder head.*

c) *Remove the battery and battery tray (see Chapter 5).*

d) *Remove the oil dipstick tube fastener and remove the tube from the oil pan.*

15   Remove the timing chain cover (see Section 8).

16   Rotate the crankshaft clockwise and place the #1 piston at TDC on the exhaust stroke. When the crankshaft is at TDC, the dimple on the crankshaft will be in line with the line made where the bearing cap meets the engine block. The front cylinder bank cam phaser arrows should be pointing toward each other and be parallel with where the cylinder head and valve cover meet. The rear side cam phaser arrows should point away from each other and the lines on the phasers should be pointing toward each other **(see illustration 8.57)**.

17   Remove the timing chain for the cylinder head or, if both cylinder heads are being removed, remove both chains (see Section 8).

18   Remove the oil control valves from the cam phasers (sprockets) (see Section 9).

19   Remove the timing chain tensioner and chain guides (see Section 8).

20   Remove the camshafts, rocker arms and lash adjusters (see Section 9).

Caution: *Once the valve covers are removed the magnetic timing wheels are exposed. The magnetic timing wheels on the camshafts must*

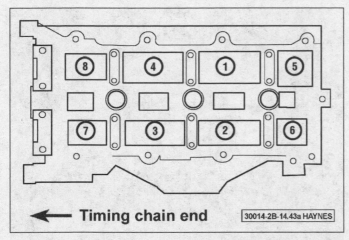

10.30a Left side cylinder head bolt TIGHTENING sequence

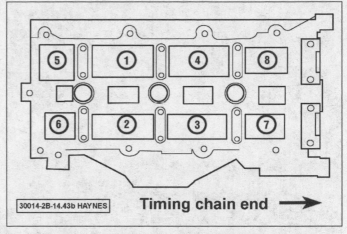

10.30b Right side cylinder head bolt TIGHTENING sequence

not come in contact with any type of magnet or magnetic field. If contact is made the timing wheels will need to be replaced (see Section 9).
Note: *Keep the rocker arms and lash adjusters in order so that they can be installed in their original locations.*

21   Loosen the cylinder head bolts in the reverse order of the tightening sequence **(see illustrations 10.30a and 10.30b)**.

22   Lift the cylinder head off the block. If resistance is felt, dislodge the cylinder head by striking it with a wood block and hammer. If prying is required, pry only on a casting protrusion - be very careful not to damage the cylinder head or block!
Caution: *Do not set the cylinder head on its gasket side; the sealing surface can be easily damaged.*

23   Have the cylinder head inspected and serviced by a qualified automotive machine shop.

Installation

24   The mating surfaces of each cylinder head and the engine block must be perfectly clean when the cylinder head is installed.

25   Carefully use a gasket scraper to remove all traces of carbon and old gasket material, then clean the mating surfaces with brake system cleaner. If there's oil on the mating surfaces when the cylinder head is installed, the gasket may not seal correctly and leaks may develop.

26   When working on the engine block, it's a good idea to cover the lifter valley with shop rags to keep debris out of the engine. Use a shop rag or vacuum cleaner to remove any debris that falls into the cylinders.

27   Check the engine block and cylinder head mating surfaces for nicks, deep scratches, and other damage. If damage is slight, it can be removed with a file; if it's excessive, machining may be the only alternative.

28   Position the new gasket over the dowel pins in the engine block. Some gaskets are marked TOP or FRONT to ensure correct installation.

11.5 Lower oil pan bolts

29   Carefully position the cylinder head on the engine block without disturbing the gasket.

30   Install NEW cylinder head bolts and tighten them in the recommended sequence **(see illustrations)** to the torque steps listed in this Chapter's Specifications.
Note: *Do not apply additional oil to the bolt threads.*
Caution: *Do not use a torque wrench for steps requiring additional rotation or turns; apply a paint mark to the bolt head or use a torque-angle gauge (available at most automotive parts stores) and a socket and breaker bar.*

31   Installation of the remaining components is the reverse of removal.

32   Change the engine oil and filter (see Chapter 1).

33   Refill the cooling system (see Chapter 1). Start the engine and check for leaks and proper operation.

11 Oil pan - removal and installation

Removal

1   Disconnect the cable from the negative terminal of the battery (see Chapter 5, Sec-

tion 3).
Caution: *On 2015 and later models, if equipped with an Intelligent Battery Sensor (IBS), disconnect the IBS connector first before disconnecting the negative battery cable.*

2   Raise the front of the vehicle and support it securely on jackstands. Apply the parking brake and block the rear wheels to keep it from rolling off the stands.

3   Drain the engine oil (see Chapter 1).

4   Remove the lower splash shield fasteners and remove the splash shield.

Lower oil pan

5   Remove the bolts, nuts and studs **(see illustration)**, then carefully separate the lower oil pan from the upper oil pan. Don't pry between the upper pan and the lower pan or damage to the sealing surfaces could occur and oil leaks may develop. Tap the pan with a soft-faced hammer to break the gasket seal. If it still sticks, slip a putty knife between the upper pan and lower pan to break the bond (but be careful not to scratch the surfaces).

Upper oil pan

6   Remove the dipstick tube bracket mounting bolt. Using a twisting motion, pull the dipstick tube out of the upper oil pan.

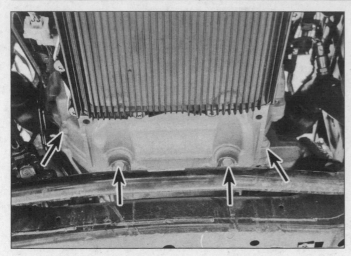

11.11 Transmission-to-oil pan bolts

11.12 Remove these plugs for access to the pan-to-rear main oil seal retainer bolts

7   Remove the steering gear (see Chapter 10).

8   Remove the colt securing the transmission fluid cooler lines.

9   Remove the lower oil pan (see Step 5). **Caution:** *Five of the upper oil pan bolts are accessible only after the lower pan has been removed.*

10   Unbolt the transmission mount from the crossmember (see Chapter 7B). Raise the transmission/engine with a floor jack and block of wood. **Caution:** *Be extremely careful not to damage the transmission fluid pan.*

11   Remove the four transmission-to-oil pan bolts **(see illustration)**.

12   Remove the rubber plugs from the bellhousing **(see illustration)**.

13   Remove the two upper pan-to-rear seal housing bolts (M6 size). **Caution:** *The oil pan-to-rear main seal bolts are hard to see and can easily be missed. If they are not removed, the rear main seal housing will be severely damaged when the pan is lowered.*

14   Remove the remaining nineteen upper oil pan bolts (M8 size) around the perimeter of the pan, then carefully separate the oil pan from the engine block. Use the two indented prying points on each side of the oil pan to carefully pry the pan free of the engine block. If it still sticks, slip a putty knife between the engine block and oil pan to break the bond (but be careful not to scratch the surfaces).

Installation

15   Clean the pan(s) with solvent and remove all old sealant and gasket material from the engine block and pan mating surfaces. Clean the mating surfaces with brake system cleaner and make sure the bolt holes in the engine block are clear. Check the oil pan flange(s) for distortion, particularly around the bolt holes. If necessary, place the pan(s) on a wood block and use a hammer to flatten and restore the gasket surface.

Upper oil pan

16   Apply a 1/8-inch wide by 1/16-inch high bead of RTV sealant to the sealing surface of the pan. Install the upper pan and the bolts, then tighten the bolts finger-tight.

17   Tighten the transaxle-to-oil pan bolts first to the torque listed in this Chapter's Specifications.

18   Tighten the remaining bolts in a circular criss-cross pattern, starting from the middle working your way outwards, to the torque listed in this Chapter's Specifications.

19   Installation of the remaining components is the reverse of removal.

20   Refill the engine with oil (see Chapter 1), Start and run the engine until normal operating temperature is reached, then check for leaks.

Lower oil pan

21   Apply a 1/8-inch wide by 1/16-inch high bead of RTV sealant to the sealing surface of the pan. Install the lower pan to the upper pan and the bolts. Then tighten the bolts in a circular pattern, starting from the middle working your way outwards, to the torque listed in this Chapter's Specifications.

22   Installation of the remaining components is the reverse of removal.

23   Refill the engine with oil (see Chapter 1). Start and run the engine until normal operating temperature is reached, then check for leaks.

12   Oil pump - removal and installation

Note: *The oil pump is sold as an assembly. The assembly includes both the pump and the solenoid. There are no serviceable components, if there is a problem with the oil pump or solenoid they must be replaced as an assembly.*

Removal

1   Disconnect the cable from the negative terminal of the battery (see Chapter 5, Section 3). **Caution:** *On 2015 and later models, if equipped with an Intelligent Battery Sensor (IBS), disconnect the IBS connector first before disconnecting the negative battery cable.*

2   Raise the front of the vehicle and support it securely on jackstands. Apply the parking brake and block the rear wheels to keep it from rolling off the stands.

3   Drain the engine oil (see Chapter 1).

4   Remove the lower splash shield fasteners and remove the splash shield.

5   Remove the lower and upper oil pans (see Section 11).

6   Remove the oil pump pick-up tube fastener, and remove the tube from the pump. Discard the pick-up tube O-ring.

7   Disconnect the oil pump solenoid electrical connector from the side of the engine then remove the connector retaining clip.

8   Working from the side of the block, depress the oil pump solenoid electrical connector locking tab and push the connector into the block. **Note:** *The connector will have to be rotated slightly clockwise and maneuvered around the tensioner mounting bolt.*

9   Remove the oil pump timing gear splash shield bolts and remove the splash shield.

10   Press the oil pump chain tensioner away from the chain until a 3 mm Allen wrench can be inserted into the housing to hold the tensioner back.

11   Using a permanent marker or paint, make reference marks on the chain and oil pump gear.

12   Hold the oil pump gear from moving, then remove the T45 Torx mounting bolt and the oil pump gear.

13   Hold the tensioner and remove the Allen wrench, allowing the tensioner to release. Remove the spring from the dowel pin and slide the tensioner from the oil pump.

14   Remove the oil pump mounting bolts and remove the pump.

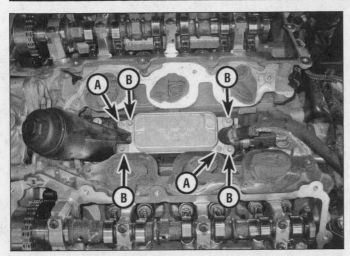

13.4 Remove the oil cooler mounting screws (A) then remove the mounting bolts (B) and cooler

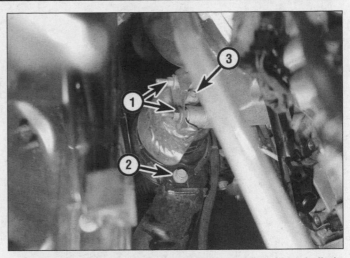

16.1 Engine mount fasteners (right side shown, left side similar)

1   Heat shield retaining nuts
2   Mount-to-frame bolt (rear bolt not visible)
3   Mount-to-bracket nut

Installation

15   Place the oil pump onto the engine block using the aligning dowels. Install the mounting bolts and tighten them to the torque listed in this Chapter's Specifications.
16   Slide the oil pump chain tensioner onto the pivot, then push the tensioner back against the spring. Insert a 3 mm Allen wrench into the tensioner to hold it in place.
17   Place the oil pump timing chain gear into the chain, center it onto the oil pump shaft and install the T45 mounting bolt. Tighten the bolt to the torque listed in this Chapter's Specifications.
Note: *Make sure the gear is facing the same way as when it was removed (see Step 11). There are no timing marks on the pump gear or chain, and no timing is necessary.*
18   Maneuver the oil pump solenoid into position and insert it through the block opening until it snaps in place.
19   Install the timing gear splash shield and bolts, then tighten the bolts to the torque listed in this Chapter's Specifications.
20   Installation of the remaining components is the reverse of removal.
21   Refill the engine with oil and change the oil filter (see Chapter 1).

13   Oil cooler - removal and installation

Warning: *Wait until the engine is completely cool before beginning this procedure.*
Note: *The oil cooler cannot be serviced separately from the oil filter housing and must be replaced as an assembly.*

Removal

1   Disconnect the cable from the negative terminal of the battery (see Chapter 5, Section 3).
Caution: *On 2015 and later models, if equipped with an Intelligent Battery Sensor (IBS), disconnect the IBS connector first before disconnecting the negative battery cable.*
2   Drain the coolant (see Chapter 1).
3   Remove the lower intake manifold (see Section 5).
4   Remove the oil cooler mounting fasteners **(see illustration)**.
5   Remove the oil cooler and discard the seals.

Installation

6   Install new seals to the oil cooler.
7   Place the oil cooler onto the block and install the two mounting screws.
8   Install the mounting bolts and tighten the screws and bolts to the torque listed in this Chapter's Specifications.
9   Installation of the remaining components is the reverse of removal.
10   Refill the cooling system (see Chapter 1). Run the engine until normal operating temperature is reached, and check for leaks.

14   Flywheel/driveplate - removal and installation

1   This procedure is essentially the same for all engines. Refer to Chapter 2A, Section 15 and follow the procedure outlined there, but use the torque listed in this Chapter's Specifications.

15   Rear main oil seal - replacement

Note: *The rear main seal and housing are serviced as one complete assembly. This procedure will require several special tools. Read the procedure carefully before starting.*
1   This procedure is essentially the same as for the 2.7L V6 engine. Refer to Chapter 2A and follow the procedure outlined there, but use the bolt torque value listed in this Chapter's Specifications.

16   Engine mounts - check and replacement

1   This procedure is essentially the same as for the 2.7L V6 engine. Refer to Chapter 2A and follow the procedure and **illustrations** outlined there, but refer to the **accompanying illustration** and use the torque values listed in this Chapter's Specifications.

Notes

Chapter 2 Part D
V8 engine

Contents

Specifications

General

Firing order	1-8-4-3-6-5-7-2
Displacement	
5.7L engine	345 cubic inches
6.1L engine	370 cubic inches
6.4L engine	392 cubic inches
Bore and stroke	
5.7L engine	3.92 x 3.58 inches
6.1L engine	4.055 x 3.58 inches
6.4L engine	4.09 x 3.72 inches
Cylinder numbers (front-to-rear)	
Left (driver's) side	1-3-5-7
Right side	2-4-6-8
Compression pressure	See Chapter 2E

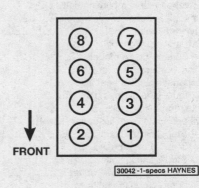

Cylinder locations

Camshaft

Journal diameters	
5.7L and 6.1L engines	
No. 1	2.290 inches
No. 2	
5.7L engines	2.280 inches
6.1L engines	2.270 inches
No. 3	2.260 inches
No. 4	2.240 inches
No. 5	1.720 inches
6.4L engines	
No. 1	2.428 inches
No. 2	2.275 inches
No. 3	2.259 inches
No. 4	2.243 inches
No. 5	1.714 inches
Journal oil clearance	
5.7L and 6.1L engines	
No. 1, 3 and 5	0.0015 to 0.0030 inch
No. 2 and 4	0.0019 to 0.0035 inch
6.4L engines	
No. 1	0.0012 to 0.0028 inch
No.2 and 4	0.0019 to 0.0035 inch
No.3 and 5	0.0015 to 0.003 inch
Endplay	0.0031 to 0.0114 inch

Torque specifications　　　　　　　　　　　　　　　　**Ft-lbs** (unless otherwise indicated)

Note: *Ono foot pound (ft-lb) of torque is equivalent to 12 inch-pounds (in-lbs) of torque. Torque values below approximately 15 ft-lbs are expressed in inch-pounds, because most foot-pound torque wrenches are not accurate at these smaller values.*

Camshaft sprocket bolt	90
Camshaft thrust plate bolts	21
Cylinder head bolts **(see illustration 10.17)**	
Step 1	
Large bolts	25
Small bolts	15
Step 2	
Large bolts	
2010 and earlier 5.7L, 6.1L and 6.4L engines	40
2011 and later 5.7L engines	45
Small bolts	
2010 and earlier 5.7L, 6.1L and 6.4L engines	15
2011 and later 5.7L engines	21
Step 3	
Large bolts	
5.7L and 6.1L engines	Tighten an additional 90 degrees
6.4L engines	45
Small bolts	
2010 and earlier 5.7L and 6.1L engines	25
Step 4	
Large bolts	
6.4L engines	Tighten an additional 90 degrees
Small bolts	
6.4L engines	21
Drivebelt tensioner mounting bolt	40
Drivebelt pulley mounting bolt	40
Driveplate bolts	70
Flywheel bolts*	55
Exhaust manifold bolts/nuts	
5.7L and 6.1L models	18
6.4L models	22
Exhaust manifold heat shield nuts	70 in-lbs
Exhaust pipe flange nuts	25
Intake manifold bolts	108 in-lbs
Oil pan bolts (see illustration 12.19)	
Step 1 Small bolts	44 in-lbs
Step 2 Large bolts	39
Step 3 Small bolts	105 in-lbs
Oil pump pick-up tube bolts	21
Oil pump mounting bolts	
5.7L and 6.1L engines	21
6.4L engines	168 in-lbs
Rear main seal retainer bolts	132 in-lbs
Rocker arm lifter rail bolts	106 in-lbs
Rocker arm shaft bolts	
2010 and earlier 5.7L and 6.1L engines	196 in-lbs
2011 and later 5.7L and 6.4L engines	
Step 1	84 in-lbs
Step 2	17
Step 3	Loosen each bolt individually 1/2 turn, then re-tighten to 17
Step 4	Tighten an additional 30 degrees
Timing chain cover bolts	21
Valve cover nuts/studs	70 in-lbs
Vibration damper-to-crankshaft bolt	
2010 and earlier 5.7L and 6.1L engines	129
2011 and later 5.7L engines	133
6.4L engines	127 to 130
Water pump-to-timing chain cover bolts	
2010 and earlier 5.7L and 6.1L engines	21
2011 and later 5.7L and 6.4L engines	18

* Use new bolts.

1   General information

1   This part of Chapter 2 is devoted to in-vehicle repair procedures for the 5.7L, 6.1L and 6.4L V8 (Hemi) engines. Information concerning engine removal and installation and engine overhaul can be found in Chapter 2E.

2   Since the repair procedures included in this Part are based on the assumption that the engine is still installed in the vehicle, if they are being used during a complete engine overhaul (with the engine already out of the vehicle and on a stand) many of the steps included here will not apply.

3   These V8 engines use a cast iron engine block with aluminum cylinder heads and intake manifold. The cylinder banks are positioned at a 90-degree angle with the camshaft mounted high in the engine block. The valvetrain includes roller lifters, with pushrods positioned between the camshaft and rocker arms in a nearly horizontal plane.

2   Repair operations possible with the engine in the vehicle

1   Many major repair operations can be accomplished without removing the engine from the vehicle.

2   Clean the engine compartment and the exterior of the engine with some type of pressure washer before any work is done. A clean engine will make the job easier and will help keep dirt out of the internal areas of the engine.

3   Depending on the components involved, it may be a good idea to remove the hood to improve access to the engine as repairs are performed (refer to Chapter 11 if necessary).

4   If oil or coolant leaks develop, indicating a need for gasket or seal replacement, the repairs can generally be made with the engine in the vehicle. The oil pan gasket, the cylinder head gaskets, intake and exhaust manifold gaskets, timing chain cover gaskets and the crankshaft front oil seal are all accessible with the engine in place.

5   Exterior engine components, such as the water pump, the starter motor, the alternator and the fuel injection components, as well as the intake and exhaust manifolds, can be removed for repair with the engine in place.

6   Since the cylinder heads can be removed without removing the engine, valve component servicing can also be accomplished with the engine in the vehicle.

7   Replacement of, repairs to or inspection of the timing chain and sprockets and the oil pump are all possible with the engine in place.

8   In extreme cases caused by a lack of necessary equipment, repair or replacement of piston rings, pistons, connecting rods and rod bearings is possible with the engine in the vehicle. However, this practice is not recommended because of the cleaning and preparation work that must be done to the components involved.

3   Top Dead Center (TDC) for number one piston - locating

1   Top Dead Center (TDC) is the highest point in the cylinder that each piston reaches as it travels up-and-down when the crankshaft turns. Each piston reaches TDC on the compression stroke and again on the exhaust stroke, but TDC generally refers to piston position on the compression stroke.

2   In order to bring any piston to TDC, the crankshaft must be turned using a breaker bar and socket on the vibration damper bolt. When looking at the front of the engine, normal crankshaft rotation is clockwise.

3   Disconnect the cable from the negative terminal of the battery (see Chapter 5, Section 1).

4   Remove one of the spark plugs from cylinder number 1, preferably the closest to the front of the engine, then thread a compression gauge into the spark plug hole **(see illustration)**. Turn the crankshaft with a socket and breaker bar attached to the large bolt that is threaded into the vibration damper **(see illustration)**. When compression registers on the gauge, the number one piston is beginning its compression stroke. Stop turning the crankshaft and remove the gauge.

5   Insert a long dowel into the number one spark plug hole until it rests on top of the piston crown. Continue rotating the crankshaft slowly until the dowel levels off (piston reaches top of travel). This will be approximate TDC for number 1 piston.

6   These engines are not equipped with components (vibration damper, flywheel, timing hole, etc.) that are marked to identify the position of number 1 TDC. Therefore the only method to pinpoint the exact position of TDC number 1 is to remove the timing chain cover to access timing chain sprockets (see Section 9), or with the use of a degree wheel on the crankshaft vibration damper and a positive stop threaded into the spark plug hole, as you would use in the process of degreeing a camshaft (this procedure is described in detail in the *Haynes Chrysler Engine Overhaul Manual*).

4   Valve covers - removal and installation

Removal

1   Disconnect the cable from the negative terminal of the battery (see Chapter 5, Section 1).

Caution: *On 2015 and later models, if equipped with an Intelligent Battery Sensor (IBS), disconnect the IBS connector first before disconnecting the negative battery cable.*

2   Remove the engine cover(s) **(see illustration)**.

Note: *6.4L engine models use two individual engine covers, one over each valve cover.*

3.4a Install a compression gauge into one of the number one spark plug holes

3.4b Use a breaker bar and socket to rotate the crankshaft

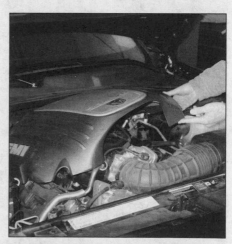

4.2 First remove the oil cap and lift the engine cover off the four support grommets, 5.7L models shown

4.4a Location of the harness clips on the right side valve cover

4.4b Location of the harness clips on the left side valve cover

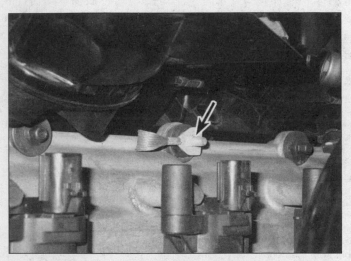

4.6 On early models, remove the ground straps from the valve cover studs

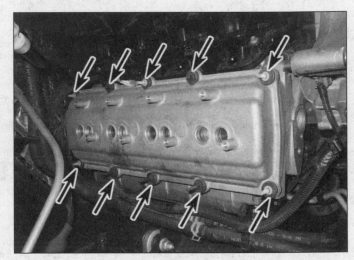

4.7a Location of the valve cover mounting bolts on the right-side valve cover

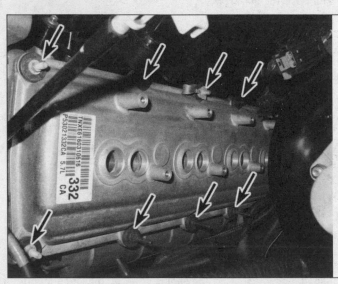

4.7b Location of the valve cover mounting bolts on the left-side valve cover (two bolts hidden from view)

3　　If you're removing the left side valve cover, it may be helpful to remove the air filter housing for better access (see Chapter 4). However, it isn't mandatory.

4　　Remove the harness clips from the valve cover studs (**see illustrations**).

5　　Remove the ignition coils and on 2005 models, the spark plug wires (see Chapter 5).

6　　Remove the ground straps from the valve cover (**see illustration**). Be sure to mark the location of each of the ground terminals for correct installation.

7　　Remove the valve cover mounting bolts (**see illustrations**).

8　　Remove the valve cover.

Note: *If the cover is stuck to the head, bump the cover with a block of wood and a hammer to release it. If it still will not come loose, try to slip a flexible putty knife between the head and cover to break the seal. Don't pry at the cover-to-head joint, as damage to the sealing surface and cover flange will result and oil leaks will develop.*

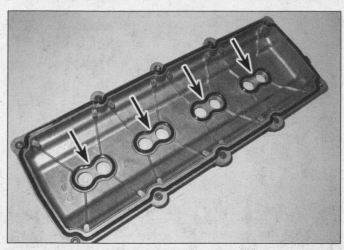

4.10 If they're damaged or hardened, replace the rubber gasket and spark plug tube seals with new ones (if they're OK, they can be re-used)

5.3 Remove the rocker arm shaft bolts - mark each rocker arm assembly and note that the upper rocker arms are the intake and the lower rocker arms are the exhaust

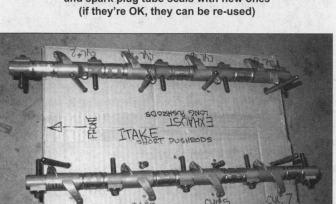

5.5 A perforated cardboard box top can be used to store the rocker arms and pushrods to ensure that they're reinstalled in their original locations - note the arrow indicating the front of the engine

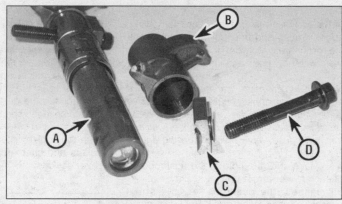

5.6 Rocker arm component details

A Rocker arm shaft	C Retainer - removed only
B Rocker arm	for illustration
	D Rocker shaft bolt

Installation

9   Clean the mating surfaces of the valve cover and cylinder head. Don't use harsh chemicals or solvent on the plastic valve covers, as they could become damaged.

10   If the valve cover gasket isn't damaged or hardened, it can be re-used. If it is in need of replacement, install a new one into the valve cover perimeter and new rubber seals into the grooves in the valve cover that seal the spark plug tubes **(see illustration)**.

11   Carefully position the cover on the head and install the bolts, making sure the bolts with the studs are in the proper locations.

12   Tighten the bolts a little at a time to the torque listed in this Chapter's Specifications. Start with the middle bolts and move to the outer bolts using a criss-cross pattern.
Caution: *DON'T over-tighten the valve cover bolts.*

13   The remaining installation steps are the reverse of removal.

14   Start the engine and check carefully for oil leaks as the engine warms up.

5   Rocker arms and pushrods - removal, inspection and installation

Removal

1   Remove the valve covers from the cylinder heads (see Section 4).

2   Insert pushrod retainer special tool #9070 onto the pushrods, then install the tool mounting bolts and tighten them securely.

3   Loosen the rocker arm shaft bolts, starting with the center bolts and working toward the outer bolts. When the bolts have been completely loosened, lift the rocker shaft assembly off the cylinder head **(see illustration)**.
Caution: *Do not remove the retainers from the rocker shaft assemblies. There are two tangs at the bottom of the retainers and if removed the tangs can break off and fall into the engine.*

4   Keep track of the rocker arm positions, since they must be returned to the same loca-

tions. Store each set of rocker components in such a way as to ensure that they're reinstalled in their original locations.

5   Remove pushrod retainer tool fasteners and remove the pushrods and store them in order as well, to make sure they don't get mixed up during installation **(see illustration)**.
Caution: *The exhaust pushrods are slightly longer than the intake pushrods - they aren't interchangeable.*

Inspection

Caution: *If the cylinder heads have been removed and the surface milled for flatness, be sure to install correct length pushrods to compensate for the reduced distances of the rocker arms-to-camshaft dimensions. Consult with the machine shop for the correct length pushrods.*

6   Check each rocker arm for wear, cracks and other damage **(see illustration)**, especially where the pushrods and valve stems contact the rocker arm.

7   Check the rocker arm shafts and bores

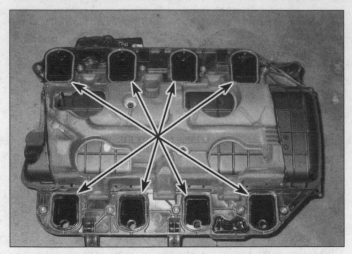

6.11 Replace the intake manifold O-rings with new ones. Make sure they seat properly in their grooves

6.13 Tightening sequence for the intake manifold bolts - 5.7L engine models shown, other models similar

of the rocker arms for wear. Look for galling, stress cracks and unusual wear patterns. If the rocker arms are worn or damaged, replace them with new ones and install new shafts as well.

Note: *Keep in mind that there is no valve adjustment on these engines, so excessive wear or damage in the valve train can easily result in excessive valve clearance, which in turn will cause valve noise when the engine is running.*

8   Make sure the hole at the pushrod end of each rocker arm is open.

9   Inspect the pushrods for cracks and excessive wear at the ends. Roll each pushrod across a piece of plate glass to see if it's bent (if it wobbles, it's bent).

Installation

10   Lubricate the lower end of each pushrod with clean engine oil or engine assembly lube and install them in their original locations. Make sure each pushrod seats completely in the lifter socket.

11   Apply engine assembly lube to the ends of the valve stems and the upper ends of the pushrods to prevent damage to the mating surfaces on initial start-up.

12   After the pushrods are reinstalled into their original locations install the pushrod retainer and tighten the fasteners securely.

13   Lubricate the rocker shafts with clean engine oil or engine assembly lube, then assemble the rocker shafts, with all of the components in their original positions. Install the rocker shafts onto the cylinder heads.

14   The rocker arm shafts must be tightened starting with the center bolt, then the center right bolt, the center left bolt, the outer right bolt and finally the outer left bolt. Follow this sequence in several steps until the torque listed in this Chapter's Specifications is reached. As the bolts are tightened, make sure the pushrods seat properly in the rocker arms.

Caution: *Do not continue tightening the rocker arms if the rocker arm bolts become tight before the shaft is seated or the pushrods are binding. Remove the rocker arm shafts and inspect all the components carefully before proceeding.*

15   Loosen the pushrod retainer fasteners, then remove the special tool.

16   Install the valve covers (see Section 4). Start the engine, listen for unusual valve train noises and check for oil leaks at the valve cover gaskets.

---

6   Intake manifold - removal and installation

Removal

1   Remove the engine cover(s) **(see illustration 4.2)**.

2   Relieve the fuel system pressure (see Chapter 4). Disconnect the cable from the negative terminal of the battery (see Chapter 5, Section 1).

Caution: *On 2015 and later models, if equipped with an Intelligent Battery Sensor (IBS), disconnect the IBS connector first before disconnecting the negative battery cable.*

3   Remove the air filter housing and the intake air duct (see Chapter 4).

4   Remove the ignition coils and, on 2005 models, the spark plug wires (see Chapter 5).

5   Disconnect the brake booster vacuum hose, canister purge hose and the PCV hose.

6   As a precaution, remove the fuel rails and injectors (see Chapter 4).

7   On 2010 and earlier 5.7L engines, remove the EGR pipe from the intake manifold (see Chapter 6).

8   Disconnect the electrical connectors from the MAP sensor (see Chapter 6) and the throttle body.

9   Starting with the outer bolts and working to the inner bolts, loosen the intake manifold

bolts in 1/4-turn increments until they can be removed by hand. Follow the reverse of the tightening sequence **(see illustration 6.13)**.

10   Remove the intake manifold. As the manifold is lifted from the engine, be sure to check for and disconnect anything still attached to the manifold.

Installation

Caution: *The mating surfaces of the cylinder heads and intake manifold must be perfectly clean when the manifold is installed, but don't use harsh chemicals or solvent on the plastic manifold, as it could become damaged.*

11   Check the O-rings on the intake manifold for damage or hardening. If they're OK, they can be re-used. If necessary, install new O-rings **(see illustration)**.

12   Carefully set the manifold in place.

Caution: *Do not disturb the rubber seals and DO NOT move the manifold fore-and-aft after it contacts the cylinder heads or the seals could be pushed out of place and the engine may develop vacuum and/or oil leaks.*

13   Install the intake manifold bolts and tighten them following the recommended sequence **(see illustration)** to the torque listed in this Chapter's Specifications.

14   The remaining installation steps are the reverse of removal. Start the engine and check carefully for vacuum leaks at the intake manifold joints.

---

7   Exhaust manifold(s) - removal and installation

Removal

Warning: *Allow the engine to cool completely before performing this procedure.*

1   Disconnect the cable from the negative terminal of the battery (see Chapter 5, Section 1). On 6.1L engine models, remove the underhood fuel/relay box fasteners and move

7.2 Front engine splash shield (1) and middle engine splash shield (2) fasteners

7.8 Location of the exhaust manifold heat shield nuts on the right-side exhaust manifold

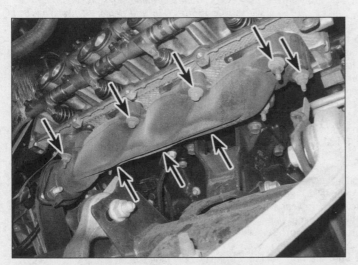

7.9a Location of the mounting bolts on the right-side exhaust manifold

7.9b Location of the mounting bolts on the left-side exhaust manifold (not all are visible in this photo)

the box out of the way.

Caution: *On 2015 and later models, if equipped with an Intelligent Battery Sensor (IBS), disconnect the IBS connector first before disconnecting the negative battery cable.*

2   Raise the front of the vehicle and support it securely on jackstands. Remove the engine splash shield **(see illustration)**.

3   Detach the exhaust pipes from the exhaust manifolds (see Chapter 4).

4   On 6.1L engine models, remove the air filter housing (see Chapter 4) and the coolant recovery bottle (see Chapter 3).

5   On 2015 and later 5.7L Challenger models and all 6.4L engine models, remove the cylinder head (see Section 10), then proceed to Step 9.

6   Connect an engine hoist to the engine (see Chapter 2E) and remove the engine mount-to-subframe fasteners (see Section 16).

7   Working from the engine compartment, disconnect the oxygen sensors' electrical connec-

tors and remove the sensors (see Chapter 6).

8   Remove the exhaust manifold heat shield mounting bolts/nuts **(see illustration)** and remove the heat shield, if equipped.

9   Remove the bolts retaining the exhaust manifold to the cylinder head **(see illustrations)**.

10   Raise the engine up enough to allow the exhaust manifold to be removed.

Installation

11   Clean the manifold and head gasket surfaces and check for cracks and flatness. Replace the exhaust manifold gaskets.

12   Install the exhaust manifold(s) and fasteners. Tighten the bolts/nuts to the torque listed in this Chapter's Specifications. Work from the center to the ends and approach the final torque in three steps. Install the heat shields.

13   Apply anti-seize compound to the exhaust pipe-to-exhaust manifold bolts and

tighten them securely.

14   Installation is otherwise the reverse of the removal procedure.

---

8   Vibration damper and front oil seal - removal and installation

Warning: *Wait until the engine is completely cool before beginning this procedure.*

1   Disconnect the cable from the negative terminal of the battery (see Chapter 5, Section 1).

Caution: *On 2015 and later models, if equipped with an Intelligent Battery Sensor (IBS), disconnect the IBS connector first before disconnecting the negative battery cable.*

2   Remove the engine drivebelt (see Chapter 1).

3   Drain the coolant (see Chapter 1), remove the upper radiator hose and remove the engine cooling fan (see Chapter 3).

8.4 A chain wrench can be used to prevent the vibration damper from turning while loosening the bolt - be sure to place a piece of drivebelt underneath the chain wrench to prevent damaging the damper

8.5 Use a three-jaw puller to remove the vibration damper from the crankshaft

4   Remove the large vibration damper-to-crankshaft bolt. To keep the crankshaft from turning, install a chain wrench or strap wrench around the circumference of the pulley. Be sure to use a piece of rubber (old drivebelt, old timing belt, etc.) under the chain wrench to protect the vibration damper from nicks or gouges **(see illustration)**.
Note: *It's a good idea to place a piece of cardboard or equivalent between the radiator and the vibration damper to protect the radiator while removing and installing the damper.*
5   Using the proper puller (commonly available from auto parts stores), detach the vibration damper **(see illustration)**.
Caution: *Do not use a puller with jaws that grip the outer edge of the pulley. The puller must be the type that utilizes bolts or arms to apply force to the pulley hub only. Also, the puller screw must not contact the threads in the nose of the crankshaft; it must use an adapter that allows the puller screw to apply force to the end of the crankshaft nose or a spacer must be inserted into the nose of the crankshaft to protect the threads.*

6   If the seal is being removed while the cover is still attached to the engine block, carefully pry the seal out of the cover with a seal removal tool or a large screwdriver **(see illustration)**.
Caution: *Be careful not to scratch, gouge or distort the area that the seal fits into or an oil leak will develop.*
7   Clean the bore to remove any old seal material and corrosion. Position the new seal in the bore with the seal lip (usually the side with the spring) facing IN (toward the engine). A small amount of oil applied to the outer edge of the new seal will make installation easier.
8   Drive the seal into the bore with a seal driver or a large socket and hammer until it's completely seated **(see illustration)**. Select a socket that's the same outside diameter as the seal and make sure the new seal is pressed into place until it bottoms against the cover flange.
9   Check the surface of the damper that the oil seal rides on. If the surface has been grooved from long-time contact with the seal, replace the vibration damper.

10   Lubricate the seal lips with engine oil and reinstall the vibration damper, aligning the Woodruff key on the nose of the crankshaft with the keyway in the damper hub. Use a special installation tool (available at most auto parts stores) to press the vibration damper onto the crankshaft.
11   Install the vibration damper bolt and tighten it to the torque listed in this Chapter's Specifications.
12   The remainder of installation is the reverse of the removal process.
13   Refill the cooling system (see Chapter 1).

9   Timing chain cover, chain and sprockets - removal, inspection and installation

Warning: *Wait until the engine is completely cool before beginning this procedure.*
Caution: *2008 and earlier 5.7L and all 6.1L engines have opposite timing chain mark locations than the 2009 and later 5.7L VVT and all 6.4L VVT engines.*

Removal

1   Disconnect the cable from the negative terminal of the battery (see Chapter 5, Section 1).
Caution: *On 2015 and later models, if equipped with an Intelligent Battery Sensor (IBS), disconnect the IBS connector first before disconnecting the negative battery cable.*
2   Drain the cooling system (see Chapter 1).
3   Drain the engine oil (see Chapter 1).
4   Remove the engine cover(s) **(see illustration 4.2)**.
5   Remove the air filter housing and the air intake duct (see Chapter 4).
6   Remove the drivebelt (see Chapter 1).
7   Remove the engine cooling fans (see Chapter 3).
8   Remove the air conditioning compressor without disconnecting the hoses (see Chapter 3). Use rope or wire to tie the compressor away from the front of the engine with the air conditioning lines attached.
9   Remove the alternator (see Chapter 5).
10   Remove the drivebelt tensioner (see Chapter 1) and the idler pulleys.
11   Disconnect the camshaft position sensor electrical connector.
Note: *It's a good idea to place a piece of cardboard or equivalent between the radiator and the vibration damper to protect the radiator while removing and installing the damper.*
12   Remove the vibration damper (see Section 8).
13   Remove the power steering pump (see Chapter 10) and set it aside without disconnecting the power steering lines.
14   Remove the dipstick tube.
15   Remove the oil pan and the pickup tube (see Section 12).

8.6 Use a seal puller to remove the seal from the timing chain cover

8.8 Use a seal driver or a large socket to drive the new seal into the cover

16 Disconnect the heater hoses and tube from the front cover.

17 Remove the water pump (see Chapter 3).

Note: *It is not absolutely necessary to remove water pump for timing cover removal but we highly recommend doing it at this time.*

18 Remove the timing chain cover bolts **(see illustration)** and the front cover.

Note: *2009 and later 5.7L VVT engine models, have five bolts that secure the oil pan to the timing chain cover.*

19 Remove the oil pump (see Section 13).

20 On 2008 and earlier 5.7L and all 6.1L engines, reinstall the bolt into the end of the crankshaft and turn the crankshaft until the crankshaft sprocket timing mark is at the 6 o'clock position (with the crankshaft key at 2 o'clock) and the camshaft timing mark at the 12 o'clock position **(see illustrations)**.

21 On 2009 and later 5.7L and all 6.4L engines, reinstall the bolt into the end of the crankshaft and turn the crankshaft until the

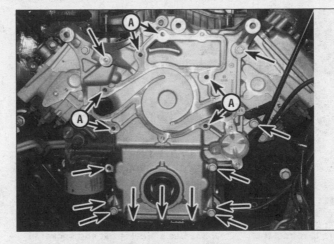

9.18 Location of the timing chain cover mounting bolts - the bolt holes marked with an (A) secure the water pump as well as the timing chain cover

crankshaft sprocket timing mark is at the 6 o'clock position (with the crankshaft key at 2 o'clock) and the camshaft timing mark at the 12 o'clock position. The two colored timing chain

links should straddle the camshaft phaser/ sprocket timing mark and the single colored timing chain link should align with the crankshaft sprocket timing dot **(see illustrations)**.

9.20a Use a breaker bar and socket to rotate the crankshaft (clockwise) to TDC number 1 position

9.20b On 2008 and earlier 5.7L and all 6.1L engines, the single colored timing chain link must align with the camshaft sprocket timing mark…

9.20c … and the two colored timing chain links (A) must straddle the crankshaft sprocket timing dot (B) at the 6 o'clock position - note that the keyway (C) should be at the 2 o'clock position

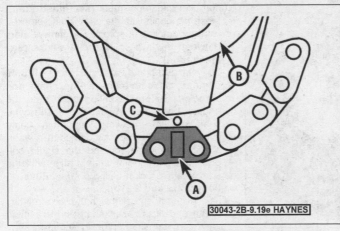

30043-2B-9.19e HAYNES

9.21a On 2009 and later 5.7L and all 6.4L engines, the single colored timing chain link (A) must align with the timing mark (C) on the crankshaft sprocket (B)…

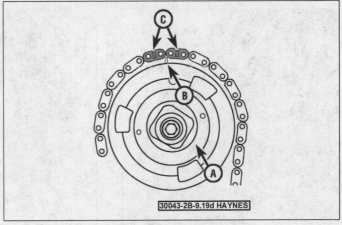

30043-2B-9.19d HAYNES

9.21b … and the two colored timing chain links must straddle the camshaft sprocket timing mark at the 6 o'clock position - note that the keyway should be at the 2 o'clock position

A Camshaft phaser
B Camshaft phaser timing mark
C Colored timing chain links

9.22 Use a large pair of pliers to retract the tensioner until the hole in the tensioner aligns with the hole in the bracket (thrust plate), then install a drill bit through the hole to keep it in the retracted position

9.23a Use a breaker bar and socket to prevent the crankshaft from rotating while loosening the camshaft sprocket bolt - 2008 and earlier models shown, 2009 and later models similar

9.23b Remove the camshaft sprocket, crankshaft sprocket and timing chain as a complete assembly - 2008 and earlier models shown, 2009 and later models similar

22   Retract the tensioner until the hole in the tensioner aligns with the hole in the camshaft tensioner thrust plate **(see illustration)**. Install a suitable size drill bit to retain the tensioner in the retracted position.
Note: *The timing chain tensioner is an integral component of the camshaft thrust plate. If necessary, the tensioner/thrust plate assembly must be replaced as a complete unit.*
23   Remove the camshaft sprocket bolt and remove the timing chain with the camshaft and crankshaft sprockets **(see illustrations)**.
Warning: *2009 and later models are equipped with Variable Valve Timing (VVT), a hydraulic timing phaser has been incorporated into the camshaft sprocket. Do not attempt to disassemble the camshaft sprocket/phaser assembly, severe engine damage could result.*
24   Remove the camshaft tensioner thrust plate mounting bolts **(see illustration 11.9)** and separate the tensioner from the engine block.

Inspection

25   Inspect the camshaft sprocket for damage or wear. The camshaft sprocket is a steel sprocket, but the teeth can become grooved or worn enough to cause a poor meshing of the sprocket and the chain.
Warning: *On 2009 and later models, do not attempt to disassemble the camshaft sprocket/phaser assembly, severe engine damage could result.*
Note: *Whenever a new timing chain is required, the entire set (chain, tensioner, camshaft/phaser sprocket and crankshaft sprocket) must be replaced as an assembly.*
26   Inspect the crankshaft sprocket for damage or wear. The crankshaft sprocket is a steel sprocket, but these teeth can also become grooved or worn enough to cause a poor meshing of the sprocket and the chain.

Installation

27   Stuff a shop rag into the opening at the front of the oil pan to keep debris out of the engine, then clean off all traces of old gasket material and sealant from the engine block. Wipe the sealing surfaces with a cloth saturated with lacquer thinner or acetone.
28   If the tensioner is not compressed (retracted position), compress the tensioner until the hole aligns with the bracket hole, then insert a suitable size drill bit through both holes to keep the tensioner locked in this position **(see illustration 9.22)**. Install the camshaft thrust plate and tighten the bolts to the torque listed in this Chapter's Specifications.
29   On 2008 and earlier models, assemble the timing chain and sprockets before installing them onto the engine. Loop the new chain over the camshaft sprocket with the single colored link (timing chain mark) aligned with the sprocket alignment mark **(see illustration)**. Mesh the chain with the crankshaft sprocket, with the crankshaft sprocket alignment mark between the two colored chain links **(see illustration)**.
30   On 2009 and later 5.7L and all 6.4L engines, assemble the timing chain and sprockets before installing them onto the engine. Loop the new chain over the camshaft sprocket, the two colored timing chain links should straddle the camshaft phaser/sprocket timing mark and the single colored timing chain link should align with the crankshaft sprocket timing dot **(see illustrations 9.21a and 9.21b)**.
31   Align the sprocket with the Woodruff key in the end of the crankshaft and assemble the chain, camshaft sprocket and crankshaft sprocket onto the crankshaft and camshaft. Tap it gently into place until it is completely seated. When the timing chain components are installed, the timing marks MUST align

9.29a On 2008 and earlier models, the camshaft sprocket timing mark(s) must align with the single colored timing chain link . . .

9.29b . . . and the two colored timing chain links must straddle the crankshaft sprocket mark

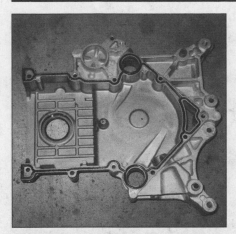

9.37 Be sure to install a new rubber gasket into the timing chain cover groove

10.10 Make sure the cylinder head surface is perfectly clean, free from old gasket material, carbon deposits and dirt - note that the dowel is part of the cylinder head stand

10.13 A die should be used to remove sealant and corrosion from the bolt threads prior to installation

as shown **in illustrations 9.20b and 9.20c or 9.21a and 9.21b** as applicable.

Caution: *If resistance is encountered, do not use excessive force to hammer the sprocket onto the crankshaft. It may eventually move onto the shaft, but it may be cracked in the process and fail later, causing extensive engine damage.*

32   Apply a thread locking compound to the camshaft sprocket bolt threads and tighten the bolt to the torque listed in this Chapter's Specifications.

33   Lubricate the chain with clean engine oil.

34   Remove the drill bit from the tensioner and bracket. Make sure the tensioner has released and is pressing against the timing chain. Verify that the timing marks are still aligned properly.

35   Install the oil pump (see Section 13).

36   Install the oil pump pick-up tube to the bottom of the oil pump. Tighten the bolt to the torque listed in this Chapter's Specifications.

37   The timing chain cover rubber gasket can be re-used if it isn't damaged or hardened, but considering the amount of work that you've done to get to this point, it's a good idea to replace it. Double-check the surface of the timing chain cover and engine block. Make sure all old gasket material is removed and the surface is clean. Install a new rubber gasket into the groove in the timing chain cover **(see illustration)**.

38   Apply a small amount of RTV sealant to the corner where the timing chain cover, engine block and oil pan meet.

39   Install the timing chain cover on the block **(see illustration 9.18)** and tighten the bolts, a little at a time, until you reach the torque listed in this Chapter's Specifications.

Note: *Be sure to install the water pump onto the timing chain cover and tighten the timing chain cover bolts and the water pump bolts at the same time (see Chapter 3).*

40   Install the oil pan (see Section 12).

41   Lubricate the oil seal contact surface of

the vibration damper hub with clean engine oil, then install the damper (see Section 8). Tighten the bolt to the torque listed in this Chapter's Specifications.

42   The remaining installation steps are the reverse of removal.

43   Add coolant and engine oil. Run the engine and check for oil and coolant leaks.

10   Cylinder head(s) - removal and installation

Warning: *Wait until the engine is completely cool before beginning this procedure.*

Removal

1   Relieve the fuel system pressure (see Chapter 4), then disconnect the cable from the negative terminal of the battery (see Chapter 5, Section 1).

Caution: *On 2015 and later models, if equipped with an Intelligent Battery Sensor (IBS), disconnect the IBS connector first before disconnecting the negative battery cable.*

2   Drain the cooling system (see Chapter 1).

3   Remove the intake manifold (see Section 6).

4   Remove the drivebelt (see Chapter 1) and on 6.1L engine models, remove the power steering pump, without disconnecting the hoses (see Chapter 10).

5   Remove the exhaust manifold(s) from the cylinder head(s) (see Section 7).

Note: *On 2015 and later Challenger models with 5.7L engines and all models with the 6.4L engine, the exhaust manifolds can only be removed after the cylinder head has been removed from the vehicle.*

6   Remove the valve covers (see Section 4).

7   Remove the rocker arms and pushrods (see Section 5).

Caution: *Again, as mentioned in Section 5,*

keep all the parts in order so they are reinstalled in the same locations.

8   Loosen the head bolts in 1/4-turn increments in a pattern opposite of the tightening sequence **(see illustration 10.17)** until they can be removed by hand.

Note: *There will be different-length head bolts for different locations, so store the bolts in order as they are removed. This will ensure that the bolts are reinstalled in their original holes.*

9   Lift the heads off the engine. If resistance is felt, do not pry between the head and block as damage to the mating surfaces will result. To dislodge the head, place a block of wood against the end of it and strike the wood block with a hammer, or lift on a casting protrusion. Store the heads on blocks of wood to prevent damage to the gasket sealing surfaces.

Installation

10   The mating surfaces of the cylinder heads and block must be perfectly clean when the heads are installed **(see illustration)**. Gasket removal solvents are available at auto parts stores and may prove helpful.

11   Use a gasket scraper to remove all traces of carbon and old gasket material, then wipe the mating surfaces with a cloth saturated with lacquer thinner or acetone. If there is oil on the mating surfaces when the heads are installed, the gaskets may not seal correctly and leaks may develop. When working on the block, cover the lifter valley with shop rags to keep debris out of the engine. Use a vacuum cleaner to remove any debris that falls into the cylinders.

12   Check the block and head mating surfaces for nicks, deep scratches and other damage. If damage is slight, it can be removed with emery cloth. If it is excessive, machining may be the only alternative.

13   Use a tap of the correct size to chase the threads in the head bolt holes in the block. Mount each bolt in a vise and run a die down the threads to remove corrosion and restore the threads **(see illustration)**. Dirt, corro-

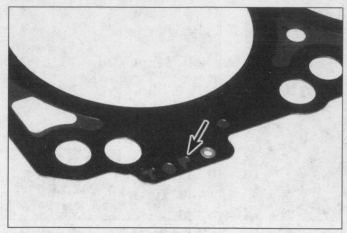

10.14 Make sure the head gaskets are installed on the correct cylinder banks, and the TOP marking is facing up

10.17 Cylinder head bolt tightening sequence

11.6 Each lifter rail is secured by one bolt

11.7a Remove the lifters from the lifter bores and install them into the lifter rail - each lifter must be installed into the same lifter bore and in the same direction (roller rotation)

11.7b Install the lifters into the lifter rail and make sure the lifters are marked in their original positions with number designations for the cylinders and letter designations for the valve each one operates (intake or exhaust)

sion, sealant and damaged threads will affect torque readings.

14   Position the new gaskets over the dowels in the block. Be sure the letter designations for the left (L) and right (R) cylinder heads and the top gasket surface (TOP/UP) are correctly oriented **(see illustration)**.

15   Carefully position the heads on the block without disturbing the gaskets.

16   Before installing the head bolts, coat the threads with a small amount of engine oil.

17   Install the bolts in their original locations and tighten them finger-tight. Following the recommended sequence **(see illustration)**, tighten the bolts in several steps to the torque listed in this Chapter's Specifications.

18   The remaining installation steps are the reverse of removal.

19   Add coolant and change the engine oil and filter (see Chapter 1). Start the engine and check for proper operation and coolant or oil leaks.

11   Camshaft and lifters - removal, inspection and installation

Warning: *Wait until the engine is completely cool before beginning this procedure.*
Warning: *If the lifter and retainer assemblies are to being reused, the lifters must be reinstalled in their original location or engine damage could result.*
Note: *Some engines are equipped with a Multiple Displacement System (MDS) that uses a specialized camshaft. Replace the camshaft ONLY with a camshaft compatible with the MDS design. Refer to Chapter 6 for a description of the MDS system.*
Note: *If the camshaft is replaced, all lifters must be replaced.*

Removal

1   Relieve the fuel system pressure (see Chapter 4), then disconnect the cable from the negative terminal of the battery (see

Chapter 5, Section 1).
Caution: *On 2015 and later models, if equipped with an Intelligent Battery Sensor (IBS), disconnect the IBS connector first before disconnecting the negative battery cable.*
2   Drain the cooling system (see Chapter 1).
3   Remove the timing chain cover, the timing chain and the sprockets (see Section 9).
4   Remove the cylinder heads (see Section 10).
5   Remove the radiator (see Chapter 3).
6   Remove the lifter rail mounting bolts **(see illustration)**.
7   Remove the lifters from the lifter bores in the engine block **(see illustration)**. Carefully place them in the correct location in the lifter rail. Each lifter must be installed into the same lifter bore and in the same direction (roller rotation) **(see illustration)**. Be sure to mark each lifter and lifter rail with a felt pen to designate the cylinder number and the type of lifter (intake or exhaust). There are several

11.9 Location of the camshaft thrust plate mounting bolts, 2008 and earlier models shown, later models have four bolts

11.12 Check the diameter of each camshaft bearing journal to pinpoint excessive wear and out-of-round conditions

11.13 Measure the camshaft lobe height (greatest dimension) with a micrometer

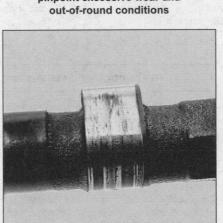

11.14 Check the cam lobes for pitting, excessive wear and scoring. If scoring is excessive, as shown here, replace the camshaft

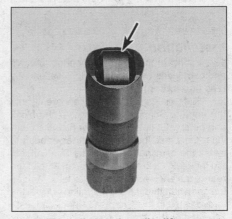

11.18 The roller on the roller lifters must turn freely - check for wear and excessive play as well

ways to extract the lifters from the bores. A special tool designed to grip and remove lifters is manufactured by many tool companies and is widely available, but it may not be required in every case. On newer engines without a lot of varnish buildup, the lifters can often be removed with a small magnet or even with your fingers. A machinist's scribe with a bent end can be used to pull the lifters out by positioning the point under the retainer ring inside the top of each lifter. Do not attempt to withdraw the camshaft with the lifters in place. **Caution:** *Do not use pliers to remove the lifters unless you intend to replace them with new ones (along with the camshaft). The pliers will damage the precision machined and hardened lifters, rendering them useless.* **Note:** *On 6.4L engines with Multi Displacement System (MDS) both standard roller lifters and "deactivating" roller lifters. The deactivating roller lifters must be used in cylinders 1, 4, 6 and 7. The deactivating lifters can be identified by the two holes in the side of the lifter body, for the latching pin locations.*

8   Before removing the camshaft, check the endplay. Mount a dial indicator so that it contacts the nose of the camshaft. Pry the camshaft forward and back using a long screwdriver with the tip taped to prevent damage to the camshaft. Record the movement of the dial indicator and compare the reading to the value given in this Chapter's Specifications. If the endplay is excessive, the camshaft must be replaced.

9   Unbolt and remove the camshaft thrust plate **(see illustration)**.

10 Thread a long bolt into the camshaft sprocket bolt hole to use as a handle when removing the camshaft from the block. Carefully pull the camshaft out. Support the cam near the block so the lobes do not nick or gouge the bearings as it is withdrawn.

Inspection

11   After the camshaft has been removed from the engine, cleaned with solvent and dried, inspect the bearing journals for uneven wear, pitting and evidence of seizure. If the

journals are damaged, the bearing inserts in the block are probably damaged as well. Both the camshaft and bearings will have to be replaced.

Note: *Camshaft bearing replacement requires special tools and expertise that place it beyond the scope of the average home mechanic. The tools for bearing removal and installation are available at stores that carry automotive tools, possibly even found at a tool rental business. It is advisable though, if bearings are bad and the procedure is beyond your ability, remove the engine block and take it to an automotive machine shop to ensure that the job is done correctly.*

12   Measure the bearing journals with a micrometer to determine if they are excessively worn or out-of-round **(see illustration)**.

13   Measure the lobe height of each cam lobe on the intake camshaft and record your measurements **(see illustration)**. Compare the measurements for excessive variations. If the lobe heights vary more than 0.005 inch (0.125 mm), replace the camshaft. Compare the lobe height measurements on the exhaust camshaft and follow the same procedure. Do not compare intake camshaft lobe heights

with exhaust camshaft lobe heights, as they are different. Only compare intake lobes with intake lobes and exhaust lobes with other exhaust lobes.

14   Check the camshaft lobes for heat discoloration, score marks, chipped areas, pitting and uneven wear **(see illustration)**. If the lobes are in good condition and if the lobe lift variation measurements recorded earlier are within the limits, the camshaft can be reused.

15   Clean the lifters with solvent and dry them thoroughly without mixing them up.

16   Check each lifter wall, pushrod seat and foot for scuffing, score marks and uneven wear. If the lifter walls are damaged or worn (which is not very likely), inspect the lifter bores in the engine block as well. If the pushrod seats are worn, check the pushrod ends.

17   If new lifters are being installed, a new camshaft must also be installed. If a new camshaft is installed, then use new lifters as well. Never install used lifters unless the original camshaft is used and the lifters can be installed in their original locations.

18   Check the rollers carefully for wear and damage and make sure they turn freely without excessive play **(see illustration)**.

11.19 Be sure to apply camshaft installation lube to the cam lobes and bearing journals before installing the camshaft

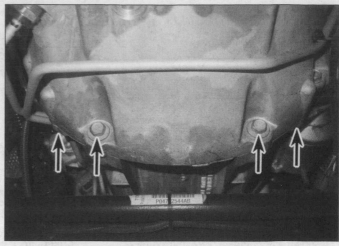

12.12 Location of the transmission-to-oil pan mounting bolts

Installation

19   Lubricate the camshaft bearing journals and cam lobes with camshaft installation lube **(see illustration)**.

20   Slide the camshaft slowly and gently into the engine. Support the cam near the block and be careful not to scrape or nick the bearings. Only install the camshaft far enough to install the camshaft thrust plate. Pushing it in too far could dislodge the camshaft plug at the rear of the engine, causing an oil leak. Tighten the camshaft thrust plate mounting bolts to the torque listed in this Chapter's Specifications.

21   Install the timing chain and sprockets (see Section 9). Align the timing marks on the crankshaft and camshaft sprockets.

22   Lubricate the lifters with clean engine oil and install them in the block. If the original lifters are being reinstalled, be sure to return them to their original locations, and with the numbers facing UP (exactly as they were removed). Install the lifter rail and mounting bolts. Tighten the lifter rail mounting bolts to the torque listed in this Chapter's Specifications.

Note: *Each lifter rail should be numbered and coincide with the correct cylinder numbers.*

23   The remaining installation steps are the reverse of removal.

24   Change the oil and install a new oil filter (see Chapter 1). Fill the cooling system with the proper type of coolant (see Chapter 1).

25   Start the engine and check for oil pressure and leaks.

Caution: *Do not run the engine above a fast idle until all the hydraulic lifters have filled with oil and become quiet again.*

26   If a new camshaft and lifters have been installed, the engine should be brought to operating temperature and run at a fast idle for 15 to 20 minutes to "break in" the new components. Change the oil and filter again after 500 miles of operation.

12   Oil pan - removal and installation

Warning: *Wait until the engine is completely cool before beginning this procedure.*

Removal

1   Disconnect the cable from the negative terminal of the battery (see Chapter 5, Section 1).

Caution: *On 2015 and later models, if equipped with an Intelligent Battery Sensor (IBS), disconnect the IBS connector first before disconnecting the negative battery cable.*

2   Drain the cooling system (see Chapter 1).

3   Raise the vehicle and support it securely on jackstands (see Chapter 1).

4   Drain the engine oil and replace the oil filter (see Chapter 1). Remove the engine oil dipstick.

5   Remove the engine splash shield **(see illustration 7.2)**. Disconnect the exhaust pipes from the exhaust manifolds (see Chapter 4).

6   Remove the intake manifold (see Section 6).

7   On 2012 and later models, with the 5.7L engines, remove the Variable Valve Timing (VVT) solenoid (see Chapter 6).

8   Remove the engine cooling fans and shroud (see Chapter 3).

9   Remove the power steering gear from the crossmember (see Chapter 10). Do not disconnect the intermediate shaft coupler or the tie rod ends. Also, do not disconnect the power steering fluid lines from the power steering gear. Position the power steering gear assembly down and away from the engine.

10   Connect an engine hoist to the engine (see Chapter 2E).

11   Remove the left and right side engine mount-to-subframe nuts and studs (see Section 16), then raise the engine slightly.

12   Remove all of the oil pan bolts **(see illustration)**, then lower the pan from the engine. The pan will probably stick to the engine, so strike the pan with a rubber mallet until it breaks the gasket seal. **Caution:** *Before using force on the oil pan, be sure all the bolts have been removed. Carefully slide the oil pan out, to the rear.* **Note:** *The M10 horizontal bolts are slightly longer in length than the other M10 vertical bolts. Be sure to mark the longer bolts and reinstall them in the correct locations.*

13   Remove the windage tray from the engine block.

Note: *The windage tray is integral to the oil pan gasket and must replaced as an assembly, whenever the oil pan is removed from the engine block.*

Installation

14   Install a new gasket and windage tray.

15   Wash out the oil pan with solvent.

16   Thoroughly clean the mounting surfaces of the oil pan and engine block of old gasket material and sealer. If the oil pan is distorted at the bolt-hole areas, straighten the flange by supporting it from below on a 1x4 wood block and tapping the bolt holes with the rounded end of a ball-peen hammer. Wipe the gasket surfaces clean with a rag soaked in lacquer thinner or acetone.

17   Apply some RTV sealant to the corners where the timing chain cover meets the block and at the rear where the rear main oil seal retainer meets the block. Then attach the one-piece oil pan gasket to the engine block with contact-cement-type gasket adhesive.

18   Make sure the alignment studs are installed in the correct locations in the engine block.

19   Lift the pan into position, slipping it over the alignment studs and being careful not to disturb the gasket, install several bolts finger tight.

20   Check that the gasket isn't sticking out anywhere around the block's perimeter. When

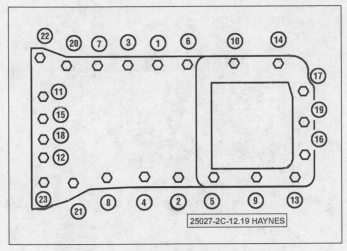

12.21 Oil pan bolt tightening sequence

13.5 Location of the oil pump pick-up tube retainer bolt (A) and
the oil pump mounting bolts (B)

all the bolts are in place, install the oil pan nuts onto the studs.

21   Following the correct torque sequence **(see illustration)**, tighten the fasteners to the torque listed in this Chapter's Specifications.

22   The remainder of the installation procedure is the reverse of removal.

23   Add the proper type and quantity of oil, change the oil filter, and refill the cooling system (see Chapter 1), start the engine and check for leaks before placing the vehicle back in service.

13   Oil pump - removal and installation

Warning: *Wait until the engine is completely cool before beginning this procedure.*

Removal

1   Disconnect the cable from the negative terminal of the battery (see Chapter 5, Section 1).

Caution: *On 2015 and later models, if equipped with an Intelligent Battery Sensor (IBS), disconnect the IBS connector first before disconnecting the negative battery cable.*

2   Drain the cooling system (see Chapter 1).

3   Remove the timing chain cover (see Section 9).

4   Remove the oil pan (see Section 12).

5   Remove the bolt from the pick-up tube assembly and lower it away from the oil pump **(see illustration)**.

6   Remove the oil pump mounting bolts and detach the pump from the engine block.

Installation

7   Position the pump on the engine. Make sure the pump rotor is aligned with the crankshaft drive. Install the oil pump mounting bolts and tighten them to the torque listed in this Chapter's Specifications.

8   Install a new O-ring onto the oil pump pick-up tube, connect the tube to the pump and tighten the bolt to the torque listed in this Chapter's Specifications.

9   The remainder of installation is the reverse of removal.

10   Fill the crankcase with the proper type and quantity of engine oil, change the oil filter and refill the cooling system (see Chapter 1).

11   Run the engine and check for oil pressure and leaks.

14   Driveplate/flywheel - removal and installation

1   Raise the vehicle and support it securely on jackstands, then refer to Chapter 7A or Chapter 7B and remove the transmission.

Warning: *The engine must be supported from above with an engine hoist or an engine support fixture before working underneath the vehicle with the transmission removed.*

2   On automatic transmissions, now would be a good time to check and replace the transmission front pump seal and on manual transmissions replace the clutch assembly (see Chapter 8).

3   Use paint or a center-punch to make alignment marks on the driveplate and crankshaft to ensure correct alignment during reinstallation.

4   Remove the bolts that secure the driveplate to the crankshaft. If the crankshaft turns, jam a large screwdriver or prybar through the driveplate to keep the crankshaft from turning, then remove the mounting bolts and bolt plate.

Caution: *On manual transmission models, discard and replace the flywheel bolts with new ones.*

5   Pull straight back on the driveplate/flywheel to detach it from the crankshaft.

6   Installation is the reverse of removal. Be sure to align the matching paint marks. Use thread locking compound on the bolt threads and tighten them to the torque listed in this Chapter's Specifications, using a criss-cross pattern.

15   Rear main oil seal - replacement

Note: *On 2007 and later models, the crankshaft rear main oil seal is integral to the crankshaft rear oil seal retainer/housing and must be replaced as an assembly. Once the crankshaft seal retainer/housing is removed it must be replaced.*

1   V8 engines use a one-piece rear main oil seal which is installed in a bolt-on housing. Replacing this seal requires removal of the transmission and driveplate/flywheel. Refer to Chapter 7A or Chapter 7B for the transmission removal procedures.

2006 and earlier models

2   On 2006 and earlier models, the seal can be removed by prying it out of the housing by inserting a screwdriver, being careful not to nick the crankshaft surface. Wrap the screwdriver tip with tape to avoid damage. Be sure to note how far it's recessed into the housing bore before removal so the new seal can be installed to the same depth. Preferably, a seal installation tool is needed to press the new seal back into place. If the proper seal installation tool is unavailable, use a large socket, section of pipe or a blunt tool and carefully drive the new seal into place. The lip is stiff so carefully work it onto the seal journal of the crankshaft. Don't rush it or you may damage the seal.

3   The rear main seal housing can also be removed to change the seal, but whenever the housing is removed from the block the oil pan must be removed (see Section 12).

15.4 Place the rear main oil seal housing on two blocks of wood and use a small punch to drive out the old seal - 2006 and earlier models only

15.5 Use a block of wood to drive the oil seal squarely into the housing - 2006 and earlier models only

4   If the housing is removed, place it on two blocks of wood and use a small punch to drive out the old seal **(see illustration)**.

5   Clean the housing thoroughly, then apply a thin coat of engine oil to the new seal. Set the seal squarely into the recess of the housing, then, using a piece of wood and a hammer, press the seal into place **(see illustration)**.

2007 and later models

6   Drain the engine oil (see Chapter 1).

7   Remove the oil pan (see Section 12).

8   Remove the crankshaft seal retainer/ housing mounting bolts then remove and discard the retainer assembly from the end of the crankshaft.

9   Carefully slide the seal over the crankshaft and bolt the seal housing to the block. Be sure to use a new gasket and tighten the bolts in a criss-cross pattern to the torque listed in this Chapter's Specifications.

10   The remainder of installation is the reverse of the removal procedure.

16   Engine mounts - check and replacement

1   There are three powertrain mounts; left and right engine mounts attached to the engine block and to the subframe and a rear mount attached to the transmission and the frame. Refer to Chapter 7A or Chapter 7B for check and replacement procedures for the transmission mount.

Check

2   During the check, the engine must be raised slightly to remove the weight from the mounts.

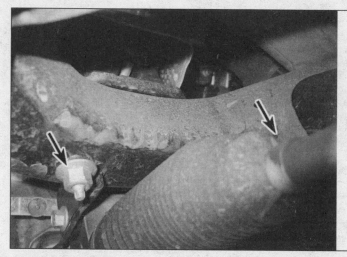

16.9 Location of the subframe-to-engine mount nuts, (one bolt hidden from view) - early models 5.7L engine shown later models similar

3   Raise the vehicle and support it securely on jackstands, then position a jack under the engine oil pan. Place a large wood block between the jack head and the oil pan, then carefully raise the engine just enough to take the weight off the mounts.

Warning: *DO NOT place any part of your body under the engine when it's supported only by a jack!*

4   Check the mount insulators to see if the rubber is cracked, hardened or separated from the metal in the center of the mount.

5   Check for relative movement between the mount and the engine or subframe. Use a large screwdriver or prybar to attempt to move the mounts.

6   If movement is noted, lower the engine and tighten the mount fasteners.

Replacement

7   Disconnect the cable from the negative terminal of the battery (see Chapter 5, Sec-

tion 1). Raise the vehicle and support it securely on jackstands (if not already done). Position a jack under the engine oil pan. Place a large wood block between the jack head and the oil pan.

Note: *On some models, there may not be clear access to the oil pan to place a jack. If this is the case, the intake manifold must be removed and the engine supported and lifted from above (see Chapter 2E).*

8   If you're removing the right side engine mount, remove the alternator support bracket (see Chapter 5).

9   Remove the engine mount-to-subframe nuts **(see illustration)** and raise the engine with the jack.

10   Remove the mount-to-engine bracket bolts.

11   Install the new mount, making sure it is correctly positioned in the bracket. Install the fasteners and tighten them securely.

Chapter 2 Part E
General engine overhaul procedures

Contents

Specifications

General

Displacement

V6 engines

2.7L	167 cubic inches
3.5L	214 cubic inches
3.6L	220 cubic inches

V8 engines

5.7L	345 cubic inches
6.1L	370 cubic inches
6.4L	392 cubic inches

Bore and stroke

V6 engines

2.7L	3.386 x 3.091 inches
3.5L	3.780 x 3.189 inches
3.6L	3.779 x 3.268 inches

V8 engines

5.7L	3.92 x 3.58 inches
6.1L	4.055 x 3.58 inches
6.4L	4.09 x 3.72 inches

Cylinder compression pressure

V6 engines

Minimum	170 psi
Maximum variation between cylinders	40 psi
V8 engines	Not available, but pressure should not vary more than 40 psi between cylinders

Oil pump pressure

V6 engines

Minimum pressure at curb idle	5 psi

Operating pressure

2.7L and 3.5L engines	45 to 105 psi at 3,000 rpm
3.6L engines	30 to 139 psi at 1,201 to 3,000 rpm

V8 engines

Minimum pressure at curb idle	4 psi
Operating pressure	25 to 110 psi at 3,000 rpm

Torque specifications

Ft-lbs (unless otherwise indicated)

Note: *One foot-pound (ft-lb) of torque is equivalent to 12 inch-pounds (in-lbs) of torque. Torque values below approximately 15 ft lbs are expressed in inch-pounds, because most foot-pound torque wrenches are not accurate at these smaller values.*

Connecting rod cap bolts
    V6 engines
        Step 1
            2.7L and 3.5L engines.. 20
            3.6L engines.. 15
        Step 2 .... Tighten an additional 90-degrees
    V8 engines
        5.7L engines
            Step 1.. 15
            Step 2.. Tighten an additional 90-degrees
        6.1L engines
            Step 1.. 33
            Step 2.. Tighten an additional 60-degrees
        6.4L engines
            Step 1.. 30
            Step 2.. Tighten an additional 90-degrees
Main bearing cap bolts
    V6 engines **(see illustration 10.19a)**
        Inner main bearing cap bolts
            Step 1.. 15
            Step 2.. Tighten an additional 90-degrees
        Outer main (windage tray) bearing cap bolts
            Step 1
                 2.7L and 3.5L engines 20
                3.6L engines.. 15
            Step 2.. Tighten an additional 90-degrees
        Main bearing cap cross (side) bolts
            2.7L and 3.5L engines... 21
            3.6L engines... 22
    3.6L Crankshaft target wheel.. 89 in-lbs
    V8 engines* **(see illustration 10.19b)**
        5.7L and 6.1L engines
            Step 1, bolts 1 through 10 15
            Step 2, bolts 1 through 10 20
            Step 3, bolts 1 through 10 .... Tighten an additional 90-degrees
            Step 4, crossbolts 11 through 20............................. 15
            Step 5, crossbolts 11 through 20............................. 21
            Step 6, crossbolts 11 through 20............................. 21 (re-check)
        6.4L engines
            Step 1, bolts 1 through 10 .... 120 in-lbs
            Step 2, bolts 1 through 10 21
            Step 3, bolts 1 through 10 .... Tighten an additional 90-degrees
            Step 4, crossbolts 11 through 20............................. 25
            Step 5, crossbolts 11 through 20............................. 25 (re-check)

* Install a new washer/seal onto each crossbolt before tightening. First, install main bearing bolts and main bearing crossbolts all finger-tight. Next tighten the main bearing bolts completely (Steps 1 and 2) before tightening the crossbolts

1.10a An engine block being bored. An engine rebuilder will use special machinery to recondition the cylinder bores

1.10b If the cylinders are bored, the machine shop will normally hone the engine on a machine like this

1 General information - engine overhaul

1   Included in this portion of Chapter 2 are general information and diagnostic testing procedures for determining the overall mechanical condition of your engine.

2   The information ranges from advice concerning preparation for an overhaul and the purchase of replacement parts and/or components to detailed, step-by-step procedures covering removal and installation.

3   The following Sections have been written to help you determine whether your engine needs to be overhauled and how to remove and install it once you've determined it needs to be rebuilt. For information concerning in-vehicle engine repair, see Chapter 2A, Chapter 2B, Chapter 2C or Chapter 2D.

4   The Specifications included in this Part are general in nature and include only those necessary for testing the oil pressure and engine compression, and bottom-end torque specifications. Refer to Chapter 2A, Chapter 2B, Chapter 2C or Chapter 2D for additional engine Specifications.

5   It's not always easy to determine when, or if, an engine should be completely overhauled, because a number of factors must be considered.

6   High mileage is not necessarily an indication that an overhaul is needed, while low mileage doesn't preclude the need for an overhaul. Frequency of servicing is probably the most important consideration. An engine that's had regular and frequent oil and filter changes, as well as other required maintenance, will most likely give many thousands of miles of reliable service. Conversely, a neglected engine may require an overhaul very early in its service life.

7   Excessive oil consumption is an indication that piston rings, valve seals and/or valve guides are in need of attention. Make sure that oil leaks aren't responsible before deciding that the rings and/or guides are bad. Perform a cylinder compression check to determine the extent of the work required (see Section 3). Also, check the vacuum readings under various conditions (see Section 4).

8   Check the oil pressure with a gauge installed (see Section 2) in place of the oil pressure sending unit and compare it to this Chapter's Specifications. If it's extremely low, the bearings and/or oil pump are probably worn out.

9   Loss of power, rough running, knocking or metallic engine noises, excessive valve train noise and high fuel consumption rates may also point to the need for an overhaul, especially if they're all present at the same time. If a complete tune-up doesn't remedy the situation, major mechanical work is the only solution.

10   An engine overhaul involves restoring the internal parts to the specifications of a new engine. During an overhaul, the piston rings are replaced and the cylinder walls are reconditioned (rebored and/or honed) **(see illustrations 1.10a and 1.10b)**. If a rebore is done by an automotive machine shop, new oversize pistons will also be installed. The main bearings, connecting rod bearings and camshaft bearings are generally replaced with new ones and, if necessary, the crankshaft may be reground to restore the journals **(see illustration 1.10c)**. Generally, the valves are serviced as well, since they're usually in less-than-perfect condition at this point. While the engine is being overhauled, other components, such as the starter and alternator, can be rebuilt as well. The end result should be similar to a new engine that will give many trouble free miles.

Note: *Critical cooling system components such as the hoses, drivebelts, thermostat and water pump should be replaced with new parts when an engine is overhauled. The radiator should be checked carefully to ensure that it isn't clogged or leaking (see Chapter 3). If you purchase a rebuilt engine or short block, some rebuilders will not warranty their engines unless the radiator has been professionally flushed. Also, we don't recommend overhauling the oil pump - always install a new one when an engine is rebuilt.*

11   Overhauling the internal components on today's engines is a difficult and time-consuming task which requires a significant amount of specialty tools and is best left to a professional engine rebuilder **(see illustrations 1.11a, 1.11b and 1.11c)**. A competent

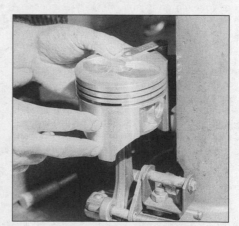

1.10c A crankshaft having a main bearing journal ground

1.11a A machinist checks for a bent connecting rod, using specialized equipment

1.11b A bore gauge being used to check a cylinder bore

1.11c Uneven piston wear like this indicates a bent connecting rod

2.2a On the 2.7L V6 engine, the oil pressure sending unit is located on the right side of the engine block

2.2b On the 3.5L V6 engine, the oil pressure sending unit is located on the left front side of the timing chain cover

2.2c On the 3.6L V6 engine, the oil pressure sending unit is located under the lower intake manifold, threaded into the oil cooler

2.2d On the 5.7L V8 engine, the oil pressure sending unit is located right above the oil filter

engine rebuilder will handle the inspection of your old parts and offer advice concerning the reconditioning or replacement of the original engine. Never purchase parts or have machine work done on other components until the block has been thoroughly inspected by a professional machine shop. As a general rule, time is the primary cost of an overhaul, especially since the vehicle may be tied up for a minimum of two weeks or more. Be aware that some engine builders only have the capability to rebuild the engine you bring them while other rebuilders have a large inventory of rebuilt exchange engines in stock. Also be aware that many machine shops could take as much as two weeks time to completely rebuild your engine depending on shop workload. Sometimes it makes more sense to simply exchange your engine for another engine that's already rebuilt to save time.

2   Oil pressure check

1   Low engine oil pressure can be a sign of an engine in need of rebuilding. A low oil pressure indicator (often called an "idiot light") is not a test of the oiling system. Such indicators only come on when the oil pressure is dangerously low. Even a factory oil pressure gauge in the instrument panel is only a relative indication, although much better for driver information than a warning light. A better test is with a mechanical (not electrical) oil pressure gauge.

2   Locate the oil pressure indicator sending unit on the engine block. The oil pressure sending unit is located in various locations depending upon engine type:

a) *On 2.7L V6 engines, the oil pressure sending unit is located on the right side of the engine block* **(see illustration)**.

b) *On 3.5L V6 engines, the oil pressure sending unit is located on the left side of the timing chain cover* **(see illustration)**.

c) *On 3.6L V6 engines, the oil pressure sending unit is located under the lower intake manifold on the oil filter housing* **(see illustration)**.

d) *On 5.7L V8 engines, the oil pressure sending unit is located on the oil filter housing above the oil filter* **(see illustration)**.

e) *On 6.1L and 6.4L V8 engines, the oil pressure sensor is located under the alternator.*

Note: *It will be necessary to remove the alternator (see Chapter 5) to access the sensor.*

3   Unscrew and remove the oil pressure sending unit, then screw in the hose for your oil pressure gauge **(see illustration)**. If necessary, install an adapter fitting. Use Teflon tape or thread sealant on the threads of the

2.3 Install an oil pressure gauge after removing the oil pressure sending unit - 5.7L V8 engine shown

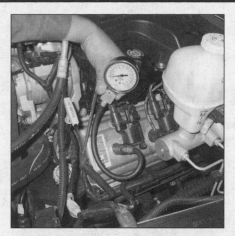

3.6 Use a compression gauge with a threaded fitting for the spark plug hole, not the type that requires hand pressure to maintain the seal

4.4 A simple vacuum gauge can be handy in diagnosing engine condition and performance

adapter and/or the fitting on the end of your gauge's hose.

4   Connect an accurate tachometer to the engine, according to the tachometer manufacturer's instructions.

5   Check the oil pressure with the engine running (normal operating temperature) at the specified engine speed, and compare it to this Chapter's Specifications. If it's extremely low, the bearings and/or oil pump are probably worn out.

3   Cylinder compression check

1   A compression check will tell you what mechanical condition the upper end of your engine (pistons, rings, valves, head gaskets) is in. Specifically, it can tell you if the compression is down due to leakage caused by worn piston rings, defective valves and seats or a blown head gasket.
Note: *The engine must be at normal operating temperature and the battery must be fully charged for this check.*

2   Begin by cleaning the area around the spark plugs before you remove them (compressed air should be used, if available). The idea is to prevent dirt from getting into the cylinders as the compression check is being done.

3   Remove the ignition coil assemblies (see Chapter 5). Also disable the fuel pump by removing the fuel pump relay (see Chapter 4, Section 3).

4   If you're working on a V8 engine, remove one spark plug from each cylinder; on all other engines, remove all of the spark plugs (see Chapter 1).

5   Block the throttle wide open.

6   Install a compression gauge in the spark plug hole **(see illustration)**.

7   Crank the engine over at least seven compression strokes and watch the gauge. The compression should build up quickly in a

healthy engine. Low compression on the first stroke, followed by gradually increasing pressure on successive strokes, indicates worn piston rings. A low compression reading on the first stroke, which doesn't build up during successive strokes, indicates leaking valves or a blown head gasket (a cracked head could also be the cause). Deposits on the undersides of the valve heads can also cause low compression. Record the highest gauge reading obtained.

8   Repeat the procedure for the remaining cylinders and compare the results to this Chapter's Specifications.

9   Add some engine oil (about three squirts from a plunger-type oil can) to each cylinder, through the spark plug hole, and repeat the test.

10   If the compression increases after the oil is added, the piston rings are definitely worn. If the compression doesn't increase significantly, the leakage is occurring at the valves or head gasket. Leakage past the valves may be caused by burned valve seats and/or faces or warped, cracked or bent valves.

11   If two adjacent cylinders have equally low compression, there's a strong possibility that the head gasket between them is blown. The appearance of coolant in the combustion chambers or the crankcase would verify this condition.

12   If one cylinder is slightly lower than the others, and the engine has a slightly rough idle, a worn lobe on the camshaft could be the cause.

13   If the compression is unusually high, the combustion chambers are probably coated with carbon deposits. If that's the case, the cylinder head(s) should be removed and decarbonized.

14   If compression is way down or varies greatly between cylinders, it would be a good idea to have a leak-down test performed by an automotive repair shop. This test will pinpoint exactly where the leakage is occurring and how severe it is.

4   Vacuum gauge diagnostic checks

1   A vacuum gauge provides inexpensive but valuable information about what is going on in the engine. You can check for worn rings or cylinder walls, leaking head or intake manifold gaskets, incorrect carburetor adjustments, restricted exhaust, stuck or burned valves, weak valve springs, improper ignition or valve timing and ignition problems.

2   Unfortunately, vacuum gauge readings are easy to misinterpret, so they should be used in conjunction with other tests to confirm the diagnosis.

3   Both the absolute readings and the rate of needle movement are important for accurate interpretation. Most gauges measure vacuum in inches of mercury (in-Hg). The following references to vacuum assume the diagnosis is being performed at sea level. As elevation increases (or atmospheric pressure decreases), the reading will decrease. For every 1,000 foot increase in elevation above approximately 2,000 feet, the gauge readings will decrease about one inch of mercury.

4   Connect the vacuum gauge directly to the intake manifold vacuum, not to ported (throttle body) vacuum **(see illustration)**. Some models are equipped with a vacuum fitting built into the brake booster vacuum hose grommet at the brake booster. Other models are equipped with a vacuum hose fitting on the intake manifold. Use a T-fitting to access the vacuum signal. Be sure no hoses are left disconnected during the test or false readings will result.

5   Before you begin the test, allow the engine to warm up completely. Block the wheels and set the parking brake. With the transmission in Park, start the engine and allow it to run at normal idle speed.
Warning: *Keep your hands and the vacuum gauge clear of the fans.*

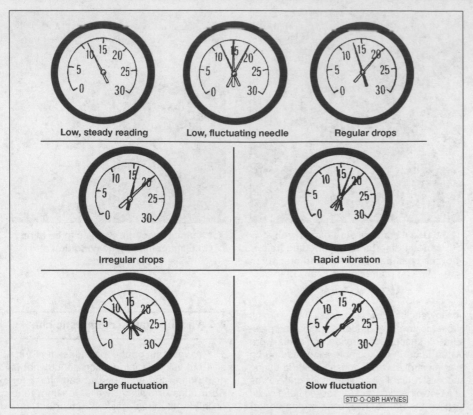

Low, steady reading   Low, fluctuating needle   Regular drops

Irregular drops   Rapid vibration

Large fluctuation   Slow fluctuation

STD-O-OBR HAYNES

4.6 Typical vacuum gauge readings

6   Read the vacuum gauge; an average, healthy engine should normally produce about 17 to 22 in-Hg with a fairly steady needle **(see illustration)**. Refer to the following vacuum gauge readings and what they indicate about the engine's condition:

7   A low, steady reading usually indicates a leaking gasket between the intake manifold and cylinder head(s) or throttle body, a leaky vacuum hose, late ignition timing or incorrect camshaft timing. Check ignition timing with a timing light and eliminate all other possible causes, utilizing the tests provided in this Chapter before you remove the timing chain cover to check the timing marks.

8   If the reading is three to eight inches below normal and it fluctuates at that low reading, suspect an intake manifold gasket leak at an intake port or a faulty fuel injector.

9   If the needle has regular drops of about two-to-four inches at a steady rate, the valves are probably leaking. Perform a compression check or leak-down test to confirm this.

10   An irregular drop or down-flick of the needle can be caused by a sticking valve or an ignition misfire. Perform a compression check or leak-down test and read the spark plugs.

11   A rapid vibration of about four in-Hg vibration at idle combined with exhaust smoke indicates worn valve guides. Perform a leak-down test to confirm this. If the rapid vibration occurs with an increase in engine speed, check for a leaking intake manifold gasket or head gasket, weak valve springs, burned

valves or ignition misfire.

12   A slight fluctuation, say one inch up and down, may mean ignition problems. Check all the usual tune-up items and, if necessary, run the engine on an ignition analyzer.

13   If there is a large fluctuation, perform a compression or leak-down test to look for a weak or dead cylinder or a blown head gasket.

14   If the needle moves slowly through a wide range, check for a clogged PCV system, incorrect idle fuel mixture, throttle body or intake manifold gasket leaks.

15   Check for a slow return after revving the engine by quickly snapping the throttle open until the engine reaches about 2,500 rpm and let it shut. Normally the reading should drop to near zero, rise above normal idle reading (about 5 in-Hg over) and then return to the previous idle reading. If the vacuum returns slowly and doesn't peak when the throttle is snapped shut, the rings may be worn. If there is a long delay, look for a restricted exhaust system (often the muffler or catalytic converter). An easy way to check this is to temporarily disconnect the exhaust ahead of the suspected part and redo the test.

5   Engine rebuilding alternatives

1   The do-it-yourselfer is faced with a number of options when purchasing a rebuilt engine. The major considerations are cost,

warranty, parts availability and the time required for the rebuilder to complete the project. The decision to replace the engine block, piston/connecting rod assemblies and crankshaft depends on the final inspection results of your engine. Only then can you make a cost effective decision whether to have your engine overhauled or simply purchase an exchange engine for your vehicle.

2   Some of the rebuilding alternatives include:

3   **Individual parts** - If the inspection procedures reveal that the engine block and most engine components are in reusable condition, purchasing individual parts and having a rebuilder rebuild your engine may be the most economical alternative. The block, crankshaft and piston/connecting rod assemblies should all be inspected carefully by a machine shop first.

4   **Short block** - A short block consists of an engine block with a crankshaft and piston/connecting rod assemblies already installed. All new bearings are incorporated and all clearances will be correct. The existing camshafts, valve train components, cylinder head and external parts can be bolted to the short block with little or no machine shop work necessary.

5   **Long block** - A long block consists of a short block plus an oil pump, oil pan, cylinder head, valve cover, camshaft and valve train components, timing sprockets and chain or gears and timing cover. All components are installed with new bearings, seals and gaskets incorporated throughout. The installation of manifolds and external parts is all that's necessary.

6   **Low mileage used engines** - Some companies now offer low mileage used engines which is a very cost effective way to get your vehicle up and running again. These engines often come from vehicles which have been in totaled in accidents or come from other countries which have a higher vehicle turn over rate. A low mileage used engine also usually has a similar warranty like the newly remanufactured engines.

7   Give careful thought to which alternative is best for you and discuss the situation with local automotive machine shops, auto parts dealers and experienced rebuilders before ordering or purchasing replacement parts.

6   Engine removal - methods and precautions

1   If you've decided that an engine must be removed for overhaul or major repair work, several preliminary steps should be taken. Read all removal and installation procedures carefully prior to committing to this job. These engines are removed by lowering the engine to the floor, along with the transmission, and then raising the vehicle sufficiently to slide the assembly out; this will require a vehicle hoist as well as an engine hoist.

2   Locating a suitable place to work is

6.3a After tightly wrapping water-vulnerable components, use a spray cleaner on everything, with particular concentration on the greasiest areas, usually around the valve cover and lower edges of the block. If one section dries out, apply more cleaner

6.3b Depending on how dirty the engine is, let the cleaner soak in according to the directions and then hose off the grime and cleaner. Get the rinse water down into every area you can get at; then dry important components with a hair dryer or paper towels

extremely important. Adequate work space, along with storage space for the vehicle, will be needed. If a shop or garage isn't available, at the very least a flat, level, clean work surface made of concrete or asphalt is required.

3   Cleaning the engine compartment and engine before beginning the removal procedure will help keep tools clean and organized **(see illustrations)**.

4   An engine hoist will also be necessary. Make sure the hoist is rated in excess of the combined weight of the engine and transmission. Safety is of primary importance, considering the potential hazards involved in removing the engine from the vehicle.

5   If you're a novice at engine removal, get at least one helper. One person cannot easily do all the things you need to do to remove a big heavy engine and transmission assembly from the engine compartment. Also helpful is to seek advice and assistance from someone who's experienced in engine removal.

6   Plan the operation ahead of time. Arrange for or obtain all of the tools and equipment you'll need prior to beginning the job **(see illustration)**. Some of the equipment necessary to perform engine removal and installation safely and with relative ease are (in addition to a vehicle hoist and an engine hoist) a heavy duty floor jack (preferably fitted with a transmission jack head adapter), complete sets of wrenches and sockets as described in the front of this manual, wooden blocks, plenty of rags and cleaning solvent for mopping up spilled oil, coolant and gasoline.

7   Plan for the vehicle to be out of use for quite a while. A machine shop can do the work that is beyond the scope of the home mechanic. Machine shops often have a busy schedule, so before removing the engine, consult the shop for an estimate of how long it will take to rebuild or repair the components that may need work.

7   Engine - removal and installation

Warning: *Gasoline is extremely flammable, so take extra precautions when you work on any part of the fuel system. Don't smoke or allow open flames or bare light bulbs near the work area, and don't work in a garage where a gas-type appliance (such as a water heater or clothes dryer) is present. Since gasoline is carcinogenic, wear fuel-resistant gloves when there's a possibility of being exposed to fuel, and, if you spill any fuel on your skin, rinse it off immediately with soap and water. Mop up any spills immediately and do not store fuel-soaked rags where they could ignite. The fuel system is under constant pressure, so, if any fuel lines are to be disconnected, the fuel pressure in the system must be relieved first (see Chapter 4 for more information). When you perform any kind of work on the fuel system, wear safety glasses and have a Class B type fire extinguisher on hand.*

Warning: *The air conditioning system is under high pressure. Do not loosen any hose fittings or remove any components until after the system has been discharged. Air conditioning refrigerant must be properly discharged into an EPA-approved recovery/recycling unit at a dealer service department or an automotive air conditioning repair facility. Always wear eye protection when disconnecting air conditioning system fittings.*

Warning: *The engine must be completely cool before beginning this procedure.*

Removal

1   Have the air conditioning system discharged by an automotive air conditioning technician.

2   Relieve the fuel system pressure (see Chapter 4).

3   Disconnect the cable from the negative

6.6 Get an engine stand sturdy enough to firmly support the engine while you're working on it. Stay away from three-wheeled models; they have a tendency to tip over more easily, so get a four-wheeled unit

terminal of the battery (see Chapter 5, Section 1).

4   Remove the hood (see Chapter 11).

5   Remove the cowl cover (see Chapter 11) and the strut support brace.

6   Remove the air filter housing and the intake air duct (see Chapter 4).

7   Remove the drivebelt (see Chapter 1).

8   Remove the alternator (see Chapter 5).

9   Remove the power steering pump, without disconnecting the hoses, and tie it out of the way (see Chapter 10).

10   Remove the PCV hose (see Chapter 6).

11   Remove the intake manifold (V8 models, see Chapter 2D) or the upper intake manifold (V6 models see Chapter 2A, Chapter 2B or Chapter 2C).

12   Remove the fuel rail (see Chapter 4).

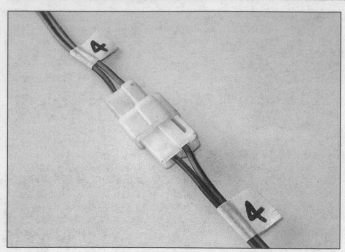

7.13 Label both ends of each wire or vacuum connection before disconnecting them

7.17 Front engine splash shield (1) and middle engine splash shield (2) fasteners

13   Label and disconnect all wires and ground straps from the engine **(see illustration)**. Masking tape and/or a touch-up paint applicator work well for marking items.
Note: *Take photos or sketch the locations of components and brackets to help with reassembly.*
14   Label and remove all vacuum lines between the engine and the firewall (or other components in the engine compartment).
15   Remove the engine oil dipstick tube.
16   Raise the vehicle and support it securely on jackstands.
17   Remove the engine splash shields **(see illustration)**.
18   Drain the cooling system (see Chapter 1).
19   On 3.5L V6 engines, disconnect the hoses from the engine oil cooler.
20   Detach the transmission cooler lines from the engine brackets and the transmission (see Chapter 7B).
21   Disconnect the engine block heater,
if equipped.
22   Disconnect the oxygen sensors and the crankshaft position sensor connector (see Chapter 6).
23   If you're working on a V6 model, disconnect the intermediate shaft coupler from the steering gear (see Chapter 10).
24   Remove the starter (see Chapter 5).
25   Drain the engine oil (see Chapter 1).
26   Disconnect the oxygen sensor connectors (see Chapter 6) and separate the exhaust pipes from the exhaust manifolds (see Chapter 4).
27   On 2.7L V6 engines, remove the oil pan-to-transmission brace (see Chapter 2A).
28   Remove the torque converter-to-driveplate bolts (see Chapter 7B).
29   On 2.7L V6 engines, remove the power steering gear mounting bolts, lower the assembly without disconnecting the power steering fluid lines, then secure the assembly using rope (see Chapter 10).
30   Remove the engine mount-to-subframe
fasteners (see Chapter 2A, Chapter 2B, Chapter 2C or Chapter 2D).
31   Remove the transmission-to-engine mounting bolts that are accessible from below.
32   Lower the vehicle.
33   Remove the engine cooling fans (see Chapter 3).
34   Remove the radiator hoses and heater hoses (see Chapter 3) and on 3.6L models, remove the heater hose tube mounting bolts and tubes.
35   Support the engine from above with a hoist. Attach the chain hoist to the engine lifting brackets. If no brackets are present, you'll have to fasten the chains to some substantial part of the engine - one that is strong enough to take the weight, but in a location that will provide good balance. If you're attaching the chain to a stud on the engine, or are using a bolt passing through the chain and into a threaded hole, place a washer between the nut or bolt head and the chain, and tighten the nut or bolt securely.
36   Support the transmission with a floor jack. Place a block of wood on the jack head to protect the transmission.
37   Remove the transmission-to-engine mounting bolts that are accessible from above.
38   Remove the ground strap from the upper right side of the transmission, if equipped **(see illustration)**.
39   Remove the harness bracket at the top of the transmission bellhousing **(see illustration)**.
40   Check to make sure everything is disconnected, then lift the engine out of the vehicle. The engine will probably need to be tilted and/or maneuvered as it's lifted out, so have an assistant handy.
Warning: *Do not place any part of your body under the engine when it is supported only by a hoist or other lifting device.*
Caution: *Be careful not to let the engine contact the radiator.*

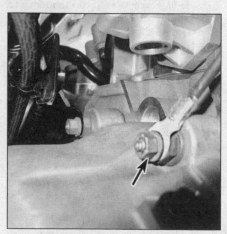

7.38 Remove the mounting nut and ground strap, then remove the transmission-to-engine stud

7.39 Remove the two mounting nuts and bracket, separate the wiring harness from the assembly and remove the two transmission-to-engine studs

41   Remove the clutch assembly if equipped (see Chapter 8) and flywheel or driveplate and mount the engine on an engine stand or set the engine on the floor and support it so it doesn't tip over. Then disconnect the engine hoist.

Installation

42   Installation is the reverse of the removal procedure, noting the following points:

a) *Check the engine mounts. If they're worn or damaged, replace them (see Chapter 2A, Chapter 2B, Chapter 2C or Chapter 2D).*

b) *Inspect the clutch disc, pressure plate and release bearing (see Chapter 8).*

c) *Inspect the front transmission fluid seal and bearing (see Chapter 7B).*

d) *Tighten the transmission-to-engine bolts and the torque converter-to-driveplate bolts to the torque listed in the Chapter 7A or Chapter 7B Specifications.* **Caution:** *Don't use the transmission-to-engine bolts to force the transmission and engine together!*

e) *Add coolant, oil, power steering and transmission fluid as needed (see Chapter 1).*

f) *Run the engine and check for proper operation and leaks. Shut off the engine and recheck the fluid levels.*

8   Engine overhaul - disassembly sequence

1   It's much easier to remove the external components if it's mounted on a portable engine stand. A stand can often be rented quite cheaply from an equipment rental yard. Before the engine is mounted on a stand, the driveplate should be removed from the engine.

2   If a stand isn't available, it's possible to remove the external engine components with it blocked up on the floor. Be extra careful not to tip or drop the engine when working without a stand.

3   If you're going to obtain a rebuilt engine, all external components must come off first, to be transferred to the replacement engine. These components include:

> Driveplate
> Ignition coils
> Emissions-related components
> Engine mounts and mount brackets
> Engine rear cover (spacer plate between driveplate and engine block)
> Intake/exhaust manifolds
> Valve covers
> Fuel injection components
> Oil filter
> Spark plugs
> Thermostat and housing assembly
> Water pump

Note: *When removing the external components from the engine, pay close attention to details that may be helpful or important during*

installation. Note the installed position of gaskets, seals, spacers, pins, brackets, washers, bolts and other small items.

4   If you're going to obtain a short block (assembled engine block, crankshaft, pistons and connecting rods), then remove the timing chain, cylinder head, oil pan, oil pump pick-up tube, oil pump and water pump from your engine so that you can turn in your old short block to the rebuilder as a core. See *Engine rebuilding alternatives* for additional information regarding the different possibilities to be considered.

9   Pistons and connecting rods - removal and installation

Removal

Note: *Prior to removing the piston/connecting rod assemblies, remove the cylinder head and oil pan (see Chapter 2A, Chapter 2B, Chapter 2C or Chapter 2D).*

1   Use your fingernail to feel if a ridge has formed at the upper limit of ring travel (about 1/4-inch down from the top of each cylinder). If carbon deposits or cylinder wear have produced ridges, they must be completely removed with a special tool **(see illustration)**. Follow the manufacturer's instructions provided with the tool. Failure to remove the ridges before attempting to remove the piston/connecting rod assemblies may result in piston breakage.

2   After the cylinder ridges have been removed, turn the engine so the crankshaft is facing up.

3   Before the main bearing cap assembly and connecting rods are removed, check the connecting rod endplay with feeler gauges. Slide them between the first connecting rod and the crankshaft throw until the play is removed **(see illustration)**. Repeat this procedure for each connecting rod. The endplay is equal to the thickness of the feeler

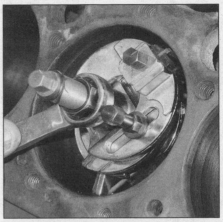

9.1 If there's a ridge at the top of the cylinder(s), use a ridge reamer to remove the raised material (ridge) from the top of the cylinders before you try to remove the piston(s)

gauge(s). Check with an automotive machine shop for the endplay service limit (a typical endplay should measure between 0.005 to 0.015 inch [0.127 to 0.396 mm]). If the play exceeds the service limit, new connecting rods will be required. If new rods (or a new crankshaft) are installed, the endplay may fall under the minimum allowable. If it does, the rods will have to be machined to restore it. If necessary, consult an automotive machine shop for advice.

4   Check the connecting rods and caps for identification marks. If they aren't plainly marked, use paint or marker **(see illustration)** to clearly identify each rod and cap (1, 2, 3, etc., depending on the cylinder they're associated with). Do not interchange the rod caps. Install the exact same rod cap onto the same connecting rod.

Caution: *Do not use a punch and hammer to mark the connecting rods or they may be damaged.*

9.3 Checking the connecting rod endplay (side clearance)

9.4 If the connecting rods or caps are not marked, use permanent ink or paint to mark the caps to the rods by cylinder number (for example, this would be number 4 cylinder connecting rod)

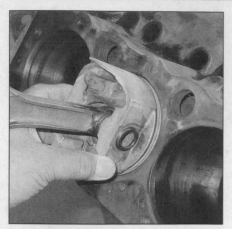

9.13 Install the piston ring into the cylinder, then push it down into position using a piston so the ring will be square in the cylinder

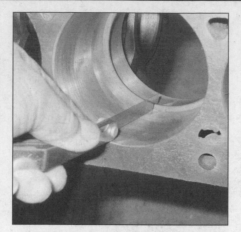

9.14 With the ring square in the cylinder, measure the ring end gap with a feeler gauge

9.15 If the ring end gap is too small, clamp a file in a vise as shown and file the piston ring ends - be sure to remove all raised material

5   Loosen each of the connecting rod cap bolts 1/4-turn at a time until they can be removed by hand.
Caution: *New connecting rod cap bolts must be used when reassembling the engine, but save the old bolts for use when checking the connecting rod bearing oil clearance.*
6   Remove the number one connecting rod cap and bearing insert. Don't drop the bearing insert out of the cap.
7   Remove the bearing insert and push the connecting rod/piston assembly out through the top of the engine. Use a wooden or plastic hammer handle to push on the upper bearing surface in the connecting rod. If resistance is felt, double-check to make sure that all of the ridge was removed from the cylinder.
8   Repeat the procedure for the remaining cylinders.
9   After removal, reassemble the connecting rod caps and bearing inserts in their respective connecting rods and install the cap bolts finger-tight. Leaving the old bearing inserts in place until reassembly will help prevent the connecting rod bearing surfaces from being accidentally nicked or gouged.

10   The pistons and connecting rods are now ready for inspection and overhaul at an automotive machine shop.

Piston ring installation

11   Before installing the new piston rings, the ring end gaps must be checked. It's assumed that the piston ring side clearance has been checked and verified correct.
12   Lay out the piston/connecting rod assemblies and the new ring sets so the ring sets will be matched with the same piston and cylinder during the end gap measurement and engine assembly.
13   Insert the top (number one) ring into the first cylinder and square it up with the cylinder walls by pushing it in with the top of the piston **(see illustration)**. The ring should be near the bottom of the cylinder, at the lower limit of ring travel.
14   To measure the end gap, slip feeler gauges between the ends of the ring until a gauge equal to the gap width is found **(see illustration)**. The feeler gauge should slide between the ring ends with a slight amount

of drag. A typical ring gap should fall between 0.010 and 0.020 inch [0.25 to 0.50 mm] for compression rings and up to 0.030 inch [0.76 mm] for the oil ring steel rails. If the gap is larger or smaller than specified, double-check to make sure you have the correct rings before proceeding.
15   If the gap is too small, it must be enlarged or the ring ends may come in contact with each other during engine operation, which can cause serious damage to the engine. If necessary, increase the end gaps by filing the ring ends very carefully with a fine file. Mount the file in a vise equipped with soft jaws, slip the ring over the file with the ends contacting the file face and slowly move the ring to remove material from the ends. When performing this operation, file only by pushing the ring from the outside end of the file towards the vise **(see illustration)**.
16   Excess end gap isn't critical unless it's greater than 0.040 inch (1.01 mm). Again, double-check to make sure you have the correct ring type.
17   Repeat the procedure for each ring that will be installed in the first cylinder and for each ring in the remaining cylinders. Remember to keep rings, pistons and cylinders matched up.
18   Once the ring end gaps have been checked/corrected, the rings can be installed on the pistons.
19   The oil control ring (lowest one on the piston) is usually installed first. It's composed of three separate components. Slip the spacer/expander into the groove **(see illustration)**. If an anti-rotation tang is used, make sure it's inserted into the drilled hole in the ring groove. Next, install the lower side rail in the same manner **(see illustration)**. Don't use a piston ring installation tool on the oil ring side rails, as they may be damaged. Instead, place one end of the side rail into the groove between the spacer/expander and the ring land, hold it firmly in place and slide a finger around the piston while pushing the rail into the groove. Finally, install the upper side rail.

9.19a Installing the spacer/expander in the oil ring groove

9.19b DO NOT use a piston ring installation tool when installing the oil ring side rails

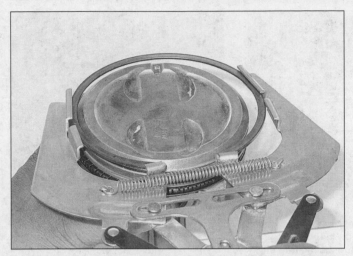

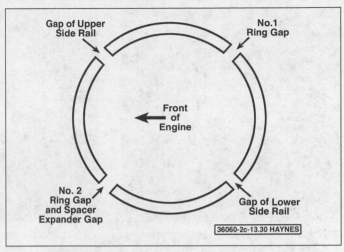

9.22 Use a piston ring installation tool to install the compression rings - on some engines the number two compression ring has a directional mark that must face toward the top of the piston

9.30 Position the piston ring end gaps as shown

20   After the three oil ring components have been installed, check to make sure that both the upper and lower side rails can be rotated smoothly inside the ring grooves.

21   The number two (middle) ring is installed next. It's usually stamped with a mark which must face up, toward the top of the piston. Do not mix up the top and middle rings, as they have different cross-sections.

Note: *Always follow the instructions printed on the ring package or box - different manufacturers may require different approaches.*

22   Use a piston ring installation tool and make sure the identification mark is facing the top of the piston, then slip the ring into the middle groove on the piston **(see illustration)**. Don't expand the ring any more than necessary to slide it over the piston.

Note: *Be careful not to confuse the number one and number two rings.*

23   Install the number one (top) ring in the same manner.

24   Repeat the procedure for the remaining pistons and rings.

Installation

25   Before installing the piston/connecting rod assemblies, the cylinder walls must be perfectly clean, the top edge of each cylinder bore must be chamfered, and the crankshaft must be in place.

26   Remove the cap from the end of the number one connecting rod (refer to the marks made during removal). Remove the original bearing inserts and wipe the bearing surfaces of the connecting rod and cap with a clean, lint-free cloth. They must be kept spotlessly clean.

Connecting rod bearing oil clearance check

27   Clean the back side of the new upper bearing insert, then lay it in place in the connecting rod.

28   Make sure the tab on the bearing fits into the recess in the rod. Don't hammer the

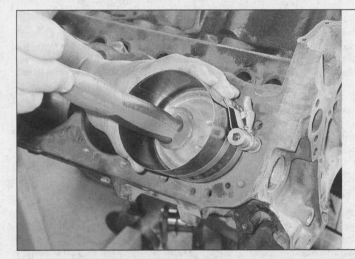

9.35 Use a plastic or wooden hammer handle to push the piston into the cylinder

bearing insert into place and be very careful not to nick or gouge the bearing face. Don't lubricate the bearing at this time.

29   Clean the back side of the other bearing insert and install it in the rod cap. Again, make sure the tab on the bearing fits into the recess in the cap, and don't apply any lubricant. It's critically important that the mating surfaces of the bearing and connecting rod are perfectly clean and oil free when they're assembled.

30   Position the piston ring gaps at the intervals around the piston as shown **(see illustration)**.

31   Lubricate the piston and rings with clean engine oil and attach a piston ring compressor to the piston. Leave the skirt protruding about 1/4-inch to guide the piston into the cylinder. The rings must be compressed until they're flush with the piston.

32   Rotate the crankshaft until the number one connecting rod journal is at BDC (bottom dead center) and apply a liberal coat of engine oil to the cylinder walls.

33   With the "front" mark (letter F or arrow) on the piston facing the front (timing chain/timing belt end) of the engine, gently insert

the piston/connecting rod assembly into the number one cylinder bore and rest the bottom edge of the ring compressor on the engine block.

Note: *Some engines have a letter "F" marking on the side of the piston near the wrist pin, others have an arrow, an "F" or a dimple or groove on the top of the piston. All of these are marks that indicate the front of the piston.*

34   Tap the top edge of the ring compressor to make sure it's contacting the block around its entire circumference.

35   Gently tap on the top of the piston with the end of a wooden or plastic hammer handle **(see illustration)** while guiding the end of the connecting rod into place on the crankshaft journal. The piston rings may try to pop out of the ring compressor just before entering the cylinder bore, so keep some downward pressure on the ring compressor. Work slowly, and if any resistance is felt as the piston enters the cylinder, stop immediately. Find out what's hanging up and fix it before proceeding. Do not, for any reason, force the piston into the cylinder - you might break a ring and/or the piston.

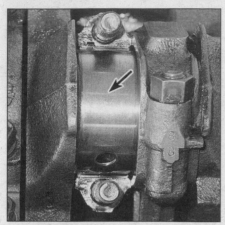

9.37 Place Plastigage on each connecting rod bearing journal parallel to the crankshaft centerline

9.41 Use the scale on the Plastigage package to determine the bearing oil clearance - be sure to measure the widest part of the Plastigage and use the correct scale; it comes with both standard and metric scales

10.1 Checking crankshaft endplay with a dial indicator

36   Once the piston/connecting rod assembly is installed, the connecting rod bearing oil clearance must be checked before the rod cap is permanently installed.

37   Cut a piece of the appropriate size Plastigage slightly shorter than the width of the connecting rod bearing and lay it in place on the number one connecting rod journal, parallel with the journal axis **(see illustration)**.

38   Clean the connecting rod cap bearing face and install the rod cap. Make sure the mating mark on the cap is on the same side as the mark on the connecting rod **(see illustration 9.4)**.

39   Install the old rod bolts at this time, and tighten them to the torque listed in this Chapter's Specifications. **Note:** *Use a thin-wall socket to avoid erroneous torque readings that can result if the socket is wedged between the rod cap and the bolt or nut. If the socket tends to wedge itself between the fastener and the cap, lift up on it slightly until it no longer contacts the cap. DO NOT rotate the crankshaft at any time during this operation.*

40   Remove the fasteners and detach the rod cap, being very careful not to disturb the Plastigage. Discard the cap bolts at this time as they cannot be reused. **Note:** *You MUST use new connecting rod bolts.*

41   Compare the width of the crushed Plastigage to the scale printed on the Plastigage envelope to obtain the oil clearance **(see illustration)**. The connecting rod bearing oil clearance is usually about 0.001 to 0.002 inch. Consult an automotive machine shop for the clearance specified for the rod bearings on your engine.

42   If the clearance is not as specified, the bearing inserts may be the wrong size (which means different ones will be required). Before deciding that different inserts are needed, make sure that no dirt or oil was between the bearing inserts and the connecting rod or cap when the clearance was measured. Also,

recheck the journal diameter. If the Plastigage was wider at one end than the other, the journal may be tapered. If the clearance still exceeds the limit specified, the bearing will have to be replaced with an undersize bearing. **Caution:** *When installing a new crankshaft always use a standard size bearing.*

Final installation

43   Carefully scrape all traces of the Plastigage material off the rod journal and/or bearing face. Be very careful not to scratch the bearing - use your fingernail or the edge of a plastic card.

44   Make sure the bearing faces are perfectly clean, then apply a uniform layer of clean moly-base grease or engine assembly lube to both of them. You'll have to push the piston into the cylinder to expose the face of the bearing insert in the connecting rod.

45   Slide the connecting rod back into place on the journal, install the rod cap, install the bolts and tighten them to the torque listed in this Chapter's Specifications. **Caution:** *Install new connecting rod cap bolts. Do NOT reuse old bolts - they have stretched and cannot be reused.*

46   Repeat the entire procedure for the remaining pistons/connecting rods.

47   The important points to remember are:

a) *Keep the back sides of the bearing inserts and the insides of the connecting rods and caps perfectly clean when assembling them.*

b) *Make sure you have the correct piston/rod assembly for each cylinder.*

c) *The mark on the piston must face the front (timing chain end) of the engine.*

d) *Lubricate the cylinder walls liberally with clean oil.*

e) *Lubricate the bearing faces when installing the rod caps after the oil clearance has been checked.*

48   After all the piston/connecting rod assemblies have been correctly installed, rotate the crankshaft a number of times by hand to check for any obvious binding.

49   As a final step, check the connecting rod endplay again. If it was correct before disassembly and the original crankshaft and rods were reinstalled, it should still be correct. If new rods or a new crankshaft were installed, the endplay may be inadequate. If so, the rods will have to be removed and taken to an automotive machine shop for resizing.

10   Crankshaft - removal and installation

Removal

Note: *The crankshaft can be removed only after the engine has been removed from the vehicle. It's assumed that the driveplate, crankshaft pulley, timing chain/belt, oil pan, oil pump, oil filter and piston/connecting rod assemblies have already been removed. The rear main oil seal retainer must be unbolted and separated from the block before proceeding with crankshaft removal.*

1   Before the crankshaft is removed, measure the endplay. Mount a dial indicator with the indicator in line with the crankshaft and touching the end of the crankshaft **(see illustration)**.

2   Pry the crankshaft all the way to the rear and zero the dial indicator. Next, pry the crankshaft to the front as far as possible and check the reading on the dial indicator. The distance traveled is the endplay. A typical crankshaft endplay will fall between 0.003 to 0.010 inch (0.076 to 0.254 mm). If it is greater than that, check the crankshaft thrust washer/bearing assembly surfaces for wear after it's removed. If no wear is evident, new main bearings should correct the endplay. Refer to

10.3 Checking crankshaft endplay with feeler gauges at the thrust bearing journal

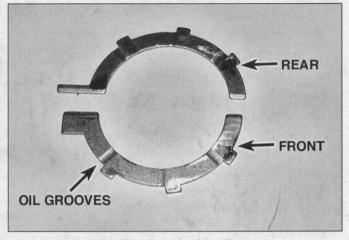

10.14a Thrust washer identification

Step 14 for the location of the thrust washer/ bearing assembly on each engine.

3   If a dial indicator isn't available, feeler gauges can be used. Gently pry the crankshaft all the way to the front of the engine. Slip feeler gauges between the crankshaft and the front face of the thrust bearing or washer to determine the clearance (see illustration).

4   Loosen the main bearing cap bolts 1/4-turn at a time each, until they can be removed by hand.

Note: The main bearing caps on the V8 engine are each secured by four bolts, two of which are accessed from the sides of the engine block.

5   Gently tap the main bearing caps with a soft-faced hammer around the perimeter of the assembly. Pull the main bearing cap assembly straight up and off the cylinder block. Try not to drop the bearing inserts if they come out with the assembly.

6   Carefully lift the crankshaft out of the engine. It may be a good idea to have an assistant available, since the crankshaft is quite heavy and awkward to handle. With the bearing inserts in place inside the engine block and main bearing caps, reinstall the main bearing cap assembly onto the engine block and tighten the bolts finger-tight. Make sure you install the main bearing cap assembly with the arrow facing the front of the engine.

Installation

7   Crankshaft installation is the first step in engine reassembly. It's assumed at this point that the engine block and crankshaft have been cleaned, inspected and repaired or reconditioned.

8   Position the engine block with the bottom facing up.

9   Remove the mounting bolts and lift off the main bearing cap assembly.

10   If they're still in place, remove the original bearing inserts from the block and from the main bearing cap assembly. Wipe the bearing surfaces of the block and main bear-

10.14b Insert the thrust washer into the machined surface between the crankshaft and the upper bearing saddle, then rotate it down into the block until it's flush with the parting line on the main bearing saddle - make sure the oil grooves on the thrust washer face the crankshaft

ing cap assembly with a clean, lint-free cloth. They must be kept spotlessly clean. This is critical for determining the correct bearing oil clearance.

Main bearing oil clearance check

11   Without mixing them up, clean the back sides of the new upper main bearing inserts (with grooves and oil holes) and lay one in each main bearing saddle in the block. Each upper bearing has an oil groove and oil hole in it. Clean the back sides of the lower main bearing inserts and lay them in the corresponding location in the main bearing cap. Make sure the tab on the bearing insert fits into the recess in the block or main bearing cap. The upper bearings with the oil holes are installed into the engine block while the lower bearings without the oil holes are installed in the caps.

Caution: The oil holes in the block must line up with the oil holes in the upper bearing inserts.

Caution: Do not hammer the bearing insert into place and don't nick or gouge the bearing faces. DO NOT apply any lubrication at this time.

12   Clean the faces of the bearing inserts in the block and the crankshaft main bearing journals with a clean, lint-free cloth.

13   Check or clean the oil holes in the crankshaft, as any dirt here can go only one way - straight through the new bearings.

14   Once you're certain the crankshaft is clean, carefully lay it in position in the cylinder block. Locate the thrust washers (see illustrations).

a) On 2.7L V6 engines, the thrust washers are located on the number 3 main journal.

b) On 3.5L and 3.6L V6 engines, the thrust washers are located on the number 2 main journal.

c) On V8 engines, the thrust washers are located on the number 3 main journal.

15   The thrust washers must be installed in the correct journal (counting from the front of the engine).

Note: Install the thrust washers with the groove in the thrust washer facing the crankshaft with the smooth sides facing the main bearing saddle.

16   Before the crankshaft can be permanently installed, the main bearing oil clearance must be checked.

ENGINE BEARING ANALYSIS

Debris

Babbitt bearing embedded with debris from machinings

Microscopic detail of debris

Microscopic detail of gouges

Overplated copper alloy bearing gouged by cast iron debris

Aluminum bearing embedded with glass beads

Microscopic detail of glass beads

Damaged lining caused by dirt left on the bearing back

Misassembly

Result of a lower half assembled as an upper - blocking the oil flow

Excessive oil clearance is indicated by a short contact arc

Polished and oil-stained backs are a result of a poor fit in the housing bore

Result of a wrong, reversed, or shifted cap

Overloading

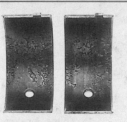

Damage from excessive idling which resulted in an oil film unable to support the load imposed

Damaged upper connecting rod bearings caused by engine lugging; the lower main bearings (not shown) were similarly affected

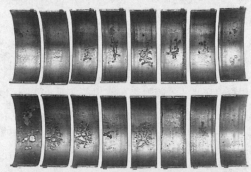

The damage shown in these upper and lower connecting rod bearings was caused by engine operation at a higher-than-rated speed under load

Misalignment

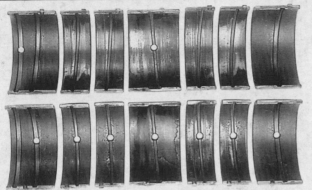

A warped crankshaft caused this pattern of severe wear in the center, diminishing toward the ends

A poorly finished crankshaft caused the equally spaced scoring shown

A tapered housing bore caused the damage along one edge of this pair

A bent connecting rod led to the damage in the "V" pattern

Lubrication

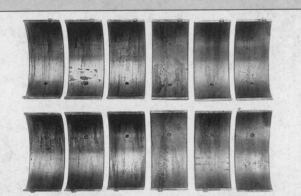

Result of dry start: The bearings on the left, farthest from the oil pump, show more damage

Result of a low oil supply or oil starvation

Severe wear as a result of inadequate oil clearance

Corrosion

Microscopic detail of corrosion

Corrosion is an acid attack on the bearing lining generally caused by inadequate maintenance, extremely hot or cold operation, or inferior oils or fuels

Microscopic detail of cavitation

Example of cavitation - a surface erosion caused by pressure changes in the oil film

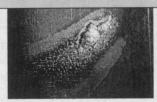

Damage from excessive thrust or insufficient axial clearance

Bearing affected by oil dilution caused by excessive blow-by or a rich mixture

10.18 Place the Plastigage onto the crankshaft bearing journal as shown

10.20a Main bearing cap bolt tightening sequence on the V6 engines

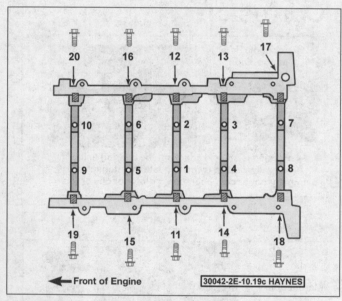

10.20b Main bearing cap and crossbolt tightening sequence on the V8 engines

10.22 Use the scale on the Plastigage package to determine the bearing oil clearance - be sure to measure the widest part of the Plastigage and use the correct scale; it comes with both standard and metric scales

17   Cut several strips of the appropriate size of Plastigage. They must be slightly shorter than the width of the main bearing journal.

18   Place one piece on each crankshaft main bearing journal, parallel with the journal axis as shown **(see illustration)**.

19   Clean the faces of the bearing inserts in the main bearing caps. Hold the bearing inserts in place and install the inserts onto the crankshaft and cylinder block. DO NOT disturb the Plastigage. Make sure you install the main bearing cap with the arrow facing the front (timing chain/timing belt end) of the engine.

20   Apply clean engine oil to all bolt threads prior to installation, then install all bolts finger-tight. Tighten main bearing cap bolts in the sequence shown **(see illustrations)** pro-gressing in steps, to the torque listed in this

Chapter's Specifications. DO NOT rotate the crankshaft at any time during this operation.

21   Remove the bolts in the reverse order of the tightening sequence and carefully lift the main bearing cap assembly straight up and off the block. Do not disturb the Plastigage or rotate the crankshaft. If the main bearing cap assembly is difficult to remove, tap it gently from side-to-side with a soft-faced hammer to loosen it.

22   Compare the width of the crushed Plasti-gage on each journal to the scale printed on the Plastigage envelope to determine the main bearing oil clearance **(see illustration)**. A typical main bearing oil clearance should fall between 0.0015 and 0.0023-inch. Check with an automotive machine shop for the clear-ance for your engine.

23   If the clearance is not as specified, the

bearing inserts may be the wrong size (which means different ones will be required). Before deciding if different inserts are needed, make sure that no dirt or oil was between the bearing inserts and the cap assembly or block when the clearance was measured. If the Plasti-gage was wider at one end than the other, the crankshaft journal may be tapered. If the clearance still exceeds the limit specified, the bearing insert(s) will have to be replaced with an undersize bearing insert(s).

Caution: *When installing a new crankshaft al-ways install a standard bearing insert set.*

24   Carefully scrape all traces of the Plasti-gage material off the main bearing journals and/or the bearing insert faces. Be sure to remove all residue from the oil holes. Use your fingernail or the edge of a plastic card - don't nick or scratch the bearing faces.

Final installation

25 Carefully lift the crankshaft out of the cylinder block.

26 Clean the bearing insert faces in the cylinder block, then apply a thin, uniform layer of moly-base grease or engine assembly lube to each of the bearing surfaces. Be sure to coat the thrust faces as well as the journal face of the thrust washers.

Note: *Install the thrust washers after the crankshaft has been installed.*

27 Make sure the crankshaft journals are clean, then lay the crankshaft back in place in the cylinder block.

28 Clean the bearing insert faces and then apply the same lubricant to them. Clean the engine block thoroughly. The surfaces must be free of oil residue. Install the thrust washers.

29 Install each main bearing cap onto the crankshaft and cylinder block. On the V8 engines, make sure the arrow on each main bearing cap faces the front of the engine.

30 Prior to installation, apply clean engine oil to all bolt threads wiping off any excess, then install all bolts finger-tight.

31 Tighten the main bearing cap bolts to the torque listed in this Chapter's Specifications (in the proper sequence) **(see illustrations 10.19a and 10.19b)**.

32 Recheck crankshaft endplay with a feeler gauge or a dial indicator. The endplay should be correct if the crankshaft thrust faces aren't worn or damaged and if new bearings have been installed.

33 Rotate the crankshaft a number of times by hand to check for any obvious binding. It should rotate with a running torque of 50 in-lbs or less. If the running torque is too high, identify and correct the problem at this time.

34 Install the new rear main oil seal (see Chapter 2A, Chapter 2B, Chapter 2C or Chapter 2D).

11 Engine overhaul - reassembly sequence

1 Before beginning engine reassembly, make sure you have all the necessary new parts, gaskets and seals as well as the following items on hand:

> *Common hand tools*
> *A 1/2-inch drive torque wrench*
> *New engine oil and filter*
> *Gasket sealant*
> *Thread locking compound*

2 If you obtained a short block it will be necessary to install the cylinder head, the oil pump and pick-up tube, the oil pan, the water pump, the timing chain and timing cover, and the valve cover (see Chapter 2A, Chapter 2B, Chapter 2C or Chapter 2D). In order to save time and avoid problems, the external components must be installed in the following general order:

> *Thermostat and housing cover*
> *Water pump*
> *Intake and exhaust manifolds*
> *Fuel injection components*
> *Emission control components*
> *Spark plugs*
> *Ignition coils*
> *Oil filter*
> *Engine mounts and mount brackets*
> *Driveplate*

12 Initial start-up and break-in after overhaul

Warning: *Have a fire extinguisher handy when starting the engine for the first time.*

1 Once the engine has been installed in the vehicle, double-check the engine oil and coolant levels.

2 With the spark plugs out of the engine and the ignition system and fuel pump disabled, crank the engine until oil pressure registers on the gauge or the light goes out.

3 Install the spark plugs and ignition coils, and reinstall the fuel pump relay.

4 Start the engine. It may take a few moments for the fuel system to build up pressure, but the engine should start without a great deal of effort.

5 After the engine starts, it should be allowed to warm up to normal operating temperature. While the engine is warming up, make a thorough check for fuel, oil and coolant leaks.

6 Shut the engine off and recheck the engine oil and coolant levels.

7 Drive the vehicle to an area with minimum traffic, accelerate from 30 to 50 mph, then allow the vehicle to slow to 30 mph with the throttle closed. Repeat the procedure 10 or 12 times. This will load the piston rings and cause them to seat properly against the cylinder walls. Check again for oil and coolant leaks.

8 Drive the vehicle gently for the first 500 miles (no sustained high speeds) and keep a constant check on the oil level. It is not unusual for an engine to use oil during the break-in period.

9 At approximately 500 to 600 miles, change the oil and filter.

10 For the next few hundred miles, drive the vehicle normally. Do not pamper it or abuse it.

11 After 2,000 miles, change the oil and filter again and consider the engine broken in.

COMMON ENGINE OVERHAUL TERMS

B

Backlash - The amount of play between two parts. Usually refers to how much one gear can be moved back and forth without moving the gear with which it's meshed.

Bearing Caps - The caps held in place by nuts or bolts which, in turn, hold the bearing surface. This space is for lubricating oil to enter.

Bearing clearance - The amount of space left between shaft and bearing surface. This space is for lubricating oil to enter.

Bearing crush - The additional height which is purposely manufactured into each bearing half to ensure complete contact of the bearing back with the housing bore when the engine is assembled.

Bearing knock - The noise created by movement of a part in a loose or worn bearing.

Blueprinting - Dismantling an engine and reassembling it to EXACT specifications.

Bore - An engine cylinder, or any cylindrical hole; also used to describe the process of enlarging or accurately refinishing a hole with a cutting tool, as to bore an engine cylinder. The bore size is the diameter of the hole.

Boring - Renewing the cylinders by cutting them out to a specified size. A boring bar is used to make the cut.

Bottom end - A term which refers collectively to the engine block, crankshaft, main bearings and the big ends of the connecting rods.

Break-in - The period of operation between installation of new or rebuilt parts and time in which parts are worn to the correct fit. Driving at reduced and varying speed for a specified mileage to permit parts to wear to the correct fit.

Bushing - A one-piece sleeve placed in a bore to serve as a bearing surface for shaft, piston pin, etc. Usually replaceable.

C

Camshaft - The shaft in the engine, on which a series of lobes are located for operating the valve mechanisms. The camshaft is driven by gears or sprockets and a timing chain. Usually referred to simply as the cam.

Carbon - Hard, or soft, black deposits found in combustion chamber, on plugs, under rings, on and under valve heads.

Cast iron - An alloy of iron and more than two percent carbon, used for engine blocks and heads because it's relatively inexpensive and easy to mold into complex shapes.

Chamfer - To bevel across (or a bevel on) the sharp edge of an object.

Chase - To repair damaged threads with a tap or die.

Combustion chamber - The space between the piston and the cylinder head, with the piston at top dead center, in which air-fuel mixture is burned.

Compression ratio - The relationship between cylinder volume (clearance volume) when the piston is at top dead center and cylinder volume when the piston is at bottom dead center.

Connecting rod - The rod that connects the crank on the crankshaft with the piston. Sometimes called a con rod.

Connecting rod cap - The part of the connecting rod assembly that attaches the rod to the crankpin.

Core plug - Soft metal plug used to plug the casting holes for the coolant passages in the block.

Crankcase - The lower part of the engine in which the crankshaft rotates; includes the lower section of the cylinder block and the oil pan.

Crank kit - A reground or reconditioned crankshaft and new main and connecting rod bearings.

Crankpin - The part of a crankshaft to which a connecting rod is attached.

Crankshaft - The main rotating member, or shaft, running the length of the crankcase, with offset throws to which the connecting rods are attached; changes the reciprocating motion of the pistons into rotating motion.

Cylinder sleeve - A replaceable sleeve, or liner, pressed into the cylinder block to form the cylinder bore.

D

Deburring - Removing the burrs (rough edges or areas) from a bearing.

Deglazer - A tool, rotated by an electric motor, used to remove glaze from cylinder walls so a new set of rings will seat.

E

Endplay - The amount of lengthwise movement between two parts. As applied to a crankshaft, the distance that the crankshaft can move forward and back in the cylinder block.

F

Face - A machinist's term that refers to removing metal from the end of a shaft or the face of a larger part, such as a flywheel.

Fatigue - A breakdown of material through a large number of loading and unloading cycles. The first signs are cracks followed shortly by breaks.

Feeler gauge - A thin strip of hardened steel, ground to an exact thickness, used to check clearances between parts.

Free height - The unloaded length or height of a spring.

Freeplay - The looseness in a linkage, or an assembly of parts, between the initial application of force and actual movement. Usually perceived as slop or slight delay.

Freeze plug - See Core plug.

G

Gallery - A large passage in the block that forms a reservoir for engine oil pressure.

Glaze - The very smooth, glassy finish that develops on cylinder walls while an engine is in service.

H

Heli-Coil - A rethreading device used when threads are worn or damaged. The device is installed in a retapped hole to reduce the thread size to the original size.

I

Installed height - The spring's measured length or height, as installed on the cylinder head. Installed height is measured from the spring seat to the underside of the spring retainer.

J

Journal - The surface of a rotating shaft which turns in a bearing.

K

Keeper - The split lock that holds the valve spring retainer in position on the valve stem.

Key - A small piece of metal inserted into matching grooves machined into two parts fitted together - such as a gear pressed onto a shaft - which prevents slippage between the two parts.

Knock - The heavy metallic engine sound, produced in the combustion chamber as a result of abnormal combustion - usually detonation. Knock is usually caused by a loose or worn bearing. Also referred to as detonation, pinging and spark knock. Connecting rod or main bearing knocks are created by too much oil clearance or insufficient lubrication.

L

Lands - The portions of metal between the piston ring grooves.

Lapping the valves - Grinding a valve face and its seat together with lapping compound.

Lash - The amount of free motion in a gear train, between gears, or in a mechanical assembly, that occurs before movement can begin. Usually refers to the lash in a valve train.

Lifter - The part that rides against the cam to transfer motion to the rest of the valve train.

M

Machining - The process of using a machine to remove metal from a metal part.

Main bearings - The plain, or babbit, bearings that support the crankshaft.

Main bearing caps - The cast iron caps, bolted to the bottom of the block, that support the main bearings.

O

O.D. - Outside diameter.

Oil gallery - A pipe or drilled passageway in the engine used to carry engine oil from one area to another.

Oil ring - The lower ring, or rings, of a piston; designed to prevent excessive amounts of oil from working up the cylinder walls and into the combustion chamber. Also called an oil-control ring.

Oil seal - A seal which keeps oil from leaking out of a compartment. Usually refers to a dynamic seal around a rotating shaft or other moving part.

O-ring - A type of sealing ring made of a special rubberlike material; in use, the O-ring is compressed into a groove to provide the sealing action.

Overhaul - To completely disassemble a unit, clean and inspect all parts, reassemble it with the original or new parts and make all adjustments necessary for proper operation.

P

Pilot bearing - A small bearing installed in the center of the flywheel (or the rear end of the crankshaft) to support the front end of the input shaft of the transmission.

Pip mark - A little dot or indentation which indicates the top side of a compression ring.

Piston - The cylindrical part, attached to the connecting rod, that moves up and down in the cylinder as the crankshaft rotates. When the fuel charge is fired, the piston transfers the force of the explosion to the connecting rod, then to the crankshaft.

Piston pin (or wrist pin) - The cylindrical and usually hollow steel pin that passes through the piston. The piston pin fastens the piston to the upper end of the connecting rod.

Piston ring - The split ring fitted to the groove in a piston. The ring contacts the sides of the ring groove and also rubs against the cylinder wall, thus sealing space between piston and wall. There are two types of rings: Compression rings seal the compression pressure in the combustion chamber; oil rings scrape excessive oil off the cylinder wall.

Piston ring groove - The slots or grooves cut in piston heads to hold piston rings in position.

Piston skirt - The portion of the piston below the rings and the piston pin hole.

Plastigage - A thin strip of plastic thread, available in different sizes, used for measuring clearances. For example, a strip of plastigage is laid across a bearing journal and mashed as parts are assembled. Then parts are disassembled and the width of the strip is measured to determine clearance between journal and bearing. Commonly used to measure crankshaft main-bearing and connecting rod bearing clearances.

Press-fit - A tight fit between two parts that requires pressure to force the parts together. Also referred to as drive, or force, fit.

Prussian blue - A blue pigment; in solution, useful in determining the area of contact between two surfaces. Prussian blue is commonly used to determine the width and location of the contact area between the valve face and the valve seat.

R

Race (bearing) - The inner or outer ring that provides a contact surface for balls or rollers in bearing.

Ream - To size, enlarge or smooth a hole by using a round cutting tool with fluted edges.

Ring job - The process of reconditioning the cylinders and installing new rings.

Runout - Wobble. The amount a shaft rotates out-of-true.

S

Saddle - The upper main bearing seat.

Scored - Scratched or grooved, as a cylinder wall may be scored by abrasive particles moved up and down by the piston rings.

Scuffing - A type of wear in which there's a transfer of material between parts moving against each other; shows up as pits or grooves in the mating surfaces.

Seat - The surface upon which another part rests or seats. For example, the valve seat is the matched surface upon which the valve face rests. Also used to refer to wearing into a good fit; for example, piston rings seat after a few miles of driving.

Short block - An engine block complete with crankshaft and piston and, usually, camshaft assemblies.

Static balance - The balance of an object while it's stationary.

Step - The wear on the lower portion of a ring land caused by excessive side and back-clearance. The height of the step indicates the ring's extra side clearance and the length of the step projecting from the back wall of the groove represents the ring's back clearance.

Stroke - The distance the piston moves when traveling from top dead center to bottom dead center, or from bottom dead center to top dead center.

Stud - A metal rod with threads on both ends.

T

Tang - A lip on the end of a plain bearing used to align the bearing during assembly.

Tap - To cut threads in a hole. Also refers to the fluted tool used to cut threads.

Taper - A gradual reduction in the width of a shaft or hole; in an engine cylinder, taper usually takes the form of uneven wear, more pronounced at the top than at the bottom.

Throws - The offset portions of the crankshaft to which the connecting rods are affixed.

Thrust bearing - The main bearing that has thrust faces to prevent excessive endplay, or forward and backward movement of the crankshaft.

Thrust washer - A bronze or hardened steel washer placed between two moving parts. The washer prevents longitudinal movement and provides a bearing surface for thrust surfaces of parts.

Tolerance - The amount of variation permitted from an exact size of measurement. Actual amount from smallest acceptable dimension to largest acceptable dimension.

U

Umbrella - An oil deflector placed near the valve tip to throw oil from the valve stem area.

Undercut - A machined groove below the normal surface.

Undersize bearings - Smaller diameter bearings used with re-ground crankshaft journals.

V

Valve grinding - Refacing a valve in a valve-refacing machine.

Valve train - The valve-operating mechanism of an engine; includes all components from the camshaft to the valve.

Vibration damper - A cylindrical weight attached to the front of the crankshaft to minimize torsional vibration (the twist-untwist actions of the crankshaft caused by the cylinder firing impulses). Also called a harmonic balancer.

W

Water jacket - The spaces around the cylinders, between the inner and outer shells of the cylinder block or head, through which coolant circulates.

Web - A supporting structure across a cavity.

Woodruff key - A key with a radiused backside (viewed from the side).

Notes

Chapter 3
Cooling, heating and air conditioning systems

Contents

Specifications

General

Coolant capacity..	See Chapter 1
Radiator/expansion tank cap pressure rating	16 to 20 psi
Refrigerant type	
2013 and earlier models...	R-134a
2014 and later models..	Either R-134a or R-1234yf (consult underhood HVAC sticker for application)

Torque specifications   Ft-lbs (unless otherwise indicated)

Note: *One foot-pound (ft-lb) of torque is equivalent to 12 inch-pounds (in-lbs) of torque. Torque values below approximately 15 ft-lbs are expressed in inch-pounds, because most foot-pound torque wrenches are not accurate at these smaller values.*

Engine oil cooler mounting bolt	
3.5L V6 engine	45
3.6L V6 engine (filter housing/cooler)	108 in-lbs
2010 and earlier 5.7L V8 engine	41
2011 and later 5.7L V8 engine	21
6.4L V8 engine	24
Thermostat housing bolts	
2.7L V6 engines	53 in-lbs
3.5L V6 engines	105 in-lbs
3.6L V6 engines	108 in-lbs
2010 and earlier 5.7L and 6.1L engines...............................	112 in-lbs
2011 and later 5.7L and 6.4L V8 engines.............................	21
Water pump mounting bolts	
2.7L and 3.5L V6 engines	105 in-lbs
3.6L engines	
2011 through 2013 models	
M6 bolts	106 in-lbs
M8 bolts	18
M10 bolts...	41
2014 and later models	
M6 bolts..	97 in-lbs
M8 bolts..	17
M10 bolts...	37
2010 and earlier 5.7L and 6.1L engines.................................	21
2011 and later 5.7L and 6.4L V8 engines..............................	18
Idler pulley mounting bolt	
V6 engines	18
V8 engines	21

1   General information

Engine cooling system

1   The main components of the cooling system consist of a radiator, an expansion tank, a pressure cap (located on the expansion tank), a thermostat, temperature-controlled electric cooling fans and a belt or chain-driven water pump.

2   The radiator cooling fans are mounted in a housing/shroud at the engine side of the radiator. It is designed to come on when the engine reaches a certain temperature, and shut off again when the engine cools down some, thereby keeping the engine in the desired operating-temperature range.

3   The expansion tank functions somewhat differently than a conventional recovery tank. It is pressurized and equipped with a pressure cap on top and it's designed to purge air from the coolant.

Warning: *Unlike a conventional coolant recovery tank, the pressure cap on the expansion tank should never be opened after the engine has warmed up, because of the danger of severe burns caused by steam or scalding coolant.*

4   Coolant circulates through the cylinder block and heads around the combustion chambers and valve seats, flows out of the cylinder head into the radiator and heater core.

5   When the engine is cold, the thermostat restricts the circulation of coolant to the radiator. When the minimum operating temperature is reached, the thermostat begins to open, allowing coolant to flow through the radiator.

6   All vehicles covered in this manual are equipped with a transmission oil cooler that is integrated with the air conditioning condenser mounted to the radiator. Hot transmission fluid is directed to the cooler by lines and hoses. Once cooled, the fluid is directed back to the transmission.

Heating system

7   The heating system consists of a blower fan and a heater core located in an air conditioning and heater housing under the dash with hoses connecting the heater core to the engine cooling system. Hot engine coolant is circulated through the heater core. When the heater is activated, a flap door opens in the housing and directs warm air to the passenger compartment. A fan switch on the heater/air conditioning control activates the blower motor, which creates the flow of warm air into the cab.

Air conditioning system

8   The air conditioning system consists of a condenser (mounted in front of the radiator), an evaporator (mounted adjacent to the heater core under the dash), a compressor (mounted on the engine), a receiver-drier (mounted to the condenser) and the plumbing that connects all of these components.

9   A blower fan forces the warmer air of the passenger's compartment through an evaporator core (sort of a radiator-in-reverse), transferring the heat from the air to the refrigerant. The liquid refrigerant boils off into low-pressure vapor, taking the heat with it when it leaves the evaporator.

2   Troubleshooting

Coolant leaks

1   A coolant leak can develop anywhere in the cooling system, but the most common causes are:

a) *A loose or weak hose clamp*
b) *A defective hose*
c) *A faulty pressure cap*
d) *A damaged radiator*
e) *A bad heater core*
f) *A faulty water pump*
g) *A leaking gasket at any joint that carries coolant*

2   Coolant leaks aren't always easy to find. Sometimes they can only be detected when the cooling system is under pressure. Here's where a cooling system pressure tester comes in handy. After the engine has cooled completely, the tester is attached in place of the pressure cap, then pumped up to the pressure value equal to that of the pressure cap rating **(see illustration)**. Now, leaks that only exist when the engine is fully warmed up will become apparent. The tester can be left connected to locate a nagging slow leak.

Coolant level drops, but no external leaks

3   If you find it necessary to keep adding coolant, but there are no external leaks, the probable causes include:

a) *A blown head gasket*
b) *A leaking intake manifold gasket (only on engines that have coolant passages in the manifold)*
c) *A cracked cylinder head or cylinder block*

4   Any of the above problems will also usually result in contamination of the engine oil, which will cause it to take on a milkshake-like appearance. A bad head gasket or cracked head or block can also result in engine oil contaminating the cooling system.

5   Combustion leak detectors (also known as block testers) are available at most auto parts stores. These work by detecting exhaust gases in the cooling system, which indicates a compression leak from a cylinder into the coolant. The tester consists of a large bulb-type syringe and bottle of test fluid **(see illustration)**. A measured amount of the fluid is added to the syringe. The syringe is placed over the cooling system filler neck and, with the engine running, the bulb is squeezed and a sample of the gases present in the cooling system are drawn up through the test fluid **(see illustration)**. If any combustion gases are present in the sample taken, the test fluid will change color.

2.2 The cooling system pressure tester is connected in place of the pressure cap, then pumped up to pressurize the system

2.5a The combustion leak detector consists of a bulb, syringe and test fluid

2.5b Place the tester over the cooling system filler neck and use the bulb to draw a sample into the tester

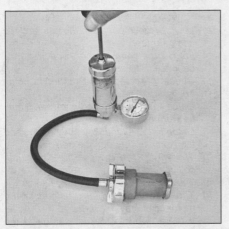

2.8 Checking the cooling system pressure cap with a cooling system pressure tester

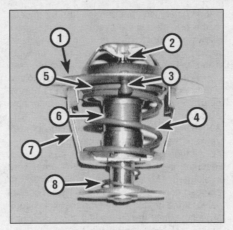

2.10 Typical thermostat:

1	Flange	5	Valve seat
2	Piston	6	Valve
3	Jiggle valve	7	Frame
4	Main coil spring	8	Secondary coil spring

2.28 The water pump weep hole is generally located on the underside of the pump

6   If the test indicates combustion gas is present in the cooling system, you can be sure that the engine has a blown head gasket or a crack in the cylinder head or block, and will require disassembly to repair.

Pressure cap

Warning: *Wait until the engine is completely cool before beginning this check.*

7   The cooling system is sealed by a spring-loaded cap, which raises the boiling point of the coolant. If the cap's seal or spring are worn out, the coolant can boil and escape past the cap. With the engine completely cool, remove the cap and check the seal; if it's cracked, hardened or deteriorated in any way, replace it with a new one.

8   Even if the seal is good, the spring might not be; this can be checked with a cooling system pressure tester **(see illustration)**. If the cap can't hold a pressure within approximately 1-1/2 lbs of its rated pressure (which is marked on the cap), replace it with a new one.

9   The cap is also equipped with a vacuum relief spring. When the engine cools off, a vacuum is created in the cooling system. The vacuum relief spring allows air back into the system, which will equalize the pressure and prevent damage to the radiator (the radiator tanks could collapse if the vacuum is great enough). If, after turning the engine off and allowing it to cool down you notice any of the cooling system hoses collapsing, replace the pressure cap with a new one.

Thermostat

10   Before assuming the thermostat **(see illustration)** is responsible for a cooling system problem, check the coolant level (see Chapter 1), drivebelt tension (see Chapter 1, Section 19) and temperature gauge (or light) operation.

11   If the engine takes a long time to warm up (as indicated by the temperature gauge or heater operation), the thermostat is probably stuck open. Replace the thermostat with a new one.

12   If the engine runs hot or overheats, a thorough test of the thermostat should be performed.

13   Definitive testing of the thermostat can only be made when it is removed from the vehicle. If the thermostat is stuck in the open position at room temperature, it is faulty and must be replaced.

Caution: *Do not drive the vehicle without a thermostat. The computer may stay in open loop and emissions and fuel economy will suffer.*

14   To test a thermostat, suspend the (closed) thermostat on a length of string or wire in a pot of cold water.

15   Heat the water on a stove while observing the thermostat. The thermostat should fully open before the water boils.

16   If the thermostat doesn't open and close as specified, or sticks in any position, replace it.

Cooling fan

Electric cooling fan

17   If the engine is overheating and the cooling fan is not coming on when the engine temperature rises to an excessive level, unplug the fan motor electrical connector(s) and connect the motor directly to the battery with fused jumper wires. If the fan motor doesn't come on, replace the motor.

18   If the fan motor is okay, but it isn't coming on when the engine gets hot, the cooling fan control module or the Powertrain Control Module (PCM) might be defective.

19   These control circuits are fairly complex, and checking them should be left to a qualified automotive technician.

20   Check all wiring and connections to the fan motor. Refer to the wiring diagrams at the end of Chapter 12.

21   If no obvious problems are found, the problem could be the Cylinder Head Temperature (CHT) sensor or the Powertrain Control Module (PCM). Have the cooling fan system and circuit diagnosed by a dealer service department or repair shop with the proper diagnostic equipment.

Belt-driven cooling fan

22   Disconnect the cable from the negative terminal of the battery and rock the fan back and forth by hand to check for excessive bearing play.

23   With the engine cold (and not running), turn the fan blades by hand. The fan should turn freely.

24   Visually inspect for substantial fluid leakage from the clutch assembly. If problems are noted, replace the clutch assembly.

25   With the engine completely warmed up, turn off the ignition switch and disconnect the negative battery cable from the battery. Turn the fan by hand. Some drag should be evident. If the fan turns easily, replace the fan clutch.

Water pump

26   A failure in the water pump can cause serious engine damage due to overheating.

Drivebelt-driven water pump

27   There are two ways to check the operation of the water pump while it's installed on the engine. If the pump is found to be defective, it should be replaced with a new or rebuilt unit.

28   Water pumps are equipped with weep (or vent) holes **(see illustration)**. If a failure occurs in the pump seal, coolant will leak from the hole.

29   If the water pump shaft bearings fail, there may be a howling sound at the pump

while it's running. Shaft wear can be felt with the drivebelt removed if the water pump pulley is rocked up and down (with the engine off). Don't mistake drivebelt slippage, which causes a squealing sound, for water pump bearing failure.

Timing chain or timing belt-driven water pump

30   Water pumps driven by the timing chain or timing belt are located underneath the timing chain or timing belt cover.

31   Checking the water pump is limited because of where it is located. However, some basic checks can be made before deciding to remove the water pump. If the pump is found to be defective, it should be replaced with a new or rebuilt unit.

32   One sign that the water pump may be failing is that the heater (climate control) may not work well. Warm the engine to normal operating temperature, confirm that the coolant level is correct, then run the heater and check for hot air coming from the ducts.

33   Check for noises coming from the water pump area. If the water pump impeller shaft or bearings are failing, there may be a howling sound at the pump while the engine is running. **Note:** *Be careful not to mistake drivebelt noise (squealing) for water pump bearing or shaft failure.*

34   It you suspect water pump failure due to noise, wear can be confirmed by feeling for play at the pump shaft. This can be done by rocking the drive sprocket on the pump shaft up and down. To do this you will need to remove the tension on the timing chain or belt as well as access the water pump.

All water pumps

35   In rare cases or on high-mileage vehicles, another sign of water pump failure may be the presence of coolant in the engine oil. This condition will adversely affect the engine in varying degrees. **Note:** *Finding coolant in the engine oil could indicate other serious issues besides a failed water pump, such as a blown head gasket or a cracked cylinder head or block.*

36   Even a pump that exhibits no outward signs of a problem, such as noise or leakage, can still be due for replacement. Removal for close examination is the only sure way to tell. Sometimes the fins on the back of the impeller can corrode to the point that cooling efficiency is diminished significantly.

Heater system

37   Little can go wrong with a heater. If the fan motor will run at all speeds, the electrical part of the system is okay. The three basic heater problems fall into the following general categories:

a) *Not enough heat*
b) *Heat all the time*
c) *No heat*

38   If there's not enough heat, the control valve or door is stuck in a partially open position, the coolant coming from the engine isn't hot enough, or the heater core is restricted.

If the coolant isn't hot enough, the thermostat in the engine cooling system is stuck open, allowing coolant to pass through the engine so rapidly that it doesn't heat up quickly enough. If the vehicle is equipped with a temperature gauge instead of a warning light, watch to see if the engine temperature rises to the normal operating range after driving for a reasonable distance.

39   If there's heat all the time, the control valve or the door is stuck wide open.

40   If there's no heat, coolant is probably not reaching the heater core, or the heater core is plugged. The likely cause is a collapsed or plugged hose, core, or a frozen heater control valve. If the heater is the type that flows coolant all the time, the cause is a stuck door or a broken or kinked control cable.

Air conditioning system

41   If the cool air output is inadequate: Inspect the condenser coils and fins to make sure they're clear.

a) *Check the compressor clutch for slippage.*
b) *Check the blower motor for proper operation.*
c) *Inspect the blower discharge passage for obstructions.*
d) *Check the system air intake filter for clogging.*

42   If the system provides intermittent cooling air:

a) *Check the circuit breaker, blower switch and blower motor for a malfunction.*
b) *Make sure the compressor clutch isn't slipping.*
c) *Inspect the plenum door to make sure it's operating properly.*
d) *Inspect the evaporator to make sure it isn't clogged.*
e) *If the unit is icing up, it may be caused by excessive moisture in the system, incorrect super heat switch adjustment or low thermostat adjustment.*

43   If the system provides no cooling air:

a) *Inspect the compressor drivebelt. Make sure it's not loose or broken.*
b) *Make sure the compressor clutch engages. If it doesn't, check for a blown fuse.*
c) *Inspect the wire harness for broken or disconnected wires.*
d) *If the compressor clutch doesn't engage, bridge the terminals of the air conditioning pressure switch(es) with a jumper wire; if the clutch now engages, and the system is properly charged, the pressure switch is bad.*
e) *Make sure the blower motor is not disconnected or burned out.*
f) *Make sure the compressor isn't partially or completely seized.*
g) *Inspect the refrigerant lines for leaks.*
h) *Check the components for leaks.*
i) *Inspect the receiver-drier/accumulator or expansion valve/tube for clogged screens.*

44   If the system is noisy:

a) *Look for loose panels in the passenger's compartment.*
b) *Inspect the compressor drivebelt. It may be loose or worn.*
c) *Check the compressor mounting bolts. They should be tight.*
d) *Listen carefully to the compressor. It may be worn out.*
e) *Listen to the idler pulley and bearing and the clutch. Either may be defective.*
f) *The winding in the compressor clutch coil or solenoid may be defective.*
g) *The compressor oil level may be low.*
h) *The blower motor fan bushing or the motor itself may be worn out.*
i) *If there is an excessive charge in the system, you'll hear a rumbling noise in the high pressure line, a thumping noise in the compressor, or see bubbles or cloudiness in the sight glass.*
j) *If there's a low charge in the system, you might hear hissing in the evaporator case at the expansion valve, or see bubbles or cloudiness in the sight glass.*

3   Air conditioning and heating system - check and maintenance

Air conditioning system

Warning: *The air conditioning system is under high pressure. Do not loosen any hose fittings or remove any components until after the system has been discharged. Air conditioning refrigerant must be properly discharged into an EPA-approved recovery/recycling unit at a dealer service department or an automotive air conditioning repair facility. Always wear eye protection when disconnecting air conditioning system fittings.*

Caution: *All models covered by this manual use either environmentally friendly R-134a or R-1234yf. These refrigerants (and their appropriate refrigerant oils) are not compatible with R-12 refrigerant system components and must never be mixed or the components will be damaged.*

Caution: *When replacing entire components, additional refrigerant oil should be added equal to the amount that is removed with the component being replaced. Be sure to read the container before adding any oil to the system, to make sure it is compatible with the type of refrigerant in the system.*

1   The following maintenance checks should be performed on a regular basis to ensure that the air conditioning continues to operate at peak efficiency.

a) *Inspect the condition of the drivebelt. If it is worn or deteriorated, replace it (see Chapter 1).*
b) *Inspect the system hoses. Look for cracks, bubbles, hardening and deterioration. Inspect the hoses and all fittings for oil bubbles or seepage. If there is any evidence of wear, damage or leakage, replace the hose(s).*

3.1 Check the evaporator housing drain tube (just above the transmission) for blockage

3.9 When feeling the pipes going to and from the evaporator, the larger-diameter pipe (A) should feel much cooler than the smaller-diameter one (B), which may even be hot)

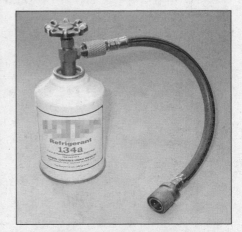

3.12 A basic charging kit is available at most auto parts stores - it must say R-134a (not R-12) or R-1234yf (as applicable) and so must the can of refrigerant

c) Inspect the condenser fins for leaves, bugs and any other foreign material that may have embedded itself in the fins. Use a "fin comb" or compressed air to remove debris from the condenser.

d) Make sure the system has the correct refrigerant charge.

e) If you hear water sloshing around in the dash area or have water dripping on the carpet, check the evaporator housing drain tube (see illustration) and insert a piece of wire into the opening to check for blockage.

2   It's a good idea to operate the system for about 10 minutes at least once a month, particularly during the winter. Long-term non-use can cause hardening, and subsequent failure, of the seals.

3   Leaks in the air conditioning system are best spotted when the system is brought up to temperature and pressure, by running the engine with the air conditioning on for five minutes. Shut the engine off and inspect the air conditioning hoses and connections. Traces of oil usually indicate refrigerant leaks.

4   Because of the complexity of the air conditioning system and the special equipment necessary to service it, in-depth troubleshooting and repairs are not included in this manual. However, simple checks and component replacement procedures are provided in this Chapter.

5   If the air conditioning system doesn't operate at all, check the fuse panel and the air conditioning relay, located in the relay box in the engine compartment (see Chapter 12).

6   The most common cause of poor cooling is simply a low system refrigerant charge. If a noticeable drop in cool air output occurs, the following quick check will help you determine if the refrigerant level is low. For more complete information on the air conditioning system, refer to the Haynes *Automotive Heating and Air Conditioning Manual.*

Checking the refrigerant charge

7   Warm the engine up to normal operating temperature.

3.15a Add refrigerant to the system at the low-pressure port - early model shown

8   Place the air conditioning temperature selector at the coldest setting and put the blower at the highest setting. Open the doors (to make sure the air conditioning system doesn't cycle off as soon as it cools the passenger compartment).

9   With the compressor engaged, the clutch will make an audible click and the center of the clutch will rotate. Feel the pipes going to and from the evaporator (see illustration); the small diameter pipe should be warm and the large diameter pipe should feel cool.

10   If the larger-diameter pipe isn't considerably cooler, the system charge is probably low. The earliest warning that a system is low on refrigerant is the air temperature coming out of the ducts inside the vehicle. If the air isn't as cold as it used to be, the system probably needs a charge.

11   Further inspection or testing of the system requires special tools and techniques and is beyond the scope of this manual.

Adding refrigerant

12   Buy an automotive charging kit at an auto parts store. A charging kit includes a can of refrigerant, a tap valve and a short section

3.15b Here's the low-side port on a later model

of hose that can be attached between the tap valve and the system low side service valve (see illustration).

Caution: *Although the system will hold more than one can of refrigerant, don't add more than one can (you could overfill the system). If more refrigerant than that is required, the system should be evacuated and leak tested.*

13   Hook up the charging kit by following the manufacturer's instructions.

Warning: *DO NOT attempt hook the charging kit hose to the system high pressure side! The fittings on the charging kit are designed to fit only on the low pressure side of the system.*

14   Back off the valve handle on the charging kit and screw the kit onto the refrigerant can, making sure first that the O-ring or rubber seal inside the threaded portion of the kit is in place.

Warning: *Wear protective eyewear when dealing with pressurized refrigerant cans.*

15   Remove the dust cap from the low-side charging port and attach the quick-connect fitting on the kit hose (see illustrations).

16   Warm up the engine and turn on the air conditioner. Keep the charging kit hose away from the fan and other moving parts. The

3.18 If you have an accurate thermometer, you can place it in the center air conditioning duct inside the vehicle

3.24 To disinfect the evaporator housing, insert the nozzle of the disinfectant can through the opening in the recirculation housing and point it toward the evaporator core (which is toward the left of the opening) - glove box removed for access

charging process requires the compressor to be running.

17   Turn the valve handle on the kit until the stem pierces the can, then back the handle out to release the refrigerant. You should be able to hear the rush of gas. Add refrigerant to the low side of the system until both the accumulator surface and the evaporator inlet pipe feel about the same temperature. Allow stabilization time between each addition.

18   If you have an accurate thermometer, you can place it in the center air conditioning duct inside the vehicle **(see illustration)** and keep track of the outlet air temperature. A charged system that is working properly should cool to approximately 40-degrees F. If the ambient (outside) air temperature is very high, say 110-degrees F, or if the relative humidity is high, the duct air temperature may be as high as 60- to 70-degrees F, but generally the air conditioning is 30- to 50-degrees F cooler than the ambient air.

19   When the can is empty, turn the valve handle to the closed position and release the connection from the low-side port. Replace the dust cap.

20   Remove the charging kit from the can and store the kit for future use with the piercing valve in the up position to prevent inadvertently piercing the can on the next use.

Eliminating air conditioning odors

21   Unpleasant odors that often develop in air conditioning systems are caused by the growth of a fungus, usually on the surface of the evaporator core. The warm, humid environment there is a perfect breeding ground for mildew to develop.

22   The evaporator core on most vehicles Is difficult to access, and factory dealerships have a lengthy, expensive process for eliminating the fungus by opening up the evaporator case and using a powerful disinfectant and

rinse on the core until the fungus is gone. You can service your own system at home, but it takes something much stronger than basic household germ-killers or deodorizers.

23   Aerosol disinfectants for automotive air conditioning systems are available in most auto parts stores, but remember when shopping for them that the most effective treatments are also the most expensive. The basic procedure for using these sprays is to start by running the system in the Recirculating mode for ten minutes with the blower on its highest speed. Use the highest heat mode to dry out the system - keep the compressor from engaging by disconnecting the wiring connector at the compressor (see Section 13).

24   Make sure that the disinfectant can comes with a long spray hose. Work the nozzle through the opening in the heater/air conditioning recirculation housing so that it protrudes inside and points toward the evaporator housing (toward the center of the instrument panel) **(see illustration)** and then spray according to the manufacturer's recommendations. Try to cover the whole surface of the evaporator core, by aiming the spray up, down and sideways. Follow the manufacturer's recommendations for the length of spray and waiting time between applications.

25   Once the evaporator has been cleaned, the best way to prevent the mildew from coming back again is to make sure your evaporator housing drain tube is clear **(see illustration 13.1)**.

Heating system

26   If the carpet under the heater core is damp, or if antifreeze vapor or steam is coming through the vents, the heater core is leaking. Remove it (see Section 12) and install a new unit (most radiator shops will not repair a leaking heater core).

27   If the air coming out of the heater vents

isn't hot, the problem could stem from any of the following causes:

a)   *The thermostat is stuck open, preventing the engine coolant from warming up enough to carry heat to the heater core. Replace the thermostat (see Section 4).*

b)   *There is a blockage in the system, preventing the flow of coolant through the heater core. Feel both heater hoses at the firewall. They should be hot. If one of them is cold, there is an obstruction in one of the hoses or in the heater core. Detach the hoses and back flush the heater core with a water hose. If the heater core is clear but circulation is still restricted, remove the two hoses and flush them out with a water hose.*

c)   *If flushing fails to remove the blockage from the heater core, the core must be replaced (see Section 12).*

4   Thermostat - replacement

Warning: *Do not remove the radiator or expansion tank cap, drain the coolant or replace the thermostat until the engine has cooled completely.*

Note: *On 2.7L and 3.5L V6 engines, the thermostat is located on the lower left side of the engine and connected to large diameter coolant hose. On 3.6L engines, the thermostat is located on the top front of the engine mounted to the coolant crossover. On V8 engines, the thermostat is located at the top front area of the engine at the intake manifold and connected to a large diameter coolant hose.*

1   Remove the engine cover (see Chapter 1) if equipped, then disconnect the cable from the negative battery terminal (see Chapter 5, Section 1).

Caution: *On 2015 and later models, if equipped with an Intelligent Battery Sensor (IBS), disconnect the IBS connector first before disconnecting the negative battery cable.*

2   Drain the cooling system (see Chapter 1). If the coolant is relatively new or in good condition, save it and reuse it (see Section 1).

3   Remove the air filter housing or the air intake duct, on models where it would interfere (see Chapter 4).

4   Squeeze the tabs on the hose clamp to loosen it from the hose(s), then reposition the clamp several inches back up the hose. Detach the hose(s) from the thermostat housing.

Note: *Special hose clamp pliers are available at most auto parts stores. If the hose is stuck, grasp it near the end with a pair of adjustable pliers and twist it to break the seal, then pull it off. If the hose is old or deteriorated, cut it off and install a new one.*

5   If the outer surface of the thermostat housing that mates with the hose is deteriorated (corroded, pitted, etc.) it may be damaged further by hose removal. If it is, the thermostat housing (or housing cover) will have to be replaced.

4.12 The thermostat housing cover fasteners on the 3.5L V6 engine - other models are similar

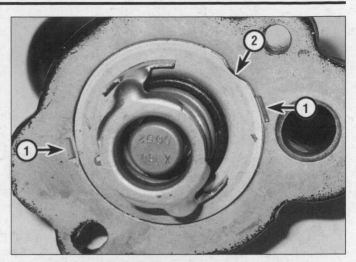

4.13a Thermostat mounting details on a 3.5L V6 engine:

1   Swaged area of thermostat housing
2   Key for thermostat and housing

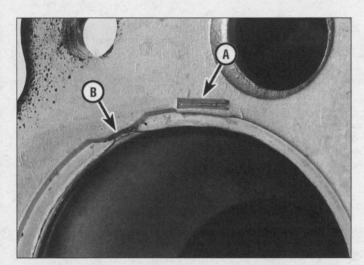

4.13b Remove the distorted metal from the swaged area of the housing (A) so the replacement thermostat will fit correctly, but leave the housing/thermostat key as shown (B)

4.15 Disconnect the radiator hose (A) then remove the thermostat housing fasteners (B) - lower fastener not visible

2.7L V6 engines

6   Loosen the bolt at the starter motor that holds the small heater tube bracket that leads to the thermostat housing.

7   Remove the heater tube flange mounting nuts from the thermostat housing, disconnect the tube and move it aside.

8   Remove the thermostat housing cover fasteners and remove the housing. Note the position of the thermostat and where the small valve is located, because it must be installed in the same position.

9   Remove the thermostat and discard the O-ring seal on the housing. If the housing is stuck, tap it with a soft-faced hammer to jar it loose. Be prepared for some coolant to spill as the gasket seal is broken.

10   Note how the thermostat is installed (which end is facing up, or out, and the position of the air bleed "jiggle valve," if equipped).

3.5L V6 engines

11   Disconnect the electrical connectors from the nearby engine oil and power steering pressure switches.

Note: *It may be necessary to raise the front of the vehicle on 3.5L engine models for access to the thermostat and surrounding components.*

12   Remove the thermostat housing cover mounting fasteners (see illustration).

13   Remove the thermostat housing cover and thermostat. The thermostat will come out with the housing cover if it is the original one installed. The original thermostat is 'staked' or 'swaged' into place and the housing and thermostat are keyed so that the thermostat fits into the housing in one position when placed in the right direction (with the large spring toward the engine) (see illustration). The

distorted metal on the housing must be carefully removed for the replacement thermostat to fit properly (see illustration). There is no need to stake or swage the replacement thermostat.

14   Remove the gasket from the engine block and discard it.

3.6L V6 engines

Thermostat

Caution: *The thermostat and housing are serviced as an assembly. Do not remove the thermostat from the housing, damage to the thermostat and the housing may occur.*

15   Disconnect the radiator hose, then remove the thermostat housing fasteners and remove the housing (see illustration).

16   Remove the thermostat housing gasket.

4.19 Disconnect the heater hose (A), then remove the crossover housing mounting bolts (B)

5.5 Location of the main cooling fan electrical connector and the fan assembly mounting fasteners

Coolant crossover housing

Removal

17   Remove the thermostat housing (see Steps 15 and 16).
18   Disconnect the heater hose from the coolant crossover housing.
19   Remove the crossover housing mounting bolts, making sure to note the bolt locations. The bolts are different lengths and must be reinstalled into their original locations **(see illustration)**.
20   Remove the crossover housing and discard the gaskets.

Installation

21   Clean all mating surfaces before beginning any reassembly.
22   Install a new gasket on to the crossover housing.
23   Place the housing onto the timing chain cover and install the mounting bolts in their original locations. Tighten the crossover hous-

5.11 The (reverse threaded) fan blade mounting nut (A) and the fan motor mounting fasteners (B) (one fastener is hidden from view)

ing bolts to the torque listed in this Chapter's Specifications.
24   Install the thermostat (see Steps 27 to 32).

5.7L, 6.1L and 6.4L V8 (Hemi) engines

25   Remove the thermostat housing cover fasteners and remove the cover.
26   Remove the thermostat and seal.
Note: *The seal and thermostat are serviced as an assembly.*

All models

27   Clean all mating surfaces before beginning any reassembly.
28   Install a new gasket or seal, as applicable.
29   The remainder of installation is the reverse of removal. Make sure that the replacement thermostat is installed in the same direction and position as the one removed. Tighten the thermostat housing cover bolts to the torque listed in this Chapter's Specifications.
30   Reattach the hose(s) to the housing cover and fitting(s), then tighten the hose clamp(s) securely.
31   Refill the cooling system (see Chapter 1).
32   Start the engine and allow it to reach normal operating temperature, then check for leaks and proper thermostat operation.

---

5   Engine cooling fans - removal and installation

Warning: *To avoid possible injury or damage, DO NOT operate the engine with a damaged fan. Do not attempt to repair fan blades - replace a damaged fan with a new one.*
1   Disconnect the cable from the negative battery terminal (see Chapter 5, Section 1).

Caution: *On 2015 and later models, if equipped with an Intelligent Battery Sensor (IBS), disconnect the IBS connector first before disconnecting the negative battery cable.*
2   Remove the air filter housing (see Chapter 4).
3   On 6.4L engine models, remove the throttle body (see Chapter 4).

All 2010 and earlier models, 2011 5.7L 300 models, 2011 Charger models, 2011 3.6L Challenger models, 2012 3.6L and all 2012 and later dual fan 5.7L/6.4L models

Warning: *Wail until the engine is completely cool before beginning this procedure.*
4   On models where the upper radiator hose would interfere with fan and shroud removal, partially drain the cooling system (see Chapter 1), then remove the upper radiator hose. On 6.4L models, also detach the oil cooler supply hose from the thermostat housing.
5   Disconnect the cooling fan electrical connector **(see illustration)**.
6   Remove the cooling fan shroud assembly mounting fasteners.
7   Guide the fan and shroud assembly up and out of the engine compartment.

All other models

8   Raise the front of the vehicle and support it securely on jackstands. Remove the under-vehicle splash shield.
9   Loosen (but do not remove) the radiator lower support bolts.
10   Push the fan shroudf upward to disengage it from the mounts, then lower the fan assembly straight down and out of the engine compartment.

Fan motor or blade replacement

11   If you are replacing a dual fan motor or fan blade, remove the fan blade mounting nut to separate it from the motor(s). If you have a single fan motor, the fan is balanced to the motor and must be replaced as an assembly **(see illustration)**.

Note: *The nut for the fan blade is reverse threaded; turn the nut clockwise to unscrew it.*

12   Remove the mounting fasteners for the motor and separate it from the shroud assembly.

All models

13   Installation is the reverse of removal. Be sure to engage the bottom of the fan assembly with the plastic mounts that hold it to the bottom of the radiator **(see illustration)**.

14   Refill the cooling system if drained (see Chapter 1).

5.13 A fan assembly lower mount at the radiator

6.2 The expansion tank mounting fasteners (A) and hoses (B) - early model shown, later models similar

---

6   Coolant expansion tank - removal and installation

Warning: *Wait until the engine is completely cool before beginning this procedure.*

1   Drain the cooling system (see Chapter 1). If the coolant is relatively new and still in good condition, it can be saved and reused.

Note: *Be careful to clean up any spilled coolant as stated in the Warning at the beginning of Section 1.*

2   Disconnect the expansion tank hoses by sliding the hose clamp away with pliers and then removing the hose from the tank fitting **(see illustration)**.

3   Remove the expansion tank mounting fasteners.

4   Clean out the tank with soapy water and a brush to remove any deposits inside.

Inspect the tank carefully for cracks. If you find a crack, replace the tank.

5   Installation is the reverse of removal. Refill the cooling system with the proper concentration of antifreeze (see Chapter 1).

---

7   Radiator - removal and installation

Warning: *Wait until the engine is completely cool before beginning this procedure.*

Removal

1   Disconnect the cable from the negative battery terminal (see Chapter 5, Section 1).

Caution: *On 2015 and later models, if equipped with an Intelligent Battery Sensor (IBS), disconnect the IBS connector first before disconnecting the negative battery cable.*

2   On 2011 and later models, remove the air filter housing (see Chapter 4).

3   Drain the cooling system (see Chapter 1). If the coolant is relatively new or in good condition, save it and reuse it.

4   Remove the upper radiator hose **(see illustration)**.

5   Remove the plastic panels between the radiator support and the top front of the front cover **(see illustration)**.

Note: *On 2011 and later V8 models, carefully remove the upper shroud cover by inserting a hooked pick under the ends of the cover until the end of the pick catches the hole in the cover and pull the cover away from the clip.*

6   Remove the cooling fan assembly (see Section 5).

7   Raise the front of the vehicle and support it securely on jackstands.

8   Remove the lower engine splash shield (see Chapter 2A).

7.4 Location of the upper radiator hose and clamps (3.5L V6 engine shown, others similar)

7.5 Pull up on the panels to release them from their retaining clips

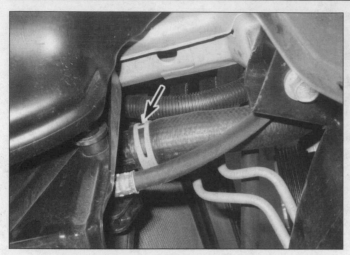

7.9 Location of the lower radiator hose clamp (3.5L V6 engine shown, others similar)

7.12 The location of the air conditioning condenser mounting fasteners (2010 and earlier models shown, front bumper cover and air deflectors removed for clarity)

9   Remove the lower radiator hose **(see illustration)**.

10   On 2011 and later models, remove the front bumper cover (see Chapter 11).

11   On 2011 and later models, remove the air deflector plastic push-pins and remove the deflectors from the sides of the radiator.

12   Remove the lower mounting fasteners that attach the air conditioning condenser to the radiator **(see illustration)**.

13   On 2011 through 2014 V8 models, remove the radiator core support-to-radiator lower mounting bolts.

14   Remove the radiator upper mounting brackets from the radiator support **(see illustration)**.

Caution: *Use hand tools on the mounting bolts because they may be treated with thread locking compound.*

15   Remove the upper mounting fasteners for the air conditioning condenser and carefully separate the condenser from the radiator **(see illustration 6.9)**.

Note: *It may be necessary to remove the fas-*

teners for the plastic air deflectors and move them aside to gain access to the upper condenser fasteners.

16   Tilt the radiator toward the engine and lift it from the engine compartment. Be careful not to damage the air-flow seals on the top and bottom of the radiator, spill coolant on the vehicle or scratch the paint.

17   Whenever the radiator is removed from the vehicle, make note of the location of all rubber mounting bushings and their condition **(see illustration)**. If they're cracked, hardened or otherwise deteriorated, replace them.

Installation

18   With the radiator removed, it can be inspected for leaks and damage. If it needs repair, have a radiator shop or dealer service department perform the work, as special tools and techniques are required.

19   Bugs and dirt can be removed from the front of the radiator with a garden hose, followed by compressed air and a soft brush. Don't bend the cooling fins as this is done.

When blowing out the core, direct the hose or air line from the engine side out only.

20   Inspect the radiator mounts for deterioration and make sure there's nothing in them when the radiator is installed.

21   Installation is the reverse of the removal procedure. Tighten the fasteners to the torque listed in this Chapter's Specifications.

22   Fill the cooling system with the proper mixture of antifreeze and water (see Chapter 1).

23   Start the engine and check for leaks. Allow the engine to reach normal operating temperature and recheck the coolant level; add more if required.

8   Water pump - removal and installation

Warning: *Wait until the engine is completely cool before beginning this procedure.*

Removal

1   Disconnect the cable from the negative battery terminal (see Chapter 5, Section 1).
Caution: *On 2015 and later models, if equipped with an Intelligent Battery Sensor (IBS), disconnect the IBS connector first before disconnecting the negative battery cable.*

2   Drain the cooling system (see Chapter 1). If the coolant is relatively new or in good condition, save it and reuse it.

3   Remove the engine cooling fan assembly (see Section 5).

4   Remove the drivebelt (see Chapter 1).

2.7L V6 engines

5   Remove the timing chain and guides (see Chapter 2A).

3.5L V6 engines

6   Remove the timing belt (see Chapter 2B).

7.14 The mounting fastener for a radiator mounting bracket (left side shown for 2010 and earlier models)

7.17 A mounting bushing for the radiator (2010 and earlier model right side shown)

8.8 Disconnect the lower radiator hose (A) and the heater hose (B) from the pump

8.15a Remove the timing chain guide mounting bolts (A) and then the water pump mounting bolts (B) on the 2.7L V6 engine

3.6L V6 engines

7   Remove the air filter housing (see Chapter 4).
8   Detach the heater hose and lower radiator hose from the water pump **(see illustration)**.
Caution: *The original spring clamps that are used to secure the coolant hoses are marked with a number or letter and, if necessary, must be replaced with a matching clamp having the same mark.*
9   Remove the idler pulley for the drivebelt located on the upper left of the water pump.

V8 engines

10   Remove the air filter housing (see Chapter 4).
11   Remove the thermostat (see Section 4).
12   Detach the radiator hose from the water pump. If a hose sticks, grasp it near the end with a pair of adjustable pliers and twist it to break the seal, then pull it off. If the hose is deteriorated, cut it off and install a new one upon reassembly.

Remove the idler pulley for the drivebelt located on the lower left of the water pump, and on later models the tensioner from the right side of the pump.
14   On 2011 and later models, remove the heater tube mounting bolt and twist the tube back and forth until the tube can be removed from the pump.
Note: *It may be necessary to spray penetrating oil on the pipe where it is inserted into the pump.*

All engines

15   Remove the bolts and detach the water pump from the engine. Note the locations of the various lengths and different types of bolts as they're removed to ensure correct installation **(see illustrations)**.

Installation

16   Clean the bolt threads and the threaded holes in the engine to remove corrosion and sealant.

17   Compare the new pump to the old one to make sure they're identical. If the old pump is being reused, check the impeller blades on the backside for corrosion. If any fins are missing or badly corroded, replace the pump with a new one.
18   Remove all traces of old gasket material or the O-ring seal from the engine and water pump (if the same one is to be installed).
19   Clean the engine and water pump mating surfaces with lacquer thinner or acetone.
20   On water pumps equipped with an O-ring seal, apply a thin coat of RTV sealant to the new water pump O-ring and position it in the groove on the back of the water pump **(see illustration)**.
Caution: *Make sure the O-ring is correctly seated in the water pump groove to avoid a coolant leak.*
21   On water pumps equipped with a gasket, apply a thin film of RTV sealant to the gasket mating surface of the new pump, and position the gasket on the pump. Apply a thin film of RTV sealant to the engine-side of the gasket

8.15b The water pump mounting bolt locations on a 3.5L V6 engine

8.20 In this example, the O-ring is seated completely in its groove (2.7L V6 shown)

10.3a The blower motor resistor electrical connector (A) and mounting fasteners (B)

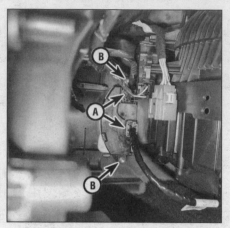

10.3b The power module electrical connectors (A) and mounting fasteners (B)

10.8 Remove the electrical connector (A) and the mounting screws (B) for the blower motor assembly

and slip a couple of bolts through the pump mounting holes to hold the gasket in place.

22   Carefully attach the pump and O-ring/gasket (as applicable) to the engine and thread the bolts into the holes finger-tight.

23   With all the bolts in place, tighten them in a criss-cross pattern, a little at a time, to the torque listed in this Chapter's Specifications. Turn the water pump by hand to make sure it rotates freely.

24   Connect the hoses to the water pump (if applicable) and reinstall all parts removed for access to the pump.

25   Refill the cooling system (see Chapter 1). Run the engine and check for leaks.

9   Coolant temperature sensor - check and replacement

Check

1   The coolant temperature indicator system consists of a warning light or a temperature gauge on the dash and a coolant temperature sending unit mounted on the engine. On the models covered by this manual, the Engine Coolant Temperature (ECT) sensor, which is an information sensor for the Powertrain Control Module (PCM), also functions as the coolant temperature sending unit (see Chapter 6).

2   If an overheating indication occurs, check the coolant level in the system and then make sure all connectors in the wiring harness between the sending unit and the indicator light or gauge are tight.

3   When the ignition switch is turned to Start and the starter motor is turning, the indicator light (if equipped) should come on. This doesn't mean the engine is overheated; it just means that the bulb is good.

4   If the light doesn't come on when the ignition key is turned to Start, the bulb might be burned out, the ignition switch might be faulty or the circuit might be open.

5   As soon as the engine starts, the indica-

tor light should go out and remain off, unless the engine overheats. If the light doesn't go out, refer to Chapter 6 and check for any stored trouble codes in the Powertrain Control Module (PCM).

6   If the engine tends to overheat easily, check the coolant to make sure it's the proper type and concentration (see Chapter 1).

Replacement

7   See Chapter 6 for the ECT sensor replacement procedure.

10   Blower motor resistor and blower motor - replacement

Warning: *The models covered by this manual are equipped with a Supplemental Restraint System (SRS), more commonly known as airbags. Always disarm the airbag system before working in the vicinity of any airbag system component to avoid the possibility of accidental deployment of the airbag, which could cause personal injury (see Chapter 12). Do not use a memory saving device to preserve the PCM's memory when working on or near airbag system components.*

Blower motor resistor/power module

Note: *Models with automatic temperature control utilize a power module instead of a blower motor resistor. It mounts the same as a blower motor resistor and is replaced in the same manner.*

1   Disconnect the cable from the negative battery terminal (see Chapter 5, Section 1).

Caution: *On 2015 and later models, if equipped with an Intelligent Battery Sensor (IBS), disconnect the IBS connector first before disconnecting the negative battery cable.*

2   Remove the lower instrument panel insulator (see Chapter 11).

3   Disconnect the electrical connector(s)

from the blower motor resistor or power module **(see illustrations)**.

4   Remove the mounting screws and remove it from the housing.

5   Installation is the reverse of removal.

Blower motor

6   Disconnect the cable from the negative battery terminal (see Chapter 5, Section 1).

7   Remove the lower instrument panel insulator (see Chapter 11).

8   Disconnect the electrical connector from the blower motor, then remove the mounting screws and lower the blower motor out of the housing **(see illustration)**.

9   If either the fan or motor is damaged, the entire unit must be replaced as an assembly.

10   Installation is the reverse of removal.

11   Heater/air conditioning control assembly - removal and installation

Warning: *The models covered by this manual are equipped with a Supplemental Restraint System (SRS), more commonly known as airbags. Always disarm the airbag system before working in the vicinity of any airbag system component to avoid the possibility of accidental deployment of the airbag, which could cause personal injury (see Chapter 12). Do not use a memory saving device to preserve the PCM's memory when working on or near airbag system components.*

1   Disconnect the cable from the negative battery terminal (see Chapter 5, Section 1).

Caution: *On 2015 and later models, if equipped with an Intelligent Battery Sensor (IBS), disconnect the IBS connector first before disconnecting the negative battery cable.*

All models except 2015 and later Challenger models

2   Detach the instrument panel center bezel (see Chapter 11).

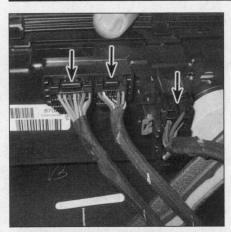

11.3 Heater/air conditioning control assembly electrical connectors (manual shown - automatic is similar)

11.4 Heater/air conditioning control assembly mounting screws

11.7 Carefully pry along the edges of the shift bezel until all the retaining clips are released

3   Disconnect the electrical connectors for the assembly (see illustration).

4   Remove the mounting screws and detach the air conditioning and heater control assembly from the bezel (see illustration). Note: *The control assembly may need calibration if replaced with a new one. This must be performed at a dealer service department or a qualified repair shop with the appropriate scan tool.*

5   Installation is the reverse of the removal procedure.

2015 and later Challenger models

6   On manual transmission models, release the shifter boot clips using a trim tool, then remove the shifter mounting bolts and shifter.

7   On automatic transmission models, use a trim tool to carefully pry along the edges of the shift bezel until all the retaining clips are released, then lift the bezel up from the floor console (see illustration).

8   Remove the shifter knob retaining screw and remove the shift knob and bezel (see illustration).

9   Using a plastic trim tool, pry along the edges of the shift console center trim panel until all the clips are released, then lift the trim panel up and disconnect the electrical connector from the shift panel (see illustration).

10   Remove the mounting screws and detach the air conditioning and heater control assembly from the shift panel (see illustration).

11   Installation is the reverse of the removal procedure.

12   Heater core - removal and installation

Warning: *The models covered by this manual are equipped with a Supplemental Restraint System (SRS), more commonly known as airbags. Always disarm the airbag system before working in the vicinity of any airbag system component to avoid the possibility of accidental deployment of the airbag, which could cause personal injury (see Chapter 12). Do not use a memory saving device to preserve*

11.8 Shifter knob retaining screw location

the PCM's memory when working on or near airbag system components.
Warning: *Wait until the engine is completely cool before beginning this procedure.*

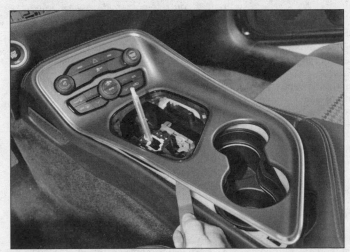

11.9 Carefully pry along the edges of the shift console center trim panel until all the clips are released

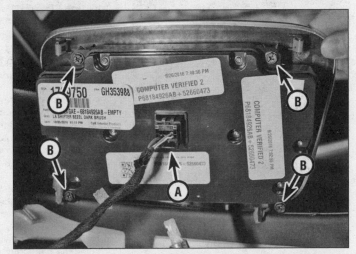

11.10 Remove the electrical connector (A) and the mounting screws (B) for the air conditioning and heater control assembly

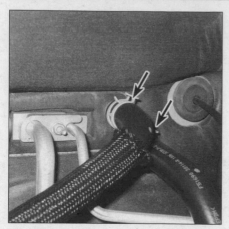

12.4 Slide the clamps away and detach the hoses from the heater core at the firewall in the engine compartment

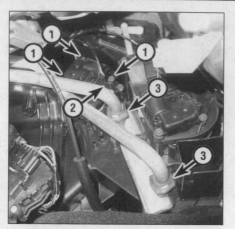

12.6 Blend door actuator mounting details and heater core fittings:

1 *Blend door actuator mounting fasteners (one hidden - vicinity shown)*
2 *Blend door actuator electrical connector*
3 *Heater core tube fittings (the shrink tubing must be cut away to remove the fitting retainer clips)*

Removal

1   On 2014 and earlier Challenger models, have the air conditioning system discharged by an automotive air conditioning technician.
2   Disconnect the cable from the negative battery terminal (see Chapter 5, Section 1).
Caution: *On 2015 and later models, if equipped with an Intelligent Battery Sensor (IBS), disconnect the IBS connector first before disconnecting the negative battery cable.*
3   Drain the cooling system (see Chapter 1).
4   Disconnect the heater hoses at the heater core in the engine compartment at the firewall **(see illustration)**. Tape or plug all openings.

2005 models

5   Remove the lower instrument panel insulator (see Chapter 11).
6   Remove the blend door actuator from the heater/air conditioning housing located just above the heater core **(see illustration)**.
7   Remove the two fasteners that hold the heater core tubes to the heater/air conditioning housing and then remove the flange from the housing **(see illustration)**.
8   Place towels underneath the heater core tubes and fittings, cut the rubber shrink tubing from around the fittings (if equipped) and then remove the retaining clips from the fittings **(see illustration 12.6 and the accompanying illustration)**.
9   On 2005 models, pull the heater core tubes directly out of the heater core and from the firewall and set them aside. Cover or plug the heater core openings.
10   Remove the retainer screw and retainer holding the heater core in the heater/air conditioning housing **(see illustration)**.

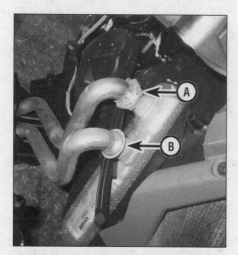

12.8 On a similar design, one heater core tube fitting retainer clip is installed (A) and the other has been removed (B) by pulling it off the fitting

12.10 The location of the heater core retainer screw

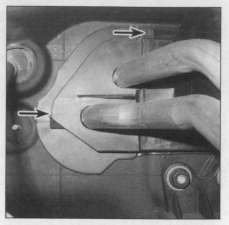

12.7 Remove the two fasteners that hold the heater core tube flange to the heater/air conditioning housing (one hidden - vicinity shown)

11   Pull the heater core out of the heater/air conditioning housing being careful not to damage any of the seals.

2006 and later models

Warning: *On 2014 and earlier Challenger models, the heater core tubes are not serviced separately from the heater core. The heater core tubes should not be moved, loosened or removed from the heater core as this can cause a coolant leak inside the vehicle which could result in severe burns caused by steam or scalding coolant.*
12   Remove the floor center console, if equipped (see Chapter 11).
13   Remove the instrument panel (see Chapter 11).
14   Using a trim tool, disengage the floor, rear floor, vent and defrost duct plastic push-pin fasteners and remove the ducts.

All models except 2014 and earlier Challenger models

15   Remove the two heater core tube clamps, then remove the tubes from the heater core. Remove and discard the heater core tube O-ring seals and cover the open heater core ports.
16   Remove the heater core bracket retaining screw from the left side of the heater/air conditioning housing and remove the bracket.
Note: *Replace the foam seal around the heater core it it has been damaged.*
17   Carefully pull the heater core straight out of the air distribution housing.

2014 and earlier Challenger models

18   Remove the cowl panel (see Chapter 11).
19   Remove the strut tower support from the engine compartment (see Chapter 10).
20   Remove the wiper motor (see Chapter 12).
21   Remove the canister purge solenoid (see Chapter 6).
22   On some models, it may be necessary to remove the Powertrain Control Model (PCM (see Chapter 6).

13.5 Air conditioning compressor details (3.5L V6 engine shown, others similar. Also, some components are removed for clarity)

1   *Compressor clutch electrical connector*
2   *High pressure line (discharge) fitting*
3   *Low pressure line (suction) fitting*
4   *Mounting fastener (some aren't visible in this photo)*

14.4 Air conditioning receiver-drier mounting details:

1   *Receiver-drier*
2   *Mounting fastener - early models shown, later models have fastener at the top of the receiver-drier*
3   *Condenser fitting*

23   Remove the interior ventilation filter (see Chapter 1), then use a trim tool to disengage the filter housing plastic push-pin fasteners. Remove the filter housing mounting nuts and remove the housing.

24   Disconnect the refrigerant lines at the firewall by removing the nut that secures the line fitting to the expansion valve **(see illustration 17.4)**. Cap or plug the line openings to prevent any dirt or moisture from entering the system.

25   Working in the engine compartment, remove the heater/air conditioning housing mounting nuts.

26   Lift the heater/air conditioning housing upwards until the drain tube clears the grommet in the floor and remove the housing from the vehicle.

27   Remove the two screws that secure the cover to the front of the heater/air conditioning housing and remove the cover.

28   Remove the screws that secure the heater core retaining bracket to the left side of the heater/air conditioning housing and remove the bracket.

29   Carefully pull the heater core tubes out from the foam seal at the front of the heater/air conditioning housing, then pull the heater core straight out of the side of the heater/air conditioning housing.

Installation

30   Before installing the heater core, make sure all foam seals are in place.

31   Carefully slide the heater core into the housing.

32   Reinstall the remaining components in the reverse order of removal.

33   Refill the cooling system (see Chapter 1) and reconnect the battery (see Chapter 5).

34   On 2006 through 2014 Challenger mod-els, have the system evacuated, recharged and leak-tested by the shop that discharged it.

35   Start the engine and check for any leaks and for proper heater operation.

13   Air conditioning compressor - removal and installation

Warning: *The air conditioning system is under high pressure. Do not loosen any hose fittings or remove any components until after the system has been discharged. Air conditioning refrigerant must be properly discharged into an EPA-approved recovery/recycling unit at a dealer service department or an automotive air conditioning repair facility. Always wear eye protection when disconnecting air conditioning system fittings.*

Note: *The receiver-drier (see Section 14) should be replaced whenever the compressor is replaced due to an internal failure.*

1   Have the air conditioning system discharged by an automotive air conditioning technician.

2   Disconnect the cable from the negative battery terminal (see Chapter 5, Section 1).

Caution: *On 2015 and later models, if equipped with an Intelligent Battery Sensor (IBS), disconnect the IBS connector first before disconnecting the negative battery cable.*

3   Remove the air filter housing (see Chapter 4).

4   Remove the drivebelt (see Chapter 1). Also, removing the power steering pump and positioning it aside (without disconnecting the hoses) will make access to the compressor easier. See Chapter 10 for the power steering pump removal procedure.

5   Disconnect the compressor clutch electrical connector **(see illustration)**.

6   Disconnect the refrigerant lines from the compressor. Plug the open fittings to prevent entry of dirt and moisture and discard the O-ring seals.

7   Raise the front of the vehicle and support it securely on jackstands.

8   Remove the front engine splash shield (see Chapter 2A).

9   Unbolt the mounting fasteners and carefully remove the compressor from the vehicle.

Note: *The compressor mounting fasteners also secure a transmission oil cooler line bracket. Be careful not to damage these lines as the compressor is removed.*

10   If a new compressor is being installed, follow the directions with the compressor regarding the draining of excess oil prior to installation.

Note: *Any replacement compressor used must be designated as compatible with R-134a refrigerant.*

11   The clutch may have to be transferred from the original to the new compressor.

12   Installation is the reverse of removal. Replace all O-rings with new ones specifically made for air conditioning system use and compatible with R-134a or R-1234yf refrigerant, as applicable. Lubricate them with refrigerant oil.

13   Have the system evacuated, recharged and leak-tested by the shop that discharged it.

14   Air conditioning receiver-drier - removal and installation

Warning: *The air conditioning system is under high pressure. Do not loosen any hose fittings or remove any components until after the system has been discharged. Air conditioning refrigerant must be properly discharged into an EPA-approved recovery/recycling unit at a dealer service department or an automotive air conditioning repair facility. Always wear eye protection when disconnecting air conditioning system fittings.*

1   Have the air conditioning system discharged by an automotive air conditioning technician.

2   Disconnect the cable from the negative battery terminal (see Chapter 5, Section 1).

Caution: *On 2015 and later models, if equipped with an Intelligent Battery Sensor (IBS), disconnect the IBS connector first before disconnecting the negative battery cable.*

3   Remove the front bumper cover (see Chapter 11).

4   Remove the receiver-drier mounting fastener **(see illustration)**.

5   Pull the receiver-drier directly up and out of the fitting for the condenser. Remove the O-ring seals and discard them.

6   Plug all open fittings to prevent entry of dirt and moisture.

7   Using new O-ring seals, install the receiver-drier into the fitting at the condenser. If a new receiver-drier is being installed, add 1 oz (30 ml) of new refrigerant oil to the new unit.

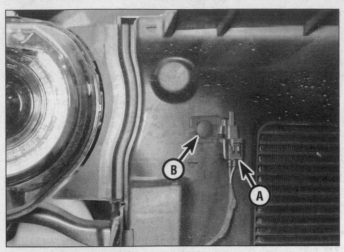

15.4a Disconnect the impact sensor electrical connector (A) then remove the plastic deflector retainer (B) - one on each side

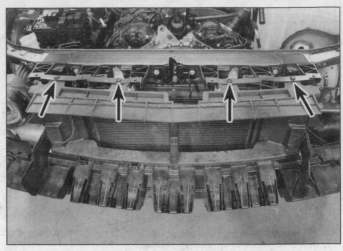

15.4b Remove the plastic trim fasteners from the top of the air deflector

15.4c Press the release tabs downward . . .

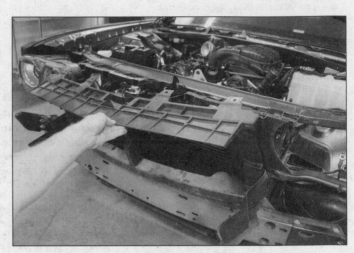

15.4d . . . then pull the air deflector out from the vehicle - later models shown

15.5 Remove the transmission cooler line manifold bolt

8   The remainder of installation is the reverse of removal.

9   Have the air conditioning system evacuated, charged and leak tested by the shop that discharged it.

---

15   Air conditioning condenser - removal and installation

Warning: *The air conditioning system is under high pressure. Do not loosen any hose fittings or remove any components until after the system has been discharged. Air conditioning refrigerant must be properly discharged into an EPA-approved recovery/recycling unit at a dealer service department or an automotive air conditioning repair facility. Always wear eye protection when disconnecting air conditioning system fittings.*

Note: *The receiver-drier (see Section 14) should be replaced whenever the condenser is replaced.*

1   Have the air conditioning system discharged by an automotive air conditioning technician.

2   Disconnect the cable from the negative battery terminal (see Chapter 5, Section 1). **Caution:** *On 2015 and later models, if equipped with an Intelligent Battery Sensor (IBS), disconnect the IBS connector first before disconnecting the negative battery cable.*

3 Remove the front bumper cover (see Chapter 11).

4   Remove the plastic air deflectors on each side of the condenser **(see illustrations)**.

5   Disconnect the transmission oil cooler lines from the condenser **(see illustration)**.

6   Remove the mounting fasteners for the power steering oil cooler (at the front of the condenser) and move the cooler aside.

15.7 Typical location of the refrigerant line fittings on the condenser

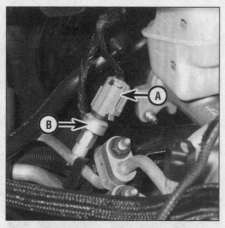

16.2 Unplug the electrical connector (A), then unscrew the switch (B)

17.3 Refrigerant line bracket at the left shock tower

7   Disconnect the refrigerant lines from the condenser **(see illustration)**. Plug the lines and fittings to prevent the entry of moisture and contaminants. Discard the O-ring seals.

8   Remove the fasteners that mount the condenser to the radiator and lower the condenser from the vehicle **(see illustration 7.12)**.

9   If the original condenser will be reinstalled, store it with the line fittings on top to prevent oil from draining out.

10   If a new condenser is being installed, pour 1 oz (30 ml) of the appropriate refrigerant oil into it prior to installation.

11   Install new O-ring seals onto the refrigerant line fittings.

12   Reinstall the components in the reverse order of removal.

13   Have the system evacuated, recharged and leak-tested by the shop that discharged it.

17.4 Remove the refrigerant line fitting nut and fittings to get to the expansion valve mounting fasteners

16   Air conditioning pressure switch - replacement

Warning: *Wear safety glasses when performing this procedure.*

Note: *It is not necessary to discharge the air conditioning system in order to replace the pressure switch (a Schrader valve is present in the fitting under the switch).*

1   Disconnect the cable from the negative battery terminal (see Chapter 5, Section 1).

Caution: *On 2015 and later models, if equipped with an Intelligent Battery Sensor (IBS), disconnect the IBS connector first before disconnecting the negative battery cable.*

2   Unplug the electrical connector from the air conditioning pressure switch **(see illustration)**.

3   Unscrew the pressure switch from the line. Be careful not to damage the refrigerant line. Discard the O-ring seal.

4   Lubricate a new R-134a or R-1234yf (as applicable) compatible O-ring seal with the

appropriate refrigerant oil and place it into position.

5   Screw the new switch onto the refrigerant line valve until hand-tight, and then tighten it securely.

6   Reconnect the electrical connector and the battery cable.

17   Air conditioning expansion valve - removal and installation

Warning: *The air conditioning system is under high pressure. Do not loosen any hose fittings or remove any components until after the system has been discharged. Air conditioning refrigerant must be properly discharged into an EPA-approved recovery/recycling unit at a dealer service department or an automotive air conditioning repair facility. Always wear eye protection when disconnecting air conditioning system fittings.*

1   Have the air conditioning system discharged by an automotive air conditioning technician.

2   Disconnect the cable from the negative battery terminal (see Chapter 5, Section 1).

Caution: *On 2015 and later models, if equipped with an Intelligent Battery Sensor (IBS), disconnect the IBS connector first before disconnecting the negative battery cable.*

3   Remove the bracket that secures the refrigerant lines to the left shock tower **(see illustration)**.

4   Disconnect the refrigerant lines at the firewall by removing the nut that secures the line fitting to the expansion valve **(see illustration)**. Cap or plug the line openings to prevent any dirt or moisture from entering the system.

5   Remove the two exposed mounting screws that secure the expansion valve.

6   Detach the expansion valve from the evaporator fitting and discard the O-rings.

7   Installation is the reverse of removal.

Note: *Use new R-134a or R-1234yf (as applicable) compatible O-rings during reassembly.*

8   Have the system evacuated, recharged and leak-tested by the shop that discharged it.

10 Engine oil cooler - removal and installation

Warning: *The engine must be completely cool before beginning this procedure.*

Note: *On 3.6L V6 models the oil cooler is located under the intake manifold. It is an integral part of the oil filter housing and not serviceable separately from the housing. See Chapter 2C for the oil cooler replacement procedure.*

1   Raise the front of the vehicle and support it securely on jackstands.

2   Drain the cooling system (see Chapter 1).

3   Remove the engine oil filter (see Chapter 1).

4   Squeeze the hose clamps and slide them back on the hoses, then detach the hoses from the oil cooler **(see illustration)**. Be prepared for coolant spillage.

5   Using a large hex bit, unscrew the bolt from the center of the oil cooler, then detach the cooler from the engine.

6   Installation is the reverse of removal, noting the following points:

18.4 Engine oil cooler mounting details (3.5L V6 engine shown)

A   Hose clamps
B   Mounting bolt

a) Clean the mating surfaces and use a new gasket.

b) When installing the cooler, align the cut-out on the cooler with the tab on the oil pan.

c) Tighten the bolt to the torque listed in this Chapter's Specifications. When tightening the bolt, hold the oil cooler to prevent it from turning.

d) Install a new oil filter, refill the cooling system, and check the oil level, adding as necessary (see Chapter 1).

Notes

Notes

Chapter 4
Fuel and exhaust systems

Contents

Specifications

General

Fuel pressure	53 to 63 psi (366 to 434 kPa)
Fuel injector resistance	
2010 and earlier models	4 to 16 ohms @ -4F (-20C)
	6 to 18 ohms @ 68F (20C)
	8 to 20 ohms @ 140F (60C)
	10 to 122 ohms @ 212F (100C)
2011 and later models	9 to 18 ohms*

* Resistance will vary depending on temperature of the injector, similar to the 2010 and earlier models.

Torque specifications   Ft-lbs (unless otherwise indicated)

Note: *One foot-pound (ft-lb) of torque is equivalent to 12 inch-pounds (in-lbs) of torque. Torque values below approximately 15 ft-lbs are expressed in inch-pounds, because most foot-pound torque wrenches are not accurate at these smaller values.*

Throttle body mounting bolts	105 in-lbs
Fuel rail mounting bolts	
2.7L, 5.7L, 6.1L and 6.4L	96 in-lbs
3.5L	21
3.6L	62 in-lbs
6.2L	89 in-lbs

1   Gonoral information and precautions

Fuel system warnings

Warning: *Gasoline is extremely flammable and repairing fuel system components can be dangerous. Consider your automotive repair knowledge and experience before attempting repairs which may be better suited for a professional mechanic.*

a) *Don't smoke or allow open flames or bare light bulbs near the work area*
b) *Don't work in a garage with a gas-type appliance (water heater, clothes dryer)*
c) *Use fuel-resistant gloves. If any fuel spills on your skin, wash it off immediately with soap and water*
d) *Clean up spills immediately*
e) *Do not store fuel-soaked rags where they could ignite*
f) *Prior to disconnecting any fuel line, you must relieve the fuel pressure (see Section 3), then disconnect the cable from the negative terminal of the battery (see Chapter 5)*
g) *Wear safety glasses*
h) *Have a proper fire extinguisher on hand*

Fuel system

1   The fuel system consists of the fuel tank, electric fuel pump/fuel level sending unit (located in the fuel tank), fuel rail, fuel injectors and, on turbocharged models, a high-pressure fuel pump mounted to the end of the cylinder head. The fuel injection system is a multi-port system; multi-port fuel injection uses timed impulses to inject the fuel directly into the intake port of each cylinder. The Powertrain Control Module (PCM) controls the injectors. The PCM monitors various engine parameters and delivers the exact amount of fuel required into the intake ports.
2   Fuel is circulated from the fuel pump to the fuel rail through fuel lines running along the underside of the vehicle. Various sections of the fuel line are either rigid metal or nylon, or flexible fuel hose. The various sections of the fuel hose are connected either by quick-connect fittings or threaded metal fittings.

Exhaust system

3   The exhaust system consists of the exhaust manifold(s), catalytic converter(s), muffler(s), tailpipe and all connecting pipes, flanges and clamps. The catalytic converters are an emission control device added to the exhaust system to reduce pollutants.

2   Troubleshooting

Electric fuel pump

1   The electric fuel pump is located inside the fuel tank. Sit inside the vehicle with the windows closed, turn the ignition key to On (not Start) and listen for the sound of the fuel pump as it's briefly activated. You will only hear the sound for a second or two, but that sound tells you that the pump is working. Alternatively, have an assistant listen at the fuel filler cap.
2   A fuel-cut off system is used in case of an accident. If there has been an accident, the restraint system sends a signal to the Fuel Pump Control Module (FPCM) and shuts the module off. To reset the FPCM, turn the ignition key to the Off position, then to the On position and start the engine.
Note: *This may take a few attempts.*
3   Check the fuel pump fuse and relay **(see illustrations)**. If the fuse and relay are okay, check the wiring back to the fuel pump. If the fuse, relay, wiring and inertia switch are okay, the fuel pump is probably defective. If the pump runs continuously with the ignition key in the On position, the Powertrain Control Module (PCM) is probably defective. Have the PCM checked by a professional mechanic.

Fuel injection system

Note: *The following procedure is based on the assumption that the fuel pump is working and the fuel pressure is adequate (see Section 4).*
4   Check all electrical connectors that are related to the system. Check the ground wire connections for tightness. Verify that the battery is fully charged (see Chapter 5).
5   Inspect the air filter element (see Chapter 1, Section 7).
6   Check all fuses related to the fuel system **(see illustrations 2.3a and 2.3b)**.
7   Check the air induction system between the throttle body and the intake manifold for air leaks. Also inspect the condition of all vacuum hoses connected to the intake manifold and to the throttle body.
8   Remove the air intake duct from the throttle body and look for dirt, carbon, varnish, or other residue in the throttle body, particularly around the throttle plate. If it's dirty, clean it with carb cleaner, a toothbrush and a clean shop towel.
9   Check to see if any trouble codes are stored in the PCM (see Chapter 6, Section 3).

3   Fuel pressure relief procedure

Warning: *Gasoline is extremely flammable. See* Fuel system warnings *in Section 1.*
1   Remove the fuel filler cap (this will relieve any pressure that has built-up in the tank).
2   The fuel pump relay is located in the rear fuse and relay box/rear power distribution center (PDC), which is located in the trunk under the spare tire access panel. Open the cover of the fuse and relay box and remove the fuel pump relay **(see illustrations 2.3a and 2.3b)**.
3   Turn the ignition key to Start and crank over the engine for several seconds. It will either start momentarily and immediately stall, or it won't start at all.
4   Turn the ignition key to the Off position.
5   Disconnect the cable from the negative battery terminal before beginning work on the fuel system (see Chapter 5, Section 1).
6   After all work on the fuel system has been completed, install the fuel pump relay.
7   After all work has been completed, the CHECK ENGINE light or Malfunction Indicator Light (MIL) might come on during operation because the engine was cranked with the fuel pump relay unplugged. The light will likely go out after a period of normal operation. If it does not go out, refer to Chapter 6 Section 3.

4   Fuel pressure - check

Warning: *Gasoline is extremely flammable. See* Fuel system warnings *in Section 1.*
Note: *In order to perform the fuel pressure test, you will need a fuel pressure gauge capable of measuring high fuel pressure. You'll also need the right fittings or adapters to attach it to the fuel rail.*
1   Relieve the fuel system pressure (see Section 3).

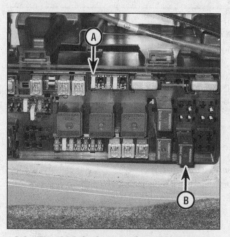

2.3a The 20A fuel pump fuse (A) and the fuel pump relay (B) are located in the fuse and relay box in the trunk (2014 and earlier model shown)

2.3b The 30A fuel pump fuse (A) and the fuel pump relay (B) are located in the fuse and relay box in the trunk (2015 and later model shown)

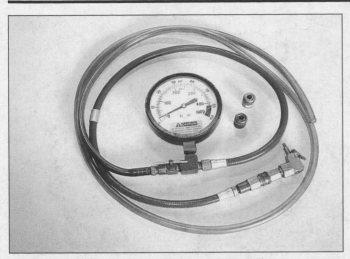

4.2 A typical fuel pressure gauge, with hoses and fittings suitable for tee-ing into the fuel system between the fuel delivery line and the fuel rail

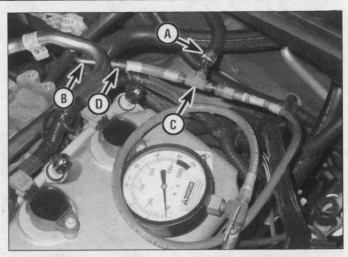

4.3a Fuel gauge connection details (3.5L V6 engine shown, other engines similar; intake manifold removed for clarity)

A Fuel feed line fitting	D Quick-connect fitting
B Fuel rail	attached to fuel rail
C Tee fitting	

2   For this check, you'll need to obtain a fuel pressure gauge with a hose and an adapter suitable for tee-ing it into the fuel system between the fuel supply line quick-connect fitting and the fuel rail inlet pipe **(see illustration)**.

3   Disconnect the fuel supply line quick-connect fitting from the fuel rail inlet pipe (see Section 5) and tee in your fuel pressure gauge **(see illustrations)**.

Note: *You must work in a very confined space to carry out a fuel pressure test. The fuel rail inlet pipe is difficult to reach on some engines because it's located behind and below the intake manifold, near the firewall. The accompanying illustration depicts a typical hook-up (on a 3.5L V6 engine) with the intake manifold removed for clarity. But you can't, of course, remove the intake manifold and still operate the engine. On some models you can make a little bit of working room by removing the cowl covers (see Chapter 11).*

4   Start the engine and check the pressure on the gauge, comparing your reading with the pressure listed in this Chapter's Specifications.

5   If the fuel pressure is not within specifications, check the following:

If the pressure is lower than specified, check for a restriction in the fuel system. Two likely suspects are the fuel inlet strainer at the base of the fuel pump/fuel level sensor module in the left fuel tank compartment and the main fuel filter in the right fuel tank compartment, or the fuel pressure regulator (see Section 9).

If the fuel pressure is higher than specified, replace the fuel pressure regulator, which is part of the auxiliary fuel pump/fuel level sensor module in the right side of the fuel tank (see Section 9).

Note: *On 2010 and earlier models, the fuel pressure regulator is a serviceable component and can be replaced separately (see Section 9). On 2011 and later models, neither of these components can be replaced separately.*

4.3b Fuel line connection at the fuel rail - 3.6L V6 engine

5   Fuel lines and fittings - general information and disconnection

Warning: *Gasoline is extremely flammable. See Fuel system warnings in Section 1.*

1   Relieve the fuel pressure before servicing fuel lines or fittings (see Section 3), then disconnect the cable from the negative battery terminal (see Chapter 5, Section 3) before proceeding.

2   The fuel supply line connects the fuel pump in the fuel tank to the fuel rail on the engine. The Evaporative Emission (EVAP) system lines connect the fuel tank to the EVAP canister and connect the canister to the intake manifold.

3   Whenever you're working under the vehicle, be sure to inspect all fuel and evaporative emission lines for leaks, kinks, dents and other damage. Always replace a damaged fuel or EVAP line immediately.

4   If you find signs of dirt in the lines during disassembly, disconnect all lines and blow them out with compressed air. Inspect the fuel strainer on the fuel pump pick-up unit for damage and deterioration.

Steel tubing

5   It is critical that the fuel lines be replaced with lines of equivalent type and specification.

6   Some steel fuel lines have threaded fittings. When loosening these fittings, hold the stationary fitting with a wrench while turning the tube nut.

Plastic tubing

7   When replacing fuel system plastic tubing, use only original equipment replacement plastic tubing.

Caution: *When removing or installing plastic fuel line tubing, be careful not to bend or twist it too much, which can damage it. Also, plastic fuel tubing is NOT heat resistant, so keep it away from excessive heat.*

Disconnecting Fuel Line Fittings

Two-tab type fitting; depress both tabs with your fingers, then pull the fuel line and the fitting apart

On this type of fitting, depress the two buttons on opposite sides of the fitting, then pull it off the fuel line

Threaded fuel line fitting; hold the stationary portion of the line or component (A) while loosening the tube nut (B) with a flare-nut wrench

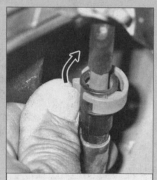

Plastic collar-type fitting; rotate the outer part of the fitting

Metal collar quick-connect fitting; pull the end of the retainer off the fuel line, and disengage the other end from the female side of the fitting . . .

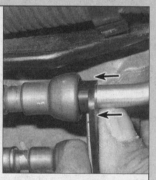

. . . insert a fuel line separator tool into the female side of the fitting, push it into the fitting until it releases the locking tabs inside the fitting, and pull the two halves of the fitting apart

Hairpin-type clip; spread the two legs of the clip apart . . .

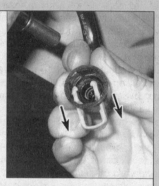

. . . pull the clip out and detach the coupling from the component (fitting detached for clarity)

Spring-lock coupling; remove the safety cover . . .

. . . install a coupling release tool and close the clamshell halves of the tool around the coupling . . .

. . . push the tool into the fitting, then pull the two lines apart

6.1 Exhaust system hangers. Inspect regularly and replace at the first sign of damage or deterioration

6.2a Identifying the collector nuts and studs

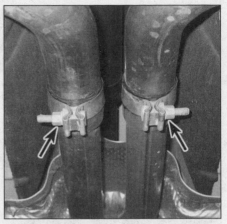

6.2b Identifying the exhaust band clamps

6.2c Identifying the exhaust ball coupling, bolts and springs

6.2d Identifying exhaust system ground strap (3.6L shown)

Flexible hoses

8 When replacing fuel system flexible hoses, use only original equipment replacements.

9 Don't route fuel hoses (or metal lines) within four inches of the exhaust system or within ten inches of the catalytic converter. Make sure that no rubber hoses are installed directly against the vehicle, particularly in places where there is any vibration. If allowed to touch some vibrating part of the vehicle, a hose can easily become chafed and it might start leaking. A good rule of thumb is to maintain a minimum of 1/4-inch clearance around a hose (or metal line) to prevent contact with the vehicle underbody.

6 Exhaust system servicing - general information

Warning: *Allow exhaust system components to cool before inspection or repair. Also, when working under the vehicle, make sure it is securely supported on jackstands.*

1 The exhaust system consists of the exhaust manifolds, catalytic converter, muffler, tailpipe and all connecting pipes, flanges and clamps. The exhaust system is isolated from the vehicle body and from chassis components by a series of rubber hangers **(see illustration)**. Periodically inspect these hangers for cracks or other signs of deterioration, replacing them as necessary.

2 Conduct regular inspections of the exhaust system to keep it safe and quiet. Look for any damaged or bent parts, open seams, holes, loose connections, excessive corrosion or other defects which could allow exhaust fumes to enter the vehicle **(see illustrations)**. Do not repair deteriorated exhaust system components; replace them with new parts.

3 If the exhaust system components are extremely corroded, or rusted together, a cutting torch is the most convenient tool for removal. Consult a properly-equipped repair shop. If a cutting torch is not available, you can use a hacksaw, or if you have compressed air, there are special pneumatic cutting chisels that can also be used. Wear

safety goggles to protect your eyes from metal chips and wear work gloves to protect your hands.

4 Here are some simple guidelines to follow when repairing the exhaust system:

a) *Work from the back to the front when removing exhaust system components.*

b) *Apply penetrating oil to the exhaust system component fasteners to make them easier to remove.*

c) *Use new gaskets, hangers and clamps.*

d) *Apply anti-seize compound to the threads of all exhaust system fasteners during reassembly.*

e) *Be sure to allow sufficient clearance between newly installed parts and all points on the underbody to avoid overheating the floor pan and possibly damaging the interior carpet and insulation. Pay particularly close attention to the catalytic converter and heat shield.*

7 Active exhaust valve actuator - replacement

Note: *The active exhaust valve is attached to the exhaust pipe, near the rear subframe, between the resonators and the mufflers.*

1 Raise and support the vehicle on jackstands.

2 Locate the active exhaust valve actuator and disconnect the electrical connector.

3 Remove the nuts attaching the actuator to the exhaust, then detach the actuator.

4 Installation is reverse of removal, making sure the actuator spring is engaged properly with the slots in the mechanical portion of the valve (the exhaust valve shaft). If the spring does not seat into the slots:

a) *Install the actuator with the nuts finger tight and the electrical connector unplugged.*

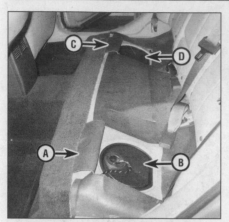

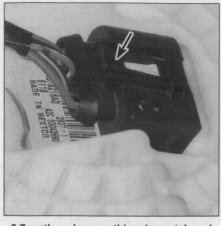

8.5 Peel back this foam pad (A) to gain access to the fuel pump/fuel level sensor module (B). Peel back this foam pad (C) to access the fuel filter/fuel pressure regulator/fuel level sensor module (D)

8.6 Carefully lift off the plastic access cover. . .

8.7. . . then depress this release tab and disconnect the electrical connector

b) Turn the ignition to the Run position (don't start the engine). On models with push-button ignition, leave your foot off the brake pedal and press the Start button two times to engage Run mode.

c) Reconnect the electrical connector; the valve actuator will cycle and the spring will seat into its slots. If it does not, the actuator will cycle every minute and the Check Engine light will turn on.

d) Turn the ignition Off. Tighten the nuts securely.

8   Fuel pump/fuel level sensor and fuel filter/fuel level sensor modules - removal and installation

Fuel pump/fuel level sensor module

Warning: Gasoline is extremely flammable. See Fuel system warnings in Section 1.

Note: This component may also be referred to as the main fuel pump module. A fuel pump/fuel level sensor module is located in the left compartment of the fuel tank. This module includes the fuel pump inlet strainer, the pump, a fuel level sensor and the fuel pressure regulator. This section covers the removal and installation of the complete module, which is removed as a complete assembly. Most replacement modules come as a complete unit with all of these components. The auxiliary fuel pump module is located in the right side of the fuel tank and includes the main fuel filter and the fuel pressure regulator. On 2010 and earlier models, the fuel pressure regulator is a serviceable component and can be replaced separately (see Section 9). On 2011 and later models, neither of these components can be replaced separately.

1   Relieve the fuel system pressure (see Section 3).

2   Disconnect the cable from the negative battery terminal (see Chapter 5, Section 1).

3   Drain the fuel tank to under 5/8-full with a hose routed through the fuel filler neck hose. Use a hard nylon tube with a 30-degree cut on the end to push open the check valve in the fuel filler neck.

Warning: If this is not done, fuel will spill into the interior of the vehicle when the fuel pump module lock-ring is loosened.

Warning: Do NOT start the siphoning action by mouth! Use a siphoning kit (available at most auto parts stores).

4   Remove the rear seat cushion (see Chapter 11).

5   Pull back the foam pad covering the fuel pump/fuel level sensor module or the pad covering the fuel filter/fuel level sensor module **(see illustration)**.

6   Pull off the plastic access cover **(see illustration)**. The wiring harness between the grommet in the access cover and the electrical connector for the fuel pump and fuel level sensor is short, so be careful.

7   Disconnect the electrical connector from the terminal on top of the fuel pump/fuel level sensor module mounting flange **(see illustration)** and place the harness safely out of the way.

8   To prevent dirt from entering the fuel tank, clean the area surrounding the mounting flange for the fuel pump/fuel level sensor module. Mark the location of the fuel pump module in relation to the fuel tank.

9   Use a fuel pump module lock-ring wrench or a hammer and a brass drift, tap on the lock-ring to turn it counterclockwise **(see illustration)**. When the lock-ring is loose, unscrew it.

Warning: Do NOT use a steel drift to loosen the lock-ring. Doing so could produce sparks, which could ignite fuel vapors from the tank.

10   Before removing the fuel pump/fuel level sensor module, note the square tab on the edge of the mounting flange, which faces forward **(see illustration)**. The flange must be oriented just like this when you install the fuel pump/fuel level sensor module.

8.9 To loosen the lock-ring, tap it counterclockwise with a hammer and a brass drift (using a steel drift can be dangerous because it might cause a spark)

8.10 Note how the square tab on the mounting flange faces forward. When you install the fuel pump/fuel level sensor module, make sure that the square tab is oriented just like this

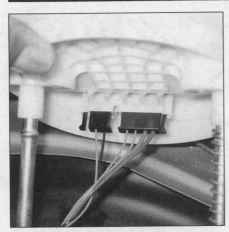

8.11 Lift up the mounting flange for the fuel pump/fuel level sensor module and disconnect these two electrical connectors

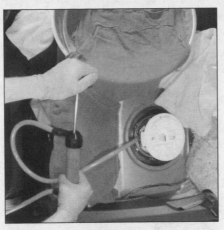

8.12 A hand-held fluid pump can be used to remove as much fuel as possible from the tank

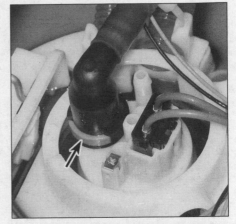

8.13 To disconnect the fuel supply line quick-connect fitting from the outlet pipe on top of the fuel pump, depress this release tab and pull off the fitting

11   Lift up the mounting flange for the fuel pump/fuel level sensor module and disconnect the two electrical connectors **(see illustration)** for the fuel level sensors and the fuel pump.

12   If the lower half of the module (the fuel pump part) is still submerged in gasoline, drain the rest of the fuel from the left fuel tank. Lift up the fuel pump module far enough to insert a siphon hose into the tank and siphon or pump out as much gas as possible **(see illustration)**.

Warning: *Once again, do NOT start the siphoning action by mouth! Use a siphoning kit (available at most auto parts stores).*

13   Disconnect the fuel supply line fitting from the outlet pipe on top of the fuel pump **(see illustration)**.

14   Disconnect the two return lines from the fuel pump module **(see illustration)**.

15   To drain the residual fuel from the reservoir inside the fuel pump module, reach through the opening and tilt the fuel pump module on its side to allow the fuel to drain out the top of the module **(see illustration)**.

16   Carefully remove the fuel pump/fuel level sensor module from the fuel tank **(see illustration)**. Tilt the pump/sensor module as necessary to protect the fuel level sensor float arm from damage.

17   If you want to replace the fuel pump module or the fuel level sensor, refer to Section 9.

18   Installation is the reverse of removal.

19   After installation, turn the ignition switch to On, but don't operate the starter. This activates the fuel pump for about two seconds, which builds up fuel pressure in the fuel lines. Repeat this step two or three times, then for fuel leakage.

Fuel filter/fuel pressure regulator/fuel level sensor module

Warning: *See the Warning in Section 1.*

Note: *This component may also be referred to as the auxiliary fuel pump module, located in the right compartment of the fuel tank. This module includes the fuel filter, the fuel pressure regulator and a fuel level sensor. This Section covers the removal and installation of the complete module, which is removed as a complete assembly. Most replacement modules come as a complete unit with all of these components. On 2010 and earlier models, the fuel pressure regulator is a serviceable component and can be replaced separately (see Section 9). On 2011 and later models, neither of these components can be replaced separately.*

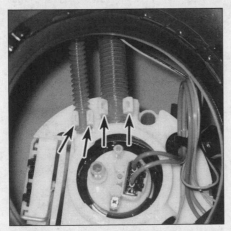

8.14 To disconnect the fuel return lines from the fuel pump module, simply lift them up and pop them loose from these retaining tangs (these two lines connect the fuel filter/fuel pressure regulator in the right fuel tank compartment to the fuel pump module)

8.15 To drain the residual fuel from the reservoir inside the fuel pump module, reach through the opening and tilt the fuel pump module on its side to allow the fuel to drain out the top of the module

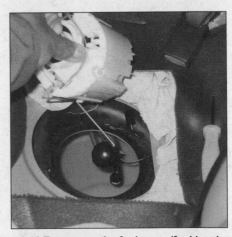

8.16 To remove the fuel pump/fuel level sensor module from the fuel tank, angle it as shown to protect the fuel level sensor float arm from damage

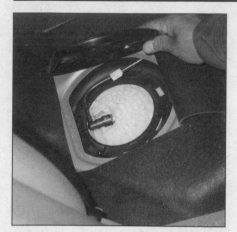

8.21 To access the fuel filter/fuel pressure regulator/fuel level sensor module, remove this plastic access cover

8.22 To disconnect the quick-connect fitting for the fuel supply line, squeeze these two tabs with a pair of needle nose pliers and pull off the fitting

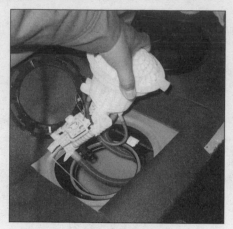

8.26 Carefully lift the fuel filter/fuel pressure regulator/fuel level sensor out of the fuel tank. Be careful not to snap the fuel return lines or the electrical wiring harnesses

20   Remove the fuel pump/fuel level sensor module from the left fuel tank compartment (see the previous procedure).

21   Pull off the plastic access cover **(see illustration)**.

22   Disconnect the fuel supply line quick-connect fitting **(see illustration)**.

23   To prevent dirt from entering the fuel tank, clean the area surrounding the mounting flange for the fuel filter/fuel pressure regulator/fuel level sensor module.

24   Use a fuel pump module lock-ring wrench or a hammer and a brass drift, tap on the lock-ring to turn it counterclockwise **(see illustration 8.9)**. When the lock-ring is loose, unscrew it.

Warning: *Do NOT use a steel drift to loosen the lock-ring. Doing so could produce sparks, which could ignite fuel vapors from the tank.*

25   Before removing the fuel filter/fuel pressure regulator/fuel level sensor module, note the square tab on the circumference of the mounting flange, which faces forward **(see**

illustration 8.10)**. The flange must be oriented just like this when you install the fuel filter/fuel pressure regulator/fuel level sensor module.

26   Lift up the mounting flange for the fuel filter/fuel pressure regulator/fuel level sensor module and carefully remove the module **(see illustration)**. The electrical harness and fuel return lines, which you have already disconnected from the fuel pump/fuel level sensor module, are permanently attached to the module that you're removing, so make sure that they don't get caught on anything.

27   If you want to replace the fuel filter, the fuel pressure regulator or the fuel level sensor, refer to Section 9.

28   Installation is the reverse of removal.

29   After installation, turn the ignition switch to On, but don't operate the starter. This activates the fuel pump for about two seconds, which builds up fuel pressure in the fuel lines. Repeat this step two or three times, then for fuel leakage.

9   Fuel pump/fuel level sensor module and fuel filter/fuel pressure regulator/fuel level sensor module - component replacement

Warning: *Gasoline is extremely flammable. See* Fuel system warnings *in Section 1.*

Left fuel level sensor and fuel pump module

Note: *This may also be referred to as the main fuel pump module.*

1   Remove the fuel pump/fuel level sensor module (see Section 8).

2   Carefully cut the wiring harness clip **(see illustration)** and disengage the fuel level sensor wires from the clip.

3   Loosen the two retaining clips **(see illustration)** and pull the fuel level sensor out of the fuel pump module.

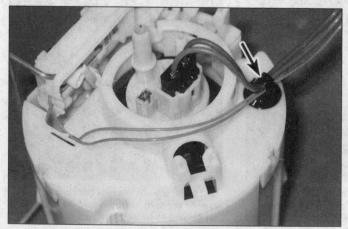

9.2 Carefully cut this wiring harness clip and disengage the fuel level sensor leads

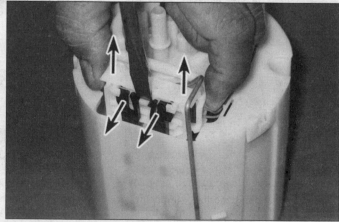

9.3 To remove the left fuel level sensor from the fuel pump module, pry these two clips loose and pull the sensor straight up

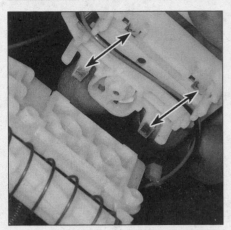

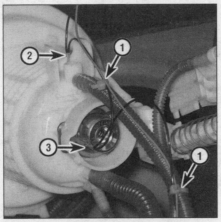

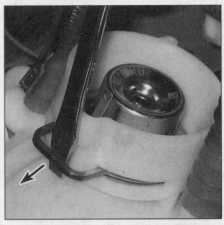

9.8 To detach the fuel level sensor from the fuel filter module, spread these four locking tangs away from the module and pull off the sensor

9.9 To remove the fuel level sensor from the fuel filter module, cut the two wire ties (1), then disconnect the ground wires from the ground terminal (2) at the check valve and from the ground terminal (3) at the fuel pressure regulator

9.11 To remove the fuel pressure regulator from the fuel filter module, pry off this retainer clip with a screwdriver

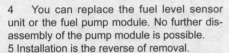

4   You can replace the fuel level sensor unit or the fuel pump module. No further disassembly of the pump module is possible.
5 Installation is the reverse of removal.
6   After installation, turn the ignition switch to On, but don't operate the starter. This activates the fuel pump for about two seconds, which builds up fuel pressure in the fuel lines. Repeat this step two or three times, then for fuel leakage.

Right fuel level sensor, fuel pressure regulator and fuel filter module

Note: *This may also be referred to as the auxiliary fuel pump module.*
Note: *On 2010 and earlier models, the fuel pressure regulator is a serviceable component and can be replaced separately. On 2011 and later models, neither of these components can be replaced separately.*

7   Remove the fuel filter/fuel pressure regulator/fuel level sensor module (see Section 8).
8   Detach the fuel level sensor from the fuel filter module (see illustration).
9   Carefully cut the two wire ties that secure the wiring to the fuel return lines, disconnect the two ground wires from the fuel filter module (see illustration) and remove the fuel level sensor.
10   If you're just replacing the fuel pressure regulator, this is as far as you need to go. Installation is the reverse of removal.
11   To remove the fuel pressure regulator from the fuel filter module, pull out the retaining clip (see illustration).
12   Inspect the condition of the fuel pressure regulator O-rings (see illustration).
13   If you're just replacing the fuel pressure regulator, this is as far as you need to go. Installation is the reverse of removal.
14   If you're replacing the fuel filter module, remove the ground ring (see illustration 6.14) from the recess for the fuel pressure regulator.

15   When installing the ground ring in the new fuel filter module, make sure that the ground ring terminal is seated correctly in the notch (see illustration).
16   Installation is otherwise the reverse of removal.
17   After installation, turn the ignition switch to ON, but don't operate the starter. This activates the fuel pump for about two seconds, which builds up fuel pressure in the fuel lines. Repeat this step two or three times, then for fuel leakage.

10   Fuel tank - removal and installation

Warning: *Gasoline is extremely flammable. See Fuel system warnings in Section 1.*
1   Relieve the fuel system pressure (see Section 3).

9.12 Inspect the condition of these two O-rings (you might find the smaller O-ring here, on the regulator, or it might be in the recess for the regulator in the underside of the fuel filter module)

2   Disconnect the cable from the negative battery terminal (see Chapter 5, Section 1).
3   If possible, drain as much fuel out of the tank through the fuel filler neck tube as you can. Use a hard nylon tube with a 30-degree cut on the end to push open the check valve in the fuel filler neck.
Warning: *Never start the siphoning action by mouth!*
Note: *The fuel tank can be drained by disconnecting the fuel supply line at the fuel rail, using a scan tool to activate the fuel pump, and draining into an approved container.*
4   Remove the lower cushion from the back seat (see Chapter 11).
5   Remove the foam pad covering the access covers for the fuel pump/fuel level sensor module and the fuel filter/fuel pressure regulator/fuel level sensor (see illustration 8.5).
6   Remove the access cover for the fuel pump/fuel level sensor module (see illustration 8.6).

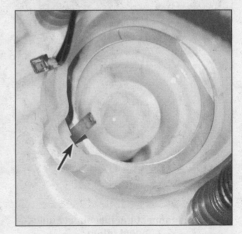

9.15 The arrow indicates the notch in which the ground ring terminal must be seated when installing the ring (the terminal has been moved out of the notch and offset a little so that you can see the notch)

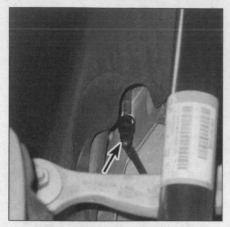

10.10 In the left rear wheel well, disconnect the quick-connect fitting for the fuel filler neck vent line

10.11 In the right rear wheel well, disconnect the quick-connect fitting for the EVAP vent line

10.12 Also in the right rear wheel well, disconnect the EVAP line quick-connect fitting

7   Disconnect the electrical connector from the fuel pump/fuel level sensor module **(see illustration 8.7)**.

8   Loosen the rear wheel lug nuts. Raise the vehicle and place it securely on jackstands. Remove the rear wheels.

9   Remove the left rear inner fender shield.

10   Disconnect the quick-connect fitting **(see illustration)** for the filler neck vent line in the left rear wheel well (see Section 5 for information on fuel line fitting disconnection)

11   Disconnect the quick-connect fitting **(see illustration)** for the vent line quick-connect fitting in the right rear wheel well.

12   Disconnect the EVAP line quick-connect fitting **(see illustration)**, which is also located in the right rear wheel well.

13   Disconnect the fuel supply line quick-connect fitting **(see illustration)**.

14   Remove the under-body splash shields **(see illustration)**.

15   Loosen the hose clamps for the fuel filler neck hose **(see illustration)** and disconnect the fuel filler neck pipe from the filler neck hose. Remove the nut attaching the fuel filler to the body and position the fuel filler neck to the side.

16   Remove the exhaust pipe(s) and the driveshaft (see Chapter 8, Section 9).

17   If equipped, remove the heat shield from under the vehicle.

18   Support the fuel tank.

19   Remove the fuel tank retaining strap bolts and remove both retaining straps.

20   Carefully lower the fuel tank just far enough to pull the fuel tank filler neck vent line through the gap between the floorpan and the left end of the rear suspension crossmember. Pull the EVAP vent line through the gap between the floorpan and the right end of the suspension crossmember. Then make a final check of the upper part of the tank for any-

thing else that still needs to be disconnected. When you have verified that EVERYTHING is disconnected, slowly lower the fuel tank.

21   Installation is the reverse of removal, noting the following points:

a)   *If you're replacing the fuel tank, remove all components from the old fuel tank and install them on the new tank. If you need help with the fuel pump/fuel level sensor module and the fuel filter/fuel pressure regulator/fuel level sensor module, refer to Sections 5 and 6. If you need help with the EVAP lines, refer to Chapter 6.*

b)   *Tighten the fuel tank strap bolts securely.*

22   After installation, turn the ignition switch to ON, but don't operate the starter. This activates the fuel pump for about two seconds, which builds up fuel pressure in the fuel lines. Repeat this step two or three times, then for fuel leakage.

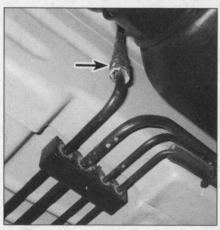

10.13 Disconnect the fuel supply quick-connect fitting

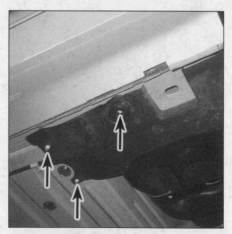

10.14 To detach the under-body splash shields, remove the two nuts at the front and the bolt in the middle, then pull the splash shield down to disengage the locator pins at the rear (left side shown)

10.15 To disconnect the fuel filler neck hose, loosen the screws for the hose clamps (A) and pull off the hose. To detach the left half of the fuel tank, remove the strap bolts (B) and remove the strap (right-side straps similar)

11.3 Air intake duct details (3.5L engine shown, 2.7L engine similar)

1 Intake Air Temperature (IAT) sensor electrical connector
2 Hose clamps

11.8a Before removing the air filter housing, disconnect the Positive Crankcase Ventilation (PCV) fresh air inlet hose (1) from the air filter housing cover, then remove the filterhousing retaining bolt (2) (2014 and earlier model shown)

11 Air filter housing and air intake duct - removal and installation

V6 models

Note: *The accompanying photos depict the air intake duct and air filter housing used on 3.5L models. However, the air intake duct and air filter housing used on 2.7L and 3.6L models are quite similar.*

1 Disconnect the negative battery cable (see Chapter 5, Section 3).
2 Remove the engine cover by pulling it straight up and off of the ballstuds.

Air intake duct

3 Disconnect the electrical connector from the Intake Air Temperature (IAT) sensor **(see illustration)**.
4 Loosen the hose clamp screws and remove the air intake duct.
Note: *The duct may be held in place with rubber grommets on ballstuds. Pull firmly on the duct to release from the studs.*
5 Inspect the condition of the air intake duct. Look for cracks, tears, deterioration and other damage. If the air intake duct is damaged in any way, replace it. A leaking air intake duct will allow the introduction of "false air" (unmetered air) into the air intake manifold, which will cause the air/fuel mixture to become excessively lean. A lean air/fuel mixture can cause rough running at idle, and even misfires if the leak is big enough.
6 Installation is the reverse of removal.

Air filter housing

Note: *The following procedure is for replacing the air filter housing, not for replacing the air filter element. If you want to replace the air filter element, refer to Chapter 1.*

7 Disconnect the air intake duct from the air filter housing **(see illustration 9.1)**.
8 Disconnect the Positive Crankcase Ventilation (PCV) fresh air inlet hose from the air filter housing **(see illustrations)**.
9 Remove the air filter housing retaining bolt **(see illustration 11.8a)** and remove the housing by lifting it straight up. The air filter housing is positioned by a single locator pin inserted into a hole in the fender and by a large inlet pipe on the bottom of the housing that fits over the outlet pipe of the air resonator box.
10 Inspect the condition of the rubber insulator that covers the locator pin **(see illustration)**. If it's cracked, torn, deteriorated or otherwise damaged, replace it.
11 Installation is the reverse of removal. Make sure that the air filter housing locator pin is aligned with its mounting hole in the fender and that the inlet pipe on the bottom of

11.8b Before removing the air filter housing, intake air duct (A), the fresh air inlet hose (B) from the air filter housing cover, then remove the filter housing retaining bolt (C) (2015 and later model shown)

11.10 Inspect the condition of the rubber insulator covering this locator pin and replace it if it's damaged or deteriorated (V6 models)

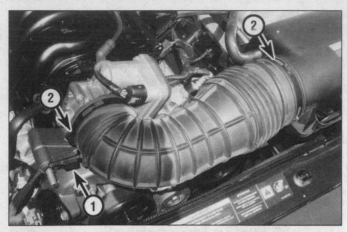

11.13 Air intake duct details (V8 engine)

1   *Intake Air Temperature (IAT) sensor electrical connector*
2   *Hose clamps*

11.16 To detach the air filter housing from a V8 model, remove the retaining bolt

the filter housing fits over the outlet pipe of the air resonator box, then install the filter housing retaining bolt.

V8 models

Note: *The accompanying photos depict the 5.7L model, but the 6.1L and 6.4L engines are similar.*

12   Remove the engine cover.

13   Disconnect the electrical connector from the Intake Air Temperature (IAT) sensor **(see illustration)**.

14   Loosen the hose clamp screws for the air intake duct at the throttle body.

15   Disconnect the Positive Crankcase Ventilation (PCV) fresh air inlet hose from the air filter housing.

16   Remove the air filter housing retaining bolt **(see illustration)** and remove the housing by lifting it straight up. The air filter housing is positioned by a pair of locator pins inserted into holes in the fender and by a large inlet pipe on the bottom of the housing that fits over the outlet pipe of the air resonator box.

Disconnect the air intake duct from the throttle body when removing the housing.

17   Inspect the condition of the rubber insulators covering the locator pins **(see illustration)**. If they're cracked, torn, deteriorated or otherwise damaged, replace them.

18   Inspect the condition of the air intake duct. Look for cracks, tears, deterioration and other damage. If the air intake duct is damaged in any way, replace it. A leaking air intake duct will allow the introduction of "false air" (unmetered air) into the air intake manifold, which will cause the air/fuel mixture to become excessively lean. A lean air/fuel mixture can cause rough running at idle, and even misfires if the leak is big enough.

19   Installation is the reverse of removal. Make sure that the air filter housing locator pins are aligned with three mounting holes in the fender and that the inlet pipe on the bottom of the filter housing fits over the outlet pipe of the air resonator box, then install the filter housing retaining bolt.

12   Throttle body - removal and installation

Caution: *Do NOT use spray carburetor cleaners or silicone lubricants on any part of the throttle body.*

V6 models

Note: *The accompanying photos depict the throttle body on a 3.5L V6, but the throttle body unit on a 2.7L and 3.6L engines are virtually identical.*

1   Disconnect the cable from the negative terminal of the battery (see Chapter 5, Section 1).

2   Remove the air intake duct (see Section 11).

3   Disconnect the electrical connector from the throttle body **(see illustration)**.

4   On 3.5L models, remove the two nuts that attach the intake manifold bracket to the throttle body **(see illustration)**, then remove the two lower bolts that attach the bracket to

11.17 Inspect the condition of the rubber insulators covering the two locator pins and replace them if they're damaged or deteriorated (V8 models)

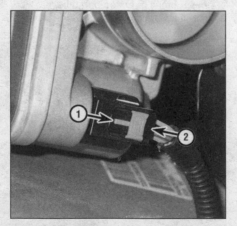

12.3 To disconnect the electrical connector from the throttle body, slide the lock (1) toward the harness, then depress the release tab (2) and pull off the connector

12.4 To remove the intake manifold bracket from a 3.5L V6 engine, remove the two upper nuts that attach the bracket to the throttle body and the two lower bolts that attach it to the cylinder head

12.5 To remove the throttle body, remove these four bolts

12.6 Remove the throttle body O-ring and inspect it for cracks, tears and deterioration. If it's damaged, replace it

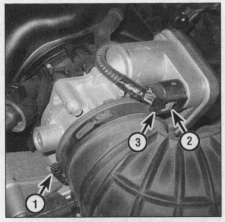

12.10 Loosen the hose clamp screw (1) and disconnect the air intake duct from the throttle body. Then slide the lock (2) toward the harness, depress the release tab (3) and pull off the connector (5.7L model shown)

the cylinder head and remove the bracket.

5   Remove the four mounting bolts **(see illustration)** and remove the throttle body.

6   Remove the throttle body O-ring **(see illustration)** and inspect it for cracks, tears and deterioration. If it's damaged, replace it.

7   Wipe off the gasket mating surfaces of the throttle body and the intake manifold.

8   Installation is the reverse of removal. Be sure to tighten the throttle body mounting bolts to the torque listed in this Chapter's Specifications. If a Diagnostic Trouble Code (DTC) is set, take the vehicle to a dealer and have the service department use a factory scan tool to re-learn parameters.

V8 models

Note: *The accompanying photos depict the 5.7L model, but the 6.1L and 6.4L engines are similar.*

9   Disconnect the cable from the negative terminal of the battery (see Chapter 5, Section 1).

10   Disconnect the air intake duct from the throttle body **(see illustration)**.

11   Disconnect the electrical connector from the throttle body.

12   Remove the four throttle body mounting bolts **(see illustration)** and remove the throttle body.

13   Remove the throttle body's O-ring type gasket **(see illustration 12.6)** inspect it for cracks, tears and deterioration. If it's damaged, replace it.

14   Wipe off the gasket mating surfaces of the throttle body and the intake manifold.

15   Installation is the reverse of removal. Be sure to tighten the throttle body mounting bolts to the torque listed in this Chapter's Specifications. If a Diagnostic Trouble Code (DTC) is set, take the vehicle to a dealer and have the

service department use a factory scan tool to re-learn parameters.

---

13 Fuel rail and injectors - test, removal and installation

Warning: *Gasoline is extremely flammable. See* Fuel system warnings *in Section 1.*

Injector resistance test

1   Remove the engine cover.

2   With the ignition off, disconnect the fuel injector connector **(see illustration)**.

3   Using a Digital Volt-Ohm Meter (DVOM) measure the resistance between the injector terminals. Compare the reading against the values listed in this Chapter's Specifications **(see illustration)**.

4   If the injector is not within specification, replace the injector.

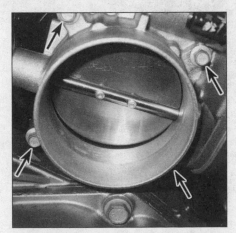

12.12 To remove the throttle body from a V8 engine, remove these four bolts (5.7L model shown)

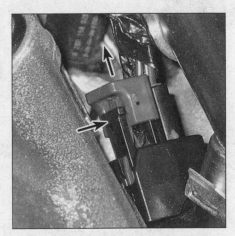

13.2 Disconnect the injector connector

13.3 Measure injector resistance using a DVOM

13.9a To disconnect the quick-connect fitting for the fuel supply line, insert the tool over the fitting, then push it into the female side of the fitting to disengage the wire spring inside. . .

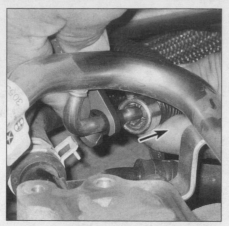

13.9b. . . and pull the female side of the fitting off the fuel supply pipe

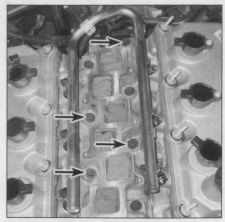

13.12 To detach the fuel rail from the engine, remove these four mounting bolts (3.5L V6 engine shown)

V6 models

Note: *The accompanying photos depict the 3.5L V6, but the 2.7L and 3.6L engines are virtually identical.*

Note: *You'll need to obtain a special fuel line disconnection tool, available at auto parts stores or from automotive tool manufacturers, to disconnect the metal-collar type fitting that connects the fuel supply line to the fuel rail.*

5   Relieve the fuel system pressure (see Section 3).

6   Disconnect the cable from the negative terminal of the battery (see Chapter 5, Section 1).

7   Remove the air intake duct (see Section 11).

8   Remove the upper intake manifold (2.7L engines, see Chapter 2A, Section 8; 3.5L engines, see Chapter 2B, Section 8; 3.6L engines, see Chapter 2C, Section 5). If equipped, remove the insulator from the left side of the engine.

9   Disconnect the fuel supply line from the fuel rail **(see illustrations)**.

Note: *You will need a special tool to disconnect this metal-collar type of quick-connect fitting. There are several versions of this tool. You can find these tools at your local auto parts store.*

10   Disconnect the electrical connectors from the fuel injectors **(see illustration 12.1)** and set the injector harness aside.

Note: *Each connector should be numbered with the corresponding cylinder number. If the number tag is obscured or missing, renumber the connectors. Also, the injector harness includes leads for other components, such as the ignition coils, but don't disconnect anything else unless it's necessary to do so in order to place the harness out of the way.*

11   Clean any debris from around the injectors so that it doesn't fall into the injector holes when you remove the fuel rail and injector assembly.

12   Remove the fuel rail mounting nuts/bolts **(see illustration)**.

13   Gently rock the fuel rail and injectors to loosen the injectors and remove the fuel rail and fuel injectors as a single assembly **(see illustration)**.

Caution: *Do not attempt to separate the left and right fuel rails. Both sides are serviced together as an assembly.*

14   Remove the safety clips and injectors from the fuel rail, then remove and discard the O-rings **(see illustrations)**.

Note: *Whether you're replacing an injector or a leaking O-ring, it's a good idea to remove all the injectors from the fuel rail and replace all the O-rings.*

15   Coat the new O-rings with clean engine oil and install them on the injector(s), then install the injector retaining clip and insert

13.13 Using a gentle rocking motion, wiggle the injectors free and lift the fuel rail and injectors out of the engine compartment as a single assembly

13.14a Using a screwdriver or pliers, remove the injector retaining clip. . .

13.14b. . . and withdraw the injector from the fuel rail

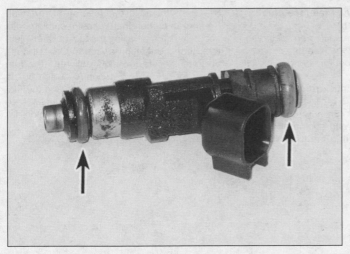

13.14c Carefully remove the O-rings from the injectors

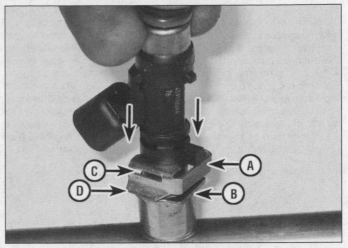

13.15 Injector installation details

A   Flat part of retainer aligned with. . .
B   . . . flat part of flange on fuel rail
C   Slots in retainer will engage with. . .
D   . . . semi-circular flanges when injector is pushed into pipe

each injector into its corresponding bore in the fuel rail until the retainer clip snaps into place **(see illustration)**.

16   Clean the injector bores on the intake manifold.

17   Install the injectors into the fuel rail and secure with the injector clips.

18   Guide the injectors/fuel rail assembly into the injector bores on the intake manifold. Make sure the injectors are fully seated, then tighten the fuel rail mounting bolts to the torque listed in this Chapter's Specifications.

19   The remainder of installation is the reverse of removal. After installing the fuel rail and injectors, turn the ignition switch to On, but don't operate the starter. This activates the fuel pump for about two seconds, which builds up fuel pressure in the fuel lines and the fuel rail. Repeat this step two or three

times, then check the fuel lines, fuel rail and injectors for fuel leakage.

V8 engines

Note: *The accompanying photos depict the 5.7L model, but the 6.1L and 6.4L engines are similar.*

Note: *You'll need to obtain a special fuel line disconnection tool, available at auto parts stores or from automotive tool manufacturers, to disconnect the metal-collar type fitting that connects the fuel supply line to the fuel rail. On 6.2L engines, there is a left and right fuel rail.*

20   Relieve the fuel system pressure (see Section 3).

21   Disconnect the cable from the negative terminal of the battery (see Chapter 5, Section 1).

22   Remove the air intake duct and housing

(see Section 11).

23   Disconnect the PCV hose.

24   Disconnect the fuel supply line fitting **(see illustration)** from the right fuel rail. If you're unfamiliar with this metal-collar type quick-connect fitting, refer to Section or to Step 6 above.

Note: *You will need a special tool to disconnect this metal collar type of quick-connect fitting. There are several versions of this tool. You can find these tools at your local auto parts store.*

25   On 6.2L models, disconnect the supply line from between the left and right fuel rails.

26   On all models, disconnect the electrical connectors from all eight fuel injectors **(see illustration)**.

27   Remove the fuel rail mounting bolts.

28   Starting with the left side of the fuel rail,

13.24 To disconnect the fuel supply line from the fuel rail on V8 engines, you'll need a tool like the one shown in illustrations 13.9a and 13.9b to disconnect the metal collar type quick-connect fitting (refer to Section 4 if you're unfamiliar with this type of fitting)

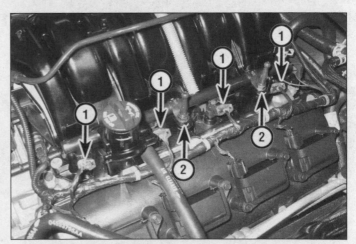

13.26 Disconnect the electrical connectors (1) from all eight fuel injectors (see illustration 13.7 for disconnection technique). Remove the mounting bolts (2) to remove the fuel rail (left fuel rail shown, right rail identical)

carefully pull on the injectors while wiggling them from side to side at the same time until all four injectors start to clear their mounting holes. Then go the right side and repeat this step. Go back and forth between the two sides of the engine, gradually working the injectors out of their mounting holes, until all eight injectors are free. When all of the injectors are free, remove the fuel rail and injectors as a single assembly.

29   To remove an injector from the fuel rail, remove the injector retainer clip and pull the injector out of the fuel rail **(see illustrations 13.14a and 13.14b)**.

Note: *Whether you're replacing an injector or a leaking O-ring, it's a good idea to remove all the injectors from the fuel rail and replace all the O-rings.*

30   Remove the O-rings from the injector **(see illustration 13.14c)**, discard them and install new O-rings. Be sure to lubricate the new O-rings with clean engine oil before installing the injectors into the fuel rail **(see illustration 13.12)**.

31   Installation is the reverse of removal. Be sure to tighten the fuel rail mounting bolts securely. After installing the fuel rail and injectors, turn the ignition switch to On, but don't operate the starter. This activates the fuel pump for about two seconds, which builds up fuel pressure in the fuel lines and the fuel rail. Repeat this step two or three times, then check the fuel lines, fuel rail and injectors for fuel leakage.

Chapter 5
Engine electrical systems

Contents

Specifications

General

Battery voltage	
Engine off	At least 12.6 volts
Engine running	13.5 to 15 volts
Firing order	
V6 engine	1-2-3-4-5-6
V8 engines	1-8-4-3-6-5-7-2
Coil resistance	
3.6L models	
Primary	0.6 to 0.9 ohms
Secondary	6.0 to 9.0 Kohms
V8 models	
Primary	0.540 to 0.660 ohms
Secondary	9.42 to 11.76 K-ohms*

* Cannot measure directly as there is a diode in the circuit.

Torque specifications   Ft-lbs (unless otherwise indicated)

Note: *One foot-pound (ft-lb) of torque is equivalent to 12 inch-pounds (in-lbs) of torque. Torque values below approximately 15 ft-lbs are expressed in inch-pounds, because most foot-pound torque wrenches are not accurate at these smaller values.*

Starter mounting	
Nuts	19
Bolts	40
Alternator mounting	
2.7L, 3.5L bolts	48
3.6L bolts	18
V8	
2010 and earlier models	
Bolts	48
Support bracket nut/bolt	30
2011 and later models	
Bolts	40
Support bracket	
Bolt	40
Nut	20.5

1   General information and precautions

Note: *The engine electrical systems include all ignition, charging and starting components. Because of their engine-related functions, these components are discussed separately from chassis electrical devices such as the lights, the instruments, etc. (which are included in Chapter 12).*

Precautions

1   Always observe the following precautions when working on the electrical systems: Be extremely careful when servicing engine electrical components. They are easily damaged if checked, connected or handled improperly.

a) *Never leave the ignition switch on for long periods of time with the engine off.*

b) *Don't disconnect the battery cables while the engine is running.*

c) *Maintain correct polarity when connecting a battery cable from another vehicle during jump-starting.*

d) *Always disconnect the negative cable first and hook it up last or the battery may be shorted by the tool being used to loosen the cable clamps.*

2   It's also a good idea to review the safety-related information regarding the engine electrical systems located in the *Safety First!* Section near the front of this manual before beginning any operation included in this Chapter.

Ignition system

3   The electronic ignition system consists of the Crankshaft Position (CKP) sensor, the Camshaft Position (CMP) sensor, the Knock Sensor (KS), the Powertrain Control Module (PCM), the ignition switch, the battery, the individual ignition coils, and the spark plugs. For more information on the CKP, CMP and KS sensors, as well as the PCM, refer to Chapter 6.

Charging system

4   The charging system includes the alternator (with an integral voltage regulator), the Powertrain Control Module (PCM), the Body Control Module (BCM), a charge indicator light on the dash, the battery, a fuse or fusible link and the wiring connecting all of these components. The charging system supplies electrical power for the ignition system, the lights, the radio, etc. The alternator is driven by a drivebelt.

5   The Intelligent Battery Sensor (IBS) is part of the negative (-) battery terminal and measures battery current, voltage and temperature. The IBS sends the battery State Of Charge (SOC) and State Of Function (SOF) to the PCM and BCM through the CAN bus, allowing the charging system to be controlled precisely so as to keep the battery properly charged at all times.

Starting system

6   The starting system consists of the battery, the ignition switch, the starter relay, the Powertrain Control Module (PCM), the Body Control Module (BCM), the Transmission Range (TR) switch, the starter motor and solenoid assembly, and the wiring connecting all of the components.

2   Troubleshooting

Ignition system

1   If a malfunction occurs in the ignition system, do not immediately assume that any particular part is causing the problem. First, check the following items:

a) *Make sure that the cable clamps at the battery terminals are clean and tight.*

b) *Test the condition of the battery (see Steps 15 through 19). If it doesn't pass all the tests, replace it.*

c) *Check the ignition coil or coil pack connections.*

d) *Check any relevant fuses in the engine compartment fuse and relay box (see Chapter 12). If they're burned, determine the cause and repair the circuit.*

Check

Warning: *Because of the high voltage generated by the ignition system, use extreme care when performing a procedure involving ignition components.*

Note: *The ignition system components on these vehicles are difficult to diagnose. In the event of ignition system failure that you can't diagnose, have the vehicle tested at a dealer service department or other qualified auto repair facility.*

Note: *You'll need a spark tester for the following test. Spark testers are available at most auto supply stores.*

2   If the engine turns over but won't start, verify that there is sufficient ignition voltage to fire the spark plugs as follows.

3   Remove a coil and install the tester between the boot at the lower end of the coil and the spark plug **(see illustrations)**.

4   Crank the engine and note whether or not the tester flashes.

Caution: *Do NOT crank the engine or allow it to run for more than five seconds; running the engine for more than five seconds may set a Diagnostic Trouble Code (DTC) for a cylinder misfire.*

5   If the tester flashes during cranking, the coil is delivering sufficient voltage to the spark plug to fire it. Repeat this test for each cylinder to verify that the other coils are OK.

6   If the tester doesn't flash, remove a coil from another cylinder and swap it for the one being tested. If the tester now flashes, you know that the original coil is bad. If the tester still doesn't flash, the PCM or wiring harness is probably defective. Have the PCM checked out by a dealer service department or other qualified repair shop (testing the PCM is beyond the scope of the do-it-yourselfer because it requires expensive special tools).

2.3a Spark tester on V6 engine

2.3b Spark tester on V8 engine companion spark plug wire (2005 models)

2.3c Spark tester on V8 engine ignition coil (2005 shown, 2006 and later models similar)

7   If the tester flashes during cranking but a misfire code (related to the cylinder being tested) has been stored, the spark plug could be fouled or defective.

Charging system

8   If a malfunction occurs in the charging system, do not automatically assume the alternator is causing the problem. First check the following items:

a) *Check the drivebelt tension and condition, as described in Chapter 1. Replace it if it's worn or deteriorated.*

b) *Make sure the alternator mounting bolts are tight.*

c) *Inspect the alternator wiring harness and the connectors at the alternator and voltage regulator. They must be in good condition, tight and have no corrosion.*

d) *Check the fusible link (if equipped) or main fuse in the underhood fuse/relay box. If it is burned, determine the cause, repair the circuit and replace the link or fuse (the vehicle will not start and/or the accessories will not work if the fusible link or main fuse is blown).*

e) *Start the engine and check the alternator for abnormal noises (a shrieking or squealing sound indicates a bad bearing).*

f) *Check the battery. Make sure it's fully charged and in good condition (one bad cell in a battery can cause overcharging by the alternator).*

g) *Disconnect the battery cables (negative first, then positive). Inspect the battery posts and the cable clamps for corrosion. Clean them thoroughly if necessary. Reconnect the cables (positive first, negative last).*

Alternator - check

9   Use a voltmeter to check the battery voltage with the engine off. It should be at least 12.6 volts **(see illustration 2.16)**.

10   Start the engine and check the battery voltage again. It should now be approximately 13.5 to 15 volts.

11   If the voltage reading is more or less than the specified charging voltage, the voltage regulator is probably defective, which will require replacement of the alternator (the voltage regulator is not replaceable separately). Remove the alternator and have it bench tested (most auto parts stores will do this for you).

12   The charging system (battery) light on the instrument cluster lights up when the ignition key is turned to On, but it should go out when the engine starts.

13   If the charging system light stays on after the engine has been started, there is a problem with the charging system. Before replacing the alternator, check the battery condition, alternator belt tension and electrical cable connections.

14   If replacing the alternator doesn't restore voltage to the specified range, have the charging system tested by a dealer service department or other qualified repair shop.

Battery - check

15   Check the battery state of charge. Visually inspect the indicator eye on the top of the battery (if equipped with one); if the indicator eye is black in color, charge the battery. Next perform an open circuit voltage test using a digital voltmeter.

Note: *The battery's surface charge must be removed before accurate voltage measurements can be made. Turn on the high beams for ten seconds, then turn them off and let the vehicle stand for two minutes.*

16   With the engine and all accessories Off, touch the negative probe of the voltmeter to the negative terminal of the battery and the positive probe to the positive terminal of the battery **(see illustration)**. The battery voltage should be 12.6 volts or slightly above. If the battery is less than the specified voltage, charge the battery before proceeding to the next test. Do not proceed with the battery load test unless the battery charge is correct.

17   Disconnect the negative battery cable, then the positive cable from the battery.

18   Perform a battery load test. An accurate check of the battery condition can only be performed with a load tester **(see illustration)**. This test evaluates the ability of the battery to operate the starter and other accessories during periods of high current draw. Connect the load tester to the battery terminals. Load test the battery according to the tool manufacturer's instructions. This tool increases the load demand (current draw) on the battery.

19   Maintain the load on the battery for 15 seconds and observe that the battery voltage does not drop below 9.6 volts. If the battery condition is weak or defective, the tool will indicate this condition immediately.

Note: *Cold temperatures will cause the minimum voltage reading to drop slightly. Follow the chart given in the manufacturer's instructions to compensate for cold climates. Minimum load voltage for freezing temperatures (32 degrees F) should be approximately 9.1 volts.*

Starting system

The starter rotates, but the engine doesn't

20   Remove the starter (see Section 10). Check the overrunning clutch and bench test the starter to make sure the drive mechanism extends fully for proper engagement with the flywheel ring gear. If it doesn't, replace the starter.

21   Check the flywheel ring gear for missing teeth and other damage. With the ignition turned off, rotate the flywheel so you can check the entire ring gear.

The starter is noisy

22   If the solenoid is making a chattering noise, first check the battery (see Steps 15 through 19). If the battery is okay, check the cables and connections.

23   If you hear a grinding, crashing metallic sound when you turn the key to Start, check for loose starter mounting bolts. If they're tight, remove the starter and inspect the teeth on the starter pinion gear and flywheel ring gear. Look for missing or damaged teeth.

24   If the starter sounds fine when you first turn the key to Start, but then stops rotating the engine and emits a zinging sound, the problem is probably a defective starter drive that's not staying engaged with the ring gear. Replace the starter.

The starter rotates slowly

25   Check the battery (see Steps 15 through 19).

26   If the battery is okay, verify all connections (at the battery, the starter solenoid and motor) are clean, corrosion-free and tight. Make sure the cables aren't frayed or damaged.

27   Check that the starter mounting bolts are tight so it grounds properly. Also check the pinion gear and flywheel ring gear for evidence of a mechanical bind (galling, deformed gear teeth or other damage).

2.16 To test the open circuit voltage of the battery, touch the black probe of the voltmeter to the negative terminal and the red probe to the positive terminal of the battery; a fully charged battery should be at least 12.6 volts

2.18 Connect a battery load tester to the battery and check the battery condition under load following the tool manufacturer's instructions

The starter does not rotate at all

28 Check the battery (see Steps 15 through 19).

29 If the battery is okay, verify all connections (at the battery, the starter solenoid and motor) are clean, corrosion-free and tight. Make sure the cables aren't frayed or damaged.

30 Check all of the fuses in the underhood fuse/relay box.

31 Check that the starter mounting bolts are tight so it grounds properly.

32 Check for voltage at the starter solenoid "S" terminal when the ignition key is turned to the start position. If voltage is present, replace the starter/solenoid assembly. If no voltage is present, the problem could be the starter relay, the Transmission Range (TR) switch (see Chapter 7B), or with an electrical connector somewhere in the circuit (see the wiring diagrams at the end of this manual). Also, on many modern vehicles, the Powertrain Control Module (PCM) and the Body Control Module (BCM) control the voltage signal to the starter solenoid; on such vehicles a special scan tool is required for diagnosis.

33 Ensure the vehicle's security system is not preventing starting of the vehicle, such as an invalid or defective ignition key or immobilizer system.

3 Battery - disconnection and reconnection

Caution: *Always disconnect the cable from the negative battery terminal FIRST and hook it up LAST or the battery may be shorted by the tool being used to loosen the cable clamps.*

Note: *The battery is located in the trunk, under the floor panel, to the right of the spare tire.*

Disconnection

1 Some systems on the vehicle require battery power to be available at all times, either to maintain continuous operation (alarm system, power door locks, etc.), or to maintain control unit memory (radio station presets, Powertrain Control Module (PCM) and other control units). When the battery is disconnected, the power that maintains these systems is cut. So, before you disconnect the battery, please note that on a vehicle with power door locks, it's a wise precaution to remove the key from the ignition and to keep it with you, so that it does not get locked inside if the power door locks should engage accidentally when the battery is reconnected!

2 Devices known as "memory-savers" can be used to avoid some of these problems. Precise details vary according to the device used. The typical memory saver is plugged into the cigarette lighter and is connected to a spare battery. Then the vehicle battery can be disconnected from the electrical system. The memory saver will provide sufficient current to maintain audio unit security codes, PCM memory, etc., and will provide power to always hot circuits such as the clock and radio memory circuits.

Warning: *Some memory savers deliver a considerable amount of current in order to keep vehicle systems operational after the main battery is disconnected. If you're using a memory saver, make sure that the circuit concerned is actually open before servicing it.*

Warning: *If you're going to work near any of the airbag system components, the battery MUST be disconnected and a memory saver must NOT be used. If a memory saver is used, power will be supplied to the airbag, which means that it could accidentally deploy and cause serious personal injury.*

Caution: *On 2015 and later models, if equipped with an Intelligent Battery Sensor (IBS), disconnect the IBS connector first before disconnecting the negative battery cable.*

3 To disconnect the battery for service procedures requiring power to be cut from the vehicle, disconnect the IBS connector (if equipped, see Section 6), then loosen the cable clamp nut and disconnect the cable from the negative battery terminal. Isolate the cable end to prevent it from coming into accidental contact with the battery terminal **(see illustration)**.

Battery reconnection procedures

4 Anytime that you disconnect the battery, you must carry out the following procedures after reconnecting the battery.

Front window express-up/express-down system

5 If the vehicle is equipped with the front window auto-up system, the door module must be re-calibrated when reconnecting the battery after servicing the power window system:

a) *Turn the ignition key to the Run position.*

b) *Lift (or depress, as applicable) the power window switch for the left (driver's side) front door window until the window reaches its fully-closed position and stops, and keep depressing the power window switch for at least two more seconds after the window stops.*

c) *Depress the power window switch for the left (driver's side) front door window until the window reaches its fully-open position and stops, and keep depressing the power window switch for at least two more seconds after the window stops.*

d) *Again, lift or depress the power window switch for the left (driver's side) front door window until the window reaches its fully-closed position and stops, and keep depressing the power window switch for at least two more seconds after the window stops.*

e) *Repeat Steps b, c and d for the right (passenger's side) front door window.*

f) *Verify that the windows are correctly calibrated by operating the express down and up features. If the windows are still not correctly calibrated, repeat this procedure until they are. If you're unable to calibrate the auto-up system using this procedure, have the system diagnosed and calibrated by a dealer service department.*

Power sunroof initialization

6 If the vehicle is equipped with a power sunroof, the Power Sunroof Module (PSM) module must be re-calibrated when reconnecting the battery after servicing the sunroof.

7 The following procedure should be used; when a new sunroof motor has been installed; the sunroof doesn't move when Vent is pressed; the express operation no longer works; the sunroof opens approximately one inch when Open is pressed and held; the sunroof closes only when Close is pressed and held:

a) *Turn the ignition to ACC or start the vehicle.*

3.3 If equipped with an Intelligent Battery Sensor (A), be sure to disconnect the sensor connector before detaching the cable

4.2 When disconnecting the battery cables, always disconnect the cable from the negative terminal (1) first, then disconnect the cable from the positive terminal (2)

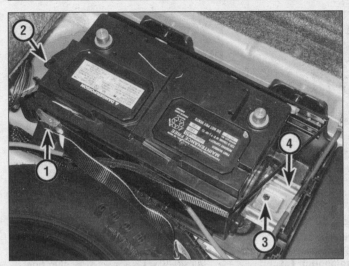

4.4 To remove the battery, loosen the retention strap buckle (1), disconnect the vent tube (2), remove the hold-down bolt (3) and remove the hold-down clamp (4)

4.13 To detach the Power Distribution Center (rear fuse and relay box) from the battery tray, disengage these two tabs with a couple of small screwdrivers

b) *Press and hold the Close switch until the sunroof moves to the vent position, then stops.*

c) *Release the Close switch, then press and hold again within five seconds. While holding the Close switch again, the sunroof should move to the closed position, then the open position, then the closed position again. If the Close switch is released too soon, the procedure must be performed again from the beginning.*

d) *Release the Close switch. Confirm proper operation of the sunroof.*

8   The following procedure should be used when the existing sunroof motor has been removed and re-installed, the closed position of the sunroof isn't right, the sunroof reverses direction during auto close, as if an object has been detected:

a) *Turn the ignition to ACC or start the vehicle.*

b) *Press and hold the Close switch until the sunroof closes. Release the switch.*

c) *Press and hold the Close switch again - DO NOT release the switch. After approximately 10 seconds, the sunroof calibration procedure begins. Continue to hold the Close switch during the calibration procedure. The sunroof moves to the vent position, then stops.*

d) *Release the Close switch, then press and hold again within five seconds. While holding the Close switch again, the sunroof should move to the closed position, then the open position, then the closed position again. If the Close switch is released too soon, the procedure must be performed again from the beginning.*

e) *Release the Close switch. Confirm proper operation of the sunroof.*

Electronic Stability Program (ESP)

9   If the vehicle is equipped with ESP, the Steering Angle Sensor (SAS) must be recali-

brated after reconnecting the battery, or after reconnecting any Anti-Lock Brake (ABS) component (the ABS system is part of the ESP). If the SAS is not recalibrated after reconnecting the battery or reconnecting an ABS component, the ESP/SAS indicator lamp will turn on after five ignition cycles, indicating the need for recalibration.

Note: *The following procedure does not apply to 2007 and later models, on which the SAS must be recalibrated with a factory scan tool. If the SAS is not recalibrated immediately after reconnecting the battery or reconnecting an ABS component, the ESP/SAS indicator lamp will flash continuously until the SAS is recalibrated, though it won't set any Diagnostic Trouble Codes (DTCs).*

a) *Start the engine.*

b) *Center the steering wheel.*

c) *Turn the steering wheel all the way to the left until it stops, then turn the wheel all the way to the right until it stops.*

d) *Center the steering wheel.*

e) *Stop the engine.*

f) *Restart the engine, drive the vehicle and verify proper operation.*

---

4   Battery and battery tray - removal and installation

Warning: *Hydrogen gas is produced by the battery, so keep open flames and lighted cigarettes away from it at all times. Always wear eye protection when working around a battery. Rinse off spilled electrolyte immediately with large amounts of water.*

Caution: *Always disconnect the negative cable first and hook it up last or you might accidentally short the battery with the tool that you're using to loosen the cable clamps.*

Battery

1   Remove the spare tire access cover.

2   Disconnect the IBS connector (if equipped) and disconnect the negative (-) battery terminal **(see illustration)** from the battery.

3   Disconnect the positive (+) battery terminal from the battery.

4   Remove the battery retention strap **(see illustration)**.

5   Disconnect the vent tube from the battery.

6   Remove the battery hold-down bolt and hold-down clamp.

7   Lift out the battery. Be careful - it's heavy.

Note: *Battery straps and handlers are available at most auto parts stores for a reasonable price. They make it easier to remove and carry the battery.*

8   While the battery is out, inspect the tray for corrosion deposits. Clean the battery tray, then use a baking soda/water solution to neutralize any deposits to prevent further oxidation. If the metal around the tray is corroded too, clean it as well and spray the area with a rust-inhibiting paint.

4.14 To detach the battery tray, remove these four bolts

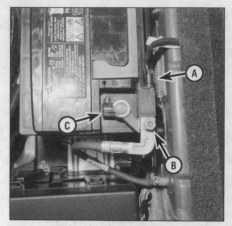

6.3 To remove the IBS, disconnect the connector (A), remove the nut (B) and disconnect the negative battery terminal (C)

7.2 To remove an ignition coil from a V6 engine, depress the release tab on top of the electrical connector (1) and disconnect the connector, remove the coil mounting screw (2), then grasp the coil firmly and pull it off the spark plug (3.5L V6 shown, 2.7L V6 similar)

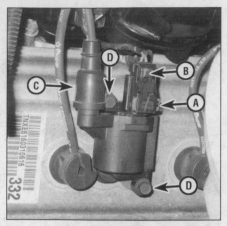

7.7 To remove an ignition coil from a 2005 V8 engine, push the slide lock (A) sideways to unlock the electrical connector, depress the release tab (B) and disconnect the connector, disconnect the spark plug wire boot (C) for the companion cylinder and remove the two coil mounting bolts (D)

9   If corrosion has leaked down past the battery tray, remove the tray (see Step 14) for further cleaning.
10   If you're replacing the battery, make sure you get an identical unit, with the same dimensions, amperage rating, cold cranking rating, etc.
11   Installation is the reverse of removal.

Battery tray

12   Remove the battery (see Steps 4 through 9).
13   Detach the Power Distribution Center (rear fuse and relay box) from the battery tray **(see illustration)**.
14   Remove the battery tray mounting bolts **(see illustration)** and remove the battery tray.
15   Installation is the reverse of removal.

5   Battery cables - replacement

1   When removing the cables, always disconnect the cable from the negative battery terminal first and hook it up last, or you might accidentally short out the battery with the tool you're using to loosen the cable clamps. Even if you're only replacing the cable for the positive terminal, be sure to disconnect the negative cable from the battery first.
2   Disconnect the old cables from the battery, then trace each of them to their opposite ends and disconnect them. Note the routing of each cable before disconnecting it to ensure correct installation.
3   If you are replacing any of the old cables, take them with you when buying new cables. It is vitally important that you replace the cables with identical parts.
Caution: *The IBS is part of the negative (-) battery terminal. Be sure to remove the nega-*

tive cable from the IBS and do not discard the IBS by accident when replacing the negative cable.
4   Clean the threads of the solenoid or ground connection with a wire brush to remove rust and corrosion. Apply a light coat of battery terminal corrosion inhibitor or petroleum jelly to the threads to prevent future corrosion.
5   Attach the cable to the solenoid or ground connection and tighten the mounting nut/bolt securely.
6   Before connecting a new cable to the battery, make sure that it reaches the battery post without having to be stretched.
7   Connect the cable to the positive battery terminal first, then connect the ground cable to the negative battery terminal.

6   Intelligent Battery Sensor (IBS) - replacement

Note: *The IBS is part of the negative (-) battery terminal and is replaced as one unit.*
1   Disconnect the IBS connector.
2   Remove the nut and disconnect the negative battery cable from the IBS terminal.
3   Disconnect the negative (-) battery terminal from the battery **(see Illustration)**.
4   Installation is reverse of removal.

7   Ignition coil - replacement

V6 engines

1   If you want to remove the rear ignition coil on the left cylinder head on 2.7L models, or any of the three coils on the right cylinder head on 3.6L models, remove the air intake

duct (see Chapter 4, Section 11). If you want to remove the rear ignition coil on the left cylinder head on 3.5L models, remove the upper intake manifold (see Chapter 2B, Section 8). If you want to remove any of the three coils on the right cylinder head on the 2.7L or 3.5L engine, remove the upper intake manifold (see Chapter 2A, Section 8 or Chapter 2B, Section 8). If you want to remove any of the three coils on the left cylinder head on the 3.6L engine, remove the upper intake manifold (see Chapter 2C, Section 5).
2   Disconnect the electrical connector from the coil **(see Illustration)**.
3   Remove the ignition coil mounting bolt and remove the coil.
4   Installation is the reverse of removal.

V8 engine

5   On some 2005 and earlier models, each cylinder has one ignition coil and a spark plug wire that goes to the companion cylinder on the opposite side of the engine. On 2006 and later models, each ignition coil provides spark for both spark plugs for each cylinder.

2005 models

6   Remove the engine cover.
7   Disconnect the electrical connector and the spark plug wire boot for the companion cylinder from the ignition coil **(see illustration)**.
8   Remove the coil mounting bolts. To remove the coil, carefully pull it up with a twisting motion to disengage the rubber spark plug boot from the plug.
9   While the coil is removed, inspect the condition of the rubber spark plug boot. If it's cracked, torn or deteriorated, replace it. Also inspect the condition of the spark plug wire connected to the coil (see Chapter 1).
10   Installation is the reverse of removal.

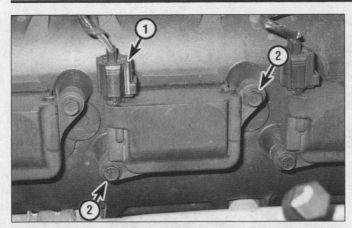

7.13 To remove an ignition coil from a 2006 and later V8 engine, disconnect the electrical connector (1), then remove the two coil mounting bolts (2)

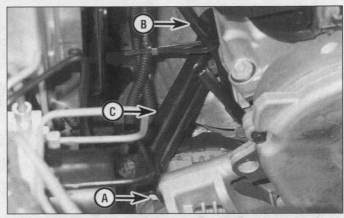

9.8a Remove the upper alternator mounting bolt (A) and the support bracket bolt (B) and remove the support bracket (C) (3.5L V6 engine)

2006 and later models

11   Remove the engine cover(s).

12   On supercharged models, remove the air cleaner housing (see Chapter 4, Section 11) and detach the engine coolant reservoir and move out of the way. To access the right side coils, detach the supercharger coolant reservoir from the bracket and position to the side.

13   On all models, disconnect the electrical connector from the ignition coil **(see illustration)**.

14   Remove the coil mounting bolts. To remove the coil, carefully pull it up with a twisting motion to disengage the rubber spark plug boots from the plugs.

Note: *For the left side, the air conditioning lines may need to be moved around to allow removal of the coils.*

15   While the coil is removed, inspect the condition of the rubber spark plug boots. If they're cracked, torn or deteriorated, replace them.

16   Installation is the reverse of removal.

Testing coils

Note: *The following procedure only applies to 3.6L and V8 models.*

17   After removing the ignition coil, use a Digital Volt-Ohm Meter (DVOM) set to measure Ohms (resistance) to measure the ignition coil primary and secondary windings.

18   To measure the primary winding resistance, connect the DVOM leads between the coil connector terminals.

19   To measure the secondary winding resistance, connect the DVOM leads between the coil terminal for the spark plug and the positive terminal at the coil connector.

20   After removing the ignition coil, use a Digital Volt-Ohm Meter (DVOM) set to measure Ohms (resistance) to measure the ignition coil primary and secondary windings. Compare the measurements against the values listed in this Chapter's Specifications.

21   After removing the ignition coil, use a Digital Volt-Ohm Meter (DVOM) set to measure Ohms (resistance) to measure the ignition coil primary and secondary windings. Com-

pare the measurements against the values listed in this Chapter's Specifications.

22   If measurements are not within specifications, replace the coil.

---

8   Ignition coil capacitor - replacement

1   Locate the ignition coil capacitor.

a) *2007 and earlier 2.7L and 3.5L models: left valve cover*

b) *2008 and later 2.7L models: one on the right valve cover, one under the intake manifold on the left side of the engine*

c) *2008 and later 3.5L models: left side of engine*

d) *All 3.6L models: rear of both cylinder heads*

e) *All V8 models: left rear corner of intake manifold*

2   Disconnect the electrical connector.

3   Remove the bolt and the capacitor.

4   Installation is reverse of removal.

---

9   Alternator - removal and installation

Note: *If you are replacing the alternator, take the old one with you when purchasing a replacement unit. Make sure the new/rebuilt unit looks identical to the old alternator. Look at the terminals - they should be the same in number, size and location as the terminals on the old alternator. Finally, look at the identification numbers - they will be stamped into the housing or printed on a tag attached to the housing. Make sure the numbers are the same on both alternators.*

Note: *Many new and remanufactured alternators do not have a pulley installed, so you may have to switch the pulley from the old unit to the new/rebuilt one. When buying an alternator, find out the shop's policy regarding pulleys; some shops will perform this service free of charge.*

2.7L V6 engine

1   Disconnect the cable from the negative terminal of the battery (see Section 4).

2   Remove the drivebelt (see Chapter 1).

3   Disconnect the battery cable from the B+ output terminal and disconnect the field wire electrical connector from the field terminal.

4   Remove the upper mounting bolt and the two lower mounting bolts and remove the alternator.

5   Installation is the reverse of removal. When you're done, check the charging voltage (see Section 2) to verify that the alternator is operating correctly.

3.5L and 3.6L V6 engines

6   Disconnect the cable from the negative terminal of the battery (see Section 4).

7   Remove the drivebelt (see Chapter 1).

8   On 3.5L engines, remove the upper mounting bolt and the support bracket bolt **(see illustration)** and remove the support bracket. On 3.6L engines, detach the B+ cable and the electrical connector, then unscrew the upper mounting bolts **(see illustration)**.

9.8b Alternator upper details - 3.6L V6 engine

1   *B+ cable (under cover)*

2   *Electrical connector*

3   *Upper mounting bolts*

9.11 Remove the nut from the battery (B+) output terminal stud (A) and disconnect the battery cable from the stud, then disconnect the electrical connector (B) for the field terminals (3.5L V6 engine)

9.12a To detach the alternator from a 3.5L V6 engine, remove these two lower mounting bolts

9.12b Alternator lower mounting bolt - 3.6L V6 engine

9   Raise the front of the vehicle and place it securely on jackstands.
10   Remove the middle engine splash shield (see Chapter 2B).
11   On 3.5L models, disconnect the battery cable from the B+ output terminal and disconnect the field wire electrical connector from the field terminal **(see illustration)**.
12   Remove the lower alternator mounting bolt(s) **(see illustrations)** and remove the alternator.
13   Installation is the reverse of removal. When you're done, check the charging voltage (see Section 2) to verify that the alternator is operating correctly.

V8 engines

Note: *The alternator is located on the lower right side of the engine, attached to the engine block.*

14   Disconnect the cable from the negative terminal of the battery (see Section 4).
15   Remove the drivebelt (see Chapter 1).
16   Raise the front of the vehicle and place it securely on jackstands.
17   Unsnap the plastic insulator cap from the B+ output terminal, remove the battery cable retaining nut from the B+ output terminal and disconnect the battery cable from the B+ output terminal. Then disconnect the field wire electrical connector from the field terminal.
18   Working underneath the engine, remove the alternator support bracket nut and bolt **(see illustration)** and remove the support bracket.
19   Remove the upper mounting bolt **(see illustration)** and the lower mounting bolt **(see illustration 10.18)** and remove the alternator.
20   Installation is the reverse of removal. When you're done, check the charging volt-

age (see Section 2) to verify that the alternator is operating correctly.

---

10   Starter motor - removal and installation

V6 engines

1   Disconnect the cable from the negative terminal of the battery (see Section 3).
2   Raise the vehicle and support it securely on jackstands.
3   Remove the middle engine splash shield (see Chapter 2B).

2010 and earlier 300, 2.7L Charger and all 2011 and later V6 models

4   Secure the steering wheel so it does not rotate freely.
Caution: *If the steering wheel is allowed to move freely with the intermediate shaft disconnected, the airbag system clockspring can be damaged.*
5   On 2.7L Charger and all 300 V6 models, remove the bolt that secures the intermediate steering shaft to the lower steering coupling shaft and separate the two shafts. On all 2011 and later V6 models, remove the intermediate shaft completely (see Chapter 10, Section 17).

2010 and earlier 3.5L Charger and Challenger 3.5L models

6   Remove the left-side catalytic converter (see Chapter 6, Section 16).

All models

7   On 3.6L models, remove the starter motor heat shield **(see illustration)**. On all models, remove the nut that secures the battery cable to the terminal stud on the starter solenoid, disconnect the cable from the terminal stud and disconnect the electrical connector from the spade terminal on the solenoid **(see illustration)**.

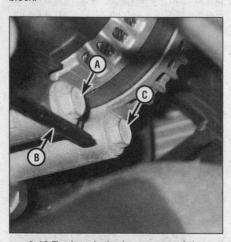

9.18 To detach the lower part of the alternator from a V8 engine, first remove this bolt (A), then remove the nut (not shown) at the rear end of the support bracket (B) and remove the support bracket, then remove the lower alternator mounting bolt (C)

9.19 To detach the upper part of the alternator from a V8 engine, remove this bolt

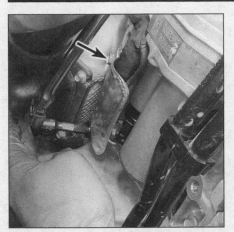

10.7a Starter heat shield bolts -
3.6L V6 engine

10.7b Disconnect the electrical connector
from the spade terminal on the solenoid (A),
then remove the nut (B) and disconnect the
battery cable from the terminal stud on the
solenoid (starter motor on 3.5L V6 shown,
starter on 2.7L similar)

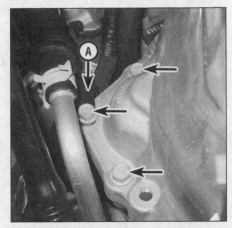

10.8 Remove the starter mounting bolts
and, if equipped the oil cooler line clip (A)
(3.5L V6 engine shown, 2.7L V6
similar. 3.6L models only have
two mounting bolts)

8   Remove the starter mounting bolts and the electrical harness/oil cooler line clip **(see illustration)**.

9   To remove the starter, pull it forward and down, work it up over the exhaust pipe, then work it past the disconnected steering shafts.

10   Installation is the reverse of removal. Don't forget to reattach the oil cooler line clip to the middle starter mounting bolt.

V8 engine

Note: *On some models, the starter needs to be unbolted from the engine and lowered to allow access to disconnect the electrical connectors.*

2014 and earlier RWD automatic transmission, all manual transmission and all 6.2L models

11   Disconnect the cable from the negative terminal of the battery. Raise the vehicle and

support it securely on jackstands.

12   Disconnect the electronic power steering connector (if equipped) for additional access to the starter.

13   Remove the starter heat shield (if equipped).

14   Remove the starter motor mounting bolts **(see illustration)**.

15   Move the starter motor assembly forward far enough for the pinion gear to clear the starter hole and the driveplate ring gear, then tilt the starter downward so that you can reach the electrical terminals on the solenoid. **Caution:** *DO NOT let the starter hang from the harness or cable.*

16   Disconnect the battery cable and electrical connector from the terminals on the starter motor solenoid **(see illustration)** and remove the starter motor.

17   Installation is the reverse of removal.

2015 and later 5.7L and 6.4L automatic transmission models

18   Disconnect the cable from the negative terminal of the battery (see Section 3). Raise the vehicle and support it securely on jackstands.

19   Disconnect the electronic power steering connector for additional access to the starter.

20   Remove the starter heat shield (if equipped).

21   Disconnect the battery cable and electrical connector from the terminals on the starter motor solenoid.

22   Remove the starter motor mounting bolts and remove the starter.

23   Installation is the reverse of removal.

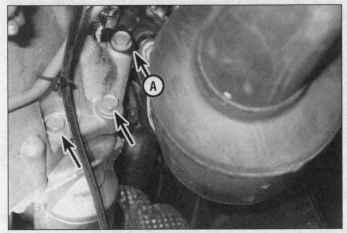

10.14 To detach the starter motor from the engine and
transmission bellhousing on a V8, remove these three mounting
bolts and the ground cable terminal from the upper bolt (A)
(automatic shown, manual similar)

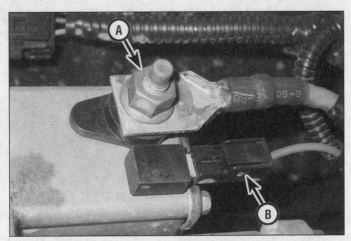

10.16 After unbolting the starter motor and moving it forward,
remove the battery cable-to-solenoid nut (A) and disconnect the
battery cable from the solenoid terminal, then disconnect the
electrical connector (B) from the solenoid terminal
(auto shown, manual similar)

Notes

Chapter 6
Emissions and engine control systems

Contents

Specifications

Torque specifications
Ft-lbs (unless otherwise indicated)

Note: *One foot-pound (ft-lb) of torque is equivalent to 12 inch-pounds (in-lbs) of torque. Torque values below approximately 15 ft-lbs are expressed in inch-pounds, because most foot-pound torque wrenches are not accurate at these smaller values.*

EGR valve (all engines except 3.6L V6)	
Lower EGR pipe mounting flange bolts (2.7L V6 only)	
At exhaust manifold	23
At EGR valve	11
Upper EGR pipe mounting flange bolts (at EGR valve)	11
EGR valve mounting bolts	23
Engine coolant temperature sensor	
V6 engines	20
V8 engine	97 in-lbs
Knock sensor	
V6 engines	
2.7L	88 in-lbs
3.5L	15
3.6L	16
V8 engines	15
Oxygen sensors	22

1   General information

1   To prevent pollution of the atmosphere from incompletely burned and evaporating gases, and to maintain good driveability and fuel economy, a number of emission control systems are incorporated.

2   The Emissions Label in the engine compartment identifies the emission controls the vehicle is equipped with as well as the emission standard the vehicle adheres to (Federal or California) **(see illustration)**. These emission systems include:

Catalytic converter

3   A catalytic converter is an emission control device in the exhaust system that reduces certain pollutants in the exhaust gas stream. There are two types of converters: oxidation converters and reduction converters.

4   Oxidation converters contain a monolithic substrate (a ceramic honeycomb) coated with the semi-precious metals platinum and palladium. An oxidation catalyst reduces unburned hydrocarbons (HC) and carbon monoxide (CO) by adding oxygen to the exhaust stream as it passes through the substrate, which, in the presence of high temperature and the catalyst materials, converts the HC and CO to water vapor (H_2O) and carbon dioxide (CO_2).

5   Reduction converters contain a monolithic substrate coated with platinum and rhodium. A reduction catalyst reduces oxides of nitrogen (NOx) by removing oxygen, which in the presence of high temperature and the catalyst material produces nitrogen (N) and carbon dioxide (CO_2).

6   Catalytic converters that combine both types of catalysts in one assembly are known as "three-way catalysts" or TWCs. A TWC can reduce all three pollutants.

Evaporative Emissions Control (EVAP) system

7   The Evaporative Emissions Control (EVAP) system prevents fuel system vapors (which contain unburned hydrocarbons) from escaping into the atmosphere. On warm days,

vapors trapped inside the fuel tank expand until the pressure reaches a certain threshold. Then the fuel vapors are routed from the fuel tank through the fuel vapor vent valve and the fuel vapor control valve to the EVAP canister, where they're stored temporarily until the next time the vehicle is operated. When the conditions are right (engine warmed up, vehicle up to speed, moderate or heavy load on the engine, etc.) the PCM opens the canister purge valve, which allows fuel vapors to be drawn from the canister into the intake manifold. Once in the intake manifold, the fuel vapors mix with incoming air before being drawn through the intake ports into the combustion chambers where they're burned up with the rest of the air/fuel mixture. The EVAP system is complex and virtually impossible to troubleshoot without the right tools and training.

Exhaust Gas Recirculation (EGR) system

Note: *This section describes the EGR system on 2010 and earlier models. 2011 and later models are not equipped with an EGR system.*

8   The EGR system reduces oxides of nitrogen by recirculating exhaust gases from the exhaust manifold, through the EGR valve and intake manifold, then back to the combustion chambers, where it mixes with the incoming air/fuel mixture before being consumed. These recirculated exhaust gases dilute the incoming air/fuel mixture, which cools the combustion chambers, thereby reducing NOx emissions.

9   The EGR system consists of the Powertrain Control Module (PCM), the EGR valve, the EGR valve position sensor and various other information sensors that the PCM uses to determine when to open the EGR valve. The degree to which the EGR valve is opened is referred to as "EGR valve lift." The PCM is programmed to produce the ideal EGR valve lift for varying operating conditions. The EGR valve position sensor, which is an integral part of the EGR valve, detects the amount of EGR valve lift and sends this information to the PCM. The PCM then compares it with the

appropriate EGR valve lift for the operating conditions. The PCM increases current flow to the EGR valve to increase valve lift and reduces the current to reduce the amount of lift. If EGR flow is inappropriate to the operating conditions (idle, cold engine, etc.) the PCM simply cuts the current to the EGR valve and the valve closes.

Powertrain Control Module (PCM)

10   The Powertrain Control Module (PCM) is the brain of the engine management system. It also controls a wide variety of other vehicle systems. In order to program the new PCM, the technician (equipped with the proper scan tool and software) needs the vehicle as well as the new PCM. If you're planning to replace the PCM with a new one, there is no point in trying to do so at home because you won't be able to program it yourself.

Positive Crankcase Ventilation (PCV) system

11   The Positive Crankcase Ventilation (PCV) system reduces hydrocarbon emissions by scavenging crankcase vapors, which are rich in unburned hydrocarbons. A PCV valve or orifice regulates the flow of gases into the intake manifold in proportion to the amount of intake vacuum available.

12   The PCV system generally consists of the fresh air inlet hose, the PCV valve or orifice and the crankcase ventilation hose (or PCV hose). The fresh air inlet hose connects the air intake duct to a pipe on the valve cover. The crankcase ventilation hose (or PCV hose) connects the PCV valve or orifice in the valve cover to the intake manifold.

2   On Board Diagnosis (OBD) system

General description

1   All models are equipped with the second generation OBD-II system. This system consists of an on-board computer known as the Powertrain Control Module (PCM), and information sensors, which monitor various functions of the engine and send data to the PCM. This system incorporates a series of diagnostic monitors that detect and identify fuel injection and emissions control system faults and store the information in the computer memory. This system also tests sensors and output actuators, diagnoses drive cycles, freezes data and clears codes.

2   The PCM is the brain of the electronically controlled fuel and emissions system. It receives data from a number of sensors and other electronic components (switches, relays, etc.). Based on the information it receives, the PCM generates output signals to control various relays, solenoids (fuel injectors) and other actuators. The PCM is specifically calibrated to optimize the emissions, fuel economy and driveability of the vehicle.

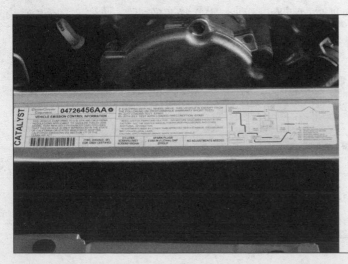

1.2 The emissions label identifies the systems the vehicle is equipped with and is located in the engine compartment

Information Sensors

Accelerator Pedal Position (APP) sensor - as you press the accelerator pedal, the APP sensor alters its voltage signal to the PCM in proportion to the angle of the pedal, and the PCM commands a motor inside the throttle body to open or close the throttle plate accordingly

Camshaft Position (CMP) sensor - produces a signal that the PCM uses to identify the number 1 cylinder and to time the firing sequence of the fuel injectors

Crankshaft Position (CKP) sensor - produces a signal that the PCM uses to calculate engine speed and crankshaft position, which enables it to synchronize ignition timing with fuel injector timing, and to detect misfires

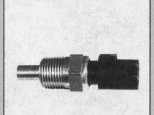

Engine Coolant Temperature (ECT) sensor - a thermistor (temperature-sensitive variable resistor) that sends a voltage signal to the PCM, which uses this data to determine the temperature of the engine coolant

Fuel tank pressure sensor - measures the fuel tank pressure and controls fuel tank pressure by signaling the EVAP system to purge the fuel tank vapors when the pressure becomes excessive

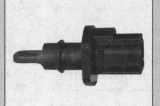

Intake Air Temperature (IAT) sensor - monitors the temperature of the air entering the engine and sends a signal to the PCM to determine injector pulse-width (the duration of each injector's on-time) and to adjust spark timing (to prevent spark knock)

Knock sensor - a piezoelectric crystal that oscillates in proportion to engine vibration which produces a voltage output that is monitored by the PCM. This retards the ignition timing when the oscillation exceeds a certain threshold

Manifold Absolute Pressure (MAP) sensor - monitors the pressure or vacuum inside the intake manifold. The PCM uses this data to determine engine load so that it can alter the ignition advance and fuel enrichment

Mass Air Flow (MAF) sensor - measures the amount of intake air drawn into the engine. It uses a hot-wire sensing element to measure the amount of air entering the engine

Oxygen sensors - generates a small variable voltage signal in proportion to the difference between the oxygen content in the exhaust stream and the oxygen content in the ambient air. The PCM uses this information to maintain the proper air/fuel ratio. A second oxygen sensor monitors the efficiency of the catalytic converter

Throttle Position (TP) sensor - a potentiometer that generates a voltage signal that varies in relation to the opening angle of the throttle plate inside the throttle body. Works with the PCM and other sensors to calculate injector pulse width (the duration of each injector's on-time)

Photos courtesy of Wells Manufacturing, except APP and MAF sensors.

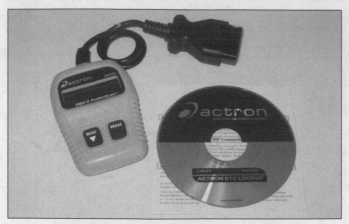

2.4a Simple code readers are an economical way to extract trouble codes when the CHECK ENGINE light comes on

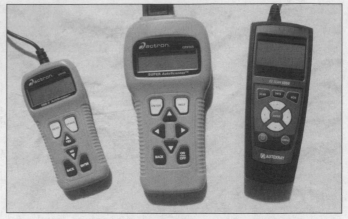

2.4b Hand-held scan tools like these can extract computer codes and also perform diagnostics

3   It isn't a good idea to attempt diagnosis or replacement of the PCM or emission control components at home while the vehicle is under warranty. Because of a Federally mandated warranty which covers the emissions system components and because any owner-induced damage to the PCM, the sensors and/or the control devices may void this warranty, take the vehicle to a dealer service department if the PCM or a system component malfunctions.

Scan tool information

4   Because extracting the Diagnostic Trouble Codes (DTCs) from an engine management system is now the first step in troubleshooting many computer-controlled systems and components, a code reader, at the very least, will be required **(see illustration)**. More powerful scan tools can also perform many of the diagnostics once associated with expensive factory scan tools **(see illustration)**. If you're planning to obtain a generic scan tool for your vehicle, make sure that it's compatible with OBD-II systems. If you don't plan to purchase a code reader or scan tool and don't have access to one, you can have the codes extracted by a dealer service department or an independent repair shop.
Note: *Some auto parts stores even provide this service.*

---

3   Obtaining and clearing Diagnostic Trouble Codes (DTCs)

---

1   All models covered by this manual are equipped with on-board diagnostics. When the PCM recognizes a malfunction in a monitored emission or engine control system, component or circuit, it turns on the Malfunction Indicator Light (MIL) on the dash. The PCM will continue to display the MIL until the problem is fixed and the Diagnostic Trouble Code (DTC) is cleared from the PCM's memory. You'll need a scan tool to access any DTCs stored in the PCM.

2   Before outputting any DTCs stored in the PCM, thoroughly inspect ALL electrical connectors and hoses. Make sure that all electrical connections are tight, clean and free of corrosion. And make sure that all hoses are correctly connected, fit tightly and are in good condition (no cracks or tears).

Accessing the DTCs

3   The Diagnostic Trouble Codes (DTCs) can only be accessed with a code reader or scan tool. Professional scan tools are expensive, but relatively inexpensive generic code readers or scan tools **(see illustrations 2.4a and 2.4b)** are available at most auto parts stores. Simply plug the connector of the scan tool into the diagnostic connector **(see illustration)**. Then follow the instructions included with the scan tool to extract the DTCs.

4   Once you have outputted all of the stored DTCs, look them up on the accompanying DTC chart.

5   After troubleshooting the source of each DTC, make any necessary repairs or replace the defective component(s).

Clearing the DTCs

6   Clear the DTCs with the code reader or scan tool in accordance with the instructions provided by the tool's manufacturer.

Diagnostic Trouble Codes

7   The accompanying tables are a list of the Diagnostic Trouble Codes (DTCs) that can be accessed by a do-it-yourselfer working at home (there are many, many more DTCs available to professional mechanics with proprietary scan tools and software, but those codes cannot be accessed by a generic scan tool). This list of DTCs may not be complete. If, after you have checked and repaired the connectors, wire harness and vacuum hoses (if applicable) for an emission-related system, component or circuit, the problem persists, have the vehicle checked by a dealer service department or other qualified repair shop.
Note: *Not all codes available are listed. Some codes may display on a scan tool but not be shown in the DTC list below.*

3.3 The Data Link Connector (DLC) is located below the steering column

OBD-II Diagnostic Trouble Codes (DTCs)

Code	Possible cause
P000B	Bank 1, camshaft 2 position, slow response
P0013	Bank 1, camshaft 2 position, actuator circuit open
P0016	Crankshaft/camshaft timing misalignment
P0031	Oxygen sensor (1/1)* heater circuit, low voltage
P0032	Oxygen sensor (1/1)* heater circuit, high voltage
P0037	Oxygen sensor (1/2)* heater circuit, low voltage
P0038	Oxygen sensor (1/2)* heater circuit, high voltage
P0051	Oxygen sensor (2/1)* heater circuit, low voltage
P0052	Oxygen sensor (2/1)* heater circuit, high voltage
P0057	Oxygen sensor (2/2)* heater circuit, low voltage
P0058	Oxygen sensor (2/2)* heater circuit, high voltage
P0071	Ambient air temperature sensor performance
P0072	Ambient air temperature sensor circuit, low voltage
P0073	Ambient air temperature sensor circuit, high voltage
P0107	Manifold Absolute Pressure (MAP) sensor circuit, low voltage
P0108	Manifold Absolute Pressure (MAP) sensor circuit, high voltage
P0111	Intake Air Temperature (IAT) sensor rationality
P0112	Intake Air Temperature (IAT) sensor circuit, low voltage
P0113	Intake Air Temperature (IAT) sensor circuit, high voltage
P0116	Engine Coolant Temperature (ECT) sensor circuit performance problem
P0117	Engine Coolant Temperature (ECT) sensor circuit, low voltage
P0118	Engine Coolant Temperature (ECT) sensor circuit, high voltage
P0121	Throttle Position (TP) sensor 1 performance
P0122	Throttle Position (TP) sensor 1 circuit, low voltage
P0123	Throttle Position (TP) sensor 1 circuit, high voltage
P0125	Insufficient coolant temperature for closed-loop fuel control
P0128	Thermostat rationality
P0129	Barometric pressure out-of-range (too low)
P0131	Oxygen sensor (1/1)* circuit, low voltage

OBD-II Diagnostic Trouble Codes (DTCs) (continued)

Code	Possible cause
P0132	Oxygen sensor (1/1)* circuit, high voltage
P0133	Oxygen sensor (1/1)* circuit, slow response
P0135	Oxygen sensor (1/1)* circuit, heater performance problem
P0137	Oxygen sensor (1/2)* circuit, low voltage
P0138	Oxygen sensor (1/2)* circuit, high voltage
P0139	Oxygen sensor (1/2)* circuit, slow response
P013A	Oxygen sensor (1/2)* circuit, slow rich-to-lean response
P013C	Oxygen sensor (2/2)* circuit, slow rich-to-lean response
P0141	Oxygen sensor (1/2)* circuit, heater performance problem
P0151	Oxygen sensor (2/1)* circuit, low voltage
P0152	Oxygen sensor (2/1)* circuit, high voltage
P0153	Oxygen sensor (2/1)* circuit, slow response
P0155	Oxygen sensor (2/1)*, heater performance problem
P0157	Oxygen sensor (2/2)* circuit, low voltage
P0158	Oxygen sensor (2/2)* circuit, high voltage
P0159	Oxygen sensor (2/2)* circuit, slow response
P0161	Oxygen sensor (2/2)* circuit, heater performance problem
P0171	Fuel system too lean (cylinder bank no. 1)
P0172	Fuel system too rich (cylinder bank no. 1)
P0174	Fuel system too lean (cylinder bank no. 2)
P0175	Fuel system too rich (cylinder bank no. 2)
P0196	Engine oil temperature sensor, circuit performance
P0197	Engine oil temperature sensor circuit, low voltage
P0198	Engine oil temperature sensor circuit, high voltage
P0201	Fuel injector circuit malfunction - cylinder no. 1
P0202	Fuel injector circuit malfunction - cylinder no. 2
P0203	Fuel injector circuit malfunction - cylinder no. 3
P0204	Fuel injector circuit malfunction - cylinder no. 4
P0205	Fuel injector circuit malfunction - cylinder no. 5

Code	Possible cause
P0206	Fuel injector circuit malfunction - cylinder no. 6
P0207	Fuel injector circuit malfunction - cylinder no. 7
P0208	Fuel injector circuit malfunction - cylinder no. 8
P0221	Throttle Position (TP) sensor 2 performance
P0222	Throttle Position (TP) sensor 2 circuit, low voltage
P0223	Throttle Position (TP) sensor 2 circuit, high voltage
P0298	Engine oil temperature too high
P0300	Multiple cylinder misfire detected
P0301	Cylinder no. 1 misfire detected
P0302	Cylinder no. 2 misfire detected
P0303	Cylinder no. 3 misfire detected
P0304	Cylinder no. 4 misfire detected
P0305	Cylinder no. 5 misfire detected
P0306	Cylinder no. 6 misfire detected
P0307	Cylinder no. 7 misfire detected
P0308	Cylinder no. 8 misfire detected
P0315	No Crankshaft Position (CKP) sensor learned
P0325	Knock sensor 1 circuit malfunction
P0330	Knock sensor 2 circuit malfunction
P0335	Crankshaft Position (CKP) sensor, circuit malfunction
P0339	Crankshaft Position (CKP) sensor, circuit intermittent
P0340	Camshaft Position (CMP) sensor circuit malfunction
P0344	Camshaft Position (CMP) sensor, circuit intermittent
P0401	Exhaust Gas Recirculation (EGR) system, performance problem
P0403	Exhaust Gas Recirculation (EGR) solenoid circuit malfunction
P0404	Exhaust Gas Recirculation (EGR) position sensor performance
P0405	Exhaust Gas Recirculation (EGR) sensor circuit, low voltage
P0406	Exhaust Gas Recirculation (EGR) sensor circuit, high voltage
P0420	Catalyst system efficiency below threshold (cylinder bank no. 1)
P0430	Catalyst system efficiency below threshold (cylinder bank no. 2)

OBD-II Diagnostic Trouble Codes (DTCs) (continued)

Code	Possible cause
P0440	General Evaporative Emission Control (EVAP) system failure
P0441	Evaporative Emission Control (EVAP) purge system performance problem
P0442	Evaporative Emission Control (EVAP) purge system, medium leak detected
P0443	Evaporative Emission Control (EVAP) purge solenoid circuit malfunction
P0452	Natural Vacuum Leak Detection (NVLD) system, pressure switch stuck closed
P0453	Natural Vacuum Leak Detection (NVLD) system, pressure switch stuck open
P0455	Evaporative Emission Control (EVAP) system, large leak
P0456	Evaporative Emission Control (EVAP) system, small leak
P0457	Loose fuel cap
P0461	Fuel level sensor 1, performance problem
P0462	Fuel level sensor 1, low voltage
P0463	Fuel level sensor 1, high voltage
P0480	Cooling fan 1, control circuit malfunction
P0481	Cooling fan 2, control circuit malfunction
P0498	Natural Vacuum Leak Detection (NVLD) canister vent valve solenoid, low voltage
P0499	Natural Vacuum Leak Detection (NVLD) canister vent valve solenoid, high voltage
P050B	Cold start ignition timing performance
P050D	Cold start rough idle
P0501	Vehicle Speed Sensor (VSS) 1, performance problem
P0503	Vehicle Speed Sensor (VSS) 1 erratic
P0506	Idle speed performance lower than expected
P0507	Idle speed performance higher than expected
P0513	Invalid Sentry Key Immobilizer System (SKIM) key
P0520	Engine oil pressure sensor circuit
P0521	Engine oil pressure sensor performance problem
P0522	Engine oil pressure sensor circuit, low voltage
P0523	Engine oil pressure sensor circuit, high voltage
P0524	Engine oil pressure too low
P0532	Air conditioning refrigerant pressure sensor circuit, low voltage

Code	Possible cause
P0533	Air conditioning refrigerant pressure sensor circuit, high voltage
P0551	Power Steering Pressure (PSP) switch, performance problem
P0562	Battery voltage low
P0563	Battery voltage high
P0571	Brake switch 1 performance problem
P0572	Brake switch 1 stuck on
P0573	Brake switch no. 1 stuck off
P0579	Speed control switch 1 performance
P0580	Speed control switch 1 circuit, low voltage
P0581	Speed control switch 1 circuit, high voltage
P0585	Speed control switch 1/2 correlation
P0591	Speed control switch performance
P0592	Speed control switch 2 circuit, low voltage
P0593	Speed control switch 2 circuit, high voltage
P0600	Serial communication link
P0601	Powertrain Control Module (PCM) internal memory checksum invalid
P0606	Internal Powertrain Control Module (PCM) processor
P060B	Electronic Throttle Control (ETC) analog-to-digital (A/D) ground performance problem
P060D	Electronic Throttle Control (ETC) level 2 Accelerator Pedal Position (APP) sensor performance problem
P060E	Electronic Throttle Control (ETC) level 2 Throttle Position (TP) sensor performance problem
P060F	Electronic Throttle Control (ETC) level 2 Engine Coolant Temperature (ECT) sensor performance problem
P061A	Electronic Throttle Control (ETC) level 2 torque performance problem
P061C	Electronic Throttle Control (ETC) level 2 rpm performance problem
P0622	Generator field control circuit
P0627	Fuel pump relay circuit
P062C	Electronic Throttle Control (ETC) level 2 mph performance problem
P0630	Vehicle Identification Number (VIN) not programmed in Powertrain Control Module (PCM)
P0632	Odometer not programmed in Powertrain Control Module (PCM)
P0633	Sentry Key Immobilizer System (SKIM) secret key not stored in Powertrain Control Module (PCM)
P0642	Sensor reference voltage no. 1 circuit low

OBD-II Diagnostic Trouble Codes (DTCs) (continued)

Code	Possible cause
P0643	Primary 5-volt supply circuit, high voltage
P0645	Air conditioning clutch relay circuit
P0652	Sensor reference voltage 2 circuit low
P0653	Sensor reference voltage 2 circuit high
P0660	Manifold Tune Valve solenoid circuit
P0685	Auto Shutdown (ASD) relay control circuit
P0688	Auto Shutdown (ASD) relay sense circuit low
P0700	Transmission control system (Malfunction Indicator Light request)
P0703	Brake switch no. 2 performance problem
P0850	Park/Neutral switch performance

Note: * (1/1) = Bank 1, upstream oxygen sensor; (1/2) = Bank 1, downstream oxygen sensor; (2/1) = Bank 2, upstream oxygen sensor; (2/2) Bank 2, downstream oxygen sensor. Bank 1 is the bank, or row of cylinders, containing cylinder no. 1. Bank 2 is the other bank of cylinders.

4   Accelerator Pedal Position Sensor (APPS) - replacement

Note: *The Accelerator Pedal Position Sensor (APPS) is an integral component of the accelerator pedal assembly - to replace the APPS, you must replace the pedal assembly.*

1   If the vehicle is not equipped with adjustable pedals, disconnect the electrical connector from the APPS **(see illustration)**.

2   If the vehicle is equipped with adjustable pedals, disconnect the electrical connector and the adjuster cable from the APPS assembly **(see illustration)**.

3   Remove the three APPS assembly mounting nuts **(see illustration)**.

4   Remove the APPS assembly.

5   Installation is the reverse of removal. Perform the ETC relearn procedure using a scan tool or DTCs will be set in the PCM. Clear any stored DTCs.

6   If the vehicle is equipped with adjustable pedals, synchronize the pedals (see Chapter 9).

5   Ambient air temperature sensor - replacement

Note: *The ambient air temperature sensor is located on the trailing edge of the front bumper, below the right horn.*

1   Raise the vehicle and place it securely on jackstands.

2   On 2014 and earlier models, remove the front splash shield that protects the area under the front part of the engine compartment, right behind the trailing edge of the front

4.1 To disconnect the electrical connector from the APPS assembly, depress this release tab and pull the connector straight up

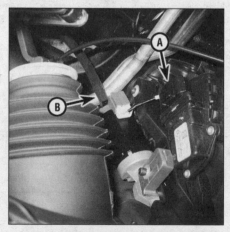

4.2 On vehicles with adjustable pedals, disconnect the electrical connector (A) and the adjuster motor drive cable (B). To disconnect the cable, simply pull it straight up

4.3 To detach the APPS assembly, remove these three mounting nuts

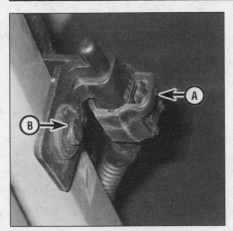

5.4a To remove the ambient air temperature sensor, depress the tab (A) and pull off the connector, then pull out the pushpin (B) and detach the sensor from the bumper (2014 and earlier model shown)

5.4b To remove the ambient air temperature sensor, disconnect the connector, and detach the sensor from the core support (2015 and later model shown)

6.2 To disconnect the electrical connector from the CMP sensor on a 2.7L V6, depress the release tab and pull off the connector. Before removing the sensor, disengage the harness clip (A), if applicable, from the sensor mounting stud by carefully prying it off

bumper cover (see Chapter 2A).

3   On 2015 and later models, remove the front grille (see Chapter 11, Section 11).

4   On all models, disconnect the electrical connector from the ambient air temperature sensor (see illustrations).

5   Detach the sensor mounting bracket pushpin from the bumper and remove the sensor.

6   Installation is the reverse of removal.

6   Camshaft Position (CMP) sensor - replacement

V6 engines

2.7L engine

Note: *The CMP sensor is located at the front end of the left cylinder head.*

1   Raise the front end of the vehicle and support it securely on jackstands.

2   Disconnect the electrical connector from the CMP sensor (see illustration).

3   Detach the wiring harness clip, if applicable, from the CMP sensor mounting stud.

4   Remove the sensor mounting bolt (see illustration) and remove the CMP sensor.

5   Installation is the reverse of removal.

6   Perform the cam/crank variation learn procedure using a scan tool.

3.5L engine

Note: *The CMP sensor is located ahead of the left cylinder head, on top of the timing belt cover.*

7   Unlock and disconnect the electrical connector from the CMP sensor (see illustration).

8   Remove the CMP sensor mounting bolt (see illustration) and remove the CMP sensor.

6.4 To detach the CMP sensor from the cylinder head on a 2.7L V6, remove this nut, then pull the sensor straight out

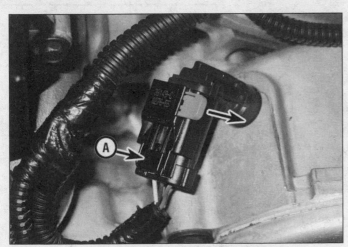

6.7 To remove the CMP sensor from the timing belt cover on a 3.5L V6, push the lock in the direction of the arrow, then depress the release tab (A) and pull off the connector

6.8 To detach the CMP sensor from the timing belt cover on a 3.5L V6, remove the sensor mounting bolt

6.16 Identifying the CMP sensor on 3.6L engines

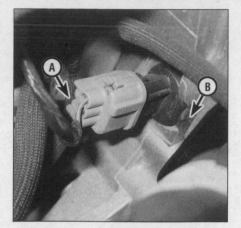

6.22 To detach the CMP sensor from the timing chain cover on a V8 engine, depress the release tab (A) and disconnect the electrical connector, then remove the sensor mounting bolt (B)

6.24 Remove the old CMP sensor O-ring and inspect it for cracks, tears and deterioration. If it's damaged, replace it (V8 engine)

9   If you're installing the old CMP sensor, scrape off the old spacer and install a new paper spacer.

10   If you're installing a new CMP sensor, it will be equipped with a new paper spacer.

11   Installation is the reverse of removal.

12   Perform the cam/crank variation learn procedure using a scan tool.

3.6L engine

13   Remove the engine cover.

14   Disconnect the negative battery cable (see Chapter 5, Section 3).

15   If removing the left (driver's side) CMP, remove the upper intake manifold (see Chapter 2C, Section 5).

16   Locate the CMP and disconnect the electrical connector **(see illustration)**.

17   Remove the fastener and pull the CMP from the valve cover.

18   Installation is reverse of removal noting

the following items:

a) *Clean the CMP sensor fastener threads in the cylinder head.*

b) *Inspect the fastener O-ring and replace if damaged.*

c) *Inspect the CMP sensor seal in the valve cover and replace as necessary.*

19   Perform the cam/crank variation learn procedure using a scan tool.

V8 engine

Note: *The CMP sensor is located on the upper right side of the timing chain cover.*

20   Remove the engine cover.

21   Disconnect the negative battery cable (see Chapter 5, Section 3).

22   Disconnect the electrical connector from the CMP sensor **(see illustration)**.

23   Remove the CMP sensor mounting bolt and remove the CMP sensor from the timing chain cover.

24   Remove the old CMP sensor O-ring **(see illustration)** and inspect it for cracks, tears and deterioration. If the old O-ring is damaged, replace it.

25   When installing the CMP sensor, apply a small dab of clean engine oil to the sensor O-ring, then use a slight rocking motion to work the O-ring into the sensor mounting bore. Do NOT use a twisting motion or you will damage the O-ring.

Caution: *If the CMP sensor is not fully seated against its mounting surface, the sensor mounting tang will be damaged when the sensor mounting bolt is tightened.*

26   Make sure that the CMP sensor mounting flange is fully seated flat against the mounting surface around the sensor mounting hole.

27   Installation is otherwise the reverse of removal.

28   Perform the cam/crank variation learn procedure using a scan tool.

---

7   Crankshaft Position (CKP) sensor - replacement

V6 engines

Note: *The CKP sensor is located on the upper right side of the transmission bellhousing.*

1   Disconnect the negative battery cable (see Chapter 5, Section 3).

2 Raise the front end of the vehicle and place it securely on jackstands.

3   Remove the under-vehicle splash shield.

4   Remove the CKP shield (if equipped).

5   Unlock and disconnect the electrical connector from the CKP sensor **(see illustrations)**.

6   Remove the CKP sensor mounting bolt and remove the sensor from the transmission bellhousing.

7   Inspect the sensor O-ring and replace if damaged.

7.5a To detach the CKP sensor from the transmission bellhousing on a V6 engine, push the lock in the direction of the arrow, depress the release tab (A), disconnect the electrical connector and remove the sensor mounting bolt (B)

7.5b Identifying the CKP on 3.6L engines

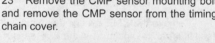

7.13 To detach the CKP sensor (A) from the (right-rear side of the) block on a V8 engine, disconnect the electrical connector (B) and remove the sensor mounting bolt (C)

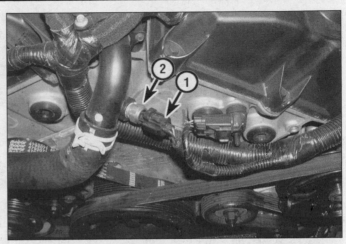

8.4a To remove the ECT sensor from a 2.7L V6, depress the release tab (1) and disconnect the electrical connector, then unscrew the ECT sensor (2) from the lower intake manifold

8   Installation is the reverse of removal.

9   Perform the cam/crank variation learn procedure using a scan tool.

V8 engine

Note: *The CKP sensor is located on the right rear side of the block.*

10   Disconnect the negative battery cable (see Chapter 5, Section 3).

11   Raise the front end of the vehicle and place it securely on jackstands.

12   Remove the starter for access to the CKP (see Chapter 5, Section 10).

13   Disconnect the electrical connector from the CKP sensor **(see illustration)**.

14   Remove the CKP sensor mounting bolt and remove the sensor from the engine.

15   Remove the CKP sensor O-ring and inspect it for cracks, tears and deterioration. If it's damaged, replace it.

16   Apply a small amount of engine oil on the O-ring and, using a slight rocking motion, push the sensor into its mounting hole in the

engine block until the sensor is fully seated.

17   Installation is otherwise the reverse of removal.

18   Perform the cam/crank variation learn procedure using a scan tool.

8   Engine Coolant Temperature (ECT) sensor - replacement

Warning: *Wait until the engine is completely cool before beginning this procedure.*

1   Partially drain the cooling system, to a level that's slightly below the cylinder heads (see Chapter 1).

V6 engines

Note: *On 2.7L and 3.5L engines, the ECT sensor is located at the front of the lower intake manifold. On 3.6L engines, the ECT sensor is located on the rear of the left (driver's) cylinder head.*

2   Disconnect the negative battery cable (see Chapter 5, Section 3).

3   Partially drain the cooling system below the level of the sensor (see Chapter 1, Section 26).

4   Disconnect the electrical connector from the ECT sensor **(see illustrations)**.

5   Using a deep socket, unscrew the ECT sensor from the lower intake manifold (2.7L and 3.5L) or left cylinder head (3.6L).

6   To prevent leakage and thread corrosion, wrap the threads of the ECT sensor with Teflon sealing tape **(see illustration)** before installing the sensor. (Seal the sensor threads whether you're installing the old sensor or a new unit.)

7   Installation is otherwise the reverse of removal. Be sure to tighten the ECT sensor to the torque listed in this Chapter's Specifications.

8   Refill the cooling system (see Chapter 1).

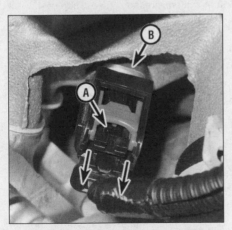

8.4b To remove the ECT sensor from a 3.5L V6, push the lock toward the harness, depress the release tab (A) and disconnect the electrical connector, then unscrew the ECT sensor (B) from the lower intake manifold

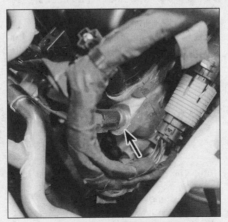

8.4c Identifying the ECT sensor on 3.6L engines

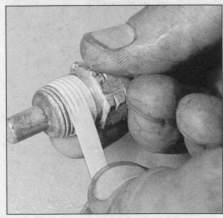

8.6 Before installing the ECT sensor, be sure to wrap the threads of the sensor with Teflon tape to prevent leaks

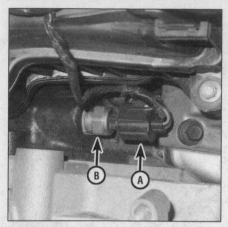

8.10 To remove the ECT sensor from the intake manifold on a V8 engine, disconnect the electrical connector (A) and unscrew the sensor (B) from the manifold

9.1a A typical IAT sensor on a 3.5L V6. To disconnect the electrical connector, push the lock toward the harness, then depress the release tab (A) and pull off the connector

9.1b A typical IAT sensor on a V8 engine. Remove it the same way you would remove the IAT sensor from a V6

V8 engine

Note: *The ECT sensor is located on the front end of the engine block, below the air conditioning compressor.*

9   Remove the accessory drivebelt (see Chapter 1), unbolt the air conditioning compressor (see Chapter 3) and set it aside. Do NOT disconnect any of the air conditioning pressure or return line fittings.

10   Disconnect the electrical connector from the ECT sensor (**see illustration**).

11   Using a deep socket, unscrew the ECT sensor from the intake manifold.

12   To prevent leakage and thread corrosion, wrap the threads of the ECT sensor with Teflon sealing tape (**see illustration 8.6**) before installing the sensor. (Seal the sensor threads whether you're installing the old sensor or a new unit.)

13   Installation is otherwise the reverse of removal. Be sure to tighten the ECT sensor

to the torque listed in this Chapter's Specifications.

14 Refill the cooling system (see Chapter 1).

9   Intake Air Temperature (IAT) sensor - replacement

Note: *The IAT sensor is located on the air intake duct on all engines. Though the exact location on the duct varies with the engine, the same IAT sensor is used on all engines.*

1   Unlock and disconnect the electrical connector from the IAT sensor (**see illustrations**).

2   Note the orientation of the IAT sensor. When you install a new (or the old) sensor, it MUST be correctly oriented in order to function correctly.

3   On 2.7L, 3.5L and 5.7L engines, to remove the IAT sensor from the air intake duct, grasp it firmly and pull it out of the rubber duct (**see illustration**).

4   On all other engines, pull up on the locking tab (if equipped) and rotate the sensor 90 degrees counterclockwise and pull from the duct. Inspect the O-ring and replace if necessary.

5   Installation is the reverse of removal.

10   Fuel Rail Pressure (FRP) sensor - replacement

Note: *The FRP is located on the right (passenger's) side fuel rail. On some models, the FRP sensor may not be servicable.*

1   Relieve fuel system pressure (see Chapter 4, Section 3).

2   Disconnect the negative battery cable (see Chapter 5, Section 3).

3   Locate the FRP and disconnect the electrical connector.

4   Disconnect the FRP sensor quick-discon-

nect and remove the sensor from the fuel rail.

5   Inspect the O-ring and replace if damaged.

6   Installation is reverse of removal.

7   Turn the key on engine off and check for fuel leaks.

11   Knock sensor - replacement

Warning: *Wait until the engine is completely cool before beginning this procedure.*

V6 engines

Note: *The knock sensor is located on top of the block, in the valley between the cylinder heads, underneath the intake manifold. On 3.6L models, two knock sensors are used; the front sensor is knock sensor 1 and the rear sensor is knock sensor 2.*

1   On 2.7L and 3.5L engines, relieve the fuel system pressure (see Chapter 4, Section 3).

2   Disconnect the negative battery cable (see Chapter 5, Section 3).

3   On 2.7L and 3.6L models, remove the upper and lower intake manifolds (2.7L engine, see Chapter 2A; 3.6L engine, see Chapter 2C). On 3.5L models, remove the upper intake manifold (see Chapter 2B).

4   On 2007 and earlier 2.7L engines, remove the right (passenger's) cylinder head (see Chapter 2A, Section 10).

5   On 2008 and later 2.7L engines, disconnect the hoses and water outlet tube that runs under the intake ports on the cylinder heads. Reposition the water outlet tube to access and remove the knock sensors.

6   On 3.6L engines, remove the oil filter housing (see Chapter 2C).

7   On all engines, disconnect the electrical connector from the knock sensor.

8   On 2007 and earlier 2.7L models, unscrew the sensor from the cylinder head.

9.3 To remove the IAT sensor, note the orientation of the sensor (it must be installed in the exact same orientation), then grasp the sensor firmly and pull it straight out of the air intake duct (3.5L V6 shown, other engines similar)

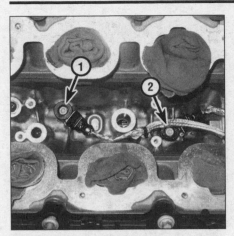

11.8 Knock sensor details - 3.6L V6 engine

1   Sensor one   2   Sensor two

11.11 The knock sensor on a V8 engine is protected by this heat shield. To remove it, grasp it firmly with a pair of pliers and pull it off

11.12 To detach a knock sensor from the block on a V8 engine, disconnect the electrical connector (A) and remove the knock sensor mounting bolt (B) (left knock sensor shown, right sensor identical)

On 3.5L and 3.6L models, remove the knock sensor mounting fastener(s) and remove the knock sensor(s) **(see illustration)**.

Caution: *Over- or under-tightening the knock sensor mounting bolts will affect knock sensor performance, which might affect the PCM's spark control ability.*

9   Installation is the reverse of removal. Be sure to tighten the knock sensor/fastener to the torque listed in this Chapter's Specifications.

V8 engine

Note: *There are two knock sensors. One is located on the left side of the block, below the exhaust manifold, and the other is located in the same place on the right side of the block.*

10   Raise the front of the vehicle and place it securely on jackstands.

11   Pull off the galvanized heat shield **(see illustration)**.

12 Disconnect the electrical connector from the knock sensor **(see illustration)**.

Note: *The foam strips on the knock sensor mounting bolt threads are used to retain the bolts during vehicle assembly at the manufacturing plant. They have no other purpose. They are not some form of adhesive, thread sealant or locking compound. Do NOT use any type of adhesive, thread sealant or locking compound when installing these bolts again.*

13   Remove the knock sensor mounting bolt and remove the knock sensor from the engine.

Caution: *Over- or under-tightening the knock sensor mounting bolts will affect knock sensor performance, which might affect the PCM's spark control ability.*

14   Installation is the reverse of removal.

Be sure to tighten the knock sensor mounting bolts to the torque listed in this Chapter's Specifications.

---

12   Manifold Absolute Pressure (MAP) sensor - replacement

Note: *The MAP sensor may also be referred to as the Manifold Air Pressure sensor.*

V6 engines

1   Remove the engine cover.

2   Unlock and disconnect the electrical connector from the MAP sensor **(see illustrations)**.

3   Rotate the MAP counterclockwise 1/4-turn **(see illustration)** and pull it straight up and out of the intake manifold.

12.2a To disconnect the electrical connector from the MAP sensor on a V6 engine, pull out the lock, then depress the release tab (A) and detach the connector (3.5L V6 shown, 2.7L and 3.6L V6 similar)

12.2b Locating the MAP sensor on 3.6L engines

12.3 To remove the MAP sensor from the intake manifold, rotate it counterclockwise 1/4-turn and pull it straight up and out of the manifold (3.5L V6 shown, 2.7L V6 similar)

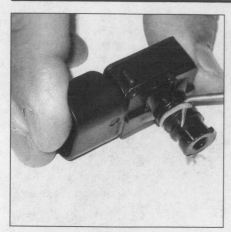

12.4 Remove the MAP sensor O-ring and inspect it for cracks, tears and deterioration; if it's damaged, replace it

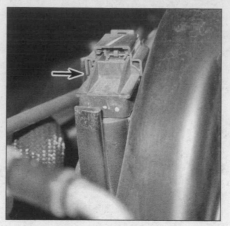

12.8 To remove the sensor from the intake manifold on a V8 engine, rotate it 1/4-turn counterclockwise and pull it out of its mounting hole

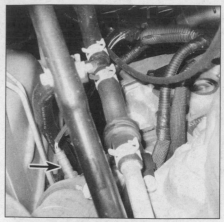

13.8a The left upstream oxygen sensor is screwed into the left exhaust manifold

4   Remove the MAP sensor O-ring **(see illustration)** and inspect it for cracks, tears and deterioration. If the O-ring is damaged, replace it.
5   Installation is the reverse of removal.

V8 engine

Note: *The MAP sensor is located on the rear part of the intake manifold, near the firewall. On some models, the sensor rotates 90 degrees to remove and install with a locking tab. Other models use a bolt to fasten the sensor to the intake manifold.*
6   Remove the engine cover.
7   Disconnect the electrical connector from the MAP sensor **(see illustration 12.2a)**.
8   Rotate the MAP sensor 1/4-turn counterclockwise and remove it from the intake manifold **(see illustration)** or remove the bolt and pull the sensor from the manifold.
9   Remove the old MAP sensor O-ring **(see illustration 12.4)** and inspect it for cracks, tears and deterioration. If it's damaged, replace it.
10   To install the MAP sensor, place it in position, insert it into the mounting hole in the intake manifold and turn it 1/4-turn clockwise, or install the bolt, as applicable.
11   Installation is otherwise the reverse of removal.

---

13   Oxygen sensors - general description and replacement

Description

1   The oxygen in the exhaust reacts with the elements inside the oxygen sensor to produce a voltage output that varies from 0.1 volt (high oxygen, lean mixture) to 0.9 volt (low oxygen, rich mixture). The pre-converter oxygen sensor (mounted in the exhaust system before the catalytic converter) provides a feedback signal to the PCM that indicates the amount of leftover oxygen in the exhaust. The PCM monitors this variable voltage continuously to determine the required fuel injector pulse width and to control the engine air/fuel ratio. A mixture ratio of 14.7 parts air to 1 part fuel is the ideal ratio for minimum exhaust emissions, as well as the best combination for fuel economy and engine performance. Based on oxygen sensor signals, the PCM tries to maintain this air/fuel ratio of 14.7:1 at all times.
2   The post-converter oxygen sensor (mounted in the exhaust system after the catalytic converter) has no effect on PCM control of the air/fuel ratio. However, the post-converter sensor is identical to the pre-converter sensor and operates in the same way. The PCM uses the post-converter signal to monitor the efficiency of the catalytic converter. A post-converter oxygen sensor will produce a slower fluctuating voltage signal that reflects the lower oxygen content in the post-catalyst exhaust.
3   There are four oxygen sensors on all vehicles. The two upstream sensors (1/1 and 2/1) are located in the exhaust manifolds and the downstream sensors (1/2 and 2/2) are located downstream from and right behind the catalytic converters.
4   An oxygen sensor produces no voltage when it is below its normal operating temperature of about 600-degrees F. During this warm-up period, the PCM operates in an open-loop fuel control mode. It does not use the oxygen sensor signal as a feedback indication of residual oxygen in the exhaust. Instead, the PCM controls fuel metering based on the inputs of other sensors and its own programs.
5   An oxygen sensor depends on four conditions in order to operate correctly:

a)   *Electrical - The low voltage generated by the sensor requires good, clean connections. Always check the connectors whenever an oxygen sensor problem is suspected or indicated.*

b)   *Outside air supply - The sensor needs air circulation to the internal portion of the sensor. Whenever the sensor is installed, make sure that the air passages are not restricted.*

c)   *Correct operating temperature - The PCM will not react to the sensor signal until the sensor reaches approximately 600-degrees F. This factor must be considered when evaluating the performance of the sensor.*

d)   *Unleaded fuel - Unleaded fuel is essential for correct sensor operation.*

6   The PCM can detect several different oxygen sensor problems and set Diagnostic Trouble Codes (DTCs) to indicate the specific fault (see Section 3). When an oxygen sensor DTC occurs, the PCM disregards the oxygen sensor signal voltage and reverts to open-loop fuel control as described previously.

Replacement

Warning: *Be careful not to burn yourself during the following procedure.*
Note: *Since the exhaust pipe contracts when cool, the oxygen sensor may be hard to loosen. To make sensor removal easier, start the engine and let it run for a minute or two, then turn it off.*

Upstream oxygen sensors

Note: *The upstream oxygen sensors are located in the exhaust manifolds, right above the flanges. Both of the connectors for the upstream sensors, and the sensors themselves, are more easily accessed from below.*
7   Raise the vehicle and place it securely on jackstands.
8   Locate the upstream sensor that you want to replace, then trace the sensor's lead to the electrical connector **(see illustrations)** and disconnect it.
9   There isn't enough room to remove an upstream oxygen sensor with a wrench; use

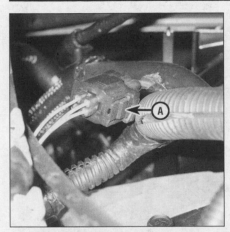

13.8b Trace the electrical lead up the connector, then depress the release tab (A) and disconnect the two halves of the connector

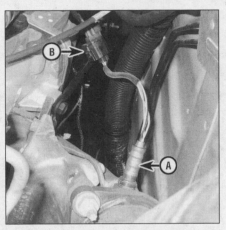

13.8c The right upstream oxygen sensor (A) and electrical connector (B) (3.5L V6 shown, other engines similar)

13.9 If there isn't enough room to put a wrench on an upstream oxygen sensor, use an oxygen sensor socket (this photo depicts an oxygen sensor socket on a downstream sensor to show you how it fits onto the sensor) (3.5L V6 shown, other engines similar)

an oxygen sensor socket **(see illustration)**. Oxygen sensor sockets are available at most auto parts stores.

10   If you're installing the old sensor, clean off the threads, then apply a coat of anti-seize compound (such as Loctite® 771-64 or a suitable equivalent) to the threads before installing the sensor. If you're installing a new sensor, do NOT apply anti-seize compound; new sensors are already coated with anti-seize.

11   Installation is otherwise the reverse of removal. Be sure to tighten the oxygen sensor to the torque listed in this Chapter's Specifications.

Downstream oxygen sensors

Note: *The downstream oxygen sensors are located right behind the catalytic converters.*

12   Raise the vehicle and place it securely on jackstands.

13   Locate the downstream sensor that you want to replace, then trace the sensor's electrical lead to the sensor electrical connector **(see illustrations)** and disconnect it.

14   All you need to remove one of the downstream sensors is a combination wrench. An oxygen sensor socket isn't really necessary.

15   If you're installing the old sensor, clean off the threads, then apply a coat of anti-seize compound (such as Loctite® 771-64 or a suitable equivalent) to the threads before installing the sensor. If you're installing a new sensor, do NOT apply anti-seize compound; new sensors are already coated with anti-seize.

16   Installation is otherwise the reverse of removal. Be sure to tighten the oxygen sensor

to the torque listed in this Chapter's Specifications.

All sensors

17   Disconnect the cable from the negative terminal of the battery and leave it disconnected for a couple of minutes (see Chapter 5, Section 3). This will erase the old operating values that the Powertrain Control Module (PCM) has learned, allowing it to learn the characteristics of the new sensor. (If this isn't done, the vehicle may exhibit driveability problems.)

Note: *Whenever the battery is disconnected, stored operating parameters may be lost from the PCM, which may cause the engine to run rough for a period of time while the PCM relearns the information.*

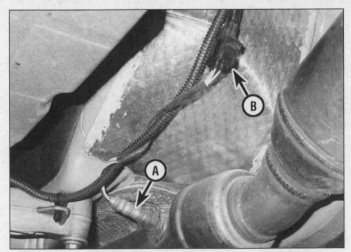

13.13a Left downstream oxygen sensor (A) and electrical connector (B) (3.5L V6 shown, other engines similar)

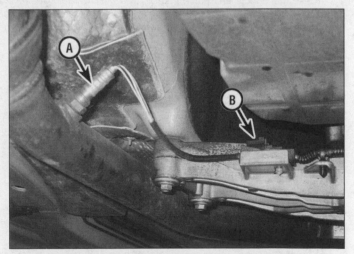

13.13b Right downstream oxygen sensor (A) and electrical connector (B) (3.5L V6 shown, other engines similar)

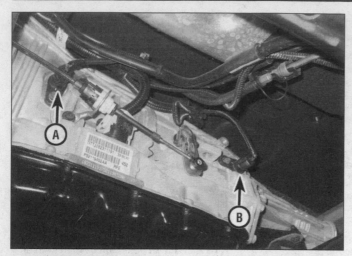

14.2 On 42RLE automatic transmissions, the transmission speed sensors are located on the left side of the transmission

A   *Input speed sensor*   B   *Output speed sensor*

14.3 To remove an input or output speed sensor from the transmission, disconnect the electrical connector (1), then remove the sensor mounting bolt (2)

14   Transmission speed sensors - replacement

Note: *This procedure applies only to the 42RLE (hydraulic, with adaptive electronics) automatic transmission. The input speed and output speed sensors are located on the left side of the transmission housing. The input speed sensor is the forward unit. The output speed sensor is the rear one. On the NAG1 (all-electronic) transmission, there are also two speed sensors, but they're located inside the control unit, which is located inside the transmission, on top of the valve body. Replacing either of these sensors is a job for a professional automatic transmission specialist. On the 8HP45/8HP45RE transmissions, these sensors are integral components of the Transmission Control Module (TCM) and are not serviceable separately.*

1   Raise the vehicle and place it securely on jackstands.

2   Locate the transmission speed sensor that you're going to replace **(see illustration)**. Place a drip pan underneath it. Some ATF will leak out when you remove the sensor.

3   Disconnect the electrical connector from the speed sensor that you want to replace **(see illustration)**.

4   Remove the mounting bolt from the speed sensor that you want to replace **(see illustration 12.3)**.

5   Installation is the reverse of removal. If you're installing the old sensor, make sure the O-ring is in good condition. Also, apply a non-hardening thread locking compound to the threads of the mounting bolt.

15   Powertrain Control Module (PCM) - removal and installation

Caution: *Avoid static electricity damage to the Powertrain Control Module (PCM) by grounding yourself to the body of the vehicle before touching the PCM and using a special anti-* static pad on which to store the PCM once it's removed.

Note: *The PCM is located in the right rear corner of the engine compartment, inside the cowl area.*

Note: *Whenever the PCM is replaced with a new unit, it must be reprogrammed with a scan tool by a dealership service department or other qualified repair shop.*

Note: *Whenever the battery is disconnected, stored operating parameters may be lost from the PCM, causing the engine to run rough for a period of time while the PCM relearns the information.*

1   Remove the engine cowl cover (see Chapter 11, Section 24).

2   Disconnect the cable from the negative terminal of the battery (see Chapter 5, Section 1). on 2011 and later models remove the right-side front top cowl cover **(see illustration)**.

3   Remove the bolt that secures the PCM mounting bracket to the cowl **(see illustration)**.

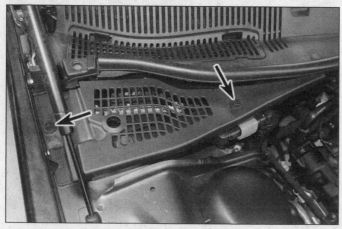

15.2 Remove these two pushpins and take off the cover (2011 and later models)

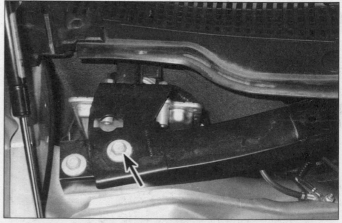

15.3 To detach the PCM mounting bracket from the cowl, remove this bolt

15.4a Lift the PCM out of the cowl area, then unlock and disconnect the electrical connectors. To do so on 2010 and earlier models, push each sliding lock toward the harness, then depress the release tab (A) and pull off the connector

15.4b On 2011 and later models, disengage the connector lever lock, then swing the lever open and disconnect the electrical connector

4   Carefully lift the PCM out of the cowl area and disconnect the electrical connectors from the PCM **(see illustrations)**.

5   If you're replacing the PCM, remove the two bolts that secure the mounting bracket to the PCM and attach the bracket to the new PCM **(see illustration)**.

6   Installation is the reverse of removal.

7   A NEW PCM must be programmed by a qualified technician or repair facility. The VIN and mileage must be entered as well as other functions performed for the vehicle to operate.

16   Catalytic converter - description, check and replacement

Note: *Because of a Federally-mandated extended warranty which covers emission-related components such as the catalytic converter, check with a dealer service department before replacing the converter at your own expense.*

General description

1   A catalytic converter (or catalyst) is an emission control device in the exhaust system that reduces certain pollutants in the exhaust gas stream. There are two types of converters. An oxidation catalyst reduces hydrocarbons (HC) and carbon monoxide (CO). A reduction catalyst reduces oxides of nitrogen (NOx). A catalyst that can reduce all three pollutants is known as a "Three-Way Catalyst" (TWC).

2   All models covered by this manual are equipped with two TWCs, one below each exhaust manifold flange. The upper end of each catalyst has a short inlet pipe with a mounting flange that bolts to the exhaust manifold flange. The lower end of each catalyst is welded to a longer outlet pipe. The left and right catalyst outlet pipes are secured to the rest of the exhaust system by large screw-type clamps.

Check

3   The test equipment for a catalytic converter (a "loaded-mode" dynamometer and a 5-gas analyzer) is expensive. If you suspect that the converter on your vehicle is malfunctioning, take it to a dealer or authorized emission inspection facility for diagnosis and repair.

4   Whenever you raise the vehicle to service underbody components, inspect the converter for leaks, corrosion, dents and other damage. Carefully inspect the welds and/or flange bolts and nuts that attach the front and rear ends of the converter to the exhaust system. If you note any damage, replace the converter.

5   Although catalytic converters don't break too often, they can become clogged or even plugged up. The easiest way to check for a restricted converter is to use a vacuum gauge to diagnose the effect of a blocked exhaust on intake vacuum.

a) *Connect a vacuum gauge to an intake manifold vacuum source (see Chapter 2D).*

b) *Warm the engine to operating temperature, place the transaxle in Park (automatic models) or Neutral (manual models) and apply the parking brake.*

c) *Note the vacuum reading at idle and write it down.*

d) *Quickly open the throttle to near its wide-open position and then quickly get off the throttle and allow it to close. Note the vacuum reading and write it down.*

e) *Do this test three more times, recording your measurement after each test.*

f) *If your fourth reading is more than one in-Hg lower than the reading that you noted at idle, the exhaust system might be restricted (the catalytic converter could be plugged, OR an exhaust pipe or muffler could be restricted).*

Replacement

Warning: *Make sure that the exhaust system is completely cooled down before proceeding. If the vehicle has just been driven, the catalytic converter can be hot enough to cause serious burns.*

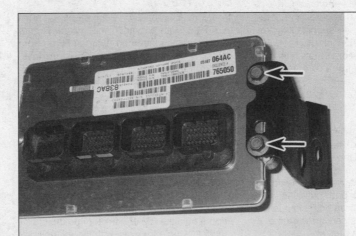

15.5 To detach the mounting bracket from the PCM, remove these two bolts. Then bolt the mounting bracket onto the new PCM unit

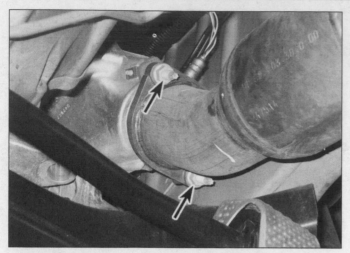

16.7a To disconnect the inlet pipe flange of either catalytic converter from the exhaust manifold flange, spray the threads of the bolts and nuts with penetrant, then remove these nuts. Replace the nut(s) and bolt(s) when installing the catalyst(s)

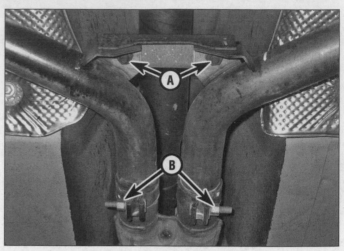

16.7b To disconnect either catalyst outlet pipe from the support bracket, remove the bolt (A) from above (not the nut). To disconnect the outlet pipe of either catalytic converter, remove the nut from the clamp bolt (B)

Note: *The photos accompanying this section depict the catalytic converters used on a 3.5L V6 model, but all models use essentially the same setup. Also, this procedure applies to both catalytic converters.*

6   Raise the vehicle and place it securely on jackstands.

7   Spray a liberal amount of penetrating oil onto the threads of the exhaust pipe-to-exhaust manifold bolts and the clamp bolt nut that connects the catalyst outlet pipe to the rest of the exhaust system. Also spray some penetrant on the two bolts and nuts that secure the small bracket between the two catalyst outlet pipes **(see illustrations)**. Wait awhile for the penetrant to loosen things up.

8   While you're waiting for the penetrant to do its work, disconnect the electrical connector for the downstream oxygen sensor and remove the downstream oxygen sensor (see Section 13).

9   Unscrew the upper exhaust pipe-to-exhaust manifold flange bolt nuts. If they're still difficult to loosen, spray the threads with some more penetrant, wait awhile and try again.

10   Remove the bolt that secures the catalyst outlet pipe to the small support bracket between the catalyst outlet pipes. If you're removing both catalytic converters, remove the both bolts and remove the support bracket.

Note: *The nuts are tack-welded to the bracket, so don't try to unscrew them. Instead, unscrew the bolts from above.*

11   Remove the nut from the clamp bolt that secures the slip joint between the catalyst outlet pipe and the rest of the exhaust system. If it's still difficult to loosen, spray the threads with some more penetrant, wait awhile and try again. If you're removing both catalytic converters, remove the nuts from both clamps.

12   Remove the catalytic converter. Remove and discard the old flange gasket.

13   Installation is the reverse of removal. Be sure to use a new flange gasket and new bolts and nuts at the exhaust manifold mounting flange. The manufacturer also recommends using a new clamp at the slip joint between the catalyst outlet pipe and the exhaust pipe. Coat the threads of the clamp bolt and the exhaust manifold bolts with anti-seize compound to facilitate future removal. Tighten the fasteners securely.

14   Once the catalytic converter is reattached to the exhaust manifold and to the rest of the exhaust system, install the small support bracket. Again, it's a good idea to use new nuts and bolts and to coat the threads of the bolts with anti-seize compound. Tighten the nuts and bolts securely.

---

17   Evaporative emissions control (EVAP) system - description and component replacement

Description

1   The Evaporative Emissions Control (EVAP) system prevents fuel system vapors (which contain unburned hydrocarbons) from escaping into the atmosphere. On warm days, vapors trapped inside the fuel tank expand. When the pressure reaches a certain threshold, these vapors are routed from the fuel tank through the fuel vapor vent valve and the fuel vapor control valve to the EVAP canister, where they're stored temporarily, until they can be consumed by the engine during normal operation. Under certain conditions (engine warmed up, vehicle up to speed, moderate or heavy loads, etc.) the Powertrain Control Module (PCM) opens the canister purge solenoid, which allows intake vacuum to pull fuel vapors from the EVAP canister into the intake manifold, where they mix with the air/fuel mix-

ture before being consumed in the combustion chambers. This system is complex and virtually impossible to troubleshoot without the right tools and training. However, the following description should give you a good idea of how the system works and where the components are located:

2   The EVAP canister, which contains activated charcoal, is the repository for storing fuel vapors produced by gasoline as it heats up inside the fuel tank. You'll have to raise the vehicle to inspect or replace the canister, but the canister is designed to be maintenance-free and should last the life of the vehicle. The EVAP canister is located in the right rear corner of the vehicle, above the right rear wheel. You'll have to raise the vehicle, remove the right rear wheel and remove the wheel well splash shield to access the EVAP canister.

3   The Natural Vacuum Leak Detection (NVLD) assembly is also located in the same spot as the EVAP canister. The NVLD unit monitors the temperature of the unburned hydrocarbon vapors inside the EVAP system. As the temperature increases, so does the pressure; as the temperature decreases, so does the pressure. But this effect can occur only if the system is fully sealed. If there's a leak, the relationship between pressure and temperature will diverge from the pressure-temperature relationship that the Powertrain Control Module (PCM) is programmed to expect. When this divergence between the pressure and the temperature of the fuel vapors inside the fuel tank and the EVAP system occurs, the PCM concludes that the system is leaking and it sets a Diagnostic Trouble Code (DTC) indicating either a small, medium or large leak in the EVAP system. Diagnosis of the NVLD assembly and detection system is beyond the scope of the home mechanic. If

17.8a To disconnect the electrical connector from the EVAP canister purge solenoid, push the lock sideways, then depress the release tab (A) and pull off the connector (3.5L V6 model shown, other models similar)

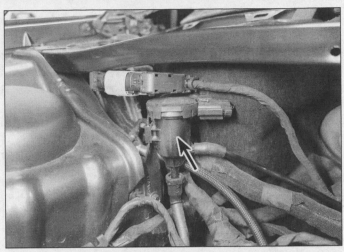

17.8b Identifying the EVAP canister purge solenoid on models with 3.6L engines

a DTC indicates a problem with the NVLD system, have a dealer service department or other qualified repair shop fix the vehicle. But a defective NVLD assembly is easy to replace, and you'll find the instructions to do so later in this section.

4   The EVAP canister purge solenoid (the manufacturer calls it a proportional purge solenoid) regulates the flow of vapors being purged from the EVAP canister into the intake manifold. The canister purge solenoid is controlled by the Powertrain Control Module (PCM). The canister purge solenoid is normally closed. It opens only when directed to do so by the PCM, which uses the availability of intake manifold vacuum and data from various information sensor inputs to determine when and how long to open the valve. The interval of time during which the purge valve is opened by the PCM is known as its "duty cycle." The canister purge solenoid is located at the right rear corner of the engine compartment on all models.

System checks

5   The most common symptom of a faulty EVAP system is a strong fuel odor (particularly during hot weather). If you smell fuel while driving or (more likely) right after you park the vehicle and turn off the engine, check the fuel filler cap first. Make sure that it's screwed onto the fuel filler neck all the way.

6   If the odor persists, inspect all EVAP hose connections, both in the engine compartment and under the vehicle. You'll have to raise the vehicle and place it securely on jackstands to inspect most of the EVAP system, since it's located under the vehicle. Be sure to inspect each hose attached to the canister for damage and leakage along its entire length. Repair or replace as necessary. Inspect the canister for damage and look for fuel leaking from the bottom. If fuel

is leaking or the canister is otherwise damaged, replace it.

7   Poor idle, stalling, and poor driveability can be caused by a defective fuel vapor vent valve or canister purge solenoid, a damaged canister, cracked hoses, or hoses connected to the wrong tubes. Fuel loss or fuel odor can be caused by fuel leaking from fuel lines or hoses, a cracked or damaged canister, or a defective vapor valve.

Component replacement
EVAP canister purge solenoid

Note: *The EVAP canister purge solenoid is located in the right rear corner of the engine compartment, on a small bracket attached to the right strut tower. On 2015 and later V8 engines, the EVAP canister purge solenoid is attached to the intake manifold.*

8   Disconnect the electrical connector from the EVAP canister purge solenoid **(see illustrations)**.

9   Clearly label the EVAP hoses **(see illus-**

tration), then disconnect them from the EVAP canister purge solenoid.

10   To disengage the EVAP canister purge solenoid from its mounting bracket, pull it straight up.

11   Installation is the reverse of removal.

EVAP canister and components (2006 and earlier models)

NVLD filter, NVLD pump and EVAP canister assembly

Note: *The NVLD filter, NVLD pump and EVAP canister are located in the upper part of the right rear wheel well, above the wheel well splash shield. The easiest way to replace any of these three components is to simply remove the entire assembly, then to remove the component that you want to replace.*

12   Loosen the right rear wheel lug nuts, raise the vehicle and place it securely on jackstands.

13   Remove the splash shield from the right rear wheel well (see Chapter 11).

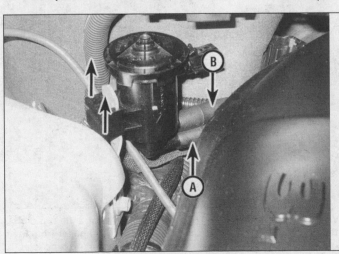

17.9 To remove the EVAP canister purge solenoid, disconnect the EVAP inlet hose (A) and the outlet hose (B), then disengage the solenoid from its mounting bracket by pulling it straight up and off the bracket (3.5L V6 model shown, other models similar)

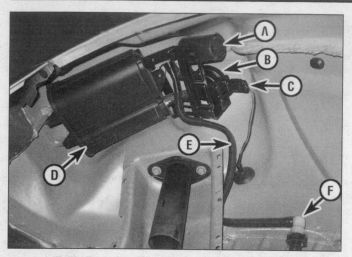

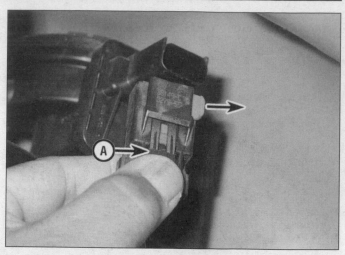

17.14a Typical EVAP canister and NVLD assembly (Chrysler 300 shown, other models similar):

A   NVLD air filter
B   Natural Vacuum Leak Detection (NVLD) pump
C   Electrical connector for NVLD pump
D   EVAP canister
E   EVAP line
F   EVAP line quick-connect fitting

17.14b To disconnect the electrical connector from the NVLD pump, push the lock to the side, then depress the release tab (A) and pull off the connector

17.15 To disconnect the EVAP line quick-connect fitting from the EVAP canister, squeeze these two tabs and pull off the fitting

14   Disconnect the NVLD electrical connector (**see illustrations**).

15   Disconnect the EVAP line from the EVAP canister (**see illustration**).

16   Remove the EVAP canister/NVLD assembly mounting nut (**see illustration**), then slide the slots in the two mounting bosses on the lower end of the canister off their corresponding mounting brackets (**see illustration**). Place the assembly on a clean workbench for further disassembly.

NVLD filter

17   Remove the EVAP canister/NVLD assembly (see Steps 12 through 15).

18   Disconnect the hose from the air filter (**see illustration**).

17.16a To detach the EVAP canister/NVLD assembly from the vehicle, remove this nut...

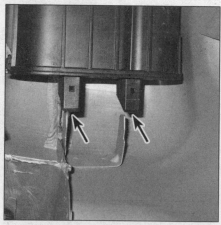

17.16b... then disengage the two slots in the lower end of the canister from their corresponding mounting brackets

17.18 Natural Vacuum Leak Detection (NVLD) assembly:

A   NVLD air filter
B   Air filter-to-NVLD pump hose
C   NVLD pump
D   NVLD pump-to-EVAP canister hose

17.19 To detach the NVLD air filter from its mounting bracket, disengage the mounting tab (A) from its slot in the mounting bracket, then swing the filter out from the bracket and unsnap the hinge (B) from its hinge pin

17.23 To detach the NVLD pump from its mounting bracket, push the mounting tab toward the pump housing to disengage the retainer tooth (A) from the bracket

17.27 To detach the NVLD filter/NVLD pump mounting bracket from the EVAP canister, remove these two nuts

19   Disengage the filter mounting tab from the mounting bracket **(see illustration)**, disengage the hinge from its hinge pin and remove the filter.

20   Installation is the reverse of removal.

NVLD pump

21   Remove the EVAP canister/NVLD assembly (see Steps 12 through 15).

22   Disconnect the air filter-to-NVLD pump hose and the NVLD pump-to-EVAP canister hose **(see illustration 17.18)**.

23   Unsnap the NVLD pump from its mounting bracket **(see illustration)**.

24   Installation is the reverse of removal.

EVAP canister

25   Remove the EVAP canister/NVLD assembly (see Steps 12 through 15).

26   Remove the NVLD filter (see Steps 18 and 19) and the NVLD pump (see Steps 22 and 23).

27   Remove the NVLD filter/NVLD pump mounting bracket nuts **(see illustration)** and remove the bracket from the EVAP canister.

28   Inspect the condition of the rubber insu-

lators inside the two canister mounting bosses **(see illustration)**.

29   Installation is the reverse of removal.

EVAP line

30   Disconnect the EVAP line from the EVAP canister **(see illustration 17.15)**.

31   Disconnect the EVAP line quick-connect fitting at the lower end of the EVAP line **(see illustration)**.

EVAP canister (2007 and later models)

Note: *The EVAP filter and Evaporative Emission System Monitor (ESIM) are part of the canister and replaced as an assembly.*

32   Disconnect the cable from the negative terminal of the battery (see Chapter 5, Section 3).

33   Loosen the gas cap to release any fuel tank pressure.

34   Loosen the right rear wheel lug nuts, then raise the rear of the vehicle and support it securely on jackstands.

35   Remove the right rear wheel.

36   Remove the right rear inner fender

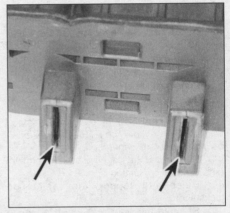

17.28 Inspect the condition of the rubber insulators inside the two EVAP canister mounting bosses

splash shield to access the canister.

37   Disconnect the EVAP tubing.

38   Disconnect the canister electrical connector **(see illustration)**.

39   Remove the canister mounting nut.

17.31 To disconnect the quick-connect fitting at the lower end of the EVAP line, squeeze the two sides together and pull it off

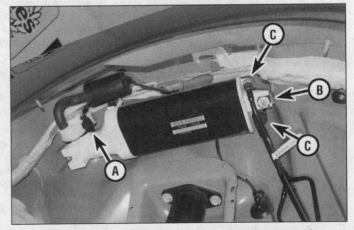

17.38 Disconnect the canister connector (A), remove the mounting nut (B) and disconnect the vapor lines (C)

17.47 Identifying the FTP sensor

18.7 To disconnect the electrical connector from the EGR valve, slide the lock sideways, then depress the release tab (A) and pull off the connector

18.12 To disconnect the lower end of the upper EGR pipe from the EGR valve, remove these two bolts (3.5L V6 shown, 2.7L V6 similar)

40   Pull the canister from the mounting tabs, using care as components are still connected.
41   Disconnect the vapor lines at the EVAP canister quick disconnect fittings.
42   If the canister is to be replaced, remove the canister from the mounting bracket.
43   Installation is reverse of removal.

Fuel Tank Pressure (FTP) sensor

Note: *2007 and later models are equipped with an FTP sensor. The FTP sensor is located in the fuel tank vent tube.*
44   Disconnect the negative battery terminal (see Chapter 5, Section 3).
45   Raise and support the vehicle on jackstands.
46   Remove the right rear inner fender to access the FTP sensor.
47   Locate the sensor and disconnect the electrical connector **(see illustration)**.
48   Clean the area around the sensor.
49   Raise the retaining tab and rotate the sensor to remove.
50   Inspect the O-ring and replace if damaged.
51   Installation is reverse of removal.

18   Exhaust Gas Recirculation (EGR) system - description and component replacement

Note: *This Section depicts the EGR system on 2010 and earlier models. 2011 and later models are not equipped with an EGR system.*

Description

1   Oxides of nitrogen, nitrogen oxide, or simply NOx, is a compound that is formed in the combustion chambers when the oxygen and nitrogen in the incoming air mix together. NOx is a natural byproduct of high combustion chamber temperatures (2500 degrees F and higher). When NOx is emitted from the tailpipe, it mixes with reactive organic compounds (ROCs), hydrocarbons (HC) and sunlight to form ozone and photochemical smog. The EGR system reduces NOx by recirculating exhaust gases from the exhaust manifold, through the EGR valve and intake manifold, then back to the combustion chambers, where it mixes with the incoming air/fuel mixture before being consumed. These recirculated exhaust gases "dilute" the incoming air/fuel mixture, which cools the combustion chambers, thereby reducing NOx emissions.
2   All engines covered by this manual use an EGR system.
3   The EGR system consists of the Powertrain Control Module (PCM), the EGR valve and various information sensors (ECT, TP, MAP, IAT, RPM and VSS sensors) that the PCM uses to determine when to open the EGR valve. When the PCM closes the power/control circuit for the EGR valve, a solenoid inside the EGR valve is energized. This creates an electromagnetic field, which causes an armature to pull up, lifting the pintle off its seat. The exhaust gas then flows from the exhaust manifold port to the intake manifold.
4   Once activated by the PCM, the EGR valve uses a position feedback circuit to control the position of the pintle valve. The feedback circuit, which functions like a potentiometer, puts out a variable output voltage signal with an operating range between 0.5 and 5.4 volts. This variable output enables the PCM to control the position of the pintle with a high degree of precision. A pintle position sensor monitors the position of the pintle, and the PCM adjusts the current to match the actual pintle position to the optimal pintle position.
5   If there is too much EGR flow at idle, cruise or during cold running conditions, the engine will stop after a cold start, stop at idle after deceleration, surge during cruising speeds or idle roughly. If there is too little EGR flow, combustion chamber temperature can become too high during acceleration or under a heavy load, which can cause spark knock (detonation) and/or engine overheating.

Component replacement

V6 engines

Note: *The EGR valve is bolted to the rear end of the right cylinder head. The lower EGR pipe connects the right exhaust manifold to the EGR valve. The upper EGR pipe connects the EGR valve to the intake manifold. You must perform the first two steps below before you can remove any of the three components.*
6   Remove the EVAP purge solenoid valve (see Section 17).
7   Disconnect the electrical connector from the EGR valve **(see illustration)** and set it aside.

Lower EGR pipe (2.7L engine only)

8   Remove the two bolts that secure the lower EGR pipe to the exhaust manifold.
9   Remove the two bolts that secure the lower EGR pipe to the EGR valve and remove the lower EGR pipe.
10   Remove the old gasket from the EGR pipe mounting flange and discard it.
11   Installation is the reverse of removal. Be sure to use a new gasket and tighten the EGR pipe mounting bolts to the torque listed in this Chapter's Specifications.

Upper EGR pipe

Note: *The photos accompanying this Section depict the EGR system on a 3.5L V6 model. Except for the lower EGR pipe (see above), the rest of the two systems are virtually identical.*
12   Remove the two bolts that secure the lower end of the upper EGR pipe to the EGR valve **(see illustration)**.
13   Disconnect the upper end of the upper EGR pipe from the intake manifold **(see illustration)** and remove the upper EGR pipe.
14   Remove the old gasket from the EGR pipe flange **(see illustration)**.
15   Remove the old rubber silicone seals from the EGR pipe **(see illustration)**. The smaller black seal is the actual O-ring; the larger blue seal is a dust seal. To ensure a

18.13 To disconnect the upper end of the upper EGR pipe from the intake manifold, grasp it firmly and pull it out

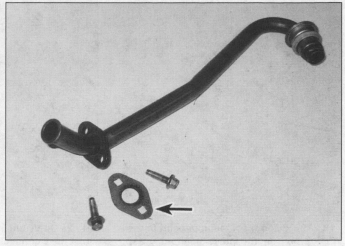

18.14 Always replace the gasket between the flange at the lower end of the upper EGR pipe and the mounting surface on the EGR valve

positive seal, you should always replace both of these seals whenever you disconnect the EGR pipe.

16   Installation is the reverse of removal. Be sure to use a new gasket and new rubber silicone seals and tighten the EGR pipe mounting flange bolts to the torque listed in this Chapter's Specifications.

EGR valve

17   Disconnect the upper EGR pipe and, on 2.7L engines, the lower EGR pipe, from the EGR valve.
18   Remove the EGR mounting bolts **(see illustration)** and remove the EGR valve from the right cylinder head.
19   Clean the EGR valve mounting surfaces on the cylinder head and on the valve itself.
20   Installation is the reverse of removal. Be sure to tighten the EGR valve mounting bolts to the torque listed in this Chapter's Specifications.

V8 engine

EGR valve

Note: *The EGR valve is located on the front of the right cylinder head.*
21   Disconnect the electrical connector from the EGR valve **(see illustration)**.
22   Remove the EGR pipe mounting flange bolts.
23   Remove the EGR valve mounting bolts and remove the EGR valve. Remove and discard the old gasket.
24   Clean the EGR valve mounting surfaces on the cylinder head and on the valve itself.
25   Installation is the reverse of removal. Be sure to use new gaskets between the EGR valve and the cylinder head and between the EGR pipe mounting flange and the EGR valve and tighten the EGR valve and EGR pipe mounting bolts to the torque listed in this Chapter's Specifications.

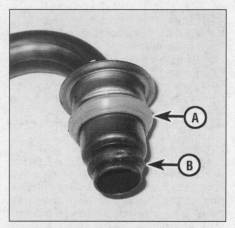

18.15 Always replace these two seals on the upper (intake manifold) end of the upper EGR pipe (3.5L V6 shown, 2.7L V6 similar):

A   Dust seal   B   O-ring

18.18 To detach the EGR valve from the cylinder, remove these two bolts (3.5L V6 shown, 2.7L V6 similar)

18.21 To remove the EGR valve from a V8 engine, slide the lock on the connector out, depress the tab (1) and detach the connector, remove the EGR pipe flange bolts (2), then remove the EGR valve bolts (3)

18.28 To disconnect the upper end of the EGR pipe from the intake manifold, grasp it firmly and pull it straight out of the manifold

19.2a On V6 engines, the PCV fresh air inlet hose connects the air filter housing to a pipe on the left valve cover (3.5L V6 shown, 2.7L V6 similar)

19.2b On V6 engines, the crankcase ventilation hose (or PCV hose) connects the PCV valve to the intake manifold (3.5L V6 shown, 2.7L V6 similar)

EGR pipe

Note: *The EGR pipe connects the EGR valve to the intake manifold.*

26   Remove the engine cover.

27   Remove the EGR pipe mounting flange bolts **(see illustration 18.21)**. Remove and discard the old EGR pipe mounting flange gasket.

28   Disconnect the EGR pipe from the intake manifold **(see illustration)**.

29   Remove the old gasket from the EGR pipe flange **(see illustration 18.14)** and replace it with a new one. Remove the old rubber silicone seals from the EGR pipe **(see illustration 18.21)**. The smaller black seal is the actual O-ring; the larger blue seal is a dust seal. To ensure a positive seal, you should always replace both of these seals whenever you disconnect the EGR pipe.

30   Clean the EGR pipe mounting surfaces on the EGR valve and on the EGR pipe mounting flange.

31   Installation is the reverse of removal. Be sure to tighten the EGR pipe mounting flange bolts to the torque listed in this Chapter's Specifications.

19 Positive Crankcase Ventilation (PCV) system - description, check and replacement

Description

1   The Positive Crankcase Ventilation (PCV) system reduces hydrocarbon emissions by scavenging crankcase vapors, which are rich in unburned hydrocarbons. A PCV valve regulates the flow of gases into the intake manifold in proportion to the amount of intake vacuum available. At idle, when intake vacuum is very high, the PCV valve restricts the flow of vapors so that the engine doesn't run poorly. As the throttle plate opens and intake vacuum begins to diminish, the PCV valve opens more to allow vapors to flow more freely.

2   On V6 engines, the PCV system consists of the fresh air inlet hose, the PCV valve and the crankcase ventilation hose (or PCV hose). The fresh air inlet hose **(see illustration)** connects the air filter housing to a pipe on top and at the front of the left valve cover. The crankcase ventilation hose (or PCV hose) connects the PCV valve on the rear end of the right valve cover to the intake manifold **(see illustration)**. On The PCV valve is located at the rear end of the right valve cover **(see illustrations)**.

3   On V8 engines, the PCV system consists of the fresh air inlet hose, the PCV valve and the crankcase ventilation hose (or PCV hose). The fresh air inlet hose **(see illustration)** connects the air filter housing to the oil filler neck on top of the left valve cover, which is on the right side of the intake manifold. The

19.2c On 2.7L and 3.5L V6 engines, the PCV valve is screwed into the rear end of the right valve cover (3.5L V6 shown, 2.7L V6 similar)

19.2d On the 3.6L V6, the valve is also located at the rear of the right valve cover, but it's retained by screws (one not visible here)

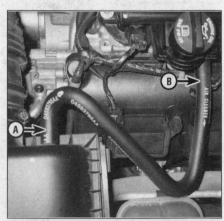

19.3a On V8 engines, the fresh air inlet hose connects the air filter housing (A) to the oil filler neck pipe (B) on top of the left valve cover

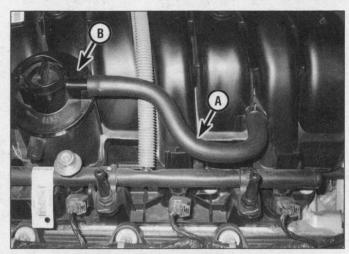

19.3b On V8 engines, the crankcase ventilation hose (A), or PCV hose, connects the PCV valve (B) to the intake manifold

19.3c On V8 engines, the PCV valve is located in the right side of the intake manifold. To remove it, disconnect the PCV hose (A), then turn it counterclockwise 90-degrees

crankcase ventilation hose, or PCV hose **(see illustration)**, connects the PCV valve to the intake manifold. The PCV valve **(see illustration)** is located in the right side of the intake manifold. To remove it, turn it 90-degrees and pull it out.

Inspection

4   An engine that is operated without a properly functioning crankcase ventilation system can be damaged. So anytime you're servicing the engine, be sure to inspect the PCV system hose(s) for cracks, tears, deterioration and other damage. Disconnect the hose(s) and inspect it/them for damage and obstructions. If a hose is clogged, clean it out. If you're unable to clean it satisfactorily, replace it.

5   A plugged PCV hose might cause any or all of the following conditions: A rough idle, stalling or a slow idle speed, oil leaks or sludge in the engine. So if the engine is running roughly, stalling and idling at a lower than normal speed, or is losing oil, or has oil in the throttle body or air intake manifold plenum, or has a build-up of sludge, a PCV system hose might be clogged. Repair or replace the hose(s) as necessary.

6   A leaking PCV hose might cause any or all of the following conditions: a rough idle, stalling or a high idle speed. So if the engine is running roughly, stalling and idling at a higher than normal speed, a PCV system hose might be leaking. Repair or replace the hose(s) as necessary.

7   Here's an easy functional check of the PCV system on a vehicle with a fresh air inlet hose and a crankcase ventilation hose with a PCV valve in it:

a) *Disconnect the crankcase ventilation hose (the crankcase ventilation hose, or simply "the PCV hose," is the hose that connects the PCV valve to the intake manifold).*

b) *Start the engine and let it warm up to its normal idle.*

c) *Verify that there is vacuum at the PCV hose. If there is no vacuum, look for a plugged hose or a clogged port or pipe on the intake manifold. Also look for a hose that collapses when it's blocked (i.e., when vacuum is applied). Replace clogged or deteriorated hoses.*

d) *Remove the engine oil dipstick and install a vacuum gauge on the upper end of the dipstick tube.*

e) *Pinch off or plug the PCV system's fresh air inlet hose.*

f) *Run the engine at 1500 rpm for 30 seconds, then read the vacuum gauge while the engine is running at 1500 rpm.*

g) *If there's vacuum present, the crankcase ventilation system is operating correctly.*

h) *If there's NO vacuum present, the engine might be drawing in outside air. The PCV system won't function correctly unless the engine is a sealed system. Inspect the valve cover(s), oil pan gasket or other sealing areas for leaks.*

i) *If the vacuum gauge indicates positive pressure, look for a plugged hose or engine blow-by.*

8   If the PCV system is functioning correctly, but there's evidence of engine oil in the throttle body or air filter housing, it could be caused by excessive crankcase pressure. Have the crankcase pressure tested by a dealer service department.

9   In the PCV system, excessive blow-by (caused by worn rings, pistons and/or cylinders, or by constant heavy loads) is discharged into the intake manifold and consumed. If you discover heavy sludge deposits or a dilution of the engine oil, even though the PCV system is functioning correctly, look for other causes (see *Troubleshooting* and Chapter 2E) and correct them as soon as possible.

Replacement

PCV valve

2.7L engine

10   The PCV valve is located at the rear of the right (passenger's) valve cover.

11   Remove the intake tube from the throttle body.

12   Remove the EGR tube (see Section 18).

13   Disconnect the hose from the PCV valve under the upper intake manifold.

14   Loosen the upper intake manifold (it does not have to be removed) and position to access the PCV valve elbow at the valve cover (see Chapter 2A, Section 8).

15   Pull the PCV valve elbow and valve from the valve cover.

16   Installation is reverse of removal.

3.5L engine

17   PCV is threaded into the valve cover.

18   Remove the PCV hose from the valve and unscrew the PCV valve from the valve cover.

19   Installation is reverse of removal.

3.6L engine

20   The PCV valve is attached the rear of the right (passenger's) valve cover facing the firewall.

21   Disconnect the hose from the intake manifold and PCV valve.

22   Remove the screws attaching the PCV valve to the valve cover (some models use two screws, others use three).

23   Inspect the O-ring and replace if damaged.

24   Installation is reverse of removal.

5.7L/6.4L engines

25   PCV valve is located on top right of intake manifold.

26   Rotate counterclockwise 90 degrees and pull to remove from intake manifold.

27   To install, apply engine oil to the O-rings,

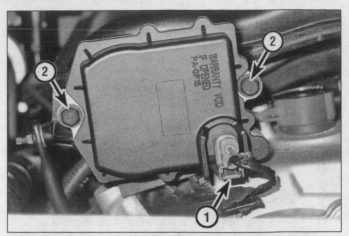

20.1 Typical Intake Manifold Tuning (IMT) system (3.5L V6 engine shown, 2.7L engine similar):

A   *Manifold tuning valve actuator (used on 2.7L and 3.5L V6 engines)*
B   *Short runner valve actuator (used only on the 3.5L engine)*

20.4 To remove the intake manifold tuning valve actuator from a V6 engine, depress the release tab (1), disconnect the electrical connector and remove the two mounting bolts (2) (3.5L engine shown, 2.7L engine similar)

insert into manifold and rotate 90 degrees clockwise.

6.1L engine

28   PCV is threaded into the right front corner of the intake manifold and faces forward.
29   Remove the PCV hose from the valve and unscrew the PCV valve from the intake manifold.
30   Installation is reverse of removal.

---

20   Intake Manifold Tuning (IMT) Valve System - description and component replacement

Description

1   The Intake Manifold Tuning (IMT) Valve System, which is used on 2.7L and 3.5L V6 engines, improves the acoustical tuning qualities of the intake manifold, which increases torque at engine speeds below 5000 rpm and horsepower at speeds above 5000 rpm. 2.7L engines are equipped with an intake manifold tuning valve actuator, which is located at the

front of the intake manifold. 3.5L engines are equipped with a manifold tuning valve actuator and a short runner valve actuator. On 3.5L engines **(see illustration)**, the manifold tuning valve actuator is located on the right front end of the intake manifold and the short runner valve actuator is located on the left front end of the manifold. Both actuators are solenoids that open or close flaps inside the manifold in response to commands from the Powertrain Control Module (PCM).
2   The intake manifold plenum on V6 engines is divided into two equal-volume upper and lower plenums by a horizontal partition running the length of the plenum. Air entering the upper half of the plenum is directed through three of the intake runners; air entering the lower half of the plenum enters the other three intake runners. The manifold tuning valve actuator controls a small valve that opens or closes a hole in the partition at the opposite end of the plenum from the throttle body. When the valve is open, the upper and lower partitions are closed off from one another, preventing the incoming air inside these two smaller ple-

nums from mixing. When the valve is closed, the upper and lower partitions are connected through this hole, allowing incoming air inside the two plenums to mix.
3   The short runner valve solenoid, which is used only on 3.5L engines, directs airflow through the longer route - the air intake plenum and six longer intake runners - or through six shorter intake runners that bypass the plenum and the long runners. Below 5000 rpm, a butterfly valve inside each runner prevents intake air from entering the short intake runners, forcing it to travel the long way around, through the intake manifold plenum, then through six longer intake runners, before it enters the combustion chambers. This gives the slow-moving mass of air time to build inertia so that it fills the combustion chambers more effectively, improving torque. Above 5000 rpm, the PCM commands the short runner valve solenoid to open the butterfly valves inside the six shorter intake runners, which allows incoming air to travel a much shorter path, bypassing the longer route through the plenum and long runners, increasing horsepower.

Component replacement

Intake manifold tuning valve actuator

Note: *On 2.7L engines, the intake manifold tuning valve actuator is located on the front end of the intake manifold. On 3.5L engines, the manifold tuning valve actuator is located on the right front end of the intake manifold.*

4   Disconnect the electrical connector from the manifold tuning valve actuator **(see illustration)**.
5   Remove the manifold tuning valve actuator mounting bolts and remove the manifold tuning valve actuator.
6   Remove the old manifold tuning valve actuator O-ring **(see illustration)** and discard it.

20.6 Be sure to remove and discard the old actuator O-ring from the intake manifold tuning valve actuator (shown) or short runner valve actuator. Use a new O-ring whether you're installing the old actuator or a new unit

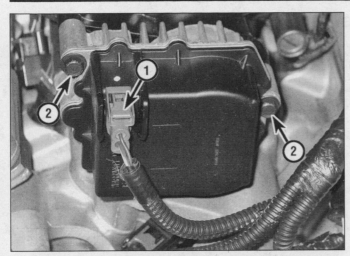

20.8 To remove the short runner valve actuator from a 3.5L engine, depress the release tab (1), disconnect the electrical connector and remove the two mounting bolts (2)

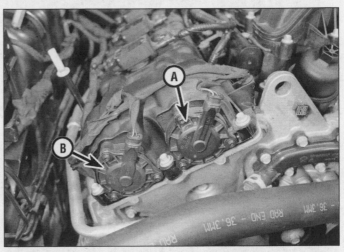

22.4 Identifying the intake (A) and exhaust (B) VVT solenoids on the right (passenger's) cylinder head on 3.6L engines

7   Installation is the reverse of removal. Be sure to use a NEW O-ring whether you're installing the old actuator or a new unit. Tighten the manifold tuning valve actuator mounting bolts to the torque listed in this Chapter's Specifications.

Short runner valve actuator

Note: *The short runner valve actuator, which is used only on 3.5L and 3.6L engines, is located on the left front end of the intake manifold.*

8   Disconnect the electrical connector from the short runner valve actuator **(see illustration)**.
9   Remove the short runner valve actuator mounting bolts and remove the short runner valve actuator.
10 Remove the old short runner valve actuator O-ring and discard it.
11 Installation is the reverse of removal. Be sure to use a NEW O-ring whether you're installing the old actuator or a new unit. Tighten the short runner valve actuator mounting bolts to the torque listed in this Chapter's Specifications.

---

21 Multi-Displacement System (MDS) - description and component replacement

---

Description

1   The Multi-Displacement System (MDS) improves the fuel economy of the V8 engine by deactivating four of the cylinders when certain conditions are met. Under acceleration or heavy loads, the engine runs on all

eight cylinders, allowing full power. But when the conditions are right - cruising at light loads or driving in city traffic, for example - the Powertrain Control Module (PCM) deactivates cylinders 1, 4, 6 and 7 (cylinders 2, 3, 5 and 8 run all the time).
2   The MDS components include an oil temperature sensor, four control valve/solenoids, special roller lifters that can be deactivated, the wiring harness that connects these components and a specially-designed camshaft.

Component replacement

Control valve/solenoid

Note: *The four solenoids are located in the valley between the cylinder heads. This procedure applies to any of the four units.*

3   Remove the intake manifold (see Chapter 2D).
4   Disconnect the electrical connector from the solenoid.
5   Remove the solenoid hold-down bolt and remove the solenoid.
6   Installation is the reverse of removal.

Lifters and camshaft

7   Refer to Chapter 2D.

---

22 Variable Valve Timing (VVT) system - component replacement

---

Solenoid

3.6L engine

Note: *The VVT solenoids are attached to the front of the valve covers. There is a solenoid*

for each intake and exhaust camshaft.

1   Disconnect the negative battery terminal (see Chapter 5, Section 3).
2   Remove the engine cover.
3   Remove the air filter housing assembly (see Chapter 4, Section 11).
4   Mark the solenoids for the intake or exhaust (and for left/right if necessary) for ease of installation **(see illustration)**.
5   Disconnect the electrical connector(s).
6   Remove the three bolts and carefully pull the solenoid from the valve cover.
7   Check the condition of the seal in the valve cover and replace if damaged.
8   Installation is reverse of removal.
9   Perform the cam/crank variation learn procedure using a scan tool.

5.7L engine (2010 and later)

Note: *The VVT solenoid is located under the intake manifold in the lifter valley, attached to the front of the engine block.*

10 Disconnect the negative battery terminal (see Chapter 5, Section 3).
11 Remove the engine cover.
12 Remove the intake manifold (see Chapter 2D, Section 6).
13 Disconnect the solenoid electrical connector.
14 Remove the retaining bolt and twist the solenoid to remove from the vehicle.
15 Installation is reverse of removal.
16 Perform the cam/crank variation learn procedure using a scan tool.

Notes

Chapter 7 Part A
Manual transmission

Contents

Specifications

General

Transmission type	Tremec TR6060 6-Speed
Lubricant type	See Chapter 1

Torque specifications

Ft-lbs (unless otherwise indicated)

Note: *One foot-pound (ft-lb) of torque is equivalent to 12 inch-pounds (in-lbs) of torque. Torque values below approximately 15 ft-lbs are expressed in inch-pounds, because most foot-pound torque wrenches are not accurate at these smaller values.*

Clutch housing bolts	25
Shift lever bolts	15
Transmission crossmember bolts	45
Transmission output flange locknut	44
Transmission-to-engine stud bolts	30
Left and right engine-to-transmission bolts	50
Transmission-to-oil pan bolts	22
Transmission mount	
Mount isloator-to-bracket bolts	18
Mount crossmember-to-chassis bolts	45
Mount isolator-to-crossmember bolts	45

1   General information

1   All vehicles covered in this manual come equipped with either a six-speed manual transmission, four-speed automatic transmission, five-speed automatic transmission or an eight-speed automatic transmission. Information on the manual transmission is included in this Chapter. Information for the automatic transmission can be found in Chapter 7B.

2   The manual transmission used is a six-speed unit known as the Tremec TR6060.

3   Due to the complexity, unavailability of replacement parts and the special tools necessary, internal repair procedures for these units are not recommended for the home mechanic. The information contained within this manual will be limited to general information and removal and installation of the transmission assembly.

4   Depending on the expense involved in having a faulty transmission overhauled, it may be an advantage to consider replacing the unit with either a new or rebuilt one. Your local dealer or transmission shop should be able to supply you with information concerning cost, availability and exchange policy. Regardless of how you decide to remedy a transmission problem, you can still save considerable expense by removing and installing the unit yourself.

2   Shift lever assembly - removal and installation

Shift knob/lever

Note: *The shift knob is part of the shift lever and replaced as an assembly.*

1   Use a trim tool to carefully pry the shift boot away from the shift knob.

2   Use a trim tool to carefully pry up the shift boot bezel and release the clips.

3   Remove the two shift lever retaining bolts and the shift knob/lever.

4   To install, place the shift lever in position and install the retaining bolts. Tighten the bolts to the torque listed in this Chapter's Specifications.

5   Install the shift lever trim bezel and boot.

Shift lever assembly

6   Disconnect the negative battery cable (see Chapter 5, Section 3).

7   Shift the transmission to neutral (N).

8   Remove the shift lever.

9   Remove the single nut at the rear of the shifter.

10   Raise the vehicle and support with jack stands.

11   Remove the driveshaft (see Chapter 8, Section 9).

12   Support the transmission with a floor jack. Place a block of wood between the floor jack pad and the transmission housing. Raise the transmission slightly to release the load on the transmission mount.

13   Remove the four transmission crossmember bolts.

14   Lower the transmission enough to access the shifter assembly.

15   Remove the center shift linkage bolt.

16   Rotate the locking pin tabs on the left and right shift linkages and remove the pins.

17   Remove the shift lever assembly from the vehicle.

18   Installation is reverse of removal.

19   Rotate the shift linkage pin tabs to lock into place.

20   Tighten fasteners to the torque listed in this Chapter's Specifications.

3   Extension housing oil seal - replacement

1   Oil leaks frequently occur due to wear of the extension housing oil seal and bushing (if equipped). Replacement is relatively easy, since the repairs can usually be performed without removing the transmission from the vehicle. The extension housing oil seal is located at the extreme rear of the transmission, where the driveshaft is attached. If leakage at the seal is suspected, raise the vehicle and support it securely on jackstands. If the seal is leaking, transmission lubricant will be built up on the front of the driveshaft and may be dripping from the rear of the transmission.

2   Remove the vehicle undercover.

3   Disconnect the driveshaft from the transmission (see Chapter 8, Section 9).

4   Use a flange holding to prevent the flange from turning, then unscrew the output flange locknut and remove the flange.

5   Using a seal removal tool or screwdriver, carefully pry the oil seal out of the rear of the transmission **(see illustration)**. Do not damage the splines on the transmission output shaft.

6   If the oil seal cannot be removed with a screwdriver or prybar, a special oil seal puller (available at auto parts stores) will be required.

7   Using a large section of pipe or a very large deep socket as a drift, install the new oil seal **(see illustration)**. Drive it into the bore squarely and make sure it's completely seated.

8   Install the output shaft flange and locknut. Tighten the locknut to the torque listed in this Chapter's Specifications.

9   Install the driveshaft (see Chapter 8).

10   Check the transmission lubricant level

3.5 Pry out the extension housing seal with a seal removal tool

3.7 Install the new extension housing seal with a seal driver or a large deep socket

and add some, if necessary, to bring it to the appropriate level (see the Chapter 1 Specifications).

4 Transmission mount - check and replacement

1   Manual transmission mount checking and replacement is essentially the same as the procedure for the automatic transmission mount. Refer to Chapter 7B, Section 9 for the mount replacment procedure, but use the torque values listed in this Chapter's Specifications.

5 Transmission cooler/thermo block - replacement

Transmission cooler

The transmission cooler and air conditionng condenser are serviced as one assembly (see Chapter 3, Section 15).

Thermo block

1   Raise the vehicle and support it securely on jackstands.
2   Locate the thermo block on the driver's side of the oil pan, towards the front of the engine.
Note: *For the next step, place a drain pan under the working area to catch lost fluid.*
3   Disconnect the four transmission cooler lines from the thermo block.
4   Remove the mounting bolt and remove the thermo block from the vehicle.
5   Installation is reverse of removal. Top off the transmission with the correct fluid to the proper level (see Chapter 1).

6 Manual transmission - removal and installation

Removal

1   Disconnect the cable from the negative terminal of the battery (see Chapter 5, Section 3).
2   Raise the vehicle and support it securely on jackstands.
3   Remove the shift lever assembly (see Section 2).
4   Remove the nut and disconnect the ground wire from the upper stud on the bellhousing.
5   Remove the single bolt and the transmission inspection cover.
6   Unbolt the starter and secure it to the side (see Chapter 5, Section 10).
7   Unplug the electrical connectors for sensors and switches on the transmission. Unclip the harness from the transmission.
8   Detach the clutch hydraulic hose retainer from the transmission

9   Disconnect the transmission cooler quick-disconnect lines at the transmission.
10   Disconnect the two transmission cooler lines at the thermal bypass valve and remove the lines.
11   Support the transmission with a jack - preferably a special jack made for this purpose. Safety chains will help steady the transmission on the jack.
12   Remove the two nuts attaching the transmission mounts to the transmission crossmember.
13   Remove the four bolts and remove the transmission crossmember.
14   Remove the four bolts and remove the transmission mount bracket from the transmission.
Note: *Use care during this next step as brake fluid will leak. Place a drain pan under the front of the transmission.*
15   Clean the clutch hydraulic line where it enters the transmission. Remove the clip and separate the lines. Plug the lines to limit the amount of brake fluid that is spilled.
16   Remove the two nuts from the studs attaching the harness to the bellhousing at the upper transaxle bolts.
17   Remove the four upper transmission-to-engine stud bolts **(see illustration)**.
18   Remove the two left and right engine-to-transmission bolts.
19   Remove the two transmission-to-oil pan bolts.
20   Make a final check that all wires and lines have been disconnected from the transmission and then move the transmission and jack toward the rear of the vehicle until the transmission input shaft is clear of the the clutch hub. Keep the transmission level as this is done.
21 Once the input shaft is clear, lower the transmission and remove it from under the vehicle.
22   Inspect the clutch components while the transmission is removed (see Chapter 8, Section 6). In general, it's always a good idea to install new clutch components any time the transmission is removed.

Installation

23   If removed, install the clutch components (see Chapter 8, Section 6).
24 With the transmission secured to the jack as on removal, raise the transmission into position, then carefully slide it forward, engaging the input shaft with the splines in the clutch hub. Do not use excessive force to install the transmission - if the input shaft does not slide into place, readjust the angle of the transmission so it is level and/or turn the input shaft so the splines engage properly with the clutch hub.
25   Install the transmission-to-engine bolts, engine-to-transmission bolts, and transmission-to-oil pan bolts. Tighten the bolts to the torque listed in this Chapter's Specifications.
26   Install the wiring harness retainer and ground strap onto the upper stud bolts.
27   Install the transmission mount and rear crossmember. Tighten all nuts and bolts to the torque listed in this Chapter's Specifications.
28   Remove the jacks supporting the transmission.
29   Attach the clutch hydraulic line and install the clip to secure. Bleed the clutch as described in Chapter 8, Section 5.
30 Reroute the electrical harnesses for the back-up light switch, the vehicle speed sensor and the oxygen sensors. Plug in all electrical connectors. Make sure the oxygen sensor connectors are clipped into the crossmember and transmission.
31   Install the starter (see Chapter 5, Section 10).
32   Install the transmission inspection cover.
33   Reconnect the oil cooler lines.
34   Install the shift lever assembly and install any other components removed as part of the shift lever assembly removal (see Section 2).
35   Fill the transmission with the recommended lubricant (see Chapter 1).
36   Remove all jackstands and lower the vehicle.
37   Connect the negative battery cable. Road test the vehicle for proper operation and check for leakage.

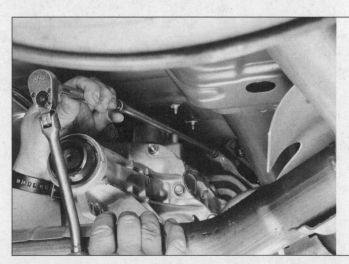

6.17 The upper transmission-to-engine bolts can be accessed with a long extension and swivel socket attached to the ratchet

7 Manual transmission overhaul - general information

1   Overhauling a manual transmission is a difficult job for the do-it-yourselfer. It involves the disassembly and reassembly of many small parts. Numerous clearances must be precisely measured and, if necessary, changed with select fit spacers and snap-rings. If transmission problems arise, you can remove and install the transmission yourself, but overhaul should be left to a transmission repair shop. Rebuilt transmissions may be available - check with your dealer parts department and auto parts stores. At any rate, the time and money involved in an overhaul is almost sure to exceed the cost of a rebuilt unit.

2   Nevertheless, it's not impossible for an inexperienced mechanic to rebuild a transmission if the special tools are available and the job is done in a deliberate step-by-step manner so nothing is overlooked.

3   The tools necessary for an overhaul include internal and external snap-ring pliers, a bearing puller, a slide hammer, a set of pin punches, a dial indicator and possibly a hydraulic press. In addition, a large, sturdy workbench and a vise or transmission stand will be required.

4   During disassembly of the transmission, make careful notes of how each piece comes off, where it fits in relation to other pieces and what holds it in place. Be sure to note how the parts are installed as you remove them; this will make it much easier to get the transmission back together.

5   Before taking the transmission apart for repair, it will help if you have some idea what area of the transmission is malfunctioning. Certain problems can be closely tied to specific areas in the transmission, which can make component examination and replacement easier. Refer to the *Troubleshooting* section at the front of this manual for information regarding possible sources of trouble.

Chapter 7 Part B
Automatic transmission

Contents

Specifications

General

Transmission type	
4-Speed	42RLE
5-Speed	NAG1, W5A580
8-Speed	845RE, 8HP45, 8HP70
Transmission fluid type	See Chapter 1

Torque specifications   Ft-lbs (unless otherwise indicated)

Note: *One foot-pound (ft-lb) of torque is equivalent to 12 inch-pounds (in-lbs) of torque. Torque values below approximately 15 ft-lbs are expressed in inch-pounds, because most foot-pound torque wrenches are not accurate at these smaller values.*

Output shaft flange nut	
42RLE/NAG1	148
8HP45/845RE/8HP70	44
Driveplate-to-torque converter bolts	
2005 and 2006	31
2007 and later	
42RLE	65
NAG1/8HP45/845RE	31
Transmission mount-to-transmission bolts	
42RLE/NAG1	35
8HP45/845RE/8HP70	45
Transmission crossmember-to-mount bolts	
42RLE/NAG1	39
8HP45/845RE	45
Transmission crossmember-to-frame bolts	50
Transmission-to-engine bolts	
NAG1 models	29
42RLE	
2005 and 2006	29
2007 and later	
Transmission-to-engine bolts	50
Oil pan-to-transmission bolts	29
8HP45/845RE	41
8HP70	33

1 General information

1   The vehicles covered in this manual are equipped with a four-speed (42RLE), five-speed (NAG1) or eight-speed (845RE, 8HP45, 8HP70) automatic transmission. All transmissions are equipped with a Torque Converter Clutch (TCC) system that increases fuel economy. The TCC engages in Drive and Overdrive modes. The TCC system consists of a solenoid, controlled by the Powertrain Control Module (PCM), that locks the torque converter when the vehicle is cruising on level ground and the engine is fully warmed up.

2   All automatic transmissions covered in this Chapter are equipped with transmission oil coolers which are mounted in front of the air conditioning condenser and serviced as a single unit with the condenser.

3   All vehicles with a four- or five-speed automatic transmission are equipped with a Brake Transmission Shift Interlock (BTSI) system that locks the shift lever in the Park position and prevents the driver from shifting out of Park unless the brake pedal is depressed.

The BTSI system also prevents the ignition key from being turned to the Lock or Accessory position unless the shift lever is in the Park position.

4   All vehicles equipped with an eight-speed transmission are equipped with an E-shifter. The E-shifter can be either a console mounted lever, similar to a cable operated shifter, or a console mounted selector knob. The E-shifter sends position to the transmission over the CAN communication network. The transmission then shifts to the selected position internally without the use of a cable between the shift lever assembly and transmission.

5   Due to the complexity of the automatic transmissions covered in this manual and the need for specialized equipment to perform most service operations, this Chapter is limited to general diagnosis, routine maintenance, adjustments and removal and installation procedures.

6   If the transmission requires major repair work, leave it to a dealer service department or a transmission repair shop. However, even if a transmission shop does the repairs, you can save some money by removing and installing

the transmission yourself. Keep in mind that a faulty transmission should not be removed before the vehicle has been diagnosed by a knowledgeable technician equipped with the proper tools, as troubleshooting must be performed with the transmission installed in the vehicle.

E-shifter manual park release

7   If the vehicle needs to be moved and cannot be shifted into neutral using the E-shifter, the following procedure instructs on releasing and engaging the manual park release.

Release (disengage)

8   Set the parking brake.
9   Remove the storage bin liner on the shift lever bezel **(see illustration)**.
10   Pull out the red release tether.
11   Insert a pocket-screwdriver into the opening in the release lever, below the tether and release the lock tab by moving to the right (passenger's) side **(see illustration)**.
12   While holding the locking tab with the screwdriver, pull up on the red release tether until the release lever locks into place (vertical) **(see illustration)**.
13   The transmission is now in the Neutral position. Release the parking brake and the vehicle can be moved.

Apply (engage)

14   While holding the tether, use a pocket-screwdriver to release the tab holding the release lever in the vertical position.
15   Continue holding the tether and let the release lever return to the horizontal position.
16   Ensure the locking tab engages with the release lever.
17   The transmission is now in the Park position.
18   Tuck the tether into the bezel and install the storage bin liner.

1.9 Remove the storage bin liner (Challenger shown)

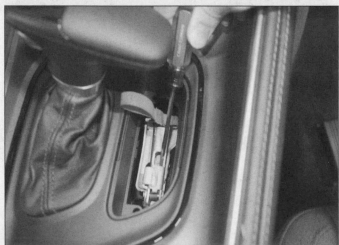

1.11 Use a screwdriver to move the lock tab to the right

1.12 Pull up on the release tether until the release lever locked is in the vertical position

Shift Lock Override feature

19   In the event the Brake Transmission Shift Interlock (BTSI) system fails and the shift lever cannot be moved out of gear, the system is equipped with an override feature.

20   Remove the bin to the right of the shift lever **(see illustration)**. Depress the brake pedal (and keep it depressed) and start the engine (or, if you just want to shift out of PARK without starting the engine, turn the ignition key to the ACC position). Locate the button alongside the right side of the shift gate **(see illustration)**, press the button and move the shifter out of the Park position.

21   Repair the BTSI system (or have it repaired) as soon as possible.

---

2   Diagnosis - general

Note: *Automatic transmission malfunctions may be caused by five general conditions: poor engine performance, incorrect adjustments, hydraulic malfunctions, mechanical malfunctions or malfunctions in the computer or its signal network. Diagnosis of these problems should always begin with a check of the easily repaired items: fluid level and condition (see Chapter 1) and shift cable adjustment. Next, perform a road test to determine if the problem has been corrected or if more diagnosis is necessary. If the problem persists after the preliminary tests and corrections are completed, additional diagnosis should be done by a dealer service department or transmission repair shop. Refer to the Troubleshooting section at the front of this manual for information on symptoms of transmission problems.*

Preliminary checks

1   Drive the vehicle to warm up the transmission to its normal operating temperature.

2   Check the fluid level as described in Chapter 1:

a) *If the fluid level is unusually low, add enough fluid to bring the level to the correct mark (see Chapter 1, Section 26), then check for external leaks (see below). The manufacturer states that routine checks of the automatic transmission fluid are not necessary; the procedure in Chapter 1 should only be used when refilling the transmission after the fluid has been drained, unless an obvious leak has been detected. Low fluid level can lead to slipping or loss of drive, while overfilling can cause foaming and loss of fluid.*

b) *If the fluid level is abnormally high, it might have been overfilled. Drain off the excess.*

c) *If the fluid is foaming, drain it and refill the transmission.*

3   Check the engine idle speed.

Note: *If the engine is malfunctioning, do not proceed with the preliminary checks until it has been repaired and runs normally.*

4   Inspect the shift cable on models so equipped (see Section 6). Make sure it's properly adjusted and that it operates smoothly.

Fluid leak diagnosis

5   Most fluid leaks are usually easy to locate because they leave a visible stain and/or wet spot. Most repairs are simply a matter of replacing a seal or gasket. If a leak is more difficult to find, the following procedure will help.

6   Identify the fluid. Make sure that it's transmission fluid, not engine oil or brake fluid.

7   Try to pinpoint the source of the leak. Drive the vehicle several miles, then park it over a large sheet of cardboard. After a minute or two, you should be able to locate the leak by determining the source of the fluid dripping onto the cardboard.

8   Make a careful visual inspection of the suspected component and the area immediately around it. Pay particular attention to gasket mating surfaces. A flashlight and mirror are often helpful for finding leaks in areas that are hard to see.

9   If you still can't find the leak, thoroughly clean the suspected area with a degreaser or solvent, then dry it off.

10   Drive the vehicle for several miles at normal operating temperature and varying speeds. After driving the vehicle, visually inspect the suspected component again.

11   Once you have located the leak, you must determine the source before you can repair it properly. For example, if you replace a pan gasket but the sealing flange is warped or bent, the new gasket won't stop the leak. The flange must first be straightened.

12   Before attempting to repair a leak, verify that the following conditions are corrected or they might cause another leak.

Note: *Some of the following conditions cannot be fixed without highly specialized tools and expertise. Such problems must be referred to a transmission repair shop or a dealer service department.*

Gasket leaks

13   Inspect the pan periodically. Make sure that the bolts are tight, that no bolts are missing, that the gasket is in good condition and that the pan is flat (dents in the pan might indicate damage to the valve body inside).

14   If the pan gasket is leaking, the fluid level or the fluid pressure might be too high, the vent might be plugged, the pan bolts might be too tight, the pan sealing flange might be warped, the sealing surface of the transmission housing might be damaged, the gasket might be damaged or the transmission casting might be cracked or porous. If sealant instead of gasket material has been used to form a seal between the pan and the transmission housing, it may be the wrong type sealant.

Seal leaks

15   If a transmission seal is leaking, the fluid level or pressure might be too high, the vent might be plugged, the seal bore might be

1.20a Lift the bin from the center console to expose the override button. . .

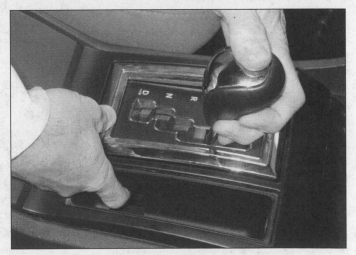

1.20b. . . and, while pressing the override button, move the console shifter out of the Park position

damaged, the seal itself might be damaged or incorrectly installed, the surface of the shaft protruding through the seal might be damaged or a loose bearing might be causing excessive shaft movement.

Case leaks

16   If the case itself appears to be leaking, the casting is porous. A porous casting must be repaired or replaced.

17   Make sure that the oil cooler hose fittings are tight and in good condition.

Fluid comes out vent pipe or fill tube

18   If this condition occurs, the transmission is overfilled, there is coolant in the fluid, the case is porous, the vent is plugged or the drain-back holes are plugged.

---

3   Shift knob - replacement

Warning: *The models covered by this manual are equipped with a Supplemental Restraint System (SRS), more commonly known as airbags. Always disable the airbag system before working in the vicinity of any airbag system component to avoid the possibility of accidental deployment of the airbag(s), which could cause personal injury (see Chapter 12). Do not use a memory saving device to preserve the PCM or radio memory when working on or near airbag system components.*

Note: *Four- and five-speed transmissions are equipped with a cable operated shift level mechanism and the shift knob can be replaced. All other transmissions, the shift knob is part of the bezel and boot.*

1   Use a trim tool to release the shift knob lower trim from the shift knob.

2   Pull firmly on the shift knob upwards to remove from the shift lever assembly.

3   Remove the lower trim from the shift lever assembly.

4   Installation is reverse of removal.

---

4   Shift lever - replacement

Warning: *The models covered by this manual are equipped with a Supplemental Restraint System (SRS), more commonly known as airbags. Always disable the airbag system before working in the vicinity of any airbag system component to avoid the possibility of accidental deployment of the airbag(s), which could cause personal injury (see Chapter 12). Do not use a memory saving device to preserve the PCM or radio memory when working on or near airbag system components.*

Note: *The following procedure applies to dash-mounted shifters.*

1   Remove the gauge panel bezel (see Chapter 11, Section 26).

2   Remove the steering column covers (see Chapter 10, Section 17).

3   Disconnect the electrical connector for the shift lever.

4   Remove the Torx screw attaching the shift lever to the shift lever assembly.

5   Remove the shift lever from the vehicle through the trim panel.

6   Installation is reverse of removal.

---

5   Shift lever assembly - replacement

Warning: *The models covered by this manual are equipped with a Supplemental Restraint System (SRS), more commonly known as airbags. Always disable the airbag system before working in the vicinity of any airbag system component to avoid the possibility of accidental deployment of the airbag(s), which could cause personal injury (see Chapter 12). Do not use a memory saving device to preserve the PCM or radio memory when working on or near airbag system components.*

Note: *42RLE, NAG1 and W5A580 transmissions are equipped with a cable operated shift level mechanism. 845RE, 8HP45, 8HP70, and 8HP90 transmissions are equipped with an E-Shifter that uses CAN communication to send*

shift lever position directly to the transmission. The transmission then shifts between selected positions internally.

Console-mounted shifter
Cable operated

1   Disconnect the cable from the negative terminal of the battery (see Chapter 5, Section 3). Remove the center console (see Chapter 11).

2   Place the shift lever into the Park position.

3   Disconnect the shift cable from the shift lever assembly.

4   Detach the cable housing from the shift lever assembly by removing the retainer.

5   On 2006 and earlier models, with the ignition switch in the LOCK position, disconnect the shift-interlock cable from the shift lever assembly cam and notch.

6   On all model years, disconnect the shift lever assembly electrical connectors.

7   Remove the shift lever assembly bolts and the shift lever assembly from the vehicle.

8   Installation is reverse of removal.

9   Adjust the shift cable (see Section 6) and the shift-interlock cable (see Section 7) if equipped.

E-Shifter lever style

10   Set the parking brake. Disconnect the cable from the negative terminal of the battery (see Chapter 5, Section 3).

11   On 2014 and earlier models, using a trim tool, carefully pry the shift knob upper cover off and remove the retaining screw.

12   Using a trim tool, carefully pry the shifter bezel assembly from the center console **(see illustration)**.

13   Disconnect the electrical connector from the bezel **(see illustration)**.

14   On 2015 and later models, remove the shift knob screw from below the bezel.

15   On all models, remove the shift knob and bezel assembly.

16   Using a trim tool, remove the center console bezel to access the shift lever assembly **(see illustration)**.

5.12 Carefully pry the shifter bezel from the center console (Challenger shown)

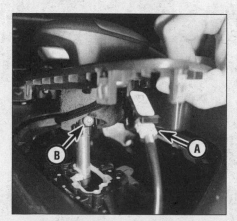

5.13 Disconnect the electrical connector (A) and remove the shift knob screw (B) to remove the shift knob, boot and bezel

5.16 Carefully pry up to detach the clips and remove the center console bezel

17   Remove the four bolts and lift the shift lever assembly to disconnect the electrical connectors (see illustration).

18   Installation is reverse of removal.

E-Shifter knob style

19   Disconnect the negative battery cable (see Chapter 5, Section 3).

20   Using a trim tool, pry the shift knob and bezel assembly from the center console.

21   Disconnect the electrical connectors.

22   Installation is reverse of removal.

Dash mounted shifter

Warning: *The models covered by this section are equipped with a Supplemental Restraint System (SRS), more commonly known as airbags. Always disable the airbag system before working in the vicinity of any airbag system component to avoid the possibility of accidental deployment of the airbags, which could cause personal injury (see Chapter 12, Section 26).*

23   Remove the shift lever (see Section 4).

24   Remove the knee bolster and knee airbag assembly (if equipped). See Chapter 11, Section 26.

25   Working under the driver's side of the dash, disconnect the shift cable from the shift lever assembly.

26   Remove the cable lock and pull the cable away from the shift lever assembly.

27   Detach the harness clip from the shift lever assembly.

28   Remove the three nuts and one screw attaching the shift lever assembly to the bracket.

29   Pull the shift lever assembly out and disconnect the connector.

30   Remove the shift lever assembly from the vehicle.

31   Installation is reverse of removal.

32   Check the shift operation and adjust the shift cable as necessary (see Section 6).

6   Shift cable - check, adjustment and replacement

Note: *Only 42RLE, NAG1, and W5A580 transmissions are equipped with a shift cable. All other transmissions use an E-shifter.*

Warning: *The models covered by this manual are equipped with a Supplemental Restraint System (SRS), more commonly known as airbags. Always disable the airbag system before working in the vicinity of any airbag system component to avoid the possibility of accidental deployment of the airbag(s), which could cause personal injury (see Chapter 12). Do not use a memory saving device to preserve the PCM or radio memory when working on or near airbag system components.*

Check

1   Firmly apply the parking brake and try to momentarily operate the starter in each shift lever position. The starter should only operate when the shift lever is in the Park or Neutral positions. If the starter operates in any position other than Park or Neutral, adjust the shift cable (see below). If, after adjustment, the starter still operates in positions other than Park or Neutral, the Transmission Range (TR) sensor may be defective (see Section 11).

Adjustment

2   Place the shift lever in the Park position.

Console mounted shifter

3   Remove the center console (see Chapter 11).

4   If equipped, remove the shield that covers the shift cable and the BTSI cable. Loosen the shift cable adjustment bolt (see illustration).

Dash mounted shifter

5   Remove the gauge bezel and steering column covers (see Chapter 11).

6   Disconnect the shift cable end from the ballstud on the shift lever assembly.

7   Remove the "C" clip near the ball-stud end of the cable.

Column mounted shifter

8   Remove the steering column covers and the knee bolster (see Chapter 11).

9   Remove the steering column cover reinforcement panel and under-dash hush panel.

10   Disconnect the shift cable end from the ballstud on the shift lever assembly.

11   Remove the "C" clip near the ball-stud end of the cable.

All shifters

12   Raise the vehicle and support it securely on jackstands.

Note: *The rear of the vehicle must also be raised, so the driveshaft can be turned in Step 8 (to verify that the transmission is completely engaged in Park).*

13   Working at the transmission, pry the shift cable end from the transmission shift lever (see illustration).

14   Verify that the manual shift lever on the transmission is all the way to the rear, in the last detent. (This is the Park position.)

15   Verify that the park lock pawl inside the transmission is engaged by trying to rotate the driveshaft. The driveshaft will not rotate if the transmission is correctly engaged in Park.

16   Reconnect the shift cable to the manual shift lever on the transmission.

17   Lower the vehicle.

18   With the parking brake firmly applied, make sure the engine starts (don't move the shift lever from Park yet).

19   On console mounted shifters, tighten the adjuster screw to the torque listed in this Chapter's Specifications.

20   On dash and column mounted shifters, snap the cable end on the ballstud and install the "C" clip back onto the shift cable.

21   If the linkage appears to be adjusted correctly, but the starter still operates in any

5.17 Remove the bolts, disconnect the connectors and remove the shift lever assembly

6.4 Location of the shift cable adjustment bolt on the shift lever assembly

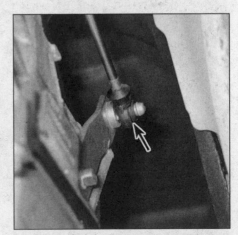

6.13 Use a screwdriver or trim panel tool to pry the shift cable off the shift lever at the transmission

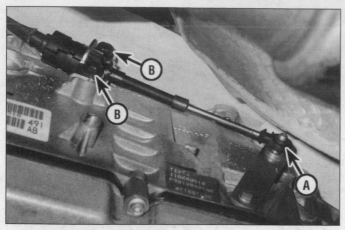

6.24a To remove the shift cable from the transmission, detach the shift cable end from the shift lever (A) and remove the two mounting bolts (B) from the cable bracket

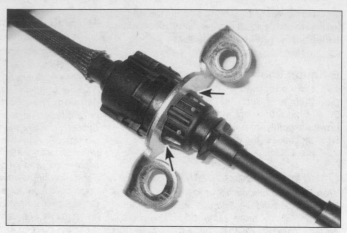

6.24b Press the two lock tabs and slide the shift cable through the bracket

other position(s) besides Park and Neutral, the Transmission Range (TR) sensor might be faulty (see Section 11).

Replacement

All shifters

22   Place the shift lever in the Park position.
23   Raise the vehicle and support it securely on jackstands.
24   Working at the transmission, pry the shift cable end from the transmission shift lever, remove the bolts from the shift cable bracket **(see illustrations)** and remove the shift cable from the transmission.

Console mounted shifter

25   Lower the vehicle and remove the center console (see Chapter 11).
26   Remove the shield that covers the shift cable and the BTSI cable, if equipped.
27   Working at the shift lever assembly, lift the shift cable from the slots in the mounting bracket, then remove the shift cable from the

shift lever stud **(see illustration)**.
28   Working in the engine compartment, disengage the rubber grommet from the firewall and pull the shift cable through the firewall **(see illustration)**.
29   Installation is the reverse of removal. When you're done, adjust the cable.

Dash and column mounted shifter

30   Lower the vehicle.
31   Remove the gauge bezel, steering column covers, knee bolster and hush panel (see Chapter 11).
32   Disconnect the shift cable end from the ballstud.
33   Remove the retainer clip and pull the cable housing from the shift lever assembly bracket.
34   Working from the engine compartment, remove the rubber grommet and pull the cable out through the engine compartment.
35   Installation is the reverse of removal. When you're done, adjust the cable.

7   Brake Transmission Shift Interlock (BTSI) system

Warning: *The models covered by this manual are equipped with a Supplemental Restraint System (SRS), more commonly known as airbags. Always disable the airbag system before working in the vicinity of any airbag system component to avoid the possibility of accidental deployment of the airbag(s), which could cause personal injury (see Chapter 12). Do not use a memory saving device to preserve the PCM or radio memory when working on or near airbag system components.*
Note: *Only 42RLE, NAG1, and W5A580 transmissions are equipped with a mechanical BTSI system.*
Note: *The shift lever and BTSI solenoid is one complete assembly. In the event of failure, replace the shift lever assembly as a complete unit. Check with a dealer parts department or other qualified automotive parts distributor.*

6.27 To remove the shift cable, disengage it from the slots (A) in the bracket, then separate the cable end from the stud (B). To adjust the shift cable, loosen the adjustment bolt (C) and allow the cable tension to release

6.28 The shift cable passes through the firewall between the heater hoses and the power brake booster

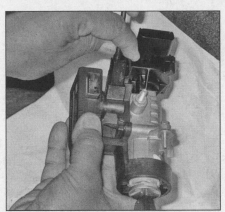

7.11 To release the Brake Transmission Shift Interlock (BTSI) cable from the key lock cylinder, place the ignition key in the On position, press the tab on the BTSI cable and pull the cable from the key lock cylinder housing (housing removed for clarity)

7.13a Push the BTSI cable back slightly and lift the cable through the notch in the BTSI solenoid

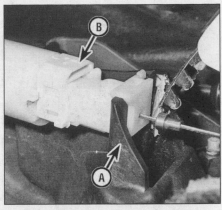

7.13b Lift the BTSI cable adjuster assembly from the slot in the bracket (A); (B) is the BTSI lock button

Description

1   The Brake Transmission Shift Interlock (BTSI) system is a solenoid-operated device, located on the shift cable, that locks the shift lever into the Park position when the ignition key is in the Lock or Accessory position. When the ignition key is in the Run position, a magnetic holding device, in line with the park lock cable, is energized. When the system is functioning correctly, the only way to unlock the shift lever and move it out of Park is to depress the brake pedal. The BTSI system also prevents the ignition key from being turned to the Lock or Accessory position unless the shift lever is fully locked into the Park position.

Check

2   Verify that the ignition key can be removed only in the Park position.
3   When the shift lever is in the Park position, you should be able to rotate the ignition key from Off to Lock. But when the shift lever is in any gear position other than Park (including Neutral), you should not be able to rotate the ignition key to the Lock position.
4   You should not be able to move the shift lever out of the Park position when the ignition key is turned to the Off position.
5   You should not be able to move the shift lever out of the Park position when the ignition key is turned to the Run or Start position until you depress the brake pedal.
6   You should not be able to move the shift lever out of the Park position when the ignition key is turned to the ACC or Lock position (brake pedal NOT depressed).
7   Once in gear, with the ignition key in the Run position, you should be able to move the shift lever between gears, or put it into Neutral or Park, without depressing the brake pedal.
8   If the BTSI system doesn't operate as described, try adjusting it (see Steps 16 through 20).

BTSI cable replacement
9   Remove the center console and the driver's knee bolster and reinforcement (see Chapter 11).
10   Put the shift lever in the Park position. Switch the ignition key to the On position.
11   Working under the steering column, remove the key lock cylinder (see Chapter 12). Disconnect the shift lock cable from the key lock cylinder (**see illustration**).
12   Working at the center console, remove the shift cable and the BTSI cable shield, if equipped.
13   Disconnect the BTSI cable end from the solenoid (**see illustration**) and lift the BTSI cable adjuster from the bracket (**see illustration**).
14   Remove the BTSI cable.
15   Installation is the reverse of removal.

BTSI cable adjustment
16   Remove the center console (see Chapter 11).
17   Put the shift lever in the Park position. Switch the ignition key to the Lock position.
18   Pull UP on the BTSI lock button (**see illustration 7.13b**).
Note: *If a new BTSI cable is being installed,*

remove the cable adjuster lock pin.
19   Check that the BTSI cable is ready for adjustment by pushing the cable to the rear and releasing.
20   Push the lock button down until it snaps into position. At this point, the lock button should be recessed approximately 1 to 2 mm into the cable adjustment housing.
21   Install the center console (see Chapter 11).

Shift Lock Override feature
22   In the event the Brake Transmission Shift Interlock (BTSI) system fails and the shift lever cannot be moved out of gear, the system is equipped with an override feature.
23   Remove the bin to the right of the shift lever (**see illustration**). Depress the brake pedal (and keep it depressed) and start the engine (or, if you just want to shift out of PARK without starting the engine, turn the ignition key to the ACC position). Locate the button alongside the right side of the shift gate (**see illustration**), press the button and move the shifter out of the PARK position.
24   Repair the BTSI system (or have it repaired) as soon as possible.

7.23a Lift the bin from the center console to expose the override button. . .

7.23b. . . and, while pressing the override button, move the console shifter out of the Park position

8.8 Use a punch or a chisel to stake the flange nut to the output shaft and be sure to repeat the procedure on the opposite side of the flange nut

8   Extension housing oil seal - replacement

1   Oil leaks frequently occur due to wear of the extension housing oil seal. Replacement of this seal is relatively easy, since the repair can be performed without removing the transmission from the vehicle.

2   If you suspect a leak at the extension housing seal, raise the vehicle and support it securely on jackstands. The extension housing seal is located at the rear end of the transmission, where the driveshaft is attached. If the extension housing seal is leaking, transmission lubricant will be evident on the front of the driveshaft and may be dripping from the rear of the transmission.

3   Remove the driveshaft (see Chapter 8).

4   Position the shift lever into Park, if not already done. Remove the output shaft flange nut. Slide the output shaft flange off the extension housing.

5   Using a screwdriver, prybar or seal removal tool, carefully pry out the extension housing seal. Do not damage the splines on the transmission output shaft.

6   Using a seal driver or a large deep socket with an outside diameter the same as that of the seal, install the new extension housing seal. Drive it into the bore squarely and make sure it's completely seated.

7   Install the output shaft flange and nut and tighten it to the torque listed in this Chapter's Specifications.

8   Use a punch to stake the flange nut to the output shaft to prevent rotation **(see illustration)**.

9   Install the driveshaft (see Chapter 8).

9   Transmission mount - check and replacement

Check

1   Raise the vehicle and support it securely on jackstands.

2   Insert a large screwdriver or prybar into the space between the transmission extension housing and the crossmember and try to pry the transmission up slightly **(see illustration)**.

3   The transmission should not move much at all and the rubber in the center of the mount should fully insulate the center of the mount from the mount bracket around it.

Replacement

4   Support the transmission with a jack, remove the bolts attaching the crossmember to the chassis and mount and the bolts attaching the mount to the transmission **(see illustrations)**.

5   Raise the transmission slightly with the jack and remove the mount.

6   Installation is the reverse of the removal procedure. Be sure to tighten all fasteners to this Chapter's Specifications.

9.2 To check the transmission mount, insert a large screwdriver between the extension housing and the crossmember and try to lever the transmission up

10   Transmission Control Module (TCM) - replacement

Warning: *The models covered by this manual are equipped with a Supplemental Restraint System (SRS), more commonly known as airbags. Always disable the airbag system before working in the vicinity of any airbag system component to avoid the possibility of accidental deployment of the airbag(s), which could cause personal injury (see Chapter 12). Do not use a memory saving device to preserve the PCM or radio memory when working on or near airbag system components.*

Caution: *The TCM is an Electro-Static Discharge (ESD) sensitive electronic device, meaning a static electricity discharge from your body could possibly damage electrical components. Be sure to properly ground yourself and the TCM before handling it. Avoid touching the electrical terminals of the TCM.*

9.4a Location of the crossmember-to-frame mounting bolts (A) and the crossmember-to-mount bolts (B)

9.4b Transmission mount-to-transmission bolts (two of four shown)

Note: *Do not interchange TCMs from different year vehicles. Whenever the TCM is replaced with a new unit, it must be programmed with a scan tool by a dealership service department or other qualified repair shop. This may require having the vehicle towed to the facility that will perform this procedure.*

NAG1 5-speed transmission

1   Disconnect the cable from the negative terminal of the battery (see Chapter 5, Section 1).
2   Remove the knee bolster (see Chapter 11).
3   Remove the knee bolster reinforcement panel (see Chapter 11).
4   Disconnect the TCM electrical connectors.
5   Remove the TCM screws **(see illustration)** and detach the TCM from the instrument panel frame.
6   Installation is the reverse of removal.

42RLE 4-speed transmission

7   On models equipped with the 42RLE transmission, the TCM is an integral part of the Powertrain Control Module (PCM). Refer to Chapter 6 for the PCM removal and installation procedure.

8-speed transmissions

8   On models equipped with an 8-speed transmission, the TCM is an integral part of the transmission valve body and not serviceable separately.

11   Transmission Range (TR) sensor - replacement

1   On the 42RLE transmission, the TR sensor is located on top of the valve body. On the NAG1 (all-electronic) transmission, there is no "TR sensor." It's been replaced by a "Park/Neutral Contact," which is hard-wired into

10.5 On models with the NAG1 transmission, the TCM is located under the left side of the instrument panel; it's secured by four screws

the electronic control unit, which is an integral part of the electro-hydraulic control unit, which is located on top of the valve body. On the 8-speed transmissions, the TR function is incorporated into the Transmsision Control Module Assembly (TCMA), which is not serviceable separately from the valve body. We don't recommend trying to replace either the TR sensor, the Park/Neutral Contact or the TCMA. Have the work performed by a dealer service department or a qualified automatic transmission shop.

12   Automatic transmission - removal and installation

Caution: *The transmission and torque converter must be removed as a single assembly. If you try to leave the torque converter attached to the driveplate, the converter driveplate, pump bushing and oil seal will be damaged. The driveplate is not designed to support the load, so none of the weight of the transmission should be allowed to rest on the driveplate during removal.*

Removal

Four- and five-speed transmissions

1   Disconnect the cable from the negative terminal of the battery (see Chapter 5, Section 1).
2   Raise the vehicle and support it securely on jackstands.
3   Remove the middle engine splash shield **(see illustration)**.
4   Remove the driveshaft (see Chapter 8).
5   Remove the starter (see Chapter 5).
6   Disconnect the oxygen sensor connectors, separate the harness connectors from the transmission brace and position the wiring harness off to the side.
7   Drain the transmission fluid, then reinstall the fluid pan (see Chapter 1).
8   Remove the torque converter access cover, if equipped **(see illustration)**.
9   On 42RLE transmissions, disconnect the output speed sensor and input speed sensor connectors (see Chapter 6).
10   On 2.7L V6 engines, remove the transmission-to-oil pan support brace (see Chapter 2A).

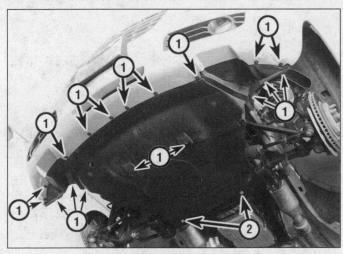

12.3 Front engine splash shield (1) and middle engine splash shield (2) fasteners

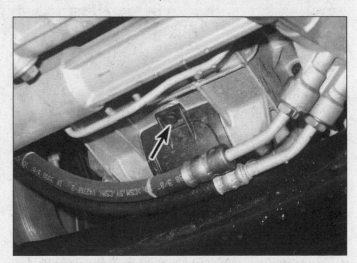

12.8 Remove this bolt and separate the torque converter cover from the transmission

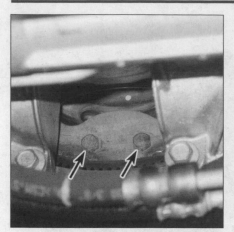

12.13 Remove the driveplate-to-torque converter bolts (3.5L V6 engine with NAG1 transmission shown)

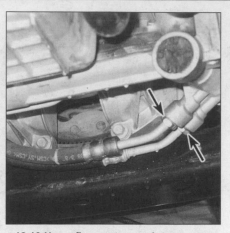

12.19 Use a flare-nut wrench to remove the transmission fluid cooler lines from the transmission

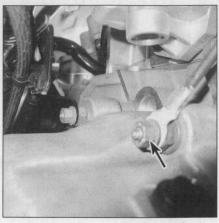

12.24 Remove the nut and detach the ground strap at the upper right transmission-to-engine bolt

11 Remove the transmission heat shield, if equipped.

12 Mark the relationship of the torque converter to the driveplate to ensure that their dynamic balance is maintained when the converter is reattached to the driveplate.

13 Remove the driveplate-to-torque converter bolts **(see illustration)**. Turn the crankshaft clockwise for access to each bolt, using a socket and breaker bar placed on the crankshaft pulley bolt.

14 Disconnect all electrical connectors from the transmission. Generally speaking, the connectors to various electrical and/or electronic devices on the transmission are different in shape, color and the number of terminals, so there's little danger of accidentally reconnecting a connector to the wrong device. The wiring harness is also designed so that each connector will only reach the device to which it's supposed to be connected. However, if any of the connectors look identical or look like they could be accidentally reconnected

to the wrong device, be sure to mark them to prevent mix-ups.

15 Disconnect the shift cable from the transmission (see Section 6).

16 Rotate the harness connector bayonet lock counterclockwise and disconnect the large wiring harness from the terminal connector at the transmission.

17 Support the transmission with a transmission jack (available at most equipment rental facilities) and secure the transmission to the jack with safety chains.

18 Unbolt and remove the transmission crossmember **(see illustration 9.4a)**.

19 Disconnect the transmission cooler lines from the transmission **(see illustration)**. Plug the ends of the lines to prevent fluid from leaking out after you disconnect them.

20 Disconnect the range sensor (see Chapter 6), the crankshaft position sensor (see Chapter 6) and all other harness connectors.

21 Remove the transmission dipstick tube. Don't lose the tube seal (unless it's damaged,

in which case you should replace it).

22 Detach the transmission fluid cooler lines from the frame (see Chapter 3).

23 Lower the jack supporting the transmission and allow it and the engine to angle down as far as possible (if you're using a transmission jack, you'll have to readjust the angle of the jack head).

24 On 3.5L V6 engines, remove the ground strap mounting nut and strap at the upper right transmission mounting bolt **(see illustration)**.

25 On 3.5L V6 engines, remove the nuts and detach the harness bracket from the upper transmission stud bolts **(see illustration)**.

26 Support the engine with a jack. Use a block of wood under the oil pan to spread the load.

27 Remove the bolts securing the transmission to the engine **(see illustration)**. A long extension and a U-joint socket will greatly simplify this step.

28 Move the transmission to the rear to dis-

12.25 Remove the two nuts and separate the wiring harness bracket from the studs

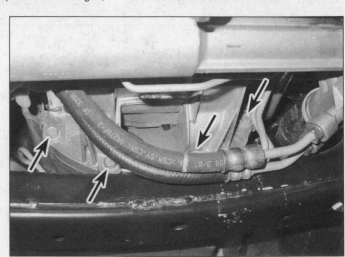

12.27 The oil pan on the 3.5L V6 engine is equipped with bolts that mount to the transmission

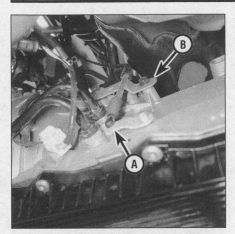

12.35 Disconnect the manual release cable from the lever (A) and remove the bolts and bracket (B) (3.6L shown, others similar)

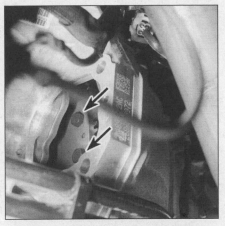

12.37 Remove the torque converter bolts through the starter opening on 3.6L models

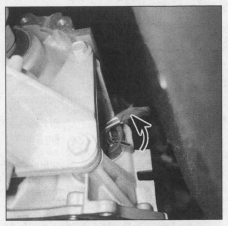

12.38 Rotate the bayonet clip counterclockwise and pull the connector from the transmission (3.6L shown, others similar)

engage it from the engine block dowel pins. Make sure the torque converter is detached from the driveplate and stays with the transmission. Lower the transmission with the jack.

Eight-speed transmissions

29   Disconnect the negative battery cable (see Chapter 5, Section 3).
30   Raise and support the vehicle on jackstands.
31   Remove the driveshaft (see Chapter 8, Section 9).
32   On 2015 and later 3.6L RWD models, remove the lower intermediate shaft from the steering column (see Chapter 10, Section 17).
33   On all models, remove the starter (see Chapter 5, Section 10).
34   On V8 models, remove the inspection cover bolt and the cover.
35   On all models, detach the manual park release cable from the lever. Remove the bolts and the manual park release bracket **(see illustration)**.
36   Mark the relationship of the torque converter to the driveplate to ensure that their

dynamic balance is maintained when the converter is reattached to the driveplate. On 3.6L models, access through the starter opening. On all V8 models, access through the inspection cover opening.
37   Lock the driveplate and remove the six torque converter bolts. On 3.6L models, access through the starter opening **(see illustration)**. On all V8 models, access through the inspection cover opening.
38   Rotate the harness connector bayonet lock counterclockwise and disconnect the large wiring harness from the terminal connector at the transmission **(see illustration)**.
39   Detach the harness clips from the transmission.
40   Remove the retainers, and spring locks and disconnect the transmission oil cooler lines from the transmission **(see illustration)**.
41   Support the transmission with a transmission jack (available at most equipment rental facilities) and secure the transmission to the jack with safety chains.
42   Unbolt and remove the transmission crossmember **(see illustration 9.4a)**.
43   Remove the left and right transmission-to-engine bolts **(see illustrations)**.

44   Remove the nuts and harness brackets from the upper transmission-to-engine bolts if equipped.
45   Remove the upper transmission-to-engine stud-bolts.
46   On V8 models, remove the two lower engine-to-transmission bolts.
47   On all models, carefully lower the transmission while holding the torque converter in the bell housing, and remove from the vehicle.
48   Remove the transmission mount bracket bolts and the bracket.

Installation

49   Installation is the reverse of the removal procedure, noting the following points:

a) *Prior to installation, make sure the torque converter is securely engaged in the pump. If you've removed the converter, spread transmission fluid on the torque converter rear hub, where the transmission front seal rides. With the front of the transmission facing up, rotate the converter back and forth. It should drop down into the transmission front pump*

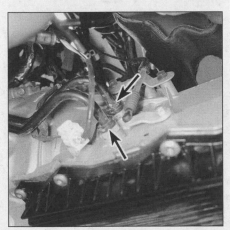

12.40 Disconnect the transmission cooler lines (3.6L shown, others similar)

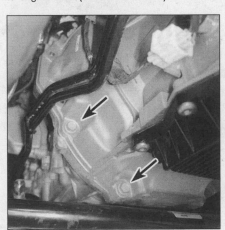

12.43a Remove the left transmission bolts . . .

12.43b . . . and the right transmission bolts (3.6L shown, others similar)

in stages. To make sure the converter is fully engaged, lay a straightedge across the transmission-to-engine mating surface and make sure the converter lugs are at least 3/4-inch below the straightedge.

b) When mating the transmission to the engine, turn the torque converter to line up the marks on the torque converter and driveplate made during removal.

c) Move the transmission forward carefully until the dowel pins and the transmission are engaged. Make sure the transmission mates with the engine with no gap. If there's a gap, make sure there are no wires or other objects pinched between the engine and transmission and also make sure the torque converter is completely engaged in the transmission front pump. Try to rotate the converter - if it doesn't rotate easily, it's probably not fully engaged in the pump. If necessary, lower the transmission and install the converter fully.

d) Install the transmission-to-engine bolts and tighten them to the torque listed in this Chapter's Specifications. As you're tightening the bolts, make sure that the engine and transmission mate completely at all points. If not, find out why. Never try to force the engine and transmission together with the bolts or you'll break the transmission case!

e) Install the driveplate-to-torque converter bolts. Tighten them to the torque listed in this Chapter's Specifications. **Note:** Install all of the bolts before tightening any of them.

f) If you're working on a 2.7L V6 engine, install the transmission-to-oil pan support brace and tighten the bolts to the torque listed in the Chapter 2A Specifications, in the sequence indicated in Chapter 2A, Section 13.

g) The remainder of installation is the reverse of removal.

h) When you're done, refill the transmission with the specified fluid (see Chapter 1, Section 27), run the engine and check for fluid leaks.

13 Automatic transmission overhaul - general information

1   In the event of a fault occurring, it will be necessary to establish whether the fault is electrical, mechanical or hydraulic in nature, before repair work can be contemplated. Diagnosis requires detailed knowledge of the transmission's operation and construction, as well as access to specialized test equipment, and so is deemed to be beyond the scope of this manual. It is therefore essential that problems with the automatic transmission are referred to a dealer service department or other qualified repair facility for assessment.

2   Note that a faulty transmission should not be removed before the vehicle has been assessed by a knowledgeable technician equipped with the proper tools, as troubleshooting must be performed with the transmission installed in the vehicle.

Chapter 8
Clutch and driveline

Contents

Specifications

General

Differential pinion shaft preload (amount of torque required
to turn the pinion shaft under no load) .... 19 to 22 inch-pounds

Torque specifications   Ft-lbs (unless otherwise indicated)

Note: *One foot-pound (ft-lb) of torque is equivalent to 12 inch-pounds (in-lbs) of torque. Torque values below approximately 15 ft-lbs are expressed in inch-pounds, because most foot-pound torque wrenches are not accurate at these smaller values.*

Coupler-to-driveshaft	
Automatic	43
Manual	67
Clutch pressure plate bolts	55
Clutch release bearing bolts	19
Differential-to-rear crossmember bolts	
2014 and earlier	162
2015 and later	111
Differential front mount insulator bolt/nut	
2014 and earlier	48
2015 and later	52
Driveaxle/hub nut	
2014 and earlier models	157
2015 and later models	170
Driveshaft center support bearing	21
Driveshaft-to-differential companion flange	
2014 and earlier	
Non-SRT8	72
Manual and SRT8	115
2015 and later	49
Driveshaft-to-transmission flange	
2014 and earlier	
Auto and Non-SRT8	72
Manual and SRT8	115
2015 and later	49
Flywheel bolts	55
Pinion nut	
2014 and earlier	180
2015 and later	Tighten per instructions (see Section 12)
Wheel lug nuts	See Chapter 1

1   General information

1   The information in this Chapter deals with the components from the rear of the engine to the rear wheels, except for the transmission, which is dealt with in the previous Chapters.

2   Some models are equipped with all-wheel drive that includes a front driveaxle and transfer case, however, these models are not covered in this manual.

3   Since nearly all the procedures covered in this Chapter involve working under the vehicle, make sure it's securely supported on sturdy jackstands or on a hoist where the vehicle can be easily raised and lowered.

2   Clutch - description and check

1   All vehicles with a manual transmissions use a dual dry plate, diaphragm spring type clutch. The clutch discs have a splined hub which allows it to slide along the splines of the transmission input shaft. The first clutch disc and pressure plate are held in contact by spring pressure exerted by the diaphragm in the pressure plate. The second clutch disc is part of the pressure plate assembly, and not serviceable, except as one unit.

2   The clutch release system is hydraulic operated. The release system includes the clutch pedal, the clutch master cylinder, and the hydraulic release bearing.

3   When the clutch pedal is depressed, it presses in on the master cylinder causing hydraulic pressure to operate the release bearing which pushes the release bearing forward. As the bearing is pushed forward, it slides forward along the input shaft, and the release bearing pushes against the fingers of the diaphragm spring in the pressure plate assembly, which releases the clutch plate.

4   Terminology can be a problem regarding the clutch components because common names have in some cases changed from that used by the manufacturer. For example, the driven plate is also called the clutch plate or disc and the clutch release bearing is sometimes called a throwout bearing.

5   Unless you're replacing components with obvious damage, perform some preliminary checks to diagnose a clutch system malfunction.

a)   *To check "clutch spin down time," run the engine at normal idle speed with the transmission in Neutral (clutch pedal up - engaged). Disengage the clutch (pedal down), wait nine seconds and shift the transmission into Reverse. No grinding noise should be heard. A grinding noise would most likely indicate a problem in the pressure plate or the clutch disc.*

b)   *To check for complete clutch release, run the engine (with the parking brake applied to prevent movement) and hold the clutch pedal approximately 1/4-inch from the floor. Shift the transmission between 1st gear and Reverse several*

times. *If the shift is not smooth, component failure is indicated.*

c)   *Visually inspect the clutch pedal bushing at the top of the clutch pedal to make sure there is no sticking or excessive wear.*

d)   *A clutch pedal that is difficult to operate is most likely caused by a faulty clutch pedal bushing, return spring or master cylinder. Check the condition of the spring and bushing. If they look good, lubricate the clutch pedal assembly with penetrating oil. If pedal operation does not improve, the clutch master cylinder may be faulty.*

3   Clutch master cylinder - replacement

1   Clamp a pair of locking pliers onto the clutch fluid feed hose, a couple of inches downstream of the brake fluid reservoir (the clutch master cylinder is supplied with fluid from the brake fluid reservoir). The pliers should be just tight enough to prevent fluid flow when the hose is disconnected.

2   Working under the instrument panel, remove the driver's side under-dash panel, steering column cover and reinforcement (see Chapter 11).

3   Working under the dash, disconnect the clutch master cylinder pushrod from the pedal by removing the clip from the clutch pedal pin.

Note: *It is recommended to use a new clip during installation.*

4   Disconnect the starter/clutch interlock switch connector.

Caution: *Don't allow brake fluid to come into contact with the paint or carpet, as it will damage the finish. Have rags handy, as some fluid will be lost as the line is removed.*

5   Separate the hydraulic line from the clutch master cylinder. Plug the line to prevent leakage.

6   Remove the master cylinder mounting nuts and detach the master cylinder.

7   Installation is reverse of removal. Bleed the clutch hydraulic system (see Section 5).

4   Clutch release cylinder/bearing - replacement

Warning: *Dust produced by clutch wear is hazardous to your health. DO NOT blow it out with compressed air and DO NOT inhale it. DO NOT use gasoline or petroleum-based solvents to remove the dust. Brake system cleaner should be used to flush the dust into a drain pan. After the clutch components are wiped clean with a rag, dispose of the contaminated rags and cleaner in a covered, marked container.*

1   Raise the vehicle and support it securely on jackstands.

2   Disconnect the hydraulic line at the

transmission. Have a small can and rags handy, as some fluid will be spilled as the line is removed. Plug the line to prevent excessive fluid loss and contamination.

3   Remove the transmission (see Chapter 7A, Section 6).

4   Remove the release cylinder mounting bolts. Remove the release cylinder.

5   Installation is reverse of removal. Bleed the clutch hydraulic system (see Section 5).

5   Clutch hydraulic system - bleeding

Note: *The clutch hydraulic system does not have a bleed screw to release trapped air.*

1   Bleed the hydraulic system whenever any part of the system has been removed or the fluid level has fallen so low that air has been drawn into the master cylinder. The bleeding procedure is very similar to bleeding a brake system.

2   Fill the brake fluid reservoir with new brake fluid conforming to DOT 3 specifications.

3   Press and release the clutch pedal until the proper pedal feel is achieved. If the pedal is hard but only for a short distance, air is in the clutch release cylinder. If the pedal is spongy, air is in the clutch master cylinder. This process can take up to 200 pedal press and release motions.

4   Check the brake fluid level again, and add some, if necessary, to bring it to the appropriate level.

6   Clutch components - removal, inspection and installation

Warning: *Dust produced by clutch wear and deposited on clutch components is hazardous to your health. DO NOT blow it out with compressed air and DO NOT inhale it. DO NOT use gasoline or petroleum-based solvents to remove the dust. Brake system cleaner should be used to flush the dust into a drain pan. After the clutch components are wiped clean with a rag, dispose of the contaminated rags and cleaner in a covered container.*

Removal

Note: *Access to the clutch components is normally accomplished by removing the transmission, leaving the engine in the vehicle. If, of course, the engine is being removed for major overhaul, then the opportunity should always be taken to check the clutch for wear and replace worn components as necessary. The following procedures assume that the engine will stay in place.*

1   Remove the transmission (see Chapter 7A, Section 6).

2   To support the clutch disc during removal, install a clutch alignment tool through the clutch disc hub.

3   Carefully inspect the flywheel and pres-

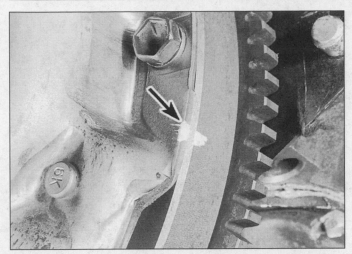

6.3 Mark the relationship of the pressure plate to the flywheel (if you're planning to re-use the old pressure plate

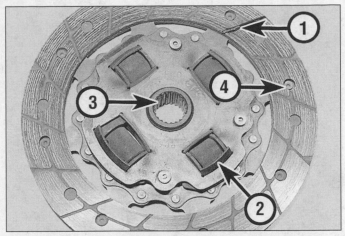

6.8 The clutch disc

1   *Lining - this will wear down in use*
2   *Rivets - these secure the lining and will damage the flywheel or pressure plate if allowed to contact the surfaces*
3   *Marks - "flywheel side" or something similar*
4   *Rivets - these secure the lining and will damage the flywheel or pressure plate if allowed to contact the surfaces*

sure plate for indexing marks. The marks are usually an X, an O or a white letter. If they cannot be found, scribe marks yourself so the pressure plate and the flywheel will be in the same alignment during installation **(see illustration)**.

4   Turning each bolt only 1/4-turn at a time, slowly loosen the pressure plate-to-flywheel bolts. Work in a diagonal pattern and loosen each bolt a little at a time until all spring pressure is relieved. Then hold the pressure plate securely and completely remove the bolts, followed by the pressure plate and clutch disc.

Inspection

5   Ordinarily, when a problem occurs in the clutch, it can be attributed to wear of the clutch driven plate assembly (clutch disc). However, all components should be inspected at this time.

6   Inspect the flywheel for cracks, heat checking, grooves or other signs of obvious defects. If the imperfections are slight, a machine shop can machine the surface flat and smooth, which is highly recommended regardless of the surface appearance. See Chapter 2D, Section 14.

7   Inspect the pilot bearing for smoothness and replace as necessary (see Section 7).

8 Inspect the lining on the clutch disc. There should be at least 1/16-inch of lining above the rivet heads. Check for loose rivets, distortion, cracks, broken springs and other obvious damage **(see illustration)**.

9   As mentioned above, ordinarily the clutch disc is replaced as a matter of course, so if in doubt about the condition, replace it with a new one. And if one disc is worn, the other disc is also worn. This requires replacement of the pressure plate along with both discs.

10   Ordinarily, the release bearing is also replaced along with the clutch disc (see Section 4).

11   Check the machined surfaces and the diaphragm spring fingers of the pressure plate. If the surface is grooved or otherwise damaged, take it to a machine shop for possible machining or replacement. Also check for obvious damage, distortion, cracking, etc. Light glazing can be removed with medium

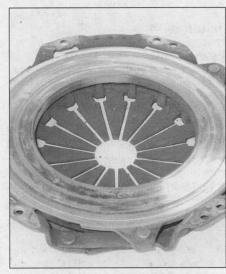

6.11a Examine the pressure plate friction surface for score marks, cracks and evidence of overheating (bluespots)

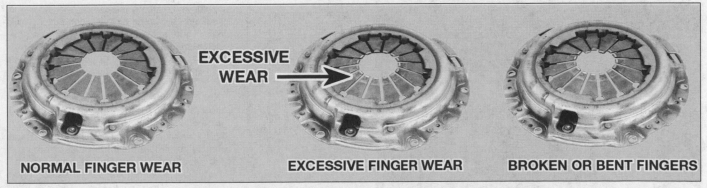

NORMAL FINGER WEAR   EXCESSIVE FINGER WEAR   BROKEN OR BENT FINGERS

EXCESSIVE WEAR ➜

6.11b Replace the pressure plate if any of these conditions are noted

6.13 Center the clutch disc in the pressure plate with a clutch alignment tool

7.4 Remove the old pilot bearing with a small slide-hammer, as shown, if you've got one

grit emery cloth. If a new pressure plate is indicated, new or factory-rebuilt units are available **(see illustrations)**.

Installation

12   Before installation, carefully wipe the flywheel and pressure plate machined surfaces clean. It's important that no oil or grease is on these surfaces or the lining of the clutch disc. Handle these parts only with clean hands.

13   Position the clutch disc and pressure plate with the clutch held in place with an alignment tool **(see illustration)**. Make sure it's installed properly (most replacement clutch plates will be marked "flywheel side" or something similar - if not marked, install the clutch disc with the damper springs toward the transmission).

14   Tighten the pressure plate-to-flywheel bolts only finger-tight, working around the pressure plate.

15   Center the clutch disc by inserting the alignment tool through the splined hub and into the pilot bearing in the crankshaft. Tighten

the pressure plate-to-flywheel bolts a little at a time, working in a crisscross pattern to prevent distorting the cover. After all of the bolts are snug, tighten them to the specified torque. Remove the alignment tool.

16   Install the transmission (see Chapter 7A).

7   Pilot bearing - inspection and replacement

1   The clutch pilot bearing is a needle roller type bearing which is pressed into the rear of the crankshaft. It is greased at the factory and does not require additional lubrication. Its primary purpose is to support the front of the transmission input shaft. The pilot bearing should be inspected whenever the clutch components are removed from the engine. Due to its inaccessibility, if you are in doubt as to its condition, replace it with a new one.

Note: *If the engine has been removed from the vehicle, disregard the following steps which do not apply.*

2   Remove the clutch components (see Section 6).

3   Inspect for any excessive wear, scoring, lack of grease, dryness or obvious damage. If any of these conditions are noted, the bearing should be replaced. A flashlight will be helpful to direct light into the recess.

4   The pilot bearing can be removed with a special puller and slide hammer **(see illustration)**.

5   To install the new bearing, lightly lubricate the outside surface with lithium-based grease, then drive it into the recess with a large socket **(see illustration)**. Make sure that the seal faces out, toward the transmission.

6   Install the clutch components, transmission and all other components removed previously, tightening all fasteners properly.

8   Clutch pedal position switch - check

1   The clutch pedal position switch is part of the clutch master cylinder and serviced as one unit (see Section 3).

2   If the engine can be started without depressing the clutch pedal, the switch may be shorted. Unplug the switch connector and hook up an ohmmeter to the connector terminals. There should be no continuity when the clutch pedal is released, and continuity when the pedal is depressed. If the switch doesn't operate as described, replace it.

3   If the engine can't be started even when the clutch pedal *is* depressed, either the switch is bad, the electrical connector is unplugged, or there's a ground, open or short in the circuit between the ignition switch and the clutch pedal position switch, or between the pedal position switch and the rest of the starter circuit. Check the switch as described in the previous step. If the switch is okay, refer to Chapter 13 and troubleshoot the circuit.

7.5 Use a large socket and hammer, or a soft-faced hammer, to install the new pilot bearing; make sure the O-ring seal faces out (toward the transmission) and make sure the bearing is fully seated

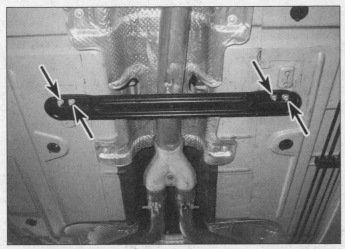

9.2 Remove the fasteners securing the crossmember brace

9.4 Mark the relationship of the driveshaft to the transmission flange

9 Driveshaft - removal and installation

Note: *Driveshaft center bearing replacement requires the use of a hydraulic press. Remove the driveshaft and have a repair shop or automotive machine shop remove the bearing.*

1   Raise the vehicle and support it securely on jackstands and place the selector lever in Neutral.

2   Remove the crossmember brace **(see illustration)**.

3   Remove the rear portion of the exhaust pipe(s) (see Chapter 4). Also remove the heat shield above the exhaust system.

4   Use chalk or a scribe to mark the relationship of the driveshaft to the transmission flange. This ensures correct alignment when the driveshaft is reinstalled **(see illustration)**.

5   Remove the fasteners securing the driveshaft to the transmission flange **(see illustration)**.

6   Use chalk or a scribe to mark the relationship of the driveshaft to the differential pinion flange **(see illustration)**.

7   Remove the fasteners securing the driveshaft flange to the differential pinion flange.

8   Remove the fasteners securing the center support bearing **(see illustration)** and remove the driveshaft.

Note: *New bolts are recommended when installing the driveshaft. If used bolts are to be installed, use thread locker.*

9   Installation is the reverse of removal. Be sure to align the marks made earlier, and tighten the fasteners to the torque listed in this Chapter's Specifications.

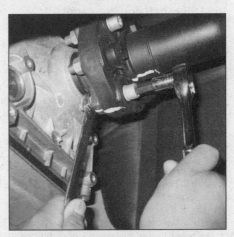

9.5 Remove the fasteners securing the driveshaft to the transmission flange (2014 and earlier model shown)

9.6 Mark the relationship of the driveshaft to the differential pinion flange

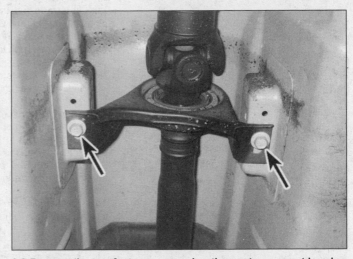

9.8 Remove the two fasteners securing the center support bearing

10.3 While supporting the driveshaft, remove the fasteners securing the coupler/damper to the driveshaft (2014 and earlier models shown)

11.5 Be sure to support the rear of the driveshaft - don't let it hang

11.6 Carefully pry the inner CV joint out of the differential just far enough to release the snap-ring

10   Driveshaft coupler/damper - removal and installation

Note: *The following procedure applies to 2014 and earlier models. On 2015 and later models, the couplers and driveshaft are one unit and not serviceable separately.*

1   Raise the vehicle and support it securely on jackstands and place the selector lever in Neutral.

2   Depending on which end of the driveshaft you're working on, disconnect the driveshaft from the differential pinion flange or the transmission flange as outlined in Section 9.

Note: *Support the driveshaft - do not let it hang from the center bearing.*

3   Remove the fasteners securing the coupler/damper to the driveshaft, then remove it from the driveshaft **(see illustration)**.

4   Installation is the reverse of removal. Tighten the fasteners to the torque listed in this Chapter's Specifications.

11.11 Install a new retaining ring on the inner stub shaft

11   Driveaxles - removal and installation

Note: *If a driveaxle boot has failed, the entire shaft assembly must be replaced. Boots are not serviceable separately.*

Removal

1   Remove the wheel cover or hub cap. Break the driveaxle/hub nut loose with a socket and large breaker bar.

2   Block the front wheels to prevent the vehicle from rolling. Loosen the wheel lug nuts, raise the rear of the vehicle and support it securely on jackstands. Remove the wheel.

3   Remove the driveaxle/hub nut and discard it.

4   Drain the gear lubricant from the differential (see Chapter 1).

5   Disconnect the driveshaft from the differential (see Section 9). Support the driveshaft - don't let it hang by the center support bearing **(see illustration)**. Remove the rear exhaust pipe and muffler (see Chapter 4).

6   Pry the inner CV joint out of the differential just far enough to release the snap-ring **(see illustration)**.

Note: *The driveaxles can't be removed from the differential until the differential has been lowered.*

7   Support the differential with a floor jack.

8   Remove the mounting fasteners securing the differential to the crossmember and differential mount (see Section 13).

9   Carefully lower and tilt the differential enough to separate the driveaxle(s) from the differential.

10   Pull the driveaxle out of the hub flange. If it sticks, tap it out with a soft-faced hammer.

Installation

11   Remove the retaining ring from the inner CV joint stub shaft and install a new one **(see illustration)**.

12   Installation is the reverse of removal,

noting the following points:

a)   *Remove the rubber isolation washer from the outer CV joint stub shaft and inspect it. Replace it if necessary.*

b)   *Tighten the differential mounting bolts to the torque listed in this Chapter's Specifications.*

c)   *Install a new driveaxle/hub nut. Tighten the hub nut securely, but don't try to tighten it to the actual torque specification until you've lowered the vehicle to the ground.*

d)   *Fill the differential with the type and amount of lubricant specified in Chapter 1.*

e)   *Install the wheels and lower the vehicle to the ground. Tighten the wheel lug nuts to the torque listed in the Chapter 1 Specifications.*

f)   *Tighten the driveaxle/hub nut to the torque listed in this Chapter's Specifications. Install the wheel cover or hub cap.*

12   Differential oil seals - replacement

1   Oil leaks frequently occur due to wear of the differential oil seals. Replacement of these seals is relatively easy, since the repair can usually be performed without removing the transmission from the vehicle.

2   Driveaxle oil seals are located at the sides of the differential, where the driveaxles are attached. The driveshaft seal is located behind the differential flange. If leakage at the seal is suspected, raise the vehicle and support it securely on jackstands. If the seal is leaking, lubricant will be found below the seals.

Driveaxle seals

3   Refer to Section 11 and remove the driveaxle(s).

4   Use a screwdriver or prybar to carefully

12.10 Use an inch-pound torque wrench to check the torque necessary to rotate the pinion shaft

12.12 A chain wrench can be used to hold the pinion flange while the nut is loosened

pry the oil seal out of the differential bore.

5   If the oil seal cannot be removed with a screwdriver or prybar, a special oil seal removal tool (available at auto parts stores) will be required.

6   Using a seal installation tool or a large deep socket (slightly smaller than the outside diameter of the seal) as a drift, install the new oil seal. Drive it into the bore squarely and make sure it's completely seated. Coat the seal lip with transmission lubricant.

7   Install the driveaxle(s). Be careful not to damage the lip of the new seal.

Pinion seal

8   Loosen the rear wheel lug nuts, raise the rear of the vehicle and support it securely on jackstands. Refer to Section 9 and remove the driveshaft.

Caution: *This procedure disturbs the pinion bearing preload adjustment. Follow the procedure very carefully to reset the pinion bearing preload during reassembly.*

9   Remove the rear wheels and brake calipers (see Chapter 9).

Note: *It is recommended that you remove the wheels and brake calipers to eliminate the added pinion shaft rotation resistance that otherwise might contribute to a false pinion shaft rotation preload torque value.*

10   Using an inch-pound torque wrench (scale from approximately 0 to 40 inch-pounds) on the drive pinion nut, measure and record the torque necessary to rotate the drive pinion in a load-free state **(see illustration).**

11   Mark the drive pinion-to-companion flange orientation for proper reassembly.

12   Using a chain wrench or a flange holding tool (available at most auto parts stores), unscrew the pinion flange locknut. Discard the nut - a new one must be used upon reassembly **(see illustration).**

13   Using a two-jaw puller, remove the companion flange from the drive pinion shaft.

Note: *Some fluid loss may occur.*

14   Remove the seal by tapping on the metal flange with a hammer and chisel.

15   Prior to installing the new seal, clean the seal mating surfaces.

16   Lubricate the lips of the new seal with high-temperature grease and tap it evenly into position with a seal installation tool or a large socket. Make sure it enters the housing squarely and is tapped in to its full depth.

17   Align the mating marks made before disassembly and install the companion flange and a *new* nut. If necessary, tighten the pinion nut to draw the flange into place. Do not try to hammer the flange into position.

18   Using a suitable holding tool, secure the companion flange while tightening the nut to the witness mark placed on the pinion nut and pinion.

Note: *Don't overtighten the nut.*

19   Measure the torque required to rotate the pinion; it should be equal to the torque recorded earlier. If too little torque is required to turn it, tighten the nut in small increments until it matches the figure recorded earlier. To compensate for the drag of the new oil seal, the nut should be tightened more until the rotational torque of the pinion exceeds the earlier recording by no more than 5 in-lbs.

Caution: *If the nut is tightened too much, resulting in a rotational torque beyond the specified range listed in this Chapter's Specifications, the differential will have to be removed and disassembled, and a new collapsible spacer installed on the pinion shaft.*

20   Using a hammer and punch, stake the pinion flange locknut.

21   The remainder of installation is the reverse of removal. Be sure to check the differential lubricant level and add if required (see Chapter 1).

---

13   Rear differential - removal and installation

Removal

1   Raise the rear of the vehicle and support it securely on jackstands.

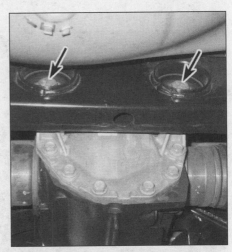

13.6a Remove the differential mounting fasteners at the rear . . .

2   Drain the lubricant from the differential (see Chapter 1).

3   Disconnect the driveshaft from the rear differential (see Section 9). Remove the rear exhaust pipe and muffler (see Chapter 4).

4   Pry the inner CV joint out of the differential just far enough to release the snap-ring **(see illustration 11.6).**

Note: *The driveaxles can't be removed from the differential until the differential has been lowered from the vehicle.*

5   Support the differential with a floor jack.

6   Remove the mounting fasteners that attach the differential to the rear crossmember **(see illustrations).**

7   Lower the differential enough to separate the driveaxles from the differential.

8   Carefully lower the differential and remove it from under the vehicle.

9   With the differential removed from the vehicle, now would be a good time to check or replace the rubber mounts for the differential mounting brackets and/or the rear crossmember.

**13.6b . . . and at the front of the differential
(2014 and earlier model shown)**

13.6c Remove the differential mounting fasteners at the rear . . .

**13.6d . . . and at
the front of the
differential (2015 and
later model shown)**

Installation

10   Place the differential on the jack head and position it directly underneath the mounting bracket and crossmember.

11   Raise the differential enough to install the driveaxles into the differential.

12   Raise the differential into position and install the rear mounting fasteners loosely. Then install the front mounting fastener. Tighten the mounting fasteners to the torque listed in this Chapter's Specifications.

13   Install the driveshaft (see Section 9).

14   Fill the differential with the type and amount of lubricant specified in Chapter 1.

15   Install the wheels, remove the jack and lower the vehicle to the ground. Tighten the wheel lug nuts to the torque listed in the Chapter 1 Specifications.

Chapter 9
Brakes

Contents

Specifications

General

Brake fluid type	See Chapter 1

Disc brakes

Minimum brake pad thickness	See Chapter 1
Disc minimum thickness	Cast into disc
Disc runout limit	
All except SRT8	0.0014 inch
SRT8	0.0012 inch
Parking brake shoe minimum thickness	1/32-inch

Torque specifications

Ft-lbs (unless otherwise indicated)

Note: *One foot pound (ft lb) of torque is equivalent to 12 inch pounds (in lbs) of torque. Torque values below approximately 15 ft lbs are expressed in inch-pounds, because most foot-pound torque wrenches are not accurate at these smaller values.*

Adjustable brake pedal assembly nuts	18
Caliper mounting bolts (all models exc. SRT8)	
2005 models	
Front	44
Rear	44
2006 and later models	
Front	44
Rear	23
Caliper mounting bolts (fixed calipers - SRT8 models)*	
Front	
Models built on or before 8/30/12	140
Models built after 8/30/12	121
Rear	98
Caliper center support bolt (2016 and later SRT8 models)	22
Caliper mounting bracket bolts (floating calipers; all models exc. SRT8)	
2012 and earlier models	
Front	70
Rear	85
2013 and 2014 models	
Models built on or before 8/30/12	140
Models built after 8/30/12	121
Rear	44
2015 and later models*	
Front	98
Rear	81
Brake hose-to-caliper banjo bolt**	
2005 models (front and rear)	32
2006 through 2014 models (front and rear)	
All except SRT8	37
SRT8	24
2015 and later models	
Front	
All except SRT8	37
SRT8	24
Rear (all models)	27
Master cylinder-to-brake booster nuts***	
2005 models	19
2006 models	18
2007 through 2012 models	
MK25 master cylinder	18
MK25E master cylinder	132 in-lbs
2013 and later models	17
Electric Vacuum Pump (EVP)-to-bracket bolts	71 in-lbs
Power brake booster mounting nuts	18
Wheel lug nuts	See Chapter 1

* *Use new fasteners*
** *Use new copper washers*
*** *Use new mounting hardware and a new O-ring seal between the master cylinder and brake booster*

1   General Information

1   The vehicles covered by this manual are equipped with hydraulically operated front and rear brake systems. The front brakes are disc type and the rear brakes are disc or drum type. Both the front and rear brakes are self adjusting. The disc brakes automatically compensate for pad wear, while the drum brakes incorporate an adjustment mechanism that is activated as the parking brake is applied.

Hydraulic system

2   The hydraulic system consists of two separate circuits. The master cylinder has separate reservoir chambers for the two circuits, and, in the event of a leak or failure in one hydraulic circuit, the other circuit will remain operative. A dynamic proportioning valve, integral with the ABS hydraulic unit, provides brake balance to each individual wheel.

Power brake booster and vacuum pump

3   The power brake booster, utilizing engine manifold vacuum, and atmospheric pressure to provide assistance to the hydraulically operated brakes, is mounted on the firewall in the engine compartment. An auxiliary vacuum pump, mounted below the power brake booster in the left side of the engine compartment, provides additional vacuum to the booster under certain operating conditions.

Parking brake

4   The parking brake operates the rear brakes only, through cable actuation. It's activated by a lever mounted in the center console.

Service

5   After completing any operation involving disassembly of any part of the brake system, always test drive the vehicle to check for proper braking performance before resuming normal driving. When testing the brakes, perform the tests on a clean, dry, flat surface. Conditions other than these can lead to inaccurate test results.

6   Test the brakes at various speeds with both light and heavy pedal pressure. The vehicle should stop evenly without pulling to one side or the other. Avoid locking the brakes, because this slides the tires and diminishes braking efficiency and control of the vehicle.

7   Tires, vehicle load and wheel alignment are factors which also affect braking performance.

Precautions

8   There are some general cautions and warnings involving the brake system on this vehicle:

a)   Use only brake fluid conforming to DOT 3 specifications.

b)   The brake pads and linings contain fibers that are hazardous to your health if inhaled. Whenever you work on brake system components, clean all parts with brake system cleaner. Do not allow the fine dust to become airborne. Also, wear an approved filtering mask.

c)   Safety should be paramount whenever any servicing of the brake components is performed. Do not use parts or fasteners that are not in perfect condition, and be sure that all clearances and torque specifications are adhered to. If you are at all unsure about a certain procedure, seek professional advice. Upon completion of any brake system work, test the brakes carefully in a controlled area before putting the vehicle into normal service. If a problem is suspected in the brake system, don't drive the vehicle until it's fixed.

d)   Used brake fluid is considered a hazardous waste and it must be disposed of in accordance with federal, state and local laws. DO NOT pour it down the sink, into septic tanks or storm drains, or on the ground. Clean up any spilled brake fluid immediately and then wash the area with large amounts of water. This is especially true for any finished or painted surfaces.

2   Troubleshooting

PROBABLE CAUSE	CORRECTIVE ACTION

No brakes - pedal travels to floor

1 Low fluid level 2 Air in system	1 and 2 Low fluid level and air in the system are symptoms of another problem a leak somewhere in the hydraulic system. Locate and repair the leak
3 Defective seals in master cylinder	3 Replace master cylinder
4 Fluid overheated and vaporized due to heavy braking	4 Bleed hydraulic system (temporary fix). Replace brake fluid (proper fix)

Brake pedal slowly travels to floor under braking or at a stop

1 Defective seals in master cylinder	1 Replace master cylinder
2 Leak in a hose, line, caliper or wheel cylinder	2 Locate and repair leak
3 Air in hydraulic system	3 Bleed the system, inspect system for a leak

Brake pedal feels spongy when depressed

1 Air in hydraulic system	1 Bleed the system, inspect system for a leak
2 Master cylinder or power booster loose	2 Tighten fasteners
3 Brake fluid overheated (beginning to boil)	3 Bleed the system (temporary fix). Replace the brake fluid (proper fix)
4 Deteriorated brake hoses (ballooning under pressure)	4 Inspect hoses, replace as necessary (it's a good idea to replace all of them if one hose shows signs of deterioration)

Troubleshooting (continued)

PROBABLE CAUSE	CORRECTIVE ACTION

Brake pedal feels hard when depressed and/or excessive effort required to stop vehicle

1 Power booster faulty	1 Replace booster
2 Engine not producing sufficient vacuum, or hose to booster clogged, collapsed or cracked	2 Check vacuum to booster with a vacuum gauge. Replace hose if cracked or clogged, repair engine if vacuum is extremely low
3 Brake linings contaminated by grease or brake fluid	3 Locate and repair source of contamination, replace brake pads or shoes
4 Brake linings glazed	4 Replace brake pads or shoes, check discs and drums for glazing, service as necessary
5 Caliper piston(s) or wheel cylinder(s) binding or frozen	5 Replace calipers or wheel cylinders
6 Brakes wet	6 Apply pedal to boil-off water (this should only be a momentary problem)
7 Kinked, clogged or internally split brake hose or line	7 Inspect lines and hoses, replace as necessary

Excessive brake pedal travel (but will pump up)

1 Drum brakes out of adjustment	1 Adjust brakes
2 Air in hydraulic system	2 Bleed system, inspect system for a leak

Excessive brake pedal travel (but will not pump up)

1 Master cylinder pushrod misadjusted	1 Adjust pushrod
2 Master cylinder seals defective	2 Replace master cylinder
3 Brake linings worn out	3 Inspect brakes, replace pads and/or shoes
4 Hydraulic system leak	4 Locate and repair leak

Brake pedal doesn't return

1 Brake pedal binding	1 Inspect pivot bushing and pushrod, repair or lubricate
2 Defective master cylinder	2 Replace master cylinder

Brake pedal pulsates during brake application

1 Brake drums out-of-round	1 Have drums machined by an automotive machine shop
2 Excessive brake disc runout or disc surfaces out-of-parallel	2 Have discs machined by an automotive machine shop
3 Loose or worn wheel bearings	3 Adjust or replace wheel bearings
4 Loose lug nuts	4 Tighten lug nuts

Brakes slow to release

1 Malfunctioning power booster	1 Replace booster
2 Pedal linkage binding	2 Inspect pedal pivot bushing and pushrod, repair/lubricate
3 Malfunctioning proportioning valve	3 Replace proportioning valve
4 Sticking caliper or wheel cylinder	4 Repair or replace calipers or wheel cylinders
5 Kinked or internally split brake hose	5 Locate and replace faulty brake hose

Brakes grab (one or more wheels)

1 Grease or brake fluid on brake lining	1 Locate and repair cause of contamination, replace lining
2 Brake lining glazed	2 Replace lining, deglaze disc or drum

PROBABLE CAUSE	CORRECTIVE ACTION

Vehicle pulls to one side during braking

PROBABLE CAUSE	CORRECTIVE ACTION
1 Grease or brake fluid on brake lining	1 Locate and repair cause of contamination, replace lining
2 Brake lining glazed	2 Deglaze or replace lining, deglaze disc or drum
3 Restricted brake line or hose	3 Repair line or replace hose
4 Tire pressures incorrect	4 Adjust tire pressures
5 Caliper or wheel cylinder sticking	5 Repair or replace calipers or wheel cylinders
6 Wheels out of alignment	6 Have wheels aligned
7 Weak suspension spring	7 Replace springs
8 Weak or broken shock absorber	8 Replace shock absorbers

Brakes drag (indicated by sluggish engine performance or wheels being very hot after driving)

PROBABLE CAUSE	CORRECTIVE ACTION
1 Brake pedal pushrod incorrectly adjusted	1 Adjust pushrod
2 Master cylinder pushrod (between booster and master cylinder)	2 Adjust pushrod incorrectly adjusted
3 Obstructed compensating port in master cylinder	3 Replace master cylinder
4 Master cylinder piston seized in bore	4 Replace master cylinder
5 Contaminated fluid causing swollen seals throughout system	5 Flush system, replace all hydraulic components
6 Clogged brake lines or internally split brake hose(s)	6 Flush hydraulic system, replace defective hose(s)
7 Sticking caliper(s) or wheel cylinder(s)	7 Replace calipers or wheel cylinders
8 Parking brake not releasing	8 Inspect parking brake linkage and parking brake mechanism, repair as required
9 Improper shoe-to-drum clearance	9 Adjust brake shoes
10 Faulty proportioning valve	10 Replace proportioning valve

Brakes fade (due to excessive heat)

PROBABLE CAUSE	CORRECTIVE ACTION
1 Brake linings excessively worn or glazed	1 Deglaze or replace brake pads and/or shoes
2 Excessive use of brakes	2 Downshift into a lower gear, maintain a constant slower speed (going down hills)
3 Vehicle overloaded	3 Reduce load
4 Brake drums or discs worn too thin	4 Measure drum diameter and disc thickness, replace drums or discs as required
5 Contaminated brake fluid	5 Flush system, replace fluid
6 Brakes drag	6 Repair cause of dragging brakes
7 Driver resting left foot on brake pedal	7 Don't ride the brakes

Brakes noisy (high-pitched squeal)

PROBABLE CAUSE	CORRECTIVE ACTION
1 Glazed lining	1 Deglaze or replace lining
2 Contaminated lining (brake fluid, grease, etc.)	2 Repair source of contamination, replace linings
3 Weak or broken brake shoe hold-down or return spring	3 Replace springs
4 Rivets securing lining to shoe or backing plate loose	4 Replace shoes or pads
5 Excessive dust buildup on brake linings	5 Wash brakes off with brake system cleaner
6 Brake drums worn too thin	6 Measure diameter of drums, replace if necessary
7 Wear indicator on disc brake pads contacting disc	7 Replace brake pads
8 Anti-squeal shims missing or installed improperly	8 Install shims correctly

Troubleshooting (continued)

PROBABLE CAUSE	CORRECTIVE ACTION

Brakes noisy (scraping sound)

1 Brake pads or shoes worn out; rivets, backing plate or brake	1 Replace linings, have discs and/or drums machined (or replace) shoe metal contacting disc or drum

Brakes chatter

1 Worn brake lining	1 Inspect brakes, replace shoes or pads as necessary
2 Glazed or scored discs or drums	2 Deglaze discs or drums with sandpaper (if glazing is severe, machining will be required)
3 Drums or discs heat checked	3 Check discs and/or drums for hard spots, heat checking, etc. Have discs/drums machined or replace them
4 Disc runout or drum out-of-round excessive	4 Measure disc runout and/or drum out-of-round, have discs or drums machined or replace them
5 Loose or worn wheel bearings	5 Adjust or replace wheel bearings
6 Loose or bent brake backing plate (drum brakes)	6 Tighten or replace backing plate
7 Grooves worn in discs or drums	7 Have discs or drums machined, if within limits (if not, replace them)
8 Brake linings contaminated (brake fluid, grease, etc.)	8 Locate and repair source of contamination, replace pads or shoes
9 Excessive dust buildup on linings	9 Wash brakes with brake system cleaner
10 Surface finish on discs or drums too rough after machining	10 Have discs or drums properly machined (especially on vehicles with sliding calipers)
11 Brake pads or shoes glazed	11 Deglaze or replace brake pads or shoes

Brake pads or shoes click

1 Shoe support pads on brake backing plate grooved or excessively worn	1 Replace brake backing plate
2 Brake pads loose in caliper	2 Loose pad retainers or anti-rattle clips
3 Also see items listed under Brakes chatter	

Brakes make groaning noise at end of stop

1 Brake pads and/or shoes worn out	1 Replace pads and/or shoes
2 Brake linings contaminated (brake fluid, grease, etc.)	2 Locate and repair cause of contamination, replace brake pads or shoes
3 Brake linings glazed	3 Deglaze or replace brake pads or shoes
4 Excessive dust buildup on linings	4 Wash brakes with brake system cleaner
5 Scored or heat-checked discs or drums	5 Inspect discs/drums, have machined if within limits (if not, replace discs or drums)
6 Broken or missing brake shoe attaching hardware	6 Inspect drum brakes, replace missing hardware

Rear brakes lock up under light brake application

1 Tire pressures too high	1 Adjust tire pressures
2 Tires excessively worn	2 Replace tires
3 Defective proportioning valve	3 Replace proportioning valve

Brake warning light on instrument panel comes on (or stays on)

1 Low fluid level in master cylinder reservoir (reservoirs with fluid level sensor)	1 Add fluid, inspect system for leak, check the thickness of the brake pads and shoes
2 Failure in one half of the hydraulic system	2 Inspect hydraulic system for a leak
3 Piston in pressure differential warning valve not centered	3 Center piston by bleeding one circuit or the other (close bleeder valve as soon as the light goes out)

PROBABLE CAUSE	CORRECTIVE ACTION

Brake warning light on instrument panel comes on (or stays on) (continued)

4 Defective pressure differential valve or warning switch	4 Replace valve or switch
5 Air in the hydraulic system	5 Bleed the system, check for leaks
6 Brake pads worn out (vehicles with electric wear sensors - small probes that fit into the brake pads and ground out on the disc when the pads get thin)	6 Replace brake pads (and sensors)

Brakes do not self adjust

Disc brakes

1 Defective caliper piston seals	1 Replace calipers. Also, possible contaminated fluid causing soft or swollen seals (flush system and fill with new fluid if in doubt)
2 Corroded caliper piston(s)	2 Same as above

Drum brakes

1 Adjuster screw frozen	1 Remove adjuster, disassemble, clean and lubricate with high-temperature grease
2 Adjuster lever does not contact star wheel or is binding	2 Inspect drum brakes, assemble correctly or clean or replace parts as required
3 Adjusters mixed up (installed on wrong wheels after brake job)	3 Reassemble correctly
4 Adjuster cable broken or installed incorrectly (cable-type adjusters)	4 Install new cable or assemble correctly

Rapid brake lining wear

1 Driver resting left foot on brake pedal	1 Don't ride the brakes
2 Surface finish on discs or drums too rough	2 Have discs or drums properly machined
3 Also see Brakes drag	

3 Anti-lock Brake System (ABS) - general information

Note: *Models equipped with an Anti-lock Brake System (ABS) will have one of two different systems installed: one is MK25 (the earlier ABS system) and can be identified by a pressure sensor mounted to the bottom of the master cylinder. The other is MK25E (the later ABS system) and the master cylinder appears typical and is not equipped with a pressure sensor on the bottom. Use this information to determine which system is on your vehicle. Knowing which system you have is important regarding part identification and torque specifications.*

1   The Anti-lock Brake System (ABS) is designed to help maintain vehicle steerability, directional stability and optimum deceleration under severe braking conditions and on most road surfaces. The ABS system is primarily designed to prevent wheel lockup during heavy or panic braking situations. It works by monitoring the rotational speed of each wheel and controlling the brake line pressure to each wheel when engaged. The data provided by the ABS wheel speed sensors is shared with another optional system called the Electronic Stability Program (ESP) which aids in vehicle control and handling. This system helps with traction control, over/under-steering and acceleration control when the base brakes are engaged or under acceleration when the system is turned on manually by the switch on the instrument panel. Another system added to vehicles with ESP is the Brake Assist System (BAS). This system works with the ABS system to maximize brake system performance in panic braking situations.

Components

Actuator assembly
2   The actuator assembly is mounted in the engine compartment and consists of an electric hydraulic pump and solenoid valves **(see illustration)**.

a) *The electric pump provides hydraulic pressure to charge the reservoirs in the actuator, which supplies pressure to the braking system. The pump and reservoirs are housed in the actuator assembly.*

b) *The solenoid valves modulate brake line pressure during ABS operation.*

Wheel speed sensors
3   These sensors are located at each wheel

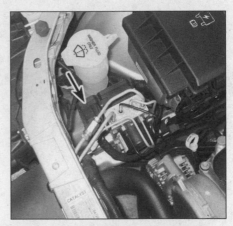

3.2 The ABS actuator assembly is mounted in the right front corner of the engine compartment

and generate small electrical pulsations when the hubs are turning, sending a signal to the electronic controller indicating wheel rotational speed.

4   The front speed sensors are mounted to the front steering knuckle in close relationship to the magnetic encoders, which are pressed into the back of the hub and bearing assemblies.

5   The rear wheel sensors are bolted to the

3.11a A wheel speed sensor wire harness and connector (front shown, rear similar)

3.11b Press the spring down to release the ABS harness connector

3.11c Challenger wheel speed sensor wire harness and connector, just slightly different length and location

A   Electrical connector
B   Pull harness out of clip to remove it

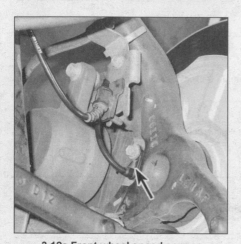

3.12a Front wheel speed sensor

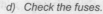

3.12b Rear wheel speed sensor

11   Trace the wiring back from the sensor, detaching all brackets and clips while noting its correct routing, then disconnect the electrical connector **(see illustrations)**.
12   Remove the mounting bolt and carefully pull the sensor out from the knuckle **(see illustrations)**.
13   Installation is the reverse of the removal procedure. Tighten the mounting fastener securely.
14   Install the wheel and lug nuts, tightening them securely. Lower the vehicle and tighten the lug nuts to the torque listed in the Chapter 1 Specifications.

rear suspension knuckles. The sensor rings are integrated with the rear hub assemblies.

ABS computer
6   The ABS computer is mounted with the actuator and is the brain for the ABS system. The function of the computer is to accept and process information received from the wheel speed sensors to control the hydraulic line pressure, avoiding wheel lock up. The computer also constantly monitors the system, even under normal driving conditions, to find faults within the system.

Diagnosis and repair
7   If a dashboard warning light comes on and stays on while the vehicle is in operation, the ABS system requires attention. Although special electronic ABS diagnostic testing tools are necessary to properly diagnose the system, you can perform a few preliminary checks before taking the vehicle to a dealer service department.

a) Check the brake fluid level in the reservoir.
b) Verify that the computer electrical connectors are securely connected.
c) Check the electrical connectors at the hydraulic control unit.

d) Check the fuses.
e) Follow the wiring harness to each wheel and verify that all connections are secure and that the wiring is undamaged.

8   If the above preliminary checks do not rectify the problem, the vehicle should be diagnosed by a dealer service department or other qualified repair shop. Due to the complexity of this system, all actual repair work must be done by a qualified automotive technician.
Warning: Do NOT try to repair an ABS wiring harness. The ABS system is sensitive to even the smallest changes in resistance. Repairing the harness could alter resistance values and cause the system to malfunction. If the ABS wiring harness is damaged in any way, it must be replaced. Make sure the ignition is turned off before unplugging or reattaching any electrical connections.

Wheel speed sensor - removal and installation
9   Loosen the wheel lug nuts, raise the vehicle and support it securely on jackstands. Remove the wheel.
10   Make sure the ignition key is turned to the Off position.

4   Disc brake pads - replacement

Warning: Disc brake pads must be replaced on both front or on both rear wheels at the same time - never replace the pads for only one wheel. Also, the dust created by the brake system is harmful to your health. Never blow it out with compressed air and don't inhale any of it. An approved filtering mask should be worn when working on the brakes. Do not, under any circumstances, use petroleum-based solvents to clean brake parts. Use brake system cleaner only!
Note: This procedure applies to both the front and rear disc brakes.
1   Remove the cap from the brake fluid reservoir.
2   Loosen the wheel lug nuts, raise the front or rear of the vehicle and support it securely on jackstands. Block the wheels at the opposite end.
3   Remove the wheels. Work on one brake assembly at a time, using the assembled brake for reference if necessary.
4   Inspect the brake disc carefully as outlined in Section 6. If machining is necessary, follow the information in that Section to

4.5 Before removing the caliper, slowly depress the piston into the caliper bore by using a large C-clamp between the outer brake pad and the back of the caliper

4.6a Always wash the brakes with brake cleaner before disassembling anything

4.6b If you're replacing the front pads, remove the lower caliper bolt while holding the caliper guide pin. On 2014 and earlier model rear calipers, remove the upper caliper bolt

4.6c If you're replacing the front pads, pivot the caliper up and secure it with a piece of wire in this position; do not allow the caliper to hang by the flexible brake hose. Be careful not to damage the guide pin boots while rotating the caliper

4.6d On 2014 and earlier model rear calipers, pivot the caliper down

4.6e On 2015 and later model rear calipers, remove both caliper mounting bolts and detach the caliper from the mounting bracket. Hang the caliper with a length of wire - don't let it hang by the hose

remove the disc, at which time the pads can be removed as well.

Floating calipers (non-SRT models)

5   Push the piston back into its bore to provide room for the new brake pads. A C-clamp can be used to accomplish this (see illustration). As the piston is depressed to the bottom of the caliper bore, the fluid in the master cylinder will rise. Make sure that it doesn't overflow. If necessary, siphon off some of the fluid.

6   Follow the accompanying photos (illustrations 4.6a through 4.6n) for the actual pad replacement procedure. Be sure to stay in order and read the caption under each illustration.

4.6f Remove inner pad. . .

4.6g. . . and the outer pad

4.6h Remove the upper and lower anti-rattle clips; make sure they fit tightly and aren't worn. Replace them if necessary

4.6i Install the clean or new anti-rattle clips

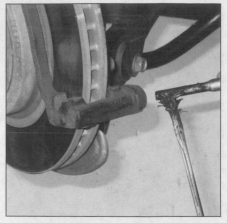

4.6j Pull out the upper and lower guide pins and clean them. Then apply a coat of high-temperature grease to the pins and reinstall them. Be careful not to damage the guide pin boots and replace any boots that are worn or damaged

4.6k Install the outer pad, making sure that the ends are seated correctly into the anti-rattle clips. . .

4.6l. . . then install the inner pad in the same way

4.6m Place the caliper back into position over the brake pads and onto the caliper mounting bracket

4.6n Install the caliper bolt(s) and tighten to the torque listed in this Chapter's Specifications

4.9a Remove the lower guide pin . . .

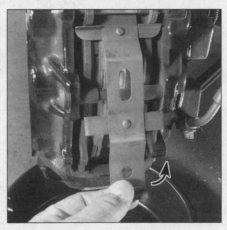

4.9b . . . then rotate the pad retainer up and remove it from the caliper

4.10 Tap the upper guide pin inward and remove it

7   When reinstalling the caliper, be sure to tighten the mounting bolt(s) to the torque listed in this Chapter's Specifications. After the job has been completed, firmly depress the brake pedal a few times to bring the pads into contact with the disc. Check the level of the brake fluid, adding some if necessary. Check the operation of the brakes carefully before placing the vehicle into normal service.

Fixed calipers (SRT models)

Warning: *Review the Warning and Note at the beginning of this Section.*

8   Perform Steps 1 through 4. Wash the caliper and surrounding components with brake system cleaner.

9   Tap the lower caliper guide pin inward, then finish driving it out with a pin punch and remove it from the caliper **(see illustration)**. Rotate the pad retainer up and remove it from the caliper **(see illustration)**.

10   Drive the upper caliper guide pin inward and remove it from the caliper **(see illustration)**.

11   On 2016 and later model calipers, remove the caliper center support bolt (if equipped), then slide the support bolt out **(see illustration)**.

12   Pull the inboard (inner) brake pad from the caliper **(see illustration)**.

Note: *Remove only one pad during this step.*

13   Using a pair of prybars or screwdrivers, slowly push the caliper pistons into the bores on the inboard side. There are two pistons on each side of the caliper for a total of four. As the pistons are depressed to the bottom of the caliper bores, the fluid in the master cylinder reservoir will rise. Continue to make sure that it doesn't overflow.

Caution: *One piston can be forced out of the caliper bore while the other is being pushed in on the same side. Use a small piece of wood or another tool to keep one piston from coming out while the other is being pushed in.*

14   Prepare each new brake pad by using a very small amount of copper-based brake paste on the edges of the pad's metal back

4.11 If equipped, remove the center support bolt

plate that contacts the caliper, and to the pad backing plates (where the shim makes contact).

Warning: *Do not get any lubricant on the pad material.*

15   Position the new inboard brake pad into the caliper.

16   Pull the inner brake pad from the caliper.

17   Repeat Step 13 to retract the outer pistons

18   Prepare the new outer brake pad as described in Step 14, then position the new outer pad into the caliper.

19   On models so equipped, install the caliper center support bolt, tightening it to the torque listed in this Chapter's Specifications.

20   Install the upper guide pin and place the spring-steel pad retainer in position.

21   Press the bottom end of the retainer and install the lower guide pin.

22   Repeat the procedure on the opposite wheel.

23   After the job is done, firmly depress the brake pedal a few times to bring the new pads into contact with the disc. Check the level of

4.12 Remove the inner pad from the caliper, then push the inner pistons back into their bores

the brake fluid, adding some if necessary. Check the operation of the brakes carefully before placing the vehicle into normal service.

---

5   Disc brake caliper - removal and installation

Warning: *Dust created by the brake system is harmful to your health. Never blow it out with compressed air and don't inhale any of it. An approved filtering mask should be worn when working on the brakes. Do not, under any circumstances, use petroleum-based solvents to clean brake parts. Use brake system cleaner only!*

Note: *Always replace the calipers in pairs - never replace just one of them.*

Removal

1   Loosen the wheel lug nuts, raise the vehicle and support it securely on jackstands. Remove the wheels.

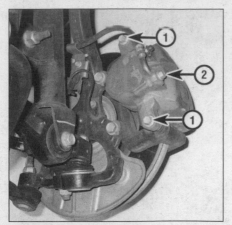

5.2a Brake caliper mounting details (front caliper shown - rear caliper similar)

1   Caliper mounting bolts
2   Brake hose banjo fitting bolt

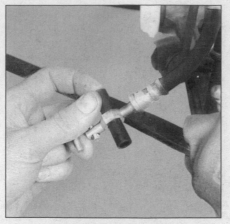

5.2b Using a piece of rubber hose of the appropriate size, plug the brake line banjo fitting to prevent brake fluid from leaking out and to prevent dirt and moisture from contaminating the fluid in the hose

6.4 The brake pads on this vehicle were obviously neglected, as they wore down completely and cut deep grooves into the disc - wear this severe means the disc must be replaced

6.5a Use a dial indicator to check disc runout; if the reading exceeds the maximum allowable runout limit, the disc will have to be machined or replaced

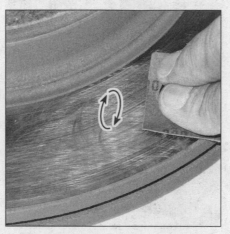

6.5b Using a swirling motion, remove the glaze from the disc surface with sandpaper or emery cloth

2   Remove the brake hose banjo bolt and disconnect the hose from the caliper. Plug the hose to keep contaminants out of the brake system and to prevent losing any more brake fluid than is necessary **(see illustrations)**. **Note:** *If you're just removing the caliper for access to other components, don't detach the hose.*

3   On non-SRT models (floating calipers), remove the caliper mounting bolts, lift the caliper from the bracket and, if the hose is still attached, support it with a length of wire.

4   On SRT models (fixed calipers), remove the caliper-to-steering knuckle bolts (front) or caliper-to-rear knuckle bolts (rear) and detach the caliper from the knuckle.

Installation

5   Install the caliper by reversing the removal procedure. Tighten the caliper mounting bolts to the torque listed in this Chapter's Specifications. If the brake hose

was detached, install *new* sealing washers, one on each side of the brake hose fitting, then tighten the banjo bolt to the torque listed in this Chapter's Specifications.

6   If the hose was disconnected, bleed the brake system (see Section 9).

7   Install the wheels and lug nuts. Lower the vehicle and tighten the lug nuts to the torque listed in the Chapter 1 Specifications.

6   Brake disc - inspection, removal and installation

Inspection

Note: *The manufacturer suggests using a specialized (on-vehicle) brake lathe that can cut the disc while it is still installed on the vehicle for disc refinishing (machining), although it is not essential. Moreover, this service requires a repair facility to have the specialized tool and to*

have the vehicle, as opposed to you taking the brake discs to a machine shop for refinishing.

1   Loosen the wheel lug nuts, raise the vehicle and support it securely on jackstands. Remove the wheel.

2   Remove the brake caliper as outlined in Section 5. It's not necessary to disconnect the brake hose for this procedure. After removing the caliper bolts, suspend the caliper out of the way with a piece of wire. Don't let the caliper hang by the hose and don't stretch or twist the hose.

3   Reinstall the lug nuts (inverted) to hold the disc against the hub. It may be necessary to install washers between the disc and the lug nuts to take up space.

4   Visually check the disc surface for score marks, cracks and other damage. Light scratches and shallow grooves are normal after use and may not always be detrimental to brake operation. Deep score marks or cracks may require disc refinishing by an automotive machine shop or disc replacement **(see illustration)**. Be sure to check both sides of the disc. If the brake pedal pulsates during brake application, suspect disc runout.

Note: *The most common symptoms of damaged or worn brake discs are pulsation in the brake pedal when the brakes are applied or loud grinding noises caused from severely worn brake pads. If these symptoms are extreme, it is very likely that the disc(s) will need to be replaced.*

5   To check disc runout, place a dial indicator at a point about 1/2-inch from the outer edge of the disc **(see illustration)**. Set the indicator to zero and turn the disc. An indicator reading that exceeds 0.003 of an inch could cause pulsation upon brake application and will require disc refinishing by an automotive machine shop or disc replacement. If disc refinishing or replacement is not necessary, you can deglaze the brake pad surface on the disc with emery cloth or sandpaper (use a swirling motion to ensure a non-directional finish) **(see illustration)**.

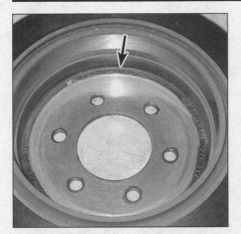

6.6a The minimum wear dimension is usually cast into the back side of the disc (typical shown)

6.6b Use a micrometer to measure disc thickness

6.7 The caliper mounting bracket is retained by these two bolts (rear shown - front similar)

6   The disc must not be machined to a thickness less than the specified minimum refinish thickness. The minimum wear (or discard) thickness is cast into either the front or backside of the disc **(see illustration)**. The disc thickness can be checked with a micrometer **(see illustration)**.

Removal and installation

7   On non-SRT models (floating calipers), remove the caliper mounting bracket **(see illustration)**.

8   Mark the disc in relation to the hub so that it can be installed in its original position on the hub and then remove the disc. If it's stuck, make sure you have removed any lug nuts installed during inspection. If the disc has never been removed, there may be wave washers on the wheel studs securing it to the hub flange. Simply cut them off and discard them **(see illustration)**.

Note: *On rear discs, make sure the parking brake is released.*

9   Clean the hub flange and the inside of the brake disc thoroughly, removing any rust or corrosion, then install the disc onto the hub assembly.

10   Install the caliper mounting bracket and tighten the bolts to the torque listed in this Chapter's Specifications.

11   Install the brake pads and caliper, tightening the bolts to the torque listed in this Chapter's Specifications.

12   Install the wheel, then lower the vehicle to the ground. Tighten the wheel lug nuts to the torque listed in the Chapter 1 Specifications. Depress the brake pedal a few times to bring the brake pads into contact with the disc. Bleeding of the system will not be necessary unless the brake hose was disconnected from the caliper. Check the operation of the brakes carefully before placing the vehicle into normal service.

7   Master cylinder - removal and installation

Removal

Caution: *Brake fluid will damage paint or finished surfaces. Cover all body parts and be careful not to spill fluid during this procedure. Clean up any spilled brake fluid immediately*

and then wash the area with large amounts of water.

Note: *Vehicle models equipped with an Antilock brake system (ABS) will have one of two different factory systems installed: one is MK25 (the earlier ABS system) and can be identified by a pressure sensor mounted to the bottom of the master cylinder. The other is MK25E (the later ABS system) and the master cylinder appears typical and is not equipped with a pressure sensor on the bottom. Use this information to determine which system is on your vehicle. Knowing which system you have is important regarding part identification and torque specifications.*

1   Disconnect the cable from the negative terminal of the battery (see Chapter 5, Section 3).

2   Firmly depress the brake pedal several times to remove all vacuum from the power brake booster.

3   Remove the master cylinder access cover in the cowl **(see illustration)**.

4   Clean the master cylinder and reservoir thoroughly with brake cleaner.

5   Remove as much fluid as possible from the reservoir with a syringe.

6.8 Cut off any retainers that may be holding the disc onto the hub and discard them (there is no need to replace them)

7.3 Lift up the access cover on the cowl

7.6 Master cylinder details

1   *Electrical connector*
2   *Brake line fittings*
3   *Mounting nut (one hidden - located on other side of mounting flange)*

7.12 Place the master cylinder in a vise by its mounting flange, attach the bleed tubes as shown and push the piston with a blunt tool several times to bench bleed the master cylinder

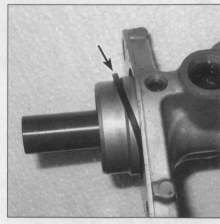

7.17 Replace the vacuum seal on the master cylinder every time the cylinder is removed (typical shown)

7.21 Loosen the fittings on the master cylinder to bleed it when it is installed on the vehicle (typical shown)

6   Unplug the electrical connector for the brake fluid level warning switch **(see illustration)**.

Note: *Some models have additional electrical connectors on the master cylinder and the power brake booster for the ESP system. Simply disconnect these connectors when necessary to remove the master cylinder.*

7   Place rags under the fittings and prepare caps or plastic bags to cover the ends of the lines once they're disconnected.

8   Loosen the fittings at the ends of the brake lines where they enter the master cylinder. To prevent rounding off the corners on the fittings, use a flare-nut wrench.

9   Carefully move the brake lines away from the master cylinder and plug the ends to prevent contamination.

10   Remove the master cylinder mounting nuts. Pull the master cylinder off the studs to remove it. Again, be careful not to spill fluid or bend the brake lines as this is done.

Installation

11   Bench bleed the new master cylinder before installing it. Because it will be necessary to depress the master cylinder piston and, at the same time, control flow from the brake line outlets, it is recommended that the master cylinder be mounted in a vise.

Note: *If the replacement master cylinder is not equipped with a reservoir, remove the reservoir from the old master cylinder and place it on the new one using new seals.*

12   Attach a pair of master cylinder bleeder tubes to the outlet ports of the master cylinder **(see illustration)**.

13   Fill the reservoir with brake fluid of the recommended type (see Chapter 1).

14   Slowly push the pistons into the master cylinder (a large Phillips screwdriver can be used for this) - air will be expelled from the pressure chambers and into the reservoir. Because the tubes are submerged in fluid, air can't be drawn back into the master cylinder when you release the pistons.

15   Repeat the procedure until no more air bubbles are present.

16   Remove the bleed tubes, one at a time, and install plugs in the open ports to prevent fluid leakage and air from entering.

17   Install a new vacuum seal on the master cylinder **(see illustration)**.

Warning: *This must be done every time the master cylinder is removed.*

18   Install the reservoir cover, then install the master cylinder over the studs on the power brake booster and tighten the attaching nuts only finger-tight at this time.

19   Thread the brake line fittings into the master cylinder. Since the master cylinder is still a bit loose, it can be moved slightly in order for the fittings to thread in easily. Do not strip the threads as the fittings are tightened.

20   Tighten the mounting nuts to the torque listed in this Chapter's Specifications and tighten the brake line fittings securely.

21   Fill the master cylinder reservoir with the correct fluid (see Chapter 1), then bleed the brake system as described in Section 9. To bleed the master cylinder on the vehicle, have an assistant pump the brake pedal several times slowly and then hold the pedal to the floor. Loosen the line fittings one at a time to allow air and fluid to escape. Repeat this procedure on both fittings until the fluid is clear of air bubbles **(see illustration)**.

Caution: *Have plenty of rags on hand to catch the fluid - brake fluid will ruin painted surfaces.*

22   The remainder of installation is the reverse of removal. Test the operation of the brake system carefully before placing the vehicle into normal service.

Warning: *Do not operate the vehicle if you are in doubt about the effectiveness of the brake system. On models equipped with ABS, it is possible for air to become trapped in the anti-lock brake system hydraulic control unit. If the pedal continues to feel spongy after repeated bleedings or if the BRAKE or ANTI-LOCK light stays on, have the vehicle towed to a dealer service department or other qualified shop to be bled with the aid of a scan tool.*

8   Brake hoses and lines - inspection and replacement

Inspection

1   About every six months, with the vehicle raised and supported securely on jackstands, the rubber hoses which connect the steel brake lines with the front and rear brake assemblies should be inspected for cracks, chafing of the outer cover, leaks, blisters and other damage. These are important and vulnerable parts of the brake system and inspection should be complete. A light and mirror will be helpful for a thorough check. If a hose exhibits any of the above conditions, replace it with a new one.

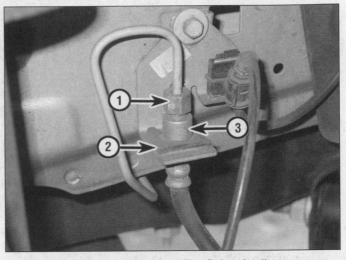

8.3a Front brake hose details

1   *Hose fitting at brake caliper*
2   *Location of steering knuckle bracket
    (actual bracket hidden behind knuckle)*
3   *Brake hose and line fitting secured to frame bracket*

8.3b Front brake hose/line fitting details

1   *Line fitting (loosen with flare-nut wrench)*
2   *U-clip (remove with pliers)*
3   *Female hose fitting (remove from frame bracket)*

Replacement

Front brake hose

2   Loosen the wheel lug nuts, raise the vehicle and support it securely on jackstands. Remove the wheel.

3   At the frame bracket, unscrew the brake line fitting from the hose **(see illustrations)**. Use a flare-nut wrench to prevent rounding off the corners.

4   Remove the U-clip from the female fitting at the frame bracket with a pair of pliers **(see illustration 7.3b)**, then pass the hose through the bracket.

5   At the caliper end of the hose, remove the banjo fitting bolt, then separate the hose from the caliper. Note that there are two copper sealing washers, one on each side of the fitting - they should be replaced with new ones during installation.

6   Remove the fastener from the hose bracket at the steering knuckle and then detach it.

Note: *The replacement hose should come equipped with a steering knuckle bracket like the original one.*

7   To install the hose, attach the hose bracket to the steering knuckle, then connect the fitting to the caliper with the banjo bolt and new sealing washers. Tighten the bolt to the torque listed in this Chapter's Specifications.

8   Push the metal support into the frame bracket and install the U-clip. Make sure the hose isn't twisted between the caliper and the steering knuckle bracket.

9   Connect the brake line fitting, starting the threads by hand, then tighten the fitting securely.

10   Bleed the caliper (see Section 9).

11   Install the wheel and lug nuts, lower the vehicle and tighten the lug nuts to the torque listed in the Chapter 1 Specifications.

Rear brake hose

12   The rear brake hose has a fitting and a bracket that is fastened to the rear frame. Otherwise, refer to the previous steps for rear brake hose replacement **(see illustration)**.

Metal brake lines

13   When replacing brake lines, be sure to use the correct parts. Don't use copper tubing for any brake system components. Purchase genuine steel brake lines from a dealer or auto parts store.

14   Prefabricated brake line, with the tube ends already flared and fittings installed, is available at auto parts stores and dealer parts departments.

15   When installing the new line, make sure it's securely supported in the brackets and has plenty of clearance between moving or hot components.

16   After installation, check the master cylinder fluid level and add fluid as necessary. Bleed the brake system (see Section 9) and test the brakes carefully before driving the vehicle in traffic.

9   Brake hydraulic system - bleeding

Warning: *Wear eye protection when bleeding the brake system. If the fluid comes in contact with your eyes, immediately rinse them with water and seek medical attention.*

Note: *Bleeding the hydraulic system is necessary to remove any air that manages to find its way into the system when it's been opened during removal and installation of a hose, line, caliper or master cylinder.*

1   You'll probably have to bleed the system at all four brakes if air has entered it due to

8.12 The rear brake hose/line fitting bracket

low fluid level, or if the brake lines have been disconnected at the master cylinder.

2   If a brake line was disconnected only at a wheel, then only that caliper must be bled. If a brake line is disconnected at a fitting located between the master cylinder and any of the brakes, that part of the system served by the disconnected line must be bled.

3   Raise the rear of the vehicle about one foot and secure it on jackstands.

Note: *This step is recommended by the manufacturer to ensure that all the air is bled from the ABS hydraulic actuator assembly or the junction block for non-ABS equipped vehicles efficiently.*

4   Remove any residual vacuum from the brake power booster by applying the brake several times with the engine off.

5   Remove the master cylinder reservoir cover and fill the reservoir with brake fluid.

9.8 When bleeding the brakes, a hose is connected to the bleeder valve at the caliper and the other end is submerged in brake fluid. Air will be seen as bubbles in the tube and container. All air must be expelled before moving to the next wheel (typical shown)

10.10a Remove the vacuum hose from the power booster by carefully pulling it off the check valve. Do not remove the check valve to remove the vacuum hose

10.10b On some models (Challenger included) there will be an additional electrical connection to be removed from the brake booster vacuum line before removing the vacuum hose from the booster. Do not remove the check valve from the booster

Reinstall the cover.
Note: *Check the fluid level often during the bleeding operation and add fluid as necessary to prevent the fluid level from falling low enough to allow air bubbles into the master cylinder.*

6   Have an assistant on hand, as well as a supply of new brake fluid, a clear container partially filled with clean brake fluid, a length of tubing to fit over the bleeder valve and a wrench to open and close the bleeder valve.

7   Beginning at the right rear wheel, loosen the bleeder valve slightly, then tighten it to a point where it's snug but can still be loosened quickly and easily.

8   Place one end of the tubing over the bleeder valve and submerge the other end in brake fluid in the container **(see illustration)**.

9   Have your assistant depress the brake pedal slowly, then hold the pedal down firmly.

10   While the pedal is held down, open the bleeder valve just enough to allow a flow of fluid to leave the valve. Watch for air bubbles to exit the submerged end of the tube. When the fluid flow slows after a couple of seconds, close the valve and have your assistant release the pedal.

11   Repeat Steps 9 and 10 until no more air is seen leaving the tube, then tighten the bleeder valve and proceed to the left rear wheel, the right front wheel and the left front wheel, in that order, and perform the same procedure. Be sure to check the fluid in the master cylinder reservoir frequently.

12   Never use old brake fluid. It contains moisture that can boil, rendering the brake system inoperative.

13   Refill the master cylinder with fluid at the end of the operation.

14   Check the operation of the brakes. The pedal should feel solid when depressed, with no sponginess. If necessary, repeat the entire process.

Warning: *Do not operate the vehicle if you are*

in doubt about the effectiveness of the brake system. It's possible for air to become trapped in the ABS hydraulic control unit. If the pedal continues to feel spongy after repeated bleedings or if the BRAKE or ANTI-LOCK (ABS) light stays on, have the vehicle towed to a dealer service department or other qualified shop to be bled with the aid of a scan tool.

---

10   Power brake booster - check, removal and installation

Note: *Vehicle models equipped with an Antilock brake system (ABS) will have one of two different factory systems installed: one is MK25 (the earlier ABS system) and can be identified by a pressure sensor mounted to the bottom of the master cylinder. The other is MK25E (the later ABS system) and the master cylinder appears typical and is not equipped with a pressure sensor on the bottom. Use this information to determine which system is on your vehicle. Knowing which system you have is important regarding part identification and torque specifications.*

Operating check

1   Depress the brake pedal several times with the engine off and make sure there's no change in the pedal reserve distance.

2   Depress the pedal and start the engine. If the pedal goes down slightly, operation is normal.

Airtightness check

3   Start the engine and turn it off after one or two minutes. Depress the brake pedal slowly several times. If the pedal depresses less each time, the booster is airtight.

4   Depress the brake pedal while the engine is running, then stop the engine with the pedal depressed. If there's no change in the pedal

reserve travel after holding the pedal for 30 seconds, the booster is airtight.

Removal

Note: *The power brake booster is not serviceable and is replaced with a new or rebuilt unit.*
Note: *Models that are equipped with ESP may have additional components integrated with power brake booster resulting in electrical connectors attached to it. The new or rebuilt unit must also have all of the original components.*

5   With the engine off, press the brake pedal several times to remove any stored vacuum in the power brake booster.

6   Move the driver's seat to the rearmost position, then disconnect the cable from the negative terminal of the battery (see Chapter 5, Section 1).

7   Remove the windshield wiper motor (see Chapter 12).

8   Remove the brake master cylinder (see Section 7).

9   Remove any electrical connectors attached to the power brake booster, if applicable.

10   Carefully disconnect the vacuum hose from the brake booster check valve **(see illustrations)**.
Warning: *Do not remove the check valve from the brake booster.*

11   Inside the vehicle, remove the brake light switch (see Section 13).

12   Pry the retaining clip off of the brake pedal pin, then disconnect the booster pushrod **(see illustration)**.

13   Remove the booster mounting fasteners. Slide the booster straight out from the firewall until the studs clear the holes and remove it from the engine compartment.

Installation

14   Installation procedures are basically the reverse of removal while noting the fol-

10.12 Power brake booster mounting details:

1   Retaining clip
2   Mounting fasteners

11.4a Turn the adjuster until it's at its shortest setting (the threads disappear into the adjuster)

11.4b Remove the adjuster spring

11.4c Remove the return spring

11.4d Remove the adjuster

11.4e Remove the lower shoe hold-down clip, pin and shoe

lowing points:

a) *Replace the seal that goes between the power booster and the firewall (over the booster's mounting studs).*

b) *Tighten the booster mounting nuts to the torque listed in this Chapter's Specifications.*

c) *Install a new retaining clip securing the booster pushrod to the brake pedal pin.*

d) *Bleed the brake system (see Section 9) and test the operation of the brakes before putting the vehicle into normal service.*

11   Parking brake shoes - replacement

Warning: *Dust created by the brake system is hazardous to your health. Never blow it out with compressed air and don't inhale any of*

it. *An approved filtering mask should be worn when working on the brakes. Do not, under any circumstances, use petroleum-based solvents to clean brake parts. Use brake system cleaner only!*

Warning: *Parking brake shoes must be replaced on both wheels at the same time - never replace the shoes on only one wheel.*

1   Remove the brake disc (see Section 6).
2   Remove the hub and bearing assembly if necessary (see Chapter 10).

Note: *It is possible to perform the shoe replacement procedure without removing the hub and bearing assembly, although working room is limited.*

3   Wash off the brake parts with brake system cleaner.
4   Follow the accompanying illustrations for the brake shoe replacement procedure (**see illustrations 11.4a through 11.4r**). Be sure to stay in order and read the caption under each illustration.

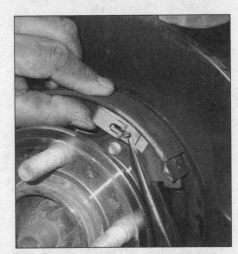

11.4f Remove the upper shoe hold-down clip, pin and shoe

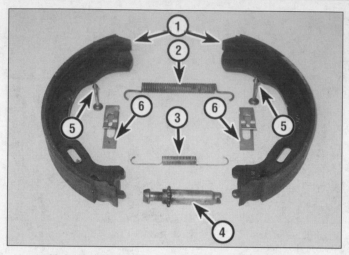

11.4g Parking brake shoes and component details:

1	Brake shoes	4	Adjuster assembly
2	Return spring	5	Hold-down pins
3	Adjuster spring	6	Hold-down clips

11.4h Clean the backing plate thoroughly, then lubricate the brake shoe contact areas on the plate with high-temperature grease

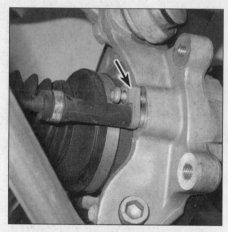

11.4i Loosen the fastener that anchors the parking brake cable to the knuckle and withdraw it enough to disengage the actuator from the end of the parking brake cable

11.4j Remove the actuator from the end of the parking brake cable and inspect it for wear. Place it back into position and then secure the parking brake cable to the knuckle

11.4k Place the upper shoe and hold-down pin in position while engaging the brake actuator

11.4l Install the upper shoe hold-down clip

11.4m Place the lower shoe and hold-down pin in position while engaging the brake actuator

11.4n Install the lower shoe hold-down clip

11.4o Lubricate the adjuster threads

11.4p Install the adjuster

11.4q Install the return spring

11.4r Install the adjuster spring

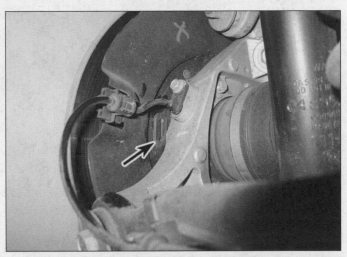

11.6 Remove the rubber plug to access the adjuster for the parking brake shoes through the backing plate

5   Install the brake disc and caliper (see Sections 4 and 5) and then install the wheel. Proceed to the shoe replacement on the opposite rear wheel.

6   With both sides complete and wheels installed, remove the rubber plug from the brake backing plate to access the parking brake adjuster **(see illustration)**.

7   Turn the adjuster star wheel with a brake adjusting tool or screwdriver until the shoes contact the drum the vehicle's wheel can't be turned. Back off the adjuster six notches to achieve the appropriate shoe-to-drum clearance, then replace the rubber plug.

8   Perform the brake shoe adjustment for both wheels.

9   Set the parking brake and confirm proper operation. Re-adjust the shoe clearance if necessary.

10   Lower the vehicle and tighten the wheel lug nuts to the torque listed in the Chapter 1 Specifications.

12   Proportioning valve - removal and installation

Note: *Models equipped with an Anti-lock Brake System (ABS) are not equipped with a proportioning valve or junction block.*
Note: *The proportioning valve is integrated with a junction block and they are replaced as an assembly.*

1   Disconnect the cable from the negative terminal of the battery (see Chapter 5, Section 1).

2   Use a suitable tool to depress and hold the brake pedal down by one inch.
Note: *Holding the pedal down by an inch will minimize fluid loss in the hydraulic circuits while the brake lines are open.*

3   Locate the junction block in the right front portion of the engine compartment. It can be recognized as a solid looking aluminum part

with several hydraulic brake lines attached to it. It is mounted to the chassis with three rubber grommets.

4   Using a flare-nut wrench, disconnect all line fittings to the junction block.

5   Pull the junction block up to remove it from the mounting grommets.

6   Installation is the reverse of removal. Tighten all line fittings securely and bleed the brake system (see Section 9). Test the operation of the brakes before putting the vehicle into normal service.

13   Brake light switch - removal and installation

1   Remove the lower instrument panel insulator below the driver's side knee bolster for access (see Chapter 11).

2   The brake light switch is mounted to a

13.2a Location of the brake light switch (early models)

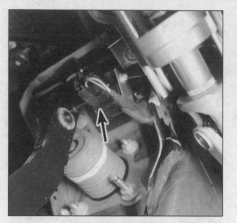

13.2b Location of the brake light switch (later models)

14.4 The drive cable attaches to the motor on the accelerator pedal assembly

14.8 Brake pedal assembly mounting nuts (two hidden)

15.3 The EVP is mounted to a bracket at the rear of the engine

1   Vacuum hose fitting
2   Mounting bolts

bracket attached to the brake pedal assembly **(see illustrations)**.
3   Depress and hold the brake pedal.
4   Rotate the switch about 30 degrees counterclockwise and remove it from its bracket.
5   Disconnect the electrical connector from the brake light switch.
6   To install the switch, pull the plunger on the switch out fully.
7   Depress the brake pedal as far as it will go and then install the switch in its bracket in the opposite manner it was removed.
8   Release the brake pedal and gently pull up on it to make sure it's fully released. The switch is now fully adjusted and installed.

14   Adjustable brake pedal - removal and installation

Note: *Not all models are equipped with adjustable brake and accelerator pedals.*
Note: *The adjustable brake pedal mechanism is driven by a cable from a motor in the accelerator pedal assembly. The motor moves the accelerator pedal up and down directly. Although the brake and accelerator pedal mechanisms*

are two separate assemblies, their movement is synchronized by a drive cable between them; the result is equal upward and downward motion of both pedals simultaneously.
1   Using the pedal adjusting switch, move the pedals to the lowest position, if possible.
2   Move the driver's seat to the rearmost position, then disconnect the cable from the negative terminal of the battery (see Chapter 5, Section 1).
3   Remove the lower instrument panel insulator below the driver's side knee bolster for access (see Chapter 11).
4   Disconnect the drive cable from the motor on the accelerator pedal assembly by gently prying up on the cable mounting flange **(see illustration)**.
5   Remove the brake light switch (see Section 13).
6   Disconnect the booster pushrod from the brake pedal (see Section 10).
7   Disconnect any electrical connectors attached to the brake pedal assembly.
8   Remove the brake pedal assembly mounting fasteners **(see illustration)**.
9   Gently move the power brake booster

forward toward the engine compartment. Be careful not to cause any damage.
10   Remove the brake pedal assembly.
11   Installation is the same as removal noting the following points:
 a) *It may be necessary to remove the drive cable and gear from the brake pedal assembly to install on a replacement assembly.*
 b) *Tighten the brake pedal assembly mounting fasteners to the torque listed in this Chapter's Specifications.*
 c) *A drill or equivalent tool attached to the end of the drive cable can be used to adjust the brake pedal mechanism to its down-most position. This is necessary to synchronize the brake pedal assembly with the accelerator assembly.*
 d) *Test the operation of the adjustable brake and accelerator pedals before driving the vehicle.*

Warning: *If the two mechanisms are not synchronized properly they can become damaged. Make certain that both mechanisms are in the same position (either in their most upward or downward position) before installing the drive cable to the accelerator pedal assembly motor.*

15   Electric Vacuum Pump (EVP) - removal and installation

1   Disconnect the cable from the negative terminal of the battery (see Chapter 5, Section 3).
2   Remove the plastic engine cover by pulling it straight up off of its ballstuds.
3   Locate the EVP at the upper center area of the rear section on the engine **(see illustration)**.
4   Disconnect the electrical connector to the EVP.
5   Disconect the vacuum hose quick-connect fitting from the EVP.
6   Remove the bolts securing the EVP to the bracket.

7   Installation is the reverse of removal.

Notes

Notes

Chapter 10
Suspension and steering

Contents

Specifications

Torque specifications

Ft-lbs (unless otherwise indicated)

Note: *One foot-pound (ft-lb) of torque is equivalent to 12 inch-pounds (in-lbs) of torque. Torque values below approximately 15 ft-lbs are expressed in inch-pounds, because most foot-pound torque wrenches are not accurate at these smaller values.*

Warning: *During installation, always use new fasteners of original part number on any of the suspension or steering systems and related components.*

Front suspension

Hub nut (2014 and earlier 300 and Charger/2013 and earlier Challenger models)	184
Hub-to-knuckle mounting bolts (2015 and later 300 and Charger/2014 and later Challenger models)	
2014 models	50
2015 and later models	74
Shock absorber	
Upper mounting nuts	20
Lower mounting nut/bolt	128
Stabilizer bar	
Link nuts	
2012 and earlier models	95
2013 and 2014 models	
Upper nut	74
Lower nut	96
2015 and later models	
Upper nut	111
Lower nut	100
Bracket bolts	
2014 and earlier models	44
2015 and later models	48

Torque specifications   Ft-lbs (unless otherwise indicated)

Note: *One foot-pound (ft-lb) of torque is equivalent to 12 inch-pounds (in-lbs) of torque. Torque values below approximately 15 ft-lbs are expressed in inch-pounds, because most foot-pound torque wrenches are not accurate at these smaller values.*

Warning: *During installation, always use new fasteners of original part number on any of the suspension or steering systems and related components.*

Front suspension (continued)

Tension strut
Strut-to-subframe bolt/nut	130
Balljoint-to-steering knuckle nut	
Step 1	50
Step 2	Tighten an additional 90 degrees

Upper control arm
Arm-to-shock tower nuts	
2014 and earlier models	55
2015 and later models	66
Balljoint-to-steering knuckle nut	
2014 and earlier models	
Step 1	35
Step 2	Tighten an additional 90 degrees
2015 and later models	
Step 1	35
Step 2	Tighten an additional 95 degrees

Lower control arm
Arm-to-subframe bolt/nut	
2014 and earlier models	130
2015 and later models	136
Balljoint-to-steering knuckle nut	
2014 and earlier models	
Step 1	50
Step 2	Tighten an additional 90 degrees
2015 and later models	
Step 1	27
Step 2	Tighten an additional 180 degrees

Rear suspension

Camber link
Link-to-subframe bolt/nut	
2014 and earlier models	63
2015 and later models	72
Link-to-knuckle bolt/nut	72

Compression link
2014 and earlier models	
Link-to-subframe bolt/nut	63
Link-to-knuckle bolt/nut	60
2015 and later models	
Link-to-subframe bolt/nut	72
Link-to-knuckle bolt/nut	72

Hub/bearing assembly mounting bolts
2014 and earlier models	50
2015 and later models	44

Shock absorber mounting fasteners
2014 and earlier models	
Upper mounting bolts	38
Lower mounting bolt/nut	53
2015 and later models	
Upper mounting bolts	46
Lower mounting bolt/nut	96

Shock absorber damper rod nut	18

Spring link
Link-to-subframe bolt/nut	80
Link-to-knuckle bolt/nut	
2012 and earlier models	102
2013 and later models	105

Stabilizer bar
Link nuts/bolts	45
Bracket bolts	45

Rear suspension (continued)

Tension link
    2014 and earlier models
        Link-to-subframe nut/bolt... 63
        Link-to-knuckle nut/bolt.. 72
    2015 and later models
        Link-to-subframe nut/bolt... 72
        Link-to-knuckle nut/bolt.. 72
Toe link
    Link-to-subframe bolt/nut 80
    Link-to-knuckle bolt/nut
        2011 and earlier models 60
        2012 and later models 70

Steering

Note: *One foot-pound (ft-lb) of torque is equivalent to 12 inch-pounds (in-lbs) of torque. Torque values below approximately 15 ft-lbs are expressed in inch-pounds, because most foot-pound torque wrenches are not accurate at these smaller values.*

Airbag module-to-steering wheel screws
    2010 and earlier models.. 89 in-lbs
    2011 through 2014 models.. 120 in-lbs
Steering wheel bolt
    2010 and earlier models.. 52
    2011 through 2014 models.. 55
    2015 and later models.. 63
Steering column mounting fasteners
    2010 and earlier models
        Upper bolts 22
        Pivot bolt 22
    2011 and later models 20
    Steering column-to-firewall boot nuts.. 62 in-lbs
Steering shaft pinch bolts
    Steering column shaft-to-intermediate shaft bolt............................. 23
    Lower coupler-to-steering gear input shaft
        2010 and earlier models 40
        2011 through 2014 models 33
        2015 and later models 42
Steering gear mounting bolts... 70
Tie-rod-to-steering knuckle (ballstud) nut
    2011 and earlier models... 63
    2012 through 2014 models
        Step 1 26
        Step 2 .... Tighten an additional 90 degrees
    2015 and later models
        Step 1 37
        Step 2 .... Tighten an additional 80 degrees
Wheel lug nuts.. See Chapter 1

1   General information

Front suspension

1   All models are equipped with an independent, long-arm/short-arm, front suspension system with upper and lower control arms, tension struts and shock absorber/coil spring assemblies. A stabilizer bar controls body roll. Each steering knuckle is positioned by a series of balljoints at the ends of the upper and lower control arms and the tension links **(see illustration)**.

Rear suspension

2   All models are equipped with an independent multi-link suspension consisting of shock absorbers, coil springs and several specialized links. Some models incorporate a stabilizer bar to control body roll **(see illustration)**.

Steering

3   The steering system consists of a rack-and-pinion steering gear and two adjustable tie-rods. Power assistance comes from either a hydraulic power steering pump or an electrically assisted steering system, depending on the model and year of the vehicle.

1.1 Front suspension and steering components

1	Stabilizer bar	4	Lower balljoint ballstud (tension strut)	7	Steering gear
2	Lower control arm	5	Tie-rod end	8	Shock absorber/coil spring assembly
3	Lower balljoint (control arm)	6	Tension strut	9	Upper control arm

1.2 Rear suspension components

1	Subframe	3	Shock absorber	5	Compression link
2	Coil spring	4	Toe link	6	Spring link

Precautions

4   Frequently, when working on the suspension or steering system components, you may come across fasteners which seem impossible to loosen. These fasteners on the underside of the vehicle are continually subjected to water, road grime, mud, etc., and can become rusted or "frozen," making them extremely difficult to remove. In order to unscrew these stubborn fasteners without damaging them (or other components), be sure to use lots of penetrating oil and allow it to soak in for a while. Using a wire brush to clean exposed threads will also ease removal of the nut or bolt and prevent damage to the threads. Sometimes a sharp blow with a hammer and punch is effective in breaking the bond between a nut and bolt threads, but care must be taken to prevent the punch from slipping off the fastener and ruining the threads. Heating the stuck fastener and surrounding area with a torch sometimes helps too, but isn't recommended because of the obvious dangers associated with fire. Long breaker bars and extension, or "cheater" pipes will increase leverage, but never use an extension pipe on a ratchet - the ratcheting mechanism could be damaged. Sometimes, turning the nut or bolt in the tightening (clockwise) direction first will help to break it loose. Fasteners that require drastic measures to unscrew should always be replaced with new ones.

5   Since most of the procedures that are dealt with in this Chapter involve jacking up the vehicle and working underneath it, a good pair of jackstands will be needed. A hydraulic floor jack is the preferred type of jack to lift the vehicle, and it can also be used to support certain components during various operations.

Warning: *Never, under any circumstances, rely on a jack to support the vehicle while working on it. Also, whenever any of the suspension or steering fasteners are loosened or removed they must be inspected and, if neces-* sary, be replaced with new ones of the same part number or of original equipment quality and design. Torque specifications must be followed for proper reassembly and component retention. Never attempt to heat or straighten any suspension or steering components. Instead, replace any bent or damaged part with a new one.

---

2   Shock absorber/coil spring assembly (front) - removal and installation

---

Warning: *Always replace shock absorbers or shock absorber/coil spring assemblies in pairs - never replace just one of them.*
Note: *These vehicle models are equipped with shock absorber/coil spring assemblies. It is possible to replace the shocks or springs individually but the unit will have to be disassembled by a qualified repair shop with the proper equipment. This will add considerable cost to the project. You can compare the cost of replacing the complete assemblies yourself to the cost of replacing individual components (with the help of a shop).*
Note: *The SRT models use a unique front shock absorber system utilizing several accelerometers to monitor the shock. This ADS "Active Damping System" works with these accelerometers through the ADCM "Active Damping Control Module" to adjust the shocks depending on the road conditions and angle of the vehicle. The ADS shock absorber is serviced in the same manner as the non-SRT models with the exception of disconnecting the three wire electrical connections from the shock.*

Removal

1   Remove the cap on the shock tower (if equipped) and remove the fasteners that attach the upper end of the shock to the frame **(see illustration)**.

2   Loosen the front wheel lug nuts. Raise the front of the vehicle and support it securely on jackstands. Remove the front wheels.
3   Disconnect the stabilizer bar link from the shock assembly (see Section 3).
4   Remove the fasteners attaching the lower end of the shock absorber to the lower control arm **(see illustration)**.
5   Pull down on the lower control arm and remove the shock absorber toward the rear of the vehicle.
6   Inspect the shock absorber for leaking fluid, dents, cracks and other damage. Inspect the coil spring for chips and cracks which could cause premature failure. Inspect the spring seats for hardness and general deterioration. If any of the components of the assembly are worn or damaged, have the unit serviced by a qualified repair shop or replace it.

Installation

Note: *Raise the lower control arm with a floor jack to simulate normal ride height before tightening lower shock absorber mounting fasteners.*
7   Installation is the reverse of removal. Tighten the mounting fasteners to the torque listed in this Chapter's Specifications.
8   Tighten the wheel lug nuts to the torque listed in the Chapter 1 Specifications.

---

3   Stabilizer bar, bushings and links (front) - removal and installation

---

1   Raise the front of the vehicle and support it securely on jackstands. Remove the front wheels.
2   Remove the engine splash shield from under the vehicle (see Chapter 2A).
Note: *The stabilizer bar link can be separated from the shock absorber assembly by removing the upper nut on the link.*

2.1 Upper shock absorber mounting fasteners

2.4 Lower shock absorber mounting fastener

3.3 Stabilizer bar mounting details

1   Link nuts
2   Link ballstud (hold with wrench while loosening or tightening link nuts)
3   Stabilizer bar bracket fasteners
4   Stabilizer bar bushing

3.4 Stabilizer bar bushing heat shield fasteners

3   Remove the lower nut from the stabilizer bar link and separate it from the bar **(see illustration)**.
4   Remove the heat shields covering the stabilizer bar brackets **(see illustration)**.
5   Remove the stabilizer bar bracket fasteners and remove the brackets.
6   Remove the stabilizer bar. Remove the rubber bushings from the stabilizer bar. Note that the slit in the bushing is facing toward the rear of the vehicle and must be installed in the same direction.
7   Clean the stabilizer bar where the bushings contact it. Inspect all rubber bushings for wear and damage. If any of the rubber parts are cracked, torn or generally deteriorated,

replace them.
8   Check each stabilizer link for signs of excessive wear and replace them as necessary.
9   Installation is the reverse of removal. Be sure to tighten all the fasteners to the torque listed in this Chapter's Specifications.

4   Tension strut - removal and installation

Note: *A special tool, available at most auto parts stores, is necessary to separate the balljoint from the steering knuckle.*

1   Loosen the front wheel lug nuts. Raise the front of the vehicle and support it securely on jackstands. Remove the front wheel.
2   Remove the engine splash shield from under the vehicle (see Chapter 2A).
Caution: *Be careful not to damage the balljoint seal during this next step.*
3   Separate the balljoint on the tension strut from the steering knuckle **(see illustrations)**.
Note: *Grooved mounting bolts may or may not be installed on the vehicle you're working on. They are installed for wheel alignment adjustment when necessary.*
4   Inspect the tension strut-to-subframe mounting bolt for grooves by looking at the

4.3a Tension strut mounting details

1   Balljoint stud nut
2   Tension strut-to-subframe pivot bolt (hold the bolt with wrench and remove the nut - do not turn the bolt)
3   Nut

4.3b To separate the strut mounted balljoint from the steering knuckle, loosen the balljoint stud nut and then install the tool as shown

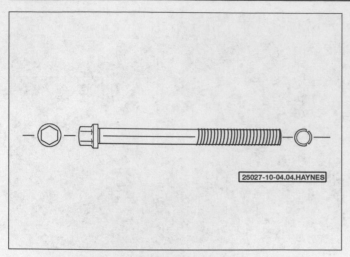

4.4 The tension strut may have a grooved mounting bolt like the one illustrated

4.6 The tension strut bushing's center area has notches to engage grooved mounting bolts. Regular (non-grooved) mounting bolts are placed in between the notches - directly in the center

end where the threads can be seen **(see illustration)**.

5   Hold the tension strut-to-subframe mounting bolt while loosening the nut **(see illustration 4.3a)**.

Caution: *DO NOT turn the tension strut mounting bolt if there are grooves in it or considerable damage will occur.*

Note: *If the bolt has no grooves in it, then it is installed in between the notches in the center of the bushing.*

6   Remove the nut and slowly withdraw the bolt while noting its placement in the notched portion of the bushing **(see illustration)**.

7   Remove the tension strut.

8   Installation is the reverse of removal. Use a new balljoint stud nut. Tighten all fasteners to the torque listed in this Chapter's Specifications.

Note: *Before tightening the tension strut-to-*

subframe nut, raise the outer end of the lower control arm with a floor jack to simulate normal ride height.

9   Install the wheel, remove the jackstands and lower the vehicle.

10   Tighten the wheel lug nuts to the torque listed in the Chapter 1 Specifications.

11   Have the front end alignment checked, and if necessary, adjusted.

---

5   Upper control arm - removal and installation

Removal

Note: *A special tool, available at most auto parts stores, is necessary to separate the balljoint from the steering knuckle.*

1   For right upper control arm removal,

move the engine compartment fuse and relay box aside for access **(see illustration)**. For left upper control arm removal, move the coolant expansion tank aside (see Chapter 3).

2   Remove the upper fasteners for the shock absorber so that it can be moved outward in order to remove the control arm fasteners (see Section 2).

3   Remove the upper control arm mounting nuts from within the engine compartment **(see illustration)**.

4   Loosen the wheel lug nuts, raise the front of the vehicle and support it securely on jackstands placed under the frame rails. Remove the wheel.

5   Remove the ABS wheel speed sensor harness from the frame bracket where the brake hose and brake line fitting are (see Chapter 9).

6   Loosen (but don't remove) the nut on

5.1 Release the tabs on the fuse and relay box to lift it off its mounting brackets and move it aside

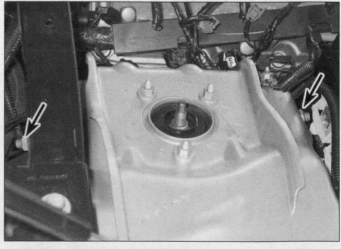

5.3 The mounting nuts for the upper control arm are on each side of the shock tower

the upper balljoint stud, then disconnect the balljoint from the steering knuckle with a balljoint removal tool **(see illustration)**.

7   Carefully secure the steering knuckle aside, then move the shock absorber outward.

8   Remove the control arm mounting flag bolts and pull the upper arm from its frame brackets **(see illustration)**.

Installation

Note: *Raise the lower control arm with a floor jack to simulate normal ride height before tightening the upper control arm mounting nuts at the shock tower.*

9   Installation is the reverse of removal. Use a new balljoint stud nut. Tighten all fasteners to the torque values listed in this Chapter's Specifications.

10   Install the wheel and lug nuts. Lower the vehicle and tighten the lug nuts to the torque listed in the Chapter 1 Specifications.

11   Have the front end alignment checked and, if necessary, adjusted.

5.6 Back off the nut a few turns, then separate the balljoint from the steering knuckle (leaving the nut on the ballstud will prevent the balljoint from separating violently)

5.8 Upper control arm flag bolt fasteners

6   Lower control arm - removal and installation

Removal

Note: *A special tool, available at most auto parts stores, is necessary to separate the balljoint from the lower control arm.*

1   Loosen the wheel lug nuts, raise the front of the vehicle and support it securely on jackstands placed under the frame rails. Remove the wheel.

2   Remove the engine splash shield from under the vehicle (see Chapter 2A).

3   Remove the stabilizer bar brackets to gain access to the lower control arm-to-subframe mounting bolt (see Section 3) **(see illustration)**.

4   Inspect the lower control arm-to-sub-frame mounting bolt for grooves by looking at the end where the threads can be seen **(see illustration 4.4)**.

Note: *Grooved mounting bolts may or may not be installed on the vehicle you're working on. They are installed for wheel alignment adjustment when necessary.*

Warning: *DO NOT turn the lower control arm mounting bolt if there are grooves in it or considerable damage will occur.*

5   Hold the lower control arm-to-subframe mounting bolt while loosening the nut **(see illustration)**.

Warning: *If the bolt has no grooves in it, then it is installed in between the notches in the center of the bushing.*

6   Remove the nut and slowly withdraw the bolt while noting its placement in the notched portion of the bushing **(see illustration)**.

7   Remove the shock absorber lower mounting fasteners **(see illustration 6.3)**.

8   Remove the ABS wheel speed sensor harness from the frame bracket where the brake hose and brake line fitting are (see Chapter 9).

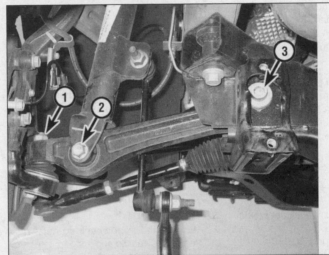

6.3 Lower control arm mounting details:

1   *Balljoint stud nut*
2   *Shock absorber mounting fastener*
3   *Lower control arm-to-subframe pivot bolt (hold the bolt with wrench and remove the nut - do not turn the bolt)*

6.5 Hold the mounting bolt while loosening the nut

6.6 The lower control arm bushing's center area has notches to engage grooved mounting bolts. Regular (non-grooved) mounting bolts are placed in between the notches - directly in the center

6.9 To separate the balljoint from the lower control arm, loosen the balljoint stud nut and install the tool as shown (leaving the nut on the ballstud will prevent the balljoint from separating violently)

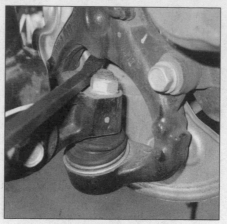

7.7 Pry up on the steering knuckle to check for wear

8.3 Carefully remove the dust cover using a sharp chisel or suitable tool

9   Loosen (but don't remove) the nut on the balljoint stud, then disconnect the balljoint from the lower control arm with a balljoint removal tool **(see illustration)**.

10   Remove the balljoint stud nut and pry the knuckle down to remove the balljoint stud from the lower control arm. Move the steering knuckle away from the control arm until they are separated.

11   Remove the lower control arm by sliding it out of the subframe bracket.

Installation

Note: *Raise the lower control arm with a floor jack to simulate normal ride height before tightening the lower control arm mounting nut at the subframe and the lower shock absorber mounting fasteners.*

12   Installation is the reverse of removal. Use a new balljoint stud nut. Be sure to tighten all fasteners to the torque values listed in this Chapter's Specifications.

13   Install the wheel and lug nuts. Lower the vehicle and tighten the lug nuts to the torque listed in the Chapter 1 Specifications.

8.4 Remove the hub nut and slide the hub off the steering knuckle spindle

14   Have the front end alignment checked and, if necessary, adjusted.

---

7   Balljoints - check and replacement

1   Inspect the control arm balljoints for looseness anytime either of them is separated from the steering knuckle. See if you can turn the ballstud in its socket with your fingers. If the balljoint is loose, or if the ballstud can be turned, replace the balljoint. You can also check the balljoints with the suspension assembled as follows.

Upper balljoint and tension strut balljoint

2   Loosen the front wheel lug nuts, raise the front of the vehicle and support it securely on jackstands, then remove the wheel. Raise the lower control arm with a floor jack to simulate normal ride height.

3   Using a large prybar inserted between the upper control arm or tension strut and the steering knuckle, pry the upper control arm or tension strut away from the knuckle. The manufacturer specifies up to 0.059-inch movement is allowed; a dial indicator can be used to check for play. If you are unable to accurately measure balljoint play, have the balljoint checked at an automotive repair shop.

4   If replacement is indicated, replace the upper control arm or tension strut (see Section 5); the balljoint cannot be replaced separately.

Lower balljoint

5   Raise the vehicle and support it securely on jackstands.

6   Place a floor jack under the lower control arm, near the outer end, and raise it until it is supporting the vehicle's weight.

7   Now, insert the prybar against the lower control arm or tension strut and pry the steering knuckle up **(see illustration)**. The manu-

facturer specifies up to 0.059-inch movement is allowed; a dial indicator can be used to check for play. If you are unable to accurately measure balljoint play, have the balljoint checked at an automotive repair shop.

8   If replacement is indicated, remove the steering knuckle or tension strut (see Section 4 or 9); the balljoint is press fit in the steering knuckle and tension strut, which necessitates the use of a special press tool and receiver cup to remove and install the balljoint. Equipment rental yards and some auto parts stores have these tools available for rent. If you don't have access to this tool, take the vehicle (or the steering knuckle) to an automotive machine shop or other qualified repair facility to have the balljoint replaced.

---

8   Hub and bearing assembly (front) - removal and installation

Warning: *The dust created by the brake system is harmful to your health. Never blow it out with compressed air and don't inhale any of it. Do not, under any circumstances, use petroleum-based solvents to clean brake parts. Use brake system cleaner only.*

Note: *The hub and bearing assembly is sealed-for-life. If any part is worn or damaged, it must be replaced as a unit.*

2014 and earlier 300 and Charger/2013 and earlier Challenger models

Warning: *A new hub nut will be required for this procedure. Don't reuse the old hub nut.*

Removal

1   Loosen the wheel lug nuts, raise the front of the vehicle and support it securely on jackstands placed under the frame rails. Remove the wheel.

2   Remove the brake disc (see Chapter 9).

3   Remove the dust cap **(see illustration)**.

4   Remove the hub nut **(see illustration)**.

5   Remove the hub from the steering knuckle spindle.

Installation

6   Installation is the reverse of removal, noting the following points:

  a) *Tighten the NEW hub nut to the torque listed in this Chapter's Specifications.*
  b) *Tighten the brake caliper mounting bracket and brake caliper mounting bolts to the torque listed in the Chapter 9 Specifications.*
  c) *Install the wheel, lower the vehicle and tighten the lug nuts to the torque listed in the Chapter 1 Specifications.*

2015 and later 300 and Charger/2014 and later Challenger models

Removal

7   Loosen the wheel lug nuts, raise the front of the vehicle and support it securely on jackstands placed under the frame rails. Remove the wheel.

8   Remove the brake disc (see Chapter 9).

9   Remove the brake hose bracket bolt from the steering knuckle to gain access to the hub assembly bolts.

10   Remove the three bolts securing the hub assembly to the steering knuckle (**see illustration**).

11   Slide hub assembly from the steering knuckle.

Note: *The hub and bearing assembly can sometimes be very stubborn to come off of the steering knuckle. Spray water liberally onto the mounting area and bolt holes of the hub assembly and wait a few minutes. If needed, lightly tap the edge of the mounting service (not hard enough to damage it) to help force the water to seep into the seam. After a few minutes it will normally come right off.*

12   Installation is the reverse of removal. Tighten the hub and bearing bolts to the torque listed in this Chapter's Specifications.

---

9   Steering knuckle - removal and installation

Warning: *The dust created by the brake system is harmful to your health. Never blow it out with compressed air and don't inhale any of it. Do not, under any circumstances, use petroleum-based solvents to clean brake parts. Use brake system cleaner only.*

1   Loosen the wheel lug nuts, raise the front of the vehicle and support it securely on jackstands. Remove the wheel.

2   Remove the brake disc, the wheel speed sensor and the small bracket that holds the brake hose and ABS wheel speed sensor harness to the knuckle (see Chapter 9).

3   Disconnect the tie-rod end from the steering knuckle (see Section 18).

4   Separate the upper control arm from the steering knuckle (see Section 5).

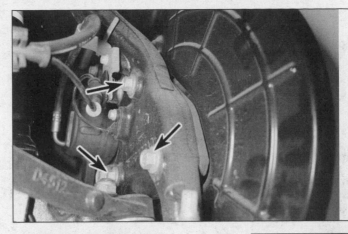

8.10 Location of the three bolts securing the hub to the steering knuckle

5   Separate the tension strut from the steering knuckle (see Section 4).

6   Separate the lower control arm from the steering knuckle (see Section 6).

7   Carefully inspect the steering knuckle for cracks, especially around the steering arm and spindle mounting area. Also inspect the balljoint stud holes. If they're elongated, or if you find any cracks in the knuckle, replace the steering knuckle.

Note: *Balljoint removal will require special tools which are available at most auto parts stores. The knuckle can also be taken to a qualified repair facility for removal.*

8   Remove the hub or hub and bearing assembly, brake dust shield or lower balljoint as necessary for replacement.

9   Installation is the reverse of removal. Be sure to tighten all suspension fasteners to the torque listed in this Chapter's Specifications. Refer to the torque values listed in the Chapter 9 Specifications for brake related fasteners.

10   Tighten the wheel lug nuts to the torque listed in the Chapter 1 Specifications.

---

10   Shock absorber (rear) - removal and installation

Note: *Some models may have optional automatic load-leveling shock absorbers. They can be identified by their considerably larger diameters. Removal and installation of these shocks requires lowering the rear subframe one side at a time. The job is a lot more difficult than changing typical shock absorbers.*

Note: *SRT models use a unique rear shock absorber system utilizing several accelerometers to monitor the shock. This Active Damping System (ADS) works with these accelerometers through the Active Damping Control Module (ADCM) to adjust the shocks depending on the road conditions and angle of the vehicle. The ADS shock absorber is serviced in the same manner as the non-SRT models with the exception of disconnecting the three wire electrical connection from the shock.*

1   Loosen the wheel lug nuts, raise the rear of the vehicle and support it securely on jackstands placed under the frame rails. Remove the wheel.

10.2 Use a wood block and a floor jack to support or raise the spring link

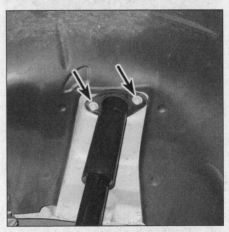

10.3 Rear shock absorber upper mounting fasteners

Regular shock absorbers

2   Support the spring link with a floor jack to keep the suspension in place when the shock is removed (**see illustration**).

3   Remove the fasteners that attach the upper end of the shock absorber to the frame (**see illustration**). If the bolts won't loosen because of rust, apply some penetrating oil and allow it to soak in for awhile.

10.4 Rear shock absorber lower mounting fasteners

10.8 Use a wood block and a floor jack to support, lower and raise the rear subframe and differential

10.11a Location of the subframe mounting bolts on the right side

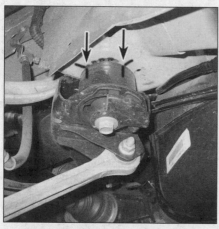

10.11b Place reference marks on the subframe and the vehicle chassis so that the subframe can be placed back into the same position when reinstalled

4   Remove the fasteners that attach the lower end of the shock to the spring link **(see illustration)**. Again, if the nut is frozen, apply some penetrating oil, wait awhile and try again.

5   Installation is the reverse of removal. Tighten the mounting fasteners to the torque listed in this Chapter's Specifications.

Note: *Raise the spring link with a floor jack to simulate normal ride height before tightening the lower shock absorber mounting fasteners (see illustration 10.2).*

Load-leveling shock absorbers

Note: *There is no need to disconnect the exhaust pipes from the resonator(s) or muffler(s); simply lower and support them.*

6   When servicing the left shock absorber, remove the rubber hose between the fuel filler neck and the gas tank, and then move the filler neck aside (see Chapter 4).

7   On vehicles with dual exhaust, or when

servicing the right shock absorber, lower the exhaust by removing it from its hangers and support it with jackstand or another jack (see Chapter 4).

8   Place a floor jack under the rear differential and use it to support the suspension and subframe **(see illustration)**.

9   Support the spring link with a floor jack and wood block **(see illustration 10.2)**. Remove the fasteners that attach the upper end of the shock absorber to the frame **(see illustration 10.3)**. If the bolt won't loosen because of rust, apply some penetrating oil and allow it to soak in for awhile. Slowly lower the floor jack supporting the spring link.

10   Remove the fasteners that attach the lower end of the shock to the spring link **(see illustration 10.4)**. Again, if the nut is frozen, apply some penetrating oil, wait awhile and try again.

Note: *Only service one side at a time to avoid changing the rear wheel alignment.*

11   Mark the subframe in relation to the chassis, then remove the front and rear subframe fasteners on the side you're working on **(see illustrations)**.

12   SLOWLY lower the differential with the floor jack just enough to remove the shock absorber.

Caution: *The other side of the subframe and suspension will flex to accommodate lowering the side of the suspension you're working on. Do not lower the subframe any more than necessary to remove the shock or damage to other suspension components will occur.*

13   Installation is the reverse of removal noting the following points:

a) *Install the lower shock absorber fasteners before raising the subframe into position, but do not tighten them yet.*

b) *When raising the subframe into position, be sure that the coil spring and insulators are seated into their original positions and that the subframe is matched to the reference marks made previously.*

c) *The subframe mounting bolts have two different lengths; the longer bolts are installed to the rear of the subframe while the shorter bolts are installed to the front.*

d) *Make sure to reinstall the brake, fuel tank, driveline and exhaust system components.*

e) *Raise the spring link with a floor jack to simulate normal ride height, then tighten the lower shock absorber mounting fasteners to the torque listed in this Chapter's Specifications (see illustration 10.2).*

f) *Be sure to tighten all suspension fasteners to the torque listed in this Chapter's Specifications.*

g) *Have the wheel alignment checked and, if necessary, adjusted.*

11   Stabilizer bar, bushings and links (rear) - removal and installation

Warning: *The removal of the rear stabilizer bar involves lowering the subframe and drivetrain components. This job is very involved and can be considered hazardous even with the proper equipment. Use extreme caution throughout the following procedures to avoid injury.*

Note: *A transmission jack or equivalent is necessary for the following procedure.*

Note: *It is recommended that additional help from an assistant be used during this procedure.*

1   Disconnect the cable from the negative battery terminal (see Chapter 5, Section 1).

2   Loosen the wheel lug nuts, raise the rear of the vehicle and support it securely on jackstands placed under the frame rails. Remove the wheel.

3   Remove the rear portion of the exhaust system (see Chapter 4).

4   Disconnect the driveshaft from the rear differential (see Chapter 8).

11.5 The parking brake equalizer is just above the rear differential

11.7 Location of the rear subframe mounting fasteners

5   Separate and detach the parking brake equalizer from the front parking brake cable **(see illustration)**.
6   Remove the bracket that holds the parking brake cable to the subframe.
Warning: *If the subframe is not returned to its original position, the rear wheel alignment will be adversely affected.*
7   Mark the subframe in relation to the chassis at all four mounting locations **(see illustration 10.11b and the accompanying illustration)**.
8   Support the spring link with a floor jack and wood block **(see illustration 10.2)**, then remove the shock absorber lower mounting fasteners (see Section 10). Slowly lower the floor jack supporting the spring link until the coil spring is extended.
9   Remove the brake calipers and secure them so that the brake hoses will not be affected when the subframe is lowered (see Chapter 9).
10   Disconnect and remove the ABS wheel speed harnesses from the chassis (see Chapter 9).
Note: *Do not secure the rear stabilizer bar along with the differential when securing it to the transmission jack.*
11   Place a transmission jack (or equivalent) under the differential and attach it securely.
12   Remove the fuel filler neck (see Chapter 4).
13   With the transmission jack in place and securing the differential and subframe, remove all four subframe bolts.
14   SLOWLY lower the subframe enough to separate the driveshaft from the differential.
15   Continue lowering the subframe until there is adequate access to the stabilizer bar brackets.
Warning: *Do not lower the subframe any more than necessary to remove the stabilizer bar.*
16   Stabilize the subframe and suspension assembly with additional jackstands or equivalent.
17   Remove   the   stabilizer   bar

bracket fasteners and remove the brackets. Remove the bushings from the bar while noting the direction of the slits.
18   Clean the stabilizer bar where the bushings contact it. Inspect the bushings for wear and damage. If they are cracked, torn or generally deteriorated, replace them.
19   Remove the stabilizer bar links and inspect them for wear or damage; replace them if necessary **(see illustration)**.
20   Installation is the reverse of removal noting the following points:

a) *Install the lower shock absorber fasteners before raising the subframe into position, but do not tighten them yet.*
b) *When raising the subframe into position, be sure that the coil spring and insulators are seated into their original positions and that the subframe is matched to the reference marks made previously. Install the driveshaft to the rear differential as the subframe is being raised. Guide the top of the shock absorbers to the chassis as the subframe nears its installed position.*
c) *The subframe mounting bolts are two different lengths; the longer bolts are*

installed to the rear of the subframe while the shorter bolts are installed to the front.
d) *Make sure to reinstall the brake, fuel tank, driveline and exhaust system components.*
e) *Raise the spring link with a floor jack to simulate normal ride height, then tighten the lower shock absorber and stabilizer bar link mounting fasteners (see illustration 10.2).*
f) *Be sure to tighten all suspension fasteners to the torque listed in this Chapter's Specifications.*
g) *Have the wheel alignment checked and, if necessary, adjusted.*

12   Suspension links (rear) - removal and installation

Note: *The subframe must be lowered one side at a time to remove the camber and tension links or the left toe link; a good floor jack, jackstands and wood blocks will be necessary.*

11.19 Rear stabilizer bar link mounting fasteners - right side shown

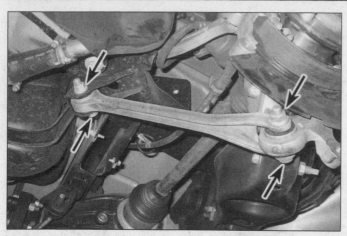

12.2 Compression link mounting fasteners

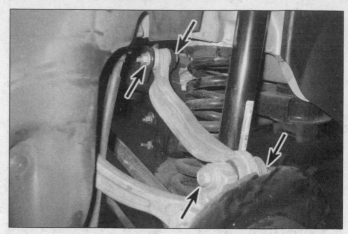

12.7 Camber link mounting fasteners

Compression link

1   Raise the rear of the vehicle and support it securely on jackstands.

2   Remove the compression link mounting fasteners from the rear knuckle **(see illustration)**.

3   Remove the compression link mounting fasteners from the rear subframe and remove the link.

4   Inspect all rubber bushings for wear and damage. If any of the rubber parts are cracked, torn or generally deteriorated, replace them.

Warning: *Raise the spring link with a floor jack to simulate normal ride height before tightening the toe link mounting fasteners (see illustration 10.2).*

5   Installation is the reverse of removal. Tighten the mounting fasteners to the torque listed in this Chapter's Specifications.

Camber link

6   Follow the procedures in Section 10 for the removal of the load-leveling type shock absorbers for the side of the vehicle you're working on. Remove the shock absorber in this manner regardless if it is the regular or load leveling type.

7   Remove the camber link mounting fas-teners at the knuckle **(see illustration)**.

8   Remove the camber link mounting fasteners at the subframe, note the position of the link and then remove it.

Note: *The thicker end of the link is attached to the subframe with the bend giving space for the coil spring.*

9   Inspect all rubber bushings for wear and damage. If any of the rubber parts are cracked, torn or generally deteriorated, replace them.

Note: *Raise the spring link with a floor jack to simulate normal ride height before tightening the toe link mounting fasteners (see illustration 10.2).*

10   For installation, refer to the same procedure in Section 10 for load-leveling shock absorbers.

Tension link

11   Follow the procedures in Section 10 for the removal of the load-leveling type shock absorbers for the side of the vehicle you're working on. Remove the shock absorber in this manner regardless if it is the regular or load leveling type.

12   Remove the tension link mounting fasteners at the knuckle **(see illustration)**.

13   Remove the tension link mounting fasten-ers at the subframe, note the position of the link and then remove it.

14   Inspect all rubber bushings for wear and damage. If any of the rubber parts are cracked, torn or generally deteriorated, replace them.

Note: *Raise the spring link with a floor jack to simulate normal ride height before tightening the toe link mounting fasteners (see illustration 10.2).*

15   For installation, refer to the same procedure in Section 10 for load-leveling shock absorbers.

Toe link

Right toe link

16   Loosen the wheel lug nuts, raise the rear of the vehicle and support it securely on jackstands placed under the frame rails. Remove the wheel.

17   Remove the ABS harness from the toe link, if equipped.

18   Mark the relationship of the toe adjusting cam bolt to the subframe **(see illustration)**.

Note: *Lower the exhaust if you need more room for access. There is no need to detach the pipes from the muffler or resonator - simply lower it and rest it on jackstands.*

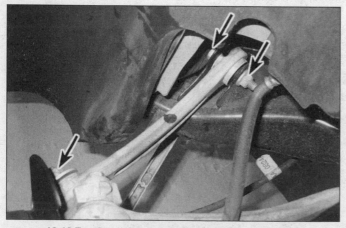

12.12 Tension link mounting fasteners - one hidden

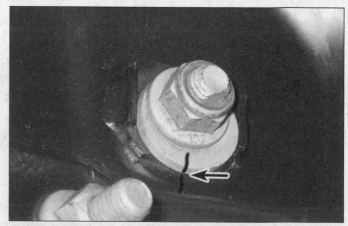

12.18 Mark the relationship of the toe adjusting cam bolt to the subframe to preserve the rear wheel alignment when the fasteners are reinstalled

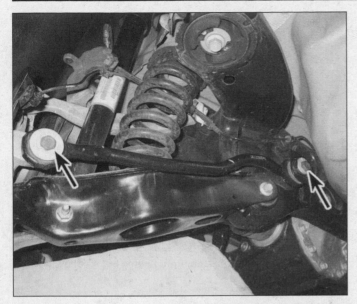

12.19 Toe link mounting fasteners

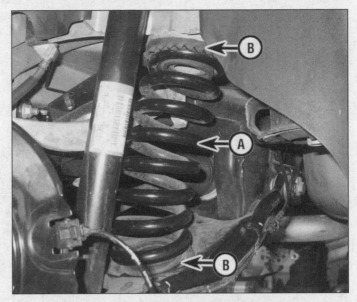

13.2 The coil spring (A) and the insulators (B)

19   Remove the toe link mounting fasteners from the rear knuckle **(see illustration)**.

20   Hold the cam bolt while loosening the toe link mounting nut from the rear subframe. Note the position of the link and then remove it.

21   Inspect all rubber bushings for wear and damage. If any of the rubber parts are cracked, torn or generally deteriorated, replace them. **Note:** *Raise the spring link with a floor jack to simulate normal ride height before tightening the toe link mounting fasteners **(see illustration 10.2)**.*

22   Installation is the reverse of removal. Tighten the mounting fasteners to the torque listed in this Chapter's Specifications.

Left toe link

23   Follow the procedures in Section 10 for the removal of the load-leveling type shock absorbers for the side of the vehicle you're working on. Remove the shock absorber in this manner regardless if it is the regular or load leveling type.

24   Remove the ABS harness from the toe link, if equipped.

25   Mark the relationship of the toe adjusting cam bolt to the subframe **(see illustration 12.18)**.

26   Remove the toe link mounting fasteners at the knuckle **(see illustration 12.19)**.

27   Hold the cam bolt while loosening the toe link mounting nut from the rear subframe. Note the position of the link and then remove it.

28   Inspect all rubber bushings for wear and damage. If any of the rubber parts are cracked, torn or generally deteriorated, replace them. **Note:** *Raise the spring link with a floor jack to simulate normal ride height before tightening the toe link mounting fasteners **(see illustration 10.2)**.*

29   For installation, refer to the same procedure in Section 10 for load-leveling shock absorbers.

Spring link

30   The bushing in the spring link, where it connects to the rear knuckle, has a metal sleeve installed in it. The metal sleeve must be withdrawn from the link before it can be separated from the knuckle. Special tools are required to remove the sleeve from the link's bushing. However, the spring link and knuckle can be removed together and taken to a dealer service department or qualified repair shop for separation. Special tools are also necessary for the link and knuckle to be rejoined with a new metal sleeve installed. Refer to Section 14 for removal of the spring link with the rear knuckle.

13   Coil spring (rear) - removal and installation

Note: *The subframe must be lowered one side at a time to remove the coil spring; a good floor jack, jackstands and wood blocks will be necessary.*

1   Follow the procedures in Section 10 for the removal of the load-leveling type shock absorbers for the side of the vehicle you're working on. Remove the shock absorber in this manner regardless if it is the regular or load leveling type. Continue to lower the subframe until the coil spring is extended.

2   Remove the coil spring and the upper and lower insulators **(see illustration)**.

3   Inspect insulators for wear and damage. If the rubber is cracked, torn or generally deteriorated, replace them. **Note:** *Raise the spring link with a floor jack to*

*simulate normal ride height before tightening the toe link mounting fasteners **(see illustration 10.2)**.*

4   For installation, refer to the same procedure in Section 10 for load-leveling shock absorbers.

14   Knuckle (rear) - removal and installation

Note: *The bushing in the spring link, where it connects to the rear knuckle, has a metal sleeve installed in it. The metal sleeve must be withdrawn from the link before it can be separated from the knuckle. Special tools are required to remove the sleeve from the link's bushing. However, the spring link and knuckle can be removed together and taken to a dealer service department or qualified repair shop for separation. Special tools are also necessary for the link and knuckle to be rejoined with a new metal sleeve installed.*

Note: *The subframe must be lowered one side at a time to remove the spring link and knuckle; a good floor jack, jackstands and wood blocks will be necessary.*

1   Loosen the wheel lug nuts, raise the rear of the vehicle and support it securely on jackstands placed under the frame rails. Remove the wheel.

2   Remove the ABS wheel speed sensor and unclip the harness from the brake backing plate (see Chapter 9).

3   Remove the hub and bearing assembly (see Section 15).

4   Remove the rear parking brake shoes and disconnect the parking brake cable from the knuckle (see Chapter 9).

5   Follow the procedures in Section 13 for the removal of the coil spring.

14.6 Spring link-to-subframe pivot bolt

15.4 Hub and bearing assembly mounting bolts - one hidden

6   Detach the spring link-to-subframe mounting fasteners **(see illustration)**.
7   Remove all links connected to the knuckle except the spring link and carefully remove the knuckle (see Section 12).
8   Installation is the reverse of removal.
Note: *Raise the spring link with a floor jack to simulate normal ride height before tightening the toe link mounting fasteners to the torque listed in this Chapter's Specifications (see illustration 10.2).*

15   Hub and bearing assembly (rear) - removal and installation

1   Loosen the driveaxle/hub nut (see Chapter 8).
2   Loosen the wheel lug nuts, raise the rear of the vehicle and support it securely on jackstands placed under the frame rails. Remove the wheel.
3   Remove the brake caliper and disc. Also remove the wheel speed sensor from the steering knuckle (see Chapter 9).
4   Remove the hub mounting bolts from the back of the steering knuckle enough to release the hub assembly **(see illustration)**.
Note: *The hub bolts can stay in the knuckle to support the brake backing plate.*
Note: *Be careful not to overextend the inner CV joint. Once the hub has been removed, support the outer end of the driveaxle with a length of wire or rope.*
5   Remove the driveaxle/hub nut and push the driveaxle through the hub splines as the hub and bearing assembly is removed. If the driveaxle sticks in the hub, it's best to push it out with a puller **(see illustration)**.
6   Installation is the reverse of removal, noting the following points:

  a) *Tighten the hub mounting bolts to the torque listed in this Chapter's Specifications.*
  b) *Tighten the brake caliper mounting bracket bolts, caliper mounting bolts and the wheel speed sensor bolt to the torque listed in the Chapter 9 Specifications.*
  c) *Tighten the driveaxle/hub nut to the torque listed in the Chapter 8 Specifications.*
  d) *Tighten the wheel lug nuts to the torque listed in the Chapter 1 Specifications.*

16   Steering wheel - removal and installation

Warning: *These models are equipped with airbags. Always disable the airbag system whenever working in the vicinity of any airbag system component to avoid the possibility of accidental airbag deployment, which could cause personal injury (see Chapter 12).*

Removal

Note: *A new steering wheel mounting bolt is necessary for installation.*
Warning: *Do NOT turn the steering shaft during or after steering wheel removal. If the shaft is turned while the steering wheel is removed, a mechanism known as the clockspring can be damaged. The clockspring, which maintains a continuous electrical circuit between the wiring harness and the airbag module, consists of a flat, ribbon-like electrically conductive tape which winds and unwinds as the steering wheel is turned.*
1   Park the vehicle with the front wheels in the straight-ahead position.
2   Disconnect the cable from the negative battery terminal, (see Chapter 5, Section 1).
Caution: *Carry the airbag module with the trim cover (upholstered side) facing away from you, and set the airbag module in a safe location with the trim cover facing up.*
3   On 2014 and earlier models, remove the two airbag module retaining screws **(see illustration)** and lift off the airbag module. On 2015 and later models, Insert a narrow blunt tool such as a punch or Allen wrench into the hole in each side of the steering wheel to dis-

15.5 If the driveaxle is stuck tightly to the hub, install a puller to remove it

16.3a Remove the airbag module retaining screws (2014 and earlier models)

16.3b On 2015 and later models, insert a punch or Allen wrench into the holes in the sides of the steering wheel and push the retainer spring inward (one side at a time) to free the airbag

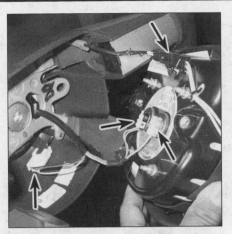

16.4 Lift the airbag off the steering wheel and unplug the electrical connectors for the airbag, horn, etc.

16.5a Remove the steering wheel bolt and discard it

16.5b After removing the bolt, note the keyed area of the steering wheel hub and its relationship of the wheel to the shaft

16.6 Clockspring mounting details

1   *Mounting screws*
2   *Alignment arrows (pointing up)*
3   *Tab that engages with steering wheel*

engage the airbag retaining spring from it corresponding hooks **(see illustration)**.

4   Unplug the electrical connectors for the airbag, horn and other components (if equipped) **(see illustration)**.

Note: *The steering wheel hub is "keyed" with one larger spline to fit the steering shaft in one position (see illustration). Slide the steering wheel off the shaft.*

5   Remove the steering wheel bolt and note the relationship of the steering wheel hub to the steering shaft before removing the steering wheel **(see illustrations)**.

6   **2010 and earlier models only:** If it is necessary to remove the clockspring, remove the mounting screws and lift the clockspring from the steering column **(see illustration)**.

Note: *On 2011 and later models there is a self-centering/locking tab that will engage when the clockspring is removed (no need to*

tape the clock spring). Also, on these models the clockspring is part of the SCCM See Chapter 12, Section 7).*

Note: *Use tape to keep the clockspring in the centered position.*

Note: *On 2011 and later models the clockspring has a self-locking mechanism (no need to tape the clockspring). Also, on these models the clockspring is part of the SCCM (see Chapter 12, Section 7).*

Installation

2010 and earlier models

7   When installing the clockspring, make absolutely sure that the airbag clockspring is centered with the arrows on the clockspring rotor and case lined up **(see illustration 16.6)**. This shouldn't be a problem as long as you have not turned the steering shaft while the

wheel was removed. If for some reason the shaft was turned, center the clockspring as follows:

a) *Rotate the clockspring clockwise until it stops (don't apply too much force, though).*
b) *Rotate the clockspring counterclockwise about 2-1/2 turns until the arrows on the clockspring rotor and case line up.*

2011 and later models

8   If the steering column has not moved while the steering wheel was off, the clockspring should still be cenetered. To confirm this, look at the inspection window on the face of the clockspring rotor; a black square on the clockspring ribbon should be visible. If it is not, the Steering Column Control Module (SCCM) must be replaced (see Chapter 12, Section 7).

17.6 The steering column shroud mounting fasteners

17.7 Lower steering column mounting fasteners

All models

9   Installation is the reverse of removal, noting the following points:

a) *Make sure the airbag clockspring is centered before installing the steering wheel (see Steps 7 and 8).*

b) *When installing the steering wheel, align the tab on the clockspring with the steering wheel hub.*

c) *Install a NEW steering wheel bolt and tighten it to the torque listed in this Chapter's Specifications.*

d) *Install the airbag module on the steering wheel. On 2014 and earlier models, tighten the mounting screws to the torque listed in this Chapter's Specifications. On 2015 and later models, push the airbag module into place until the retaining spring engages its hooks on the steering wheel hub.*

e) *Enable the airbag system (see Chapter 12).*

17   Steering column - removal and installation

Warning: *These models are equipped with airbags. Always disable the airbag system whenever working in the vicinity of any airbag system component to avoid the possibility of accidental airbag deployment, which could cause personal injury (see Chapter 12).*

Note: *The manufacturer recommends using new mounting fasteners for installation.*

Removal

1   Park the vehicle with the wheels pointing straight ahead. Extend the column all the way out. Disconnect the cable from the negative battery terminal (see Chapter 5, Section 1). Wait at least two minutes before proceeding (to allow the backup power supply for the airbag system to become depleted).

Note: *On models equipped with electric telescoping columns, place the column in the mid-tilt position.*

2   Remove the steering wheel (see Section 16), then turn the ignition key to the Lock position to prevent the steering shaft from turning.

Warning: *If this is not done, the airbag clockspring could be damaged.*

3   **2010 and earlier models:** Loosen (but don't remove) one of the clockspring mounting fasteners to keep the clockspring centered **(see illustration 16.6)**.

4   Remove the knee bolster and reinforcement panel underneath it (see Chapter 11).

5   Remove the Steering Column Control Module (SCCM) (see Chapter 12).

6   Remove the mounting screws for the plastic shroud at the end of the column and detach the tilt adjuster handle, if equipped **(see illustration)**.

7   Remove the lower steering column mounting fasteners at the firewall **(see illustration)**.

8   On 2010 and earlier models, remove the instrument cluster (see Chapter 12).

9   Detach the electrical wiring harness for the column **(see illustration)**.

10   Raise the front of the vehicle and support it securely on jackstands.

11   From under the vehicle, mark the steering shaft to the lower coupling shaft and remove the pinch bolt **(see illustration)**. On 2011 and later models, also detach the lower end of the intermediate shaft from the steering gear (see Sections 20, 21, **illustration 20.3**). Separate the shafts.

12   On vehicles equipped with electric telescoping columns, remove the brace attached to dash support and the column (inside the vehicle and under the dash on the left side of the column).

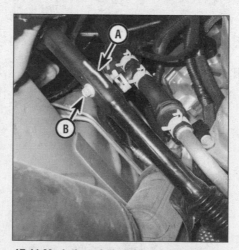

17.9 Remove the wiring harness by releasing the tabs that hold the connector to the column

17.11 Mark the relationship of the steering shaft to the lower coupling shaft (A) and remove the coupler pinch bolt (B)

17.13 Column pivot bolt (2010 and earlier models)

17.14 Steering column upper mounting bolts (2010 and earlier models)

17.15 Steering column mounting bolts (A) and nuts (B) (2011 and later models)

2010 and earlier models

13   Remove the pivot bolt on the column **(see illustration)**.

14   Remove the upper steering column mounting bolts and remove the column **(see illustration)**.

2011 and later models

15   Remove the column forward mounting bolts, followed by the rear mounting nuts **(see illustration)**.

Installation

Note: *Make sure the lower boot is properly seated on the column before placing it into position through the firewall.*

16   Guide the steering column into position and install the top and pivot mounting fasteners, but don't tighten them yet.

17   Install the brace removed in Step 12 if equipped.

18   The remainder of installation is the reverse of removal.

19   Tighten the mounting fasteners to the torque listed in this Chapter's Specifications.

20   On models with electrically assisted power steering, perform the steering gear calibration procedure (see Section 21).

---

18   Tie-rod ends - removal and installation

Removal

1   Loosen the wheel lug nuts, raise the front of the vehicle and support it securely on jackstands. Apply the parking brake. Remove the wheel.

2   Loosen the tie-rod end jam nut **(see illustration)**.

3   Mark the relationship of the tie-rod end to the threaded portion of the tie-rod. This will ensure the toe-in setting is restored when reassembled **(see illustration)**.

4   Loosen (but don't remove) the nut on the tie-rod end ballstud and disconnect the tie-rod end from the steering knuckle arm with a puller **(see illustration)**.

Note: *If the ballstud turns while loosening the nut, hold the ballstud with a wrench while removing the nut.*

5   If you're replacing the tie-rod end, unscrew the tie-rod end from the tie-rod, then thread the new tie-rod end onto the tie-rod to the marked position.

Installation

6   Connect the tie-rod end to the steering knuckle arm. Install the nut on the ballstud and tighten it to the torque listed in this Chapter's Specifications. Install the wheel. Lower the vehicle and tighten the lug nuts to the torque listed in the Chapter 1 Specifications.

7   On models with electrically assisted power steering, perform the steering gear calibration procedure (see Section 21).

8   Have the front end alignment checked and, if necessary, adjusted.

18.2 Hold the tie-rod with a wrench while loosening the jam nut

18.3 Mark the position of the tie-rod end in relation to the threads

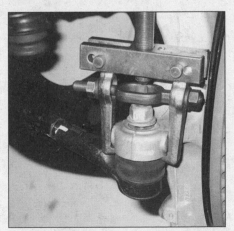

18.4 Back-off the ballstud nut a few turns, then separate the tie-rod end from the steering knuckle with a puller (leaving the nut on the ballstud will prevent the tie-rod end from separating violently)

19.3a Remove both clamps to remove the steering gear boot

19.3b The easiest way to remove the old clamps is to shear off the crimped section with a pair of side cutters

19   Steering gear boots - replacement

1   Loosen the wheel lug nuts, raise the front of the vehicle and support it securely on jackstands. Remove the wheel.

2   Remove the tie-rod end and jam nut (see Section 18).

3   Remove the steering gear boot clamps and slide the boot off **(see illustrations)**.
Note: *Check for the presence of power steering fluid in the boot. If there is a substantial amount, it means the steering gear seals are leaking and the power steering gear should be replaced with a new or rebuilt unit.*

4   Before installing the new boot, wrap the threads and serrations on the end of the steering rod with a layer of tape so the small end of the new boot isn't damaged.

5   Slide the new boot into position on the steering gear until it seats in the grooves, then install new clamps.

6   Remove the tape and install the tie-rod end (see Section 18).

7   Install the wheel and lug nuts. Lower the vehicle and tighten the lug nuts to the torque listed in the Chapter 1 Specifications.

8   Have the front end alignment checked and, if necessary, adjusted.

20   Power steering gear (hydraulic) - removal and installation

Warning: *Make sure the steering column shaft is not turned while the steering gear is removed or you could damage the airbag system clockspring. To prevent the shaft from turning, turn the ignition key to the lock position before beginning work, and run the seat belt through the steering wheel and clip it into its latch.*

1   Park the vehicle with the wheels pointing straight ahead. Loosen the front wheel lug nuts, raise the front of the vehicle and support it securely on jackstands. Apply the parking brake. Remove the wheels.

2   Detach the tie-rod ends from the steering knuckles (see Section 18).

3   Mark the relationship of the intermediate shaft coupler to the steering gear input shaft. Remove the pinch bolt and separate the coupler from the input shaft **(see illustration)**.

4   Position a drain pan under the steering gear. Using a flare-nut wrench, disconnect the power steering pressure and return lines from the steering gear **(see illustration)**. Cap the lines to prevent leakage.

5   Remove the mounting fasteners and then remove the steering gear **(see illustration)**.

6   Inspect all rubber bushings for wear and damage. If any of the rubber parts are cracked, torn or generally deteriorated, replace them.

7   Installation is the reverse of removal,

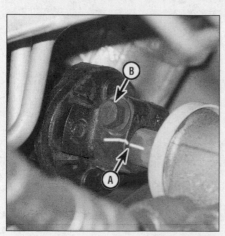

20.3 Mark the relationship of the steering shaft coupler to the steering input shaft (A) and then remove the lower coupler pinch bolt (B)

20.4 Use a flare-nut wrench to remove the line fittings from the steering gear

20.5 Steering gear mounting fasteners

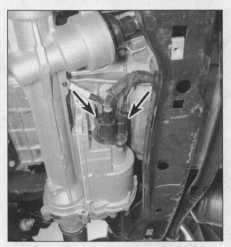

21.9 To disconnect the electrical connectors, first raise the red safety tab, then depress the locking tab to the main connector. Now the connectors can be unplugged

21.11 Steering gear mounting bolt locations

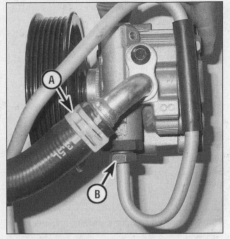

22.5 Detach the supply hose (A) and the pressure line fitting (B)

noting the following points:

a) *Tighten the steering gear mounting bolts and the lower coupler pinch bolt to the torque values listed in this Chapter's Specifications.*

b) *Tighten the wheel lug nuts to the torque listed in the Chapter 1 Specifications.*

c) *Check the power steering fluid level and add some, if necessary (see Chapter 1) and then bleed the system as described in Section 22.*

d) *Re-check the power steering fluid level.*

21 Power steering gear (electric) - removal, installation and calibration

Note: *The only serviceable parts of the steering gear are the tie-rods, steering gear boots and tie-rod ends.*

Removal and installation

Note: *Vehicles equipped with EPA (Electronic Power Assisted) steering systems utilize a rack and pinion type steering mechanism mounted on the front suspension cradle. The only serviceable parts are the inner and outer tie rods and the dust bellows. The EPA is serviced as one complete assembly.*

Note: *If the steering gear is replaced, the gear will need to be programmed to the vehicle. This will require a professional-grade scan tool with the proper software.*

Warning: *Before servicing the steering system be sure to follow the procedures for disarming the air bag system (see Chapter 12).*

Warning: *The Steering Column Control Module (SCCM) remembers and knows the exact center point of the steering system. Failure to hold the steering column and steering gear in the exact same location and relationship can*

result in damage to the SCCM.

1   Park the vehicle on a flat and level surface with the wheels pointing straight ahead.

2   Disconnect the cable from the negative terminal of the battery (see Chapter 5, Section 3).

3   If equipped with a telescopic steering system, fully extend the column.

4   Loosen the front wheel lug nuts. Raise the front of the vehicle and support it securely on jackstands.

5   Remove the wheels.

6   Separate the tie-rod ends from the steering knuckles (see Section 18).

7   Mark the relationship of the steering intermediate shaft coupler to the steering gear, then remove the pinch bolt from the coupler (see Section 20, **illustration 20.3**).

8   Stabilize the steering column shaft from turning, then separate the intermediate shaft coupler from the steering gear.

Warning: *If this is not done, the airbag clockspring could be damaged.*

Note: *One way to do this is to pass the seat belt through the steering wheel and click it into its latch.*

9   Disconnect the two electrical connections at the steering gear (**see illustration**).

10   Remove the heat shields protecting the steering gear boots.

11   Remove the steering gear-to-subframe bolts (**see illustration**) and remove the steering gear.

Note: *New replacement steering gears generally do not come with the heat shield. Drill out the three rivets from the old steering gear, then transfer the heat shield to the replacement unit. Self-tapping screws come with the new replacement steering gear for reinstalling the heat shield.*

Installation is the reverse of removal, except you will have to calibrate the steering gear to the vehicle.

Steering gear calibration

Warning: *Calibration is NOT programming. Programming (which will require a professional-grade scan tool with the proper software) is required if you are replacing the assembly.*

Note: *Read and follow all the directions before starting the procedure.*

12   After reinstalling the steering gear, steering column, or replacing the tie-rods you must calibrate the system.

13   After reconnecting the battery, turn the key to On with the engine Off.

14   Rotate the steering wheel from full right to full left position, being sure that the steering wheel reaches its maximum "lock" position in each direction.

15   Return the steering wheel to the center position and turn the key to Off.

16   The system is now calibrated. Start the vehicle and check for any related service codes (if any).

22 Power steering pump - removal and installation

1   Disconnect the cable from the negative battery terminal, (see Chapter 5, Section 1).

2   Using a large syringe or suction gun, suck as much fluid out of the power steering fluid reservoir as possible.

3   Remove the air filter housing (see Chapter 4).

4   Remove the drivebelt (see Chapter 1).

5   Position a drain pan under the power steering pump, then disconnect the pressure line and the supply hose (**see illustration**). Plug the openings to prevent excessive fluid loss and the entry of contaminants.

6   Remove the pump mounting bolts. The bolts can be accessed through the holes in

the power steering pump pulley **(see illustration)**.

7   Lift the pump from the engine compartment, being careful not to let any power steering fluid drip on the vehicle's paint.

Note: *The pulley can be removed from the pump if necessary. Special tools for this are available at most auto parts stores.*

8   Installation is the reverse of removal. Be sure to tighten all fasteners securely. Fill the power steering reservoir with the recommended fluid (see Chapter 1) and bleed the system following the procedure described in Section 23. Re-check the power steering fluid level.

22.6 Use a socket through the pulley to remove the power steering pump mounting bolts

23   Power steering system - bleeding

1   The power steering system must be bled whenever a line is disconnected. Bubbles can be seen in power steering fluid that has air in it and the fluid will often have a milky appearance. Low fluid level can cause air to mix with the fluid, resulting in a noisy pump as well as foaming of the fluid.

2   Open the hood and check the fluid level in the reservoir, adding the specified fluid necessary to bring it up to the proper level (see Chapter 1).

3   Start the engine and slowly turn the steering wheel several times from left-to-right and back again. Do not turn the wheel completely from lock-to-lock. Check the fluid level, topping it up as necessary until it remains steady and no more bubbles are visible.

24   Wheels and tires - general information

1   All models covered by this manual are equipped with radial tires **(see illustration)**. Use of other size or type of tires may affect the ride and handling of the vehicle. Don't mix different types of tires, such as radials and bias belted tires, on the same vehicle - handling may be seriously affected. It's recommended that tires be replaced in pairs on the same axle, but if only one tire is being replaced, be sure it's the same size, structure and tread design as the other tire on the same axle.

2   Because tire pressure has a substantial effect on handling and wear, the pressure of all tires should be checked at least once a month or before any extended trips are taken (see Chapter 1).

3   Wheels must be replaced if they are bent, dented, leak air, have elongated bolt holes, are heavily rusted, out of vertical symmetry or if the lug nuts won't stay tight. Wheel repairs that use welding or peening are not recommended.

4   Tire and wheel balance are important to the overall handling, braking and performance of the vehicle. Unbalanced wheels can adversely affect handling and ride characteristics as well as tire life. Whenever a tire is

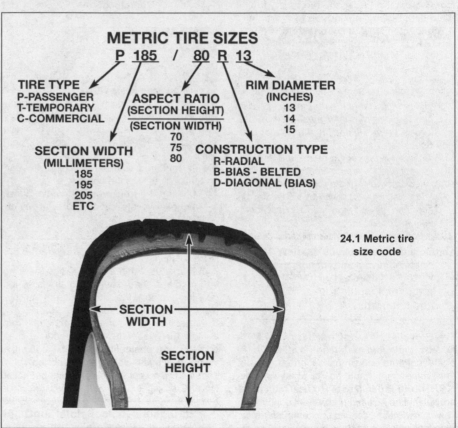

24.1 Metric tire size code

installed on a wheel, the tire and wheel should be balanced by a shop with the proper equipment and expertise.

25   Wheel alignment - general information

Note: *Since wheel alignment requires special equipment and techniques, it is beyond the scope of this manual. This Section is intended only to familiarize the reader with the basic terms used and procedures followed during a typical wheel alignment.*

1   The three basic checks made when

aligning a vehicle's front wheels are camber, caster and toe-in **(see illustration)**.

2   Camber and caster are the angles at which the wheels and suspension are inclined in relation to a vertical centerline. Camber is the angle of the wheel in the lateral, or side-to-side plane, while caster is the tilt between the steering axis and the vertical plane, as viewed from the side. Camber angle affects the amount of tire tread which contacts the road and compensates for changes in suspension geometry as the vehicle travels around curves and over bumps. Caster angle affects the self-centering action of the steering, which governs straight-line stability.

3   Toe-in is the amount the front wheels

are angled in relationship to the center line of the vehicle. For example, in a vehicle with zero toe-in, the distance measured between the front edges of the wheels and the distance measured between the rear edges of the wheels are the same. In other words, the wheels are running parallel with the centerline of the vehicle. Toe-in is adjusted by lengthening or shortening the tie-rods. Incorrect toe-in will cause the tires to wear improperly by allowing them to "scrub" against the road surface.

4   Proper wheel alignment is essential for safe steering and even tire wear. Symptoms of alignment problems are pulling of the steering to one side or the other and uneven tire wear. If these symptoms are present, check for the following before having the alignment adjusted:

 a) *Loose steering gear mounting bolts*
 b) *Damaged or worn steering gear mounts*
 c) *Worn or damaged wheel bearings*
 d) *Bent tie-rods*
 e) *Worn balljoints*
 f) *Improper tire pressures*
 g) *Mixing tires of different construction*

5   Front wheel alignment should be left to an alignment shop with the proper equipment and experienced personnel.

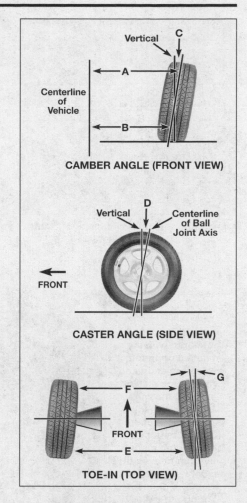

CAMBER ANGLE (FRONT VIEW)

CASTER ANGLE (SIDE VIEW)

TOE-IN (TOP VIEW)

25.1 Wheel alignment details

A minus B = C (degrees camber)
D = degrees caster
E minus F = toe-in
(measured in inches)
G = toe-in (expressed in degrees)

Notes

Chapter 11
Body

Contents

Specifications

Torque specifications
Ft-lbs (unless otherwise indicated)

Note: *One foot-pound (ft-lb) of torque is equivalent to 12 inch-pounds (in-lbs) of torque. Torque values below approximately 15 ft-lbs are expressed in inch-pounds, because most foot-pound torque wrenches are not accurate at these smaller values.*

Seat mounting bolts/nuts	
Front seat back-to-bottom section bolts	37
Front seat-to-floor pan bolts	52
Rear seat nuts	41
60/40 rear seat back bolts	42
Window regulator assembly mounting bolts/nuts	96 in-lbs
Door glass-to-regulator bolts	120 in-lbs

1   General Information

Warning: *The models covered by this manual are equipped with Supplemental Restraint Systems (SRS), more commonly known as airbags. Always disable the airbag system before working in the vicinity of any airbag system components to avoid the possibility of accidental deployment of the airbags, which could cause personal injury (see Chapter 12).*

1   Certain body components are particularly vulnerable to accident damage and can be unbolted and repaired or replaced. Among these parts are the hood, doors, tailgate, liftgate, bumpers and front fenders.

2   Only general body maintenance practices and body panel repair procedures within the scope of the do-it-yourselfer are included in this Chapter.

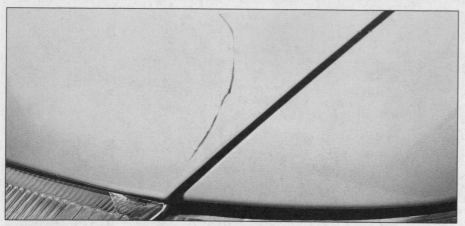

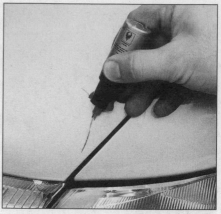

Make sure the damaged area is perfectly clean and rust free. If the touch-up kit has a wire brush, use it to clean the scratch or chip. Or use fine steel wool wrapped around the end of a pencil. Clean the scratched or chipped surface only, not the good paint surrounding it. Rinse the area with water and allow it to dry thoroughly

Thoroughly mix the paint, then apply a small amount with the touch-up kit brush or a very fine artist's brush. Brush in one direction as you fill the scratch area. Do not build up the paint higher than the surrounding paint

2   Repair minor paint scratches

1   No matter how hard you try to keep your vehicle looking like new, it will inevitably be scratched, chipped or dented at some point. If the metal is actually dented, seek the advice of a professional. But you can fix minor scratches and chips yourself. Buy a touch-up paint kit from a dealer service department or an auto parts store. To ensure that you get the right color, you'll need to have the specific make, model and year of your vehicle and, ideally, the paint code, which is located on a special metal plate under the hood or in the door jamb.

3   Body repair - minor damage

Plastic body panels

1   The following repair procedures are for minor scratches and gouges. Repair of more serious damage should be left to a dealer service department or qualified auto body shop. Below is a list of the equipment and materials necessary to perform the following repair procedures on plastic body panels.

> *Wax, grease and silicone removing solvent*
> *Cloth-backed body tape*
> *Sanding discs*
> *Drill motor with three-inch disc holder*
> *Hand sanding block*
> *Rubber squeegees*
> *Sandpaper*
> *Non-porous mixing palette*
> *Wood paddle or putty knife*
> *Wood paddle or putty knife*
> *Curved-tooth body file*
> *Flexible parts repair material*

Flexible panels (bumper trim)

2   Remove the damaged panel, if necessary or desirable. In most cases, repairs can be carried out with the panel installed.
3   Clean the area(s) to be repaired with a

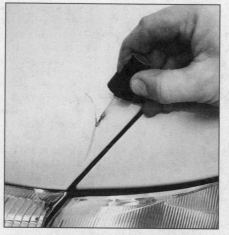

If the vehicle has a two-coat finish, apply the clear coat after the color coat has dried

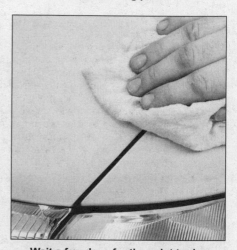

Wait a few days for the paint to dry thoroughly, then rub out the repainted area with a polishing compound to blend the new paint with the surrounding area. When you're happy with your work, wash and polish the area

wax, grease and silicone removing solvent applied with a water-dampened cloth.
4   If the damage is structural, that is, if it extends through the panel, clean the backside of the panel area to be repaired as well. Wipe dry.
5   Sand the rear surface about 1-1/2 inches beyond the break.
6   Cut two pieces of fiberglass cloth large enough to overlap the break by about 1-1/2 inches. Cut only to the required length.
7   Mix the adhesive from the repair kit according to the instructions included with the kit, and apply a layer of the mixture approximately 1/8-inch thick on the backside of the panel. Overlap the break by at least 1-1/2 inches.
8   Apply one piece of fiberglass cloth to the adhesive and cover the cloth with additional adhesive. Apply a second piece of fiberglass cloth to the adhesive and immediately cover the cloth with additional adhesive in sufficient quantity to fill the weave.
9   Allow the repair to cure for 20 to 30 min-

utes at 60-degrees to 80-degrees F.
10   If necessary, trim the excess repair material at the edge.
11   Remove all of the paint film over and around the area(s) to be repaired. The repair material should not overlap the painted surface.
12   With a drill motor and a sanding disc (or a rotary file), cut a "V" along the break line approximately 1/2-inch wide. Remove all dust and loose particles from the repair area.
13   Mix and apply the repair material. Apply a light coat first over the damaged area; then continue applying material until it reaches a level slightly higher than the surrounding finish.
14   Cure the mixture for 20 to 30 minutes at 60-degrees to 80-degrees F.
15   Roughly establish the contour of the area being repaired with a body file. If low areas or

pits remain, mix and apply additional adhesive.

16   Block sand the damaged area with sandpaper to establish the actual contour of the surrounding surface.

17   If desired, the repaired area can be temporarily protected with several light coats of primer. Because of the special paints and techniques required for flexible body panels, it is recommended that the vehicle be taken to a paint shop for completion of the body repair.

Steel body panels

Repairing simple dents

18   When repairing dents, the first job is to pull the dent out until the affected area is as close as possible to its original shape. There is no point in trying to restore the original shape completely as the metal in the damaged area will have stretched on impact and cannot be restored to its original contours. It is better to bring the level of the dent up to a point that is about 1/8-inch below the level of the surrounding metal. In cases where the dent is very shallow, it is not worth trying to pull it out at all.

19   If the backside of the dent is accessible, it can be hammered out gently from behind using a soft-face hammer. While doing this, hold a block of wood firmly against the opposite side of the metal to absorb the hammer blows and prevent the metal from being stretched.

20   If the dent is in a section of the body which has double layers, or some other factor makes it inaccessible from behind, a different technique is required. Drill several small holes through the metal inside the damaged area, particularly in the deeper sections. Screw long, self-tapping screws into the holes just enough for them to get a good grip in the metal. Now pulling on the protruding heads of the screws with locking pliers can pull out the dent.

21   The next stage of repair is the removal of paint from the damaged area and from an inch or so of the surrounding metal. This is easily done with a wire brush or sanding disk in a drill motor, although it can be done just as effectively by hand with sandpaper. To complete the preparation for filling, score the surface of the bare metal with a screwdriver or the tang of a file or drill small holes in the affected area. This will provide a good grip for the filler material. To complete the repair, see the Section on filling and painting.

Repair of rust holes or gashes

22   Remove all paint from the affected area and from an inch or so of the surrounding metal using a sanding disk or wire brush mounted in a drill motor. If these are not available, a few sheets of sandpaper will do the job just as effectively.

23   With the paint removed, you will be able to determine the severity of the corrosion and decide whether to replace the whole panel, if possible, or repair the affected area. New body panels are not as expensive as most people think and it is often quicker to install a new panel than to repair large areas of rust.

24   Remove all trim pieces from the affected area except those which will act as a guide to the original shape of the damaged body, such as headlight shells, etc. Using metal snips or a hacksaw blade, remove all loose metal and any other metal that is badly affected by rust. Hammer the edges of the hole in to create a slight depression for the filler material.

25   Wire-brush the affected area to remove the powdery rust from the surface of the metal. If the back of the rusted area is accessible, treat it with rust inhibiting paint.

26   Before filling is done, block the hole in some way. This can be done with sheet metal riveted or screwed into place, or by stuffing the hole with wire mesh.

27   Once the hole is blocked off, the affected area can be filled and painted. See the following subsection on filling and painting.

Filling and painting

28   Many types of body fillers are available, but generally speaking, body repair kits which contain filler paste and a tube of resin hardener are best for this type of repair work. A wide, flexible plastic or nylon applicator will be necessary for imparting a smooth and contoured finish to the surface of the filler material. Mix up a small amount of filler on a clean piece of wood or cardboard (use the hardener sparingly). Follow the manufacturer's instructions on the package, otherwise the filler will set incorrectly.

29   Using the applicator, apply the filler paste to the prepared area. Draw the applicator across the surface of the filler to achieve the desired contour and to level the filler surface. As soon as a contour that approximates the original one is achieved, stop working the paste. If you continue, the paste will begin to stick to the applicator. Continue to add thin layers of paste at 20-minute intervals until the level of the filler is just above the surrounding metal.

30   Once the filler has hardened, the excess can be removed with a body file. From then on, progressively finer grades of sandpaper should be used, starting with a 180-grit paper and finishing with 600-grit wet-or-dry paper. Always wrap the sandpaper around a flat rubber or wooden block, otherwise the surface of the filler will not be completely flat. During the sanding of the filler surface, the wet-or-dry paper should be periodically rinsed in water. This will ensure that a very smooth finish is produced in the final stage.

31   At this point, the repair area should be surrounded by a ring of bare metal, which in turn should be encircled by the finely feathered edge of good paint. Rinse the repair area with clean water until all of the dust produced by the sanding operation is gone.

32   Spray the entire area with a light coat of primer. This will reveal any imperfections in the surface of the filler. Repair the imperfections with fresh filler paste or glaze filler and once more smooth the surface with sandpaper. Repeat this spray-and-repair procedure until you are satisfied that the surface of the filler and the feathered edge of the paint are perfect. Rinse the area with clean water and allow it to dry completely.

33   The repair area is now ready for painting. Spray painting must be carried out in a warm, dry, windless and dust free atmosphere. These conditions can be created if you have access to a large indoor work area, but if you are forced to work in the open, you will have to pick the day very carefully. If you are working indoors, dousing the floor in the work area with water will help settle the dust that would otherwise be in the air. If the repair area is confined to one body panel, mask off the surrounding panels. This will help minimize the effects of a slight mismatch in paint color. Trim pieces such as chrome strips, door handles, etc., will also need to be masked off or removed. Use masking tape and several thickness of newspaper for the masking operations.

34   Before spraying, shake the paint can thoroughly, then spray a test area until the spray painting technique is mastered. Cover the repair area with a thick coat of primer. The thickness should be built up using several thin layers of primer rather than one thick one. Using 600-grit wet-or-dry sandpaper, rub down the surface of the primer until it is very smooth. While doing this, the work area should be thoroughly rinsed with water and the wet-or-dry sandpaper periodically rinsed as well. Allow the primer to dry before spraying additional coats.

35   Spray on the top coat, again building up the thickness by using several thin layers of paint. Begin spraying in the center of the repair area and then, using a circular motion, work out until the whole repair area and about two inches of the surrounding original paint is covered. Remove all masking material 10 to 15 minutes after spraying on the final coat of paint. Allow the new paint at least two weeks to harden, then use a very fine rubbing compound to blend the edges of the new paint into the existing paint. Finally, apply a coat of wax

4   Body repair - major damage

1   Major damage must be repaired by an auto body shop specifically equipped to perform body and frame repairs. These shops have the specialized equipment required to do the job properly.

2   If the damage is extensive, the frame must be checked for proper alignment or the vehicle's handling characteristics may be adversely affected and other components may wear at an accelerated rate.

3   Due to the fact that all of the major body components (hood, fenders, etc.) are separate and replaceable units, any seriously damaged components should be replaced rather than repaired. Sometimes the components can be found in a wrecking yard that specializes in used vehicle components, often at considerable savings over the cost of new parts.

These photos illustrate a method of repairing simple dents. They are intended to supplement *Body repair - minor damage* in this Chapter and should not be used as the sole instructions for body repair on these vehicles.

1 If you can't access the backside of the body panel to hammer out the dent, pull it out with a slide-hammer-type dent puller. Tap with a hammer near the edge of the dent to help 'pop' the metal back to its original shape, about 1/8-inch below the surface of the surrounding metal

2 Using coarse-grit sandpaper, remove the paint down to the bare metal. Clean the repair area with wax/silicone remover.

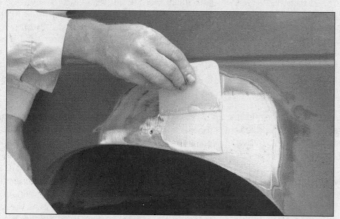

3 Following label instructions, mix up a batch of plastic filler and hardener, then quickly press it into the metal with a plastic applicator. Work the filler until it matches the original contour and is slightly above the surrounding metal

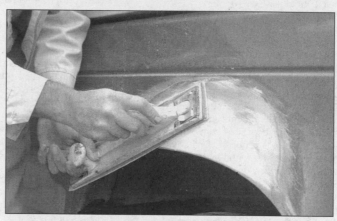

4 Let the filler harden until you can just dent it with your fingernail. File, then sand the filler down until it's smooth and even. Work down to finer grits of sandpaper - always using a board or block - ending up with 360 or 400 grit

5 When the area is smooth to the touch, clean the area and mask around it. Apply several layers of primer to the area. A professional-type spray gun is being used here, but aerosol spray primer works fine

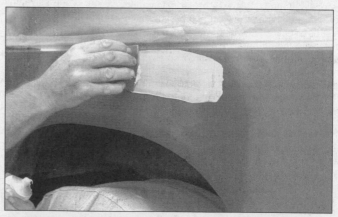

6 Fill imperfections or scratches with glazing compound. Sand with 360 or 400-grit and re-spray. Finish sand the primer with 600 grit, clean thoroughly, then apply the finish coat. Don't attempt to rub out or wax the repair area until the paint has dried completely (at least two weeks)

5   Fastener and trim removal

1   There is a variety of plastic fasteners used to hold trim panels, splash shields and other parts in place in addition to typical screws, nuts and bolts. Once you are familiar with them, they can usually be removed without too much difficulty.

2   The proper tools and approach can prevent added time and expense to a project by minimizing the number of broken fasteners and/or parts.

3   The following illustration shows various types of fasteners that are typically used on most vehicles and how to remove and install them (see illustration). Replacement fasteners are commonly found at most auto parts stores, if necessary.

4   Trim panels are typically made of plastic and their flexibility can help during removal. The key to their removal is to use a tool to pry the panel near its retainers to release it without damaging surrounding areas or breaking-off any retainers. The retainers will usually snap out of their designated slot or hole after force is applied to them. Stiff plastic tools designed for prying on trim panels are available at most auto parts stores (see illustration). Tools that are tapered and wrapped in protective tape, such as a screwdriver or small pry tool, are also very effective when used with care.

6   Upholstery, carpets and vinyl trim - maintenance

Upholstery and carpets

1   Every three months remove the floormats and clean the interior of the vehicle (more frequently if necessary). Use a stiff whiskbroom to brush the carpeting and loosen dirt and dust, then vacuum the upholstery and carpets thoroughly, especially along seams and crevices.

2   Dirt and stains can be removed from carpeting with basic household or automotive carpet shampoos available in spray cans.

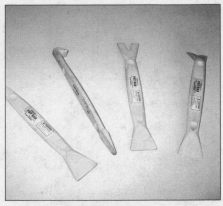

5.4 These small plastic pry tools are ideal for prying off trim panels

Follow the directions and vacuum again, then use a stiff brush to bring back the "nap" of the carpet.

3   Most interiors have cloth or vinyl upholstery, either of which can be cleaned and maintained with a number of material-

Fasteners

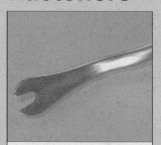

This tool is designed to remove special fasteners. A small pry tool used for removing nails will also work well in place of this tool

A Phillips head screwdriver can be used to release the center portion, but light pressure must be used because the plastic is easily damaged. Once the center is up, the fastener can easily be pried from its hole

Here is a view with the center portion fully released. Install the fastener as shown, then press the center in to set it

This fastener is used for exterior panels and shields. The center portion must be pried up to release the fastener. Install the fastener with the center up, then press the center in to set it

This type of fastener is used commonly for interior panels. Use a small blunt tool to press the small pin at the center in to release it . . .

. . . the pin will stay with the fastener in the released position

Reset the fastener for installation by moving the pin out. Install the fastener, then press the pin flush with the fastener to set it

This fastener is used for exterior and interior panels. It has no moving parts. Simply pry the fastener from its hole like the claw of a hammer removes a nail. Without a tool that can get under the top of the fastener, it can be very difficult to remove

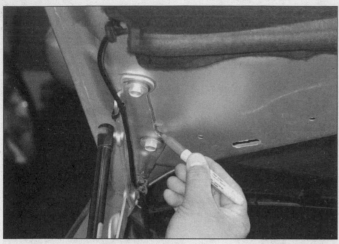

9.2 Before removing the hood, draw a mark around the hinge plate

9.3 Use a small screwdriver to remove the ball and socket retainer from the hood support strut. Be prepared to catch the clip or it may fly off

specific cleaners or shampoos available in auto supply stores. Follow the directions on the product for usage, and always spot-test any upholstery cleaner on an inconspicuous area (bottom edge of a backseat cushion) to ensure that it doesn't cause a color shift in the material.

4   After cleaning, vinyl upholstery should be treated with a protectant.
Note: *Make sure the protectant container indicates the product can be used on seats - some products may make a seat too slippery.*
Caution: *Do not use protectant on vinyl-covered steering wheels.*

5   Leather upholstery requires special care. It should be cleaned regularly with saddle-soap or leather cleaner. Never use alcohol, gasoline, nail polish remover or thinner to clean leather upholstery.

6   After cleaning, regularly treat leather upholstery with a leather conditioner, rubbed in with a soft cotton cloth. Never use car wax on leather upholstery.

7   In areas where the interior of the vehicle is subject to bright sunlight, cover leather seating areas of the seats with a sheet if the vehicle is to be left out for any length of time.

Vinyl trim

8   Don't clean vinyl trim with detergents, caustic soap or petroleum-based cleaners. Plain soap and water works just fine, with a soft brush to clean dirt that may be ingrained. Wash the vinyl as frequently as the rest of the vehicle.

9   After cleaning, application of a high-quality rubber and vinyl protectant will help prevent oxidation and cracks. The protectant can also be applied to weather-stripping, vacuum lines and rubber hoses, which often fail as a result of chemical degradation, and to the tires.

7   Hinges and locks - maintenance

1   Once every 3000 miles, or every three months, the hinges and latch assemblies on the doors, hood and trunk should be given a few drops of light oil or lock lubricant. The door latch strikers should also be lubricated with a thin coat of grease to reduce wear and ensure free movement. Lubricate the door and trunk locks with spray-on graphite lubricant.

8   Windshield and fixed glass - replacement

1   Replacement of the windshield and fixed glass requires the use of special fast-setting adhesive/caulk materials and some specialized tools. It is recommended that these operations be left to a dealer or a shop specializing in glass work.

9   Hood - removal, installation and adjustment

Note: *The hood is heavy and somewhat awkward to remove and install - at least two people should perform this procedure.*

Removal and installation

1   Use blankets or pads to cover the cowl area of the body and fenders. This will protect the body and paint as the hood is lifted off.
Note: *Have a grease pencil, marking paint, or a scribe handy for the next step.*

2   Make marks or scribe a line around the hood hinge to ensure proper alignment during installation **(see illustration)**.

3   If equipped, disconnect any cables, wires or windshield washer hoses that will interfere with removal of the hood, and the hood support struts **(see illustration)**.
Note: *With the hood support struts removed, have your assistant hold the hood open. Then proceed with the remainder of the removal.*

4   Have an assistant support one side of the hood while you support the other. Simultaneously remove the hinge-to-hood nuts **(see illustration)**.

5   Lift off the hood.
Note: *If you are removing the hood to replace the hood hinge, be sure to mark the hinge bolts to the fender area before removing the bolts. Use these marks to align the hood hinge back into the correct position.*

6   Installation is the reverse of removal.

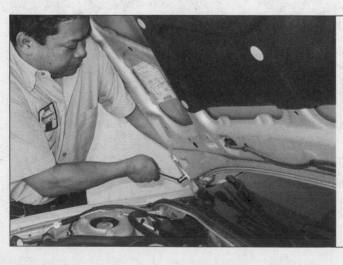

9.4 Support the hood with your shoulder while removing the hood bolts

9.12 Make a mark around the latch to use as a reference point. To adjust the hood latch, loosen the retaining bolts, move the latch and retighten bolts, then close the hood to check the fit

9.13 Adjust the hood closing height by turning the hood bumpers in or out

10.9 Work the cable off of the inside handle by aligning the cable end to the opening in the latch handle. Then pry the cable out of the handle

Hood air induction bezels

7   The air induction bezels are held in place by plastic clips. Use a flat trim tool to release the clips and remove the bezel trim plates.
8   To install, simply line up the bezel to the opening and press into place.

Adjustment

9   Fore-and-aft and side-to-side adjustment of the hood is done by moving the hinge plate slot after loosening the bolts or nuts.
10   Mark around the each hinge plate so you can determine the amount of movement **(see illustration 9.2).**
11   Loosen the bolts or nuts and move the hood into correct alignment. Move it only a little at a time. Tighten the hinge bolts and carefully lower the hood to check the position.
12   If necessary after installation, the hood latch can be adjusted up-and-down as well as from side-to-side on the radiator support so the hood closes securely and flush with the fenders. To make the adjustment, scribe a line or mark around the hood latch mounting bolts to provide a reference point, then loosen them and reposition the latch, as necessary **(see illustration).** Following adjustment, retighten the mounting bolts.
13   Finally, adjust the hood bumpers on the radiator support so the hood, when closed, is flush with the fenders **(see illustration).**
14   The hood latch assembly, as well as the hinges, should be periodically lubricated with white, lithium-base grease to prevent binding and wear.

10   Hood latch and release cable - removal and installation

Latch

1   Scribe a line around the latch to aid alignment when reinstalling the latch assembly.

2   Remove the latch retaining bolts securing the latch to the radiator support and remove the latch.
3   Disconnect the hood release cable by disengaging the cable from the back of the latch assembly.
4   Installation is the reverse of the removal procedure.
Note: *Adjust the latch so the hood engages securely when closed and the hood bumpers are slightly compressed.*

Cable

5   Remove the hood latch as described earlier in this Section, then detach the cable from the latch.
6   Remove the left-side headlight housing (see Chapter 12).
7   Attach a length of wire to the end of the cable (in the engine compartment). This will be used to pull the new cable back into the engine compartment.
8   Working in the engine compartment, detach the cable from all of its retaining clips. It may be necessary to cut some of the clips to free the cable.
9   Working under the instrument panel,

remove the screws and detach the hood release handle **(see illustration).** Dislodge the grommet and pull the cable through the firewall and into the cab.
10   Detach the wire from the old cable, then attach it to the end of the new cable.
Note: *Make sure the new cable is equipped with a grommet.*
11   Pull the new cable through the firewall and into the engine compartment. Seat the grommet in the firewall.
12   The remainder of installation is the reverse of removal.

11   Radiator grille - removal and installation

1   Remove the bumper cover (see Section 12).

300/Charger/Magnum models

2   Remove the fasteners securing the grille to the bumper cover and separate the grille from the cover **(see illustration).**
3   Installation is the reverse of removal.

11.2 Radiator grille mounting fasteners (300 model shown, Charger and Magnum models similar)

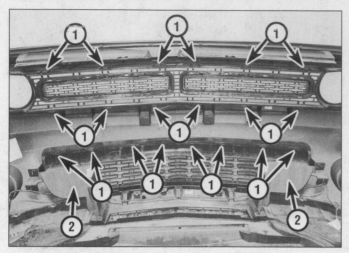

11.5 Radiator grille details (Challenger models)

1   *Retaining tabs*   2   *Close-out panels*

12.3 Remove the fasteners securing the ends of the bumper cover to the inner fender splash shield

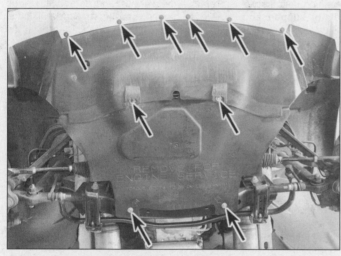

12.4 Remove the fasteners securing the engine splash shield (300 model shown)

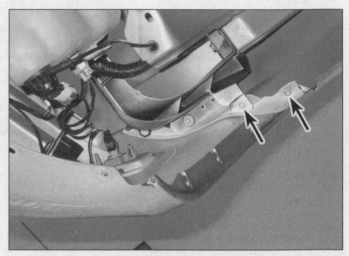

12.5a Remove the fasteners securing the bumper cover to the fender (300 model shown)

Challenger models

Upper grille

4   If present, remove any plastic rivets retaining the top of the grille to the bumper cover.

5   Using a screwdriver or a flat plastic trim tool, pry the retaining tabs free and separate the grille from the bumper cover **(see illustration)**.

6 Installation is the reverse of removal.

Lower grille

7   Remove the screws from the underside and the front of the grille.

8   Remove the plastic rivets and detach the close-out panels from each side of the grille **(see illustration 11.5)**.

9   Using a screwdriver or a flat plastic trim tool, pry the retaining tabs free and separate the grille from the bumper cover.

10   Installation is the reverse of removal.

12   Bumper covers - removal and installation

1   Bumpers on all models are composed of a plastic fascia, or bumper cover, fascia support and a structural beam.

Front bumper cover

2   Raise the vehicle and support it securely on jackstands,

3   Working in the wheelwell, remove the fasteners securing the ends of the bumper cover to the inner fender splash shield **(see illustration)**.

Note: *On some models you may find rivets instead of plastic retaining pins (see Section 13). Plastic push-pin fasteners are available from most auto parts stores and dealership parts departments for replacement.*

4   Remove the engine splash shield **(see illustration)**.

5   Working in the wheelwell, remove the fasteners securing the bumper cover to the fender and bumper cover support **(see illustrations)**.

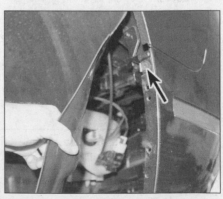

12.5b The Challenger model fasteners are in a slightly different location

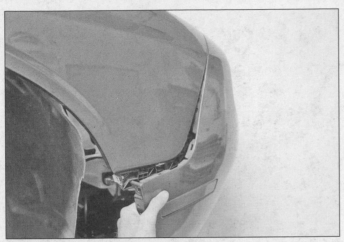

12.5c Separate the bumper cover from the fender

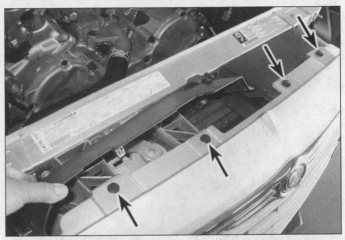

12.6a Bumper cover upper mounting fasteners (300 model shown)

12.6b On Challenger models, remove the close-out panels above the radiator support by pulling straight up, then remove the upper bumper cover-to-fender nuts above the headlight housings on each side

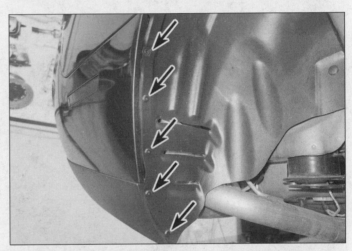

12.9 On some models there will be rivets instead of plastic push-pins. These must be drilled out

6   Remove the close-out panel above the radiator support, then remove the bumper cover upper mounting fasteners and remove the bumper cover **(see illustrations)**.

7   Disconnect the fog lamps, if equipped, then carefully remove the bumper cover from the vehicle and disconnect the TPM (Tire Pressure Monitor) connector.

8   Installation is the reverse of removal.

Rear bumper cover

9   Raise the vehicle and support it securely on jackstands. Remove the fasteners that secure the inner-fender splash shield to the rear bumper cover **(see illustration)**.

Note: *Replacement push-pin fasteners are available at most auto parts stores and dealership parts departments. These can be used instead of plastic rivets.*

10   Challenger models: Remove the rear tail light assemblies (see Chapter 12).

11   Working in the rear wheelwell, remove the fasteners securing the bumper cover to the rear quarter panel **(see illustration)**.

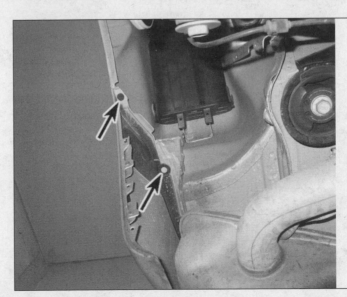

12.11 Remove the fasteners securing the bumper cover to the rear quarter panel

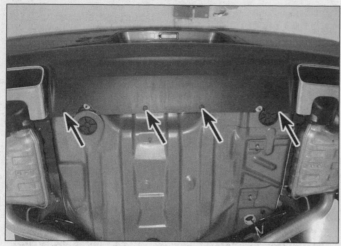

12.12 Bumper cover lower mounting fasteners (Challenger shown)

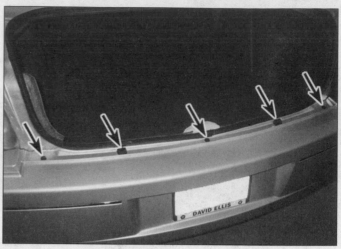

12.13a Bumper cover upper fasteners (300)

12.13b Bumper cover upper fasteners (Challenger)

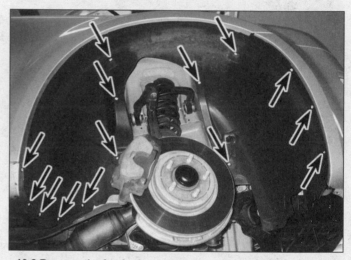

13.2 Remove the fender inner splash shield mounting fasteners

13.3 Carefully drill out the plastic rivets, if equipped

12   Remove the fasteners from the lower part of the rear bumper fascia **(see illustration)**.

13   Remove the remaining bumper cover mounting fasteners and remove the bumper cover **(see illustrations)**.

14   Installation is the reverse of removal.

13   Front fender - removal and installation

1   Loosen the wheel lug nuts, then raise the vehicle and support it securely on jackstands and remove the front wheel.

2   Remove the fasteners retaining the fender inner splash shield **(see illustration)**.

3   Some models use plastic rivets for some of the inner fender fasteners - these must be drilled out **(see illustration)**. Replacement push-pin fasteners are available at most auto parts stores and dealership parts departments.

4   Remove the front bumper cover (see Section 12).

5   Remove the headlight housings (see Chapter 12).

6   Remove the fender mounting bolts **(see illustrations)**.

7   Detach the fender. It's a good idea to have an assistant support the fender while it's being moved away from the vehicle to prevent damage to the surrounding body panels.

8   Installation is the reverse of removal.

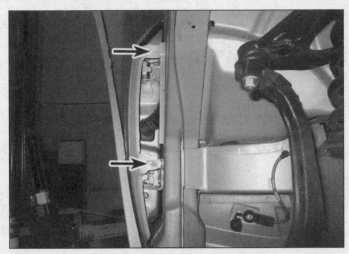

13.6a Fender-to-door pillar fasteners

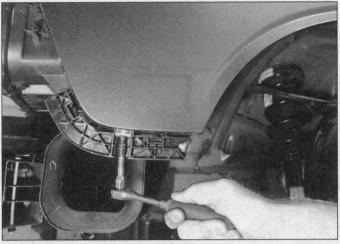

13.6b Remove the fastener securing the fender to the support

13.6c With the bolt removed, pry the retaining bracket off. Do not discard. This will need to be reinstalled

13.6d Fender lower corner mounting bolts

13.6e On some models, pry back the sill plate guard to expose the hidden lower fasteners

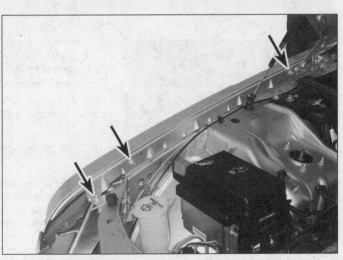

13.6f Fender upper mounting bolts

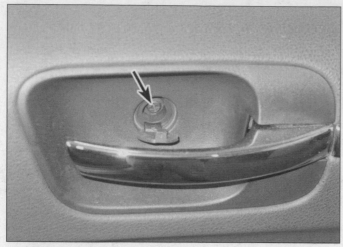

14.2a Pry open the trim cap and remove the handle retaining screw (on later models a panel is used instead of a cap)

14.2b Door panel retaining fasteners locations. Some of the screws will have a trim cap over the screw. Remove the trim cap by prying it off with a trim tool or a pocket screwdriver

14.3 Rotate the plastic retaining clip off the handle link rod and detach the rod from the handle

14.4 From the corner, carefully peel back the plastic watershield

14.7 Use a flat trim tool to ease the window switch out of the door panel

14   Door trim panels - removal and installation

1   Disconnect the cable from the negative terminal of the battery (see Chapter 5, Section 3).

All models except Challenger

Removal

2   Remove all door trim panel retaining fasteners **(see illustrations)**.

Note: *On later models there is also a screw under a panel in the door armrest* **(see illustration 14.8)**, *and the perimeter of the panel is secured to the door by push retainers which must be pried out with a trim panel tool.*

3   Disconnect the handle link rod **(see illustration)**, disconnect any wiring harness connectors, then remove the trim panel from the vehicle.

4   For access to the inner door, carefully peel back the plastic watershield **(see illustration)**.

Installation

5   Connect the wiring harness connectors and place the panel in position on the door. Connect the handle link rod.

6   The remainder of installation is the reverse of removal.

Challenger

7   Pry out the window switch and disconnect the electrical connector **(see illustration)**.

14.8 With the trim cover removed you can now get to the hidden fastener

14.9 A flat trim tool or a small pocket screwdriver works well to remove the trim cover

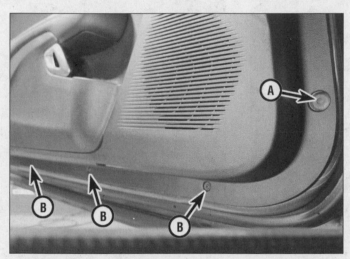

14.10 Depress the center of the push-pin fastener then pry it out (A), then remove the door panel retaining screws (B)

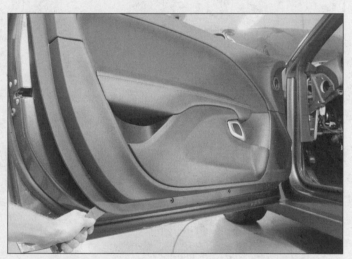

14.11 Pry the plastic panel retainers from the door

8   Remove the trim cover from the interior handle to expose the hidden fastener (**see illustration**).

9   Remove the trim cover from below the door release handle, then remove the fastener hidden by the trim cover (**see illustration**).

10   Remove the plastic fastener and the screws that secure the trim panel to the door (**see illustration**).

11   Using a plastic trim tool, pry out the door panel retainers around the perimeter of the panel (**see illustration**).

12   Disconnect the release cables from the inner handle assembly by sliding the cable end out of the locking slot (**see illustration**).

13   Remove the panel.

14   Installation is the reverse of removal.

14.12 Pry the cable from the housing, then remove the cable ball end from the lever

15.1 Pull out the boot and disconnect the door wire harness electrical connector at the A-pillar

15.5 The latch striker on the door jamb can be adjusted slightly up/down or in/out

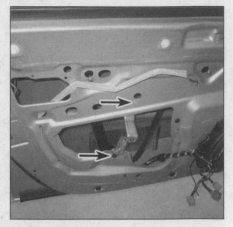

16.3 Remove the fasteners securing the pull cup bracket

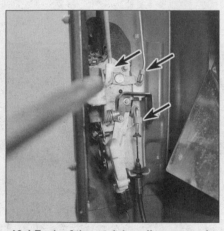

16.4 Each of the retaining clips are made to be rotated off of the rod, then pull the rod out of the clip. No need in trying to remove the plastic retainer until the latch has been removed from the door

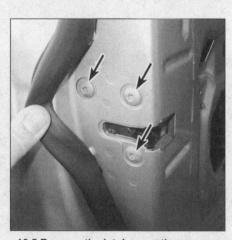

16.5 Remove the latch mounting screws

16.10 With the outside handle out of the way, the door lock cylinder can be removed

15　Door - removal, installation and adjustment

1　Open the door and disconnect the door wire harness electrical connector at the A-pillar **(see illustration)**.
2　Place a jack under the door or have an assistant on hand to support it when the hinge fasteners are removed.
Note: *If a jack is used, place a few rags between it and the door to protect the door's painted surfaces.*
3　Scribe around the hinges with a marking pen, remove the fasteners and carefully lift off the door.
4　Installation is the reverse of removal, making sure to align the hinge with the marks made during removal before tightening the fasteners.
5　Following installation of the door, check the alignment and adjust the hinges, if necessary. Adjust the door lock striker, centering it

in the door latch **(see illustration)**.
6　Rear doors are removed and installed as described above for front doors.

16　Door latch, lock cylinder and handle - removal and installation

Latch

1　Raise the window completely and remove the door trim panel and watershield (see Section 14).
2　Remove the door speaker (see Chapter 12).
3　Remove the fasteners at the rear glass run channel and move the channel away from the latch **(see illustration)**. This provides a little extra working room in the latch area inside the door.
4　Rotate the plastic retaining clips off the rods **(see illustration)**, then detach the latch links. Disconnect the electrical connector.

5　Remove the three mounting screws (it may be necessary to use an impact-type screwdriver to loosen them), then remove the latch from the door **(see illustration)**.
6　Place the latch in position and install the screws. Tighten the screws securely.
7　Connect the link rods and electrical connector to the latch.
8　The remainder of installation is the reverse of removal.

Lock cylinder

9　Remove the outside door handle (see Step 12).
10　Withdraw the lock cylinder from the door handle **(see illustration)**.
11　Installation is the reverse of removal.

Outside handle

12　Remove the door trim panel (see Section 14).
13　Remove the door speaker (see Chapter 12).

16.15 Door handle mounting fasteners

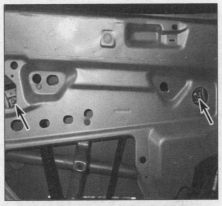

17.4 Align the window so that the fasteners can be removed from the glass track

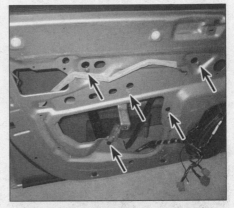

18.3a Door window regulator mounting bolt locations

14   Remove the fasteners at the rear glass run channel and move the channel away from the latch. This provides a little extra working room in the latch area inside the door.
15   Disconnect the link rods from the outside handle, remove the mounting fasteners and carefully detach the handle from the door **(see illustration)**.
16   Place the handle in position, attach the link and install the nuts. Tighten the fasteners securely.
17   The remainder of installation is the reverse of removal.

17   Door window glass - removal and installation

Front

1   Remove the door trim panel and watershield (see Section 14).
2   Remove the door speaker (see Chapter 12).
3   Pry the inner weather seal out of the door glass opening.
4   Lower the window to access the glass retaining fasteners, then remove the fasteners **(see illustration)**.
Note: *On the Challenger, lower the window to the bottom and use the speaker opening as one of the access points for the window glass fasteners.*
5   Remove the window by tilting it forward, then lifting it out of the door.
6   To install, lower the glass into the door, slide it into position and install the nuts.
7   The remainder of installation is the reverse of removal.

Rear

8   The rear doors are serviced similarly to the front doors.

18   Door window glass regulator - removal and installation

1   Remove the door trim panel and watershield (see Section 14).
2   Unbolt the window glass from the regulator (see Section 17). Push the glass all the way up and tape it to the door frame.
3   Remove the window regulator-to-door and track mounting fasteners **(see illustrations)**.
4   Unplug the electrical connector.
5   Remove the regulator from the door.
6   Installation is the reverse of removal. Tighten the glass retaining fasteners to the torque listed in this Chapter's Specifications.

Window regulator re-learn procedure

7   At any time the regulator has been removed or replaced (or the battery has been disconnected see Chapter 5) a relearn procedure must be performed in order for the DDM or PDM (Driver's door module - Passenger's

18.3b The Challenger window regulator mounting fastener positions are shown, which are slightly different than the Charger and 300 models. Notice the two slotted holes for the regulator motor. No need in removing the bolts, just loosen them so that they clear the door

door module) to properly operate the window.
Caution: *Failure to follow these procedures may result in the window not functioning correctly and/or not opening or closing completely.*

a) *The battery must be fully charged or the vehicle must be running at the time of the relearn procedure*
b) *All doors must be completely closed before proceeding.*
c) *If the regulator has been replaced, use a scan tool to activate an ECU reset for the appropriate door module (DDM or PDM). This will set a calibration missing DTC for that particular module. The calibration code should show that it is "active". (This is needed to erase the existing calibration so that the replacement regulator can be registered into the DDM or PDM).*
d) *These next steps are for both a replacement regulator or reinstalling the original regulator.*
e) *Lower the window glass to its lowest possible position (fully open). Then, without letting go of the down position button, hold the button for an additional two seconds.*

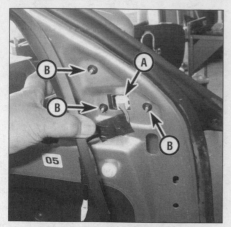

19.2a Disconnect the electrical connector (A), then remove the mirror mounting fasteners (B) (300 and Charger models)

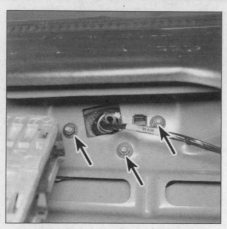

19.2b Challenger mirror mounting nuts

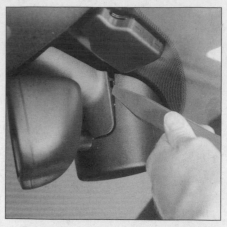

19.4a Use your flat trim tool to pry the cover off and rotate the upper trim downward to remove it . . .

19.4b . . . then pry the other half of the cover off

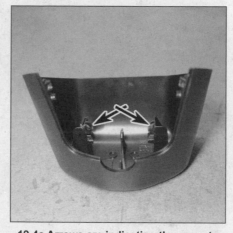

19.4c Arrows are indicating the areas to press outward to remove the cover

19.4d With the covers removed, disconnect the electrical connection(s)

f) Raise the window to its highest position (fully closed). Now, without letting go of the window switch up position, hold the button for an additional two seconds.

g) Check window operation.

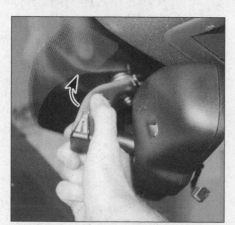

19.5 Grasp the mirror as shown, then rotate 90 degrees clockwise to remove the assembly

h) If the regulator was replaced, check for any stored DTC's. The DTC for missing calibration should have changed from "active" to "stored." Clear the code and check window operation.

Note: *It's a good idea to check for any stored service codes in the BCM whether or not you are installing a replacement regulator or the original regulator. Some codes are stored because the regulator has been disconnected while the key was in the On position or the battery was disconnected. Clear the codes and verify the window operation.*

19   Mirrors - removal and installation

Outside mirrors

1   Remove the door trim panel (see Section 14).

2   Pull back the window trim molding, on power mirrors, unplug the electrical connector, then remove the fasteners and detach the mirror from the door **(see illustrations)**.

3   Installation is the reverse of removal.

Inside mirror

Electronic type

Note: *It's a good idea when disconnecting any electrical system in the vehicle to always disconnect and isolate the negative battery terminal.*

4   On models with electrochromic, rain sensing, optional automatic day/night mirror, or telematic (hands free phone) options, remove the covers to expose the wire connections **(see illustrations)**.

5   With covers removed, and electrical connections disconnected, grasp the base of the mirror assembly and rotate it 90 degrees clockwise to remove the mirror assembly from the windshield **(see illustration)**.

6   Installation is the reverse of removal.

Note: *If the support base for the mirror has to come off the windshield, it can be reattached with a special mirror adhesive kit available at most auto parts stores. Clean the glass and support base thoroughly as described on the package and follow the directions for applying the adhesive.*

Inside mirror - with electrochromic and or telematics features only or no features at all

7   On models with electrochromic (automatic dimming) and telematics (hands free phone) only, or on models equipped with a standard-type inside mirror, loosen or remove the setscrew holding the mirror assembly to the mirror base that is mounted (glued) on the windshield. With the setscrew loose, remove the mirror from the mirror bracket by grasping the mirror base and slide it upward off of the bracket. Be sure to disconnect any electrical connectors from the back of the mirror before removing the assembly.

Mirror support base

8   If the support base for the mirror has come off the windshield, it can be reattached with a special mirror adhesive kit available at auto parts stores. Clean the glass and support base thoroughly and follow the directions on the adhesive package.

---

20   Liftgate (Magnum models) - removal and installation

---

1   Open the liftgate and cover the upper body area around the opening with pads or cloths to protect the painted surfaces when the liftgate is removed.
2   Using a trim stick, remove the upper headliner trim panel, then disconnect the body-to-liftgate electrical connectors.
3   Make alignment marks around the liftgate hinge flanges.
4   While an assistant supports the liftgate, detach the support struts (see Section 22).
5   Remove the hinge bolts and detach the liftgate from the vehicle.
6   Installation is the reverse of removal.
7   After installation, close the liftgate and make sure it's in proper alignment with the surrounding body panels.
8   If the liftgate needs to be adjusted, loosen the hinge bolts slightly, gently close the liftgate and verify that it's centered (the striker should center it). Then carefully open the liftgate and retighten the hinge bolts.

---

21   Liftgate panels, outside handle, latch and support struts (Magnum models) - removal and installation

---

Interior trim panels

1   Remove the fasteners securing the lower trim panel inside the pull handle area.
2   Using a trim stick, carefully pry around the lower trim panel to release the clips securing the trim panel.
3   With the lower panel pulled away from the liftgate, disconnect the electrical connectors at the courtesy lights.
4   Remove the two fasteners securing the

22.2 Use a small screwdriver to detach the retaining clip at the top end of the support strut

upper trim panel.
5   Using a trim stick, carefully pry around the upper trim panel to release the clips securing the trim panel.
6   Remove the upper trim panel by pulling it outward carefully to free it from the spring clips.
7   Installation is the reverse of the removal procedure.

Outside handle

8   Remove the interior lower liftgate panel.
9   Working through the liftgate access hole, disconnect the handle electrical connector, then remove the handle mounting fasteners.
10   Installation is the reverse of the removal procedure.

Latch

11   Remove the interior lower liftgate panel.
12   Working through the liftgate access hole, disconnect the latch electrical connector, then remove the latch mounting fasteners.
13   Installation is the reverse of the removal procedure.

Support struts

14   Open the liftgate and prop it securely in the full open position.
15   Using a small screwdriver, detach the retaining clip at the ends of the support strut. Then pry or pull sharply to detach it from the vehicle.
Note: *Have an assistant hold the liftgate as you remove the support struts. One support strut is not enough to hold the liftgate open.*
16 Installation is the reverse of the removal procedure.

---

22   Trunk lid - removal, installation and adjustment

Note: *The trunk lid is heavy and somewhat awkward to remove and install - at least two people should perform this procedure.*

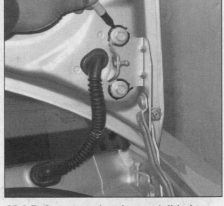

22.3 Before removing the trunk lid, draw a mark around the hinge plate

Removal and installation

1   Open the trunk lid and cover the edges of the trunk compartment with pads or cloths to protect the painted surfaces when the lid is removed.
2   Disconnect any cables or wire harness connectors attached to the trunk lid that would interfere with removal. Remove the trunk lid support struts **(see illustration)**.
3   Make alignment marks around the hinge mounting bolts with a marking pen, then while an assistant supports the trunk lid, remove the lid-to-hinge bolts **(see illustration)** on both sides of the trunk and lift it off.
4   Installation is the reverse of removal.
Note: *When reinstalling the trunk lid, align the lid-to-hinge bolts with the marks made during removal.*

Adjustment

5   Fore-and-aft and side-to-side adjustment of the trunk lid is accomplished by moving the lid in relation to the hinge after loosening the bolts or nuts.
6   Scribe a line around the entire hinge plate as described earlier in this Section so you can determine the amount of movement.
7   Loosen the bolts or nuts and move the trunk lid into correct alignment. Move it only a little at a time. Tighten the hinge bolts or nuts and carefully lower the trunk lid to check the alignment.
8   If necessary after installation, the entire trunk lid striker assembly can be adjusted up and down as well as from side to side on the trunk lid so the lid closes securely and is flush with the rear quarter panels. To do this, scribe a line around the trunk lid striker assembly to provide a reference point. Then loosen the bolts and reposition the striker as necessary. Following adjustment, retighten the mounting bolts.
9   The trunk lid latch assembly, as well as the hinges, should be periodically lubricated with white lithium-base grease to prevent sticking and wear.

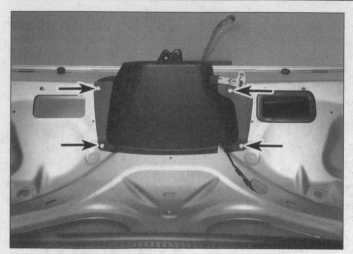

23.1a Remove the latch trim cover fasteners

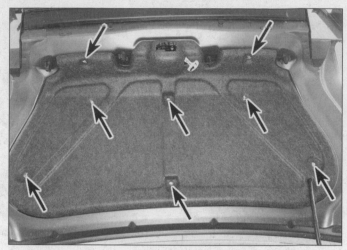

23.1b To gain access to the trunk latch assembly on models with a trunk lid liner, remove these push pins

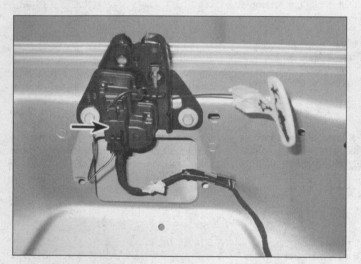

23.3 Remove the latch electrical connector

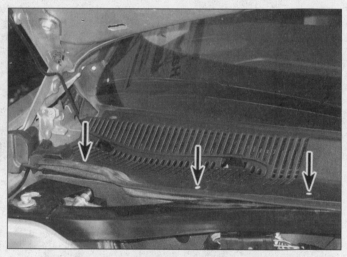

24.2 Remove the fasteners securing the right side of the cowl cover

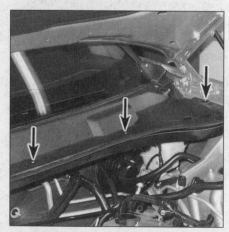

24.3 Remove the fasteners from the left side of the cowl cover

23 Trunk lid latch - removal and installation

Trunk lid latch

1   Open the trunk and remove the latch trim cover **(see illustrations)**.
2   Scribe a line around the trunk lid latch assembly for a reference point to aid the installation procedure.
3   Disconnect the electrical connector **(see illustration)**.
4   Detach the mounting fasteners and remove the latch.
5   Installation is the reverse of removal.

24 Cowl cover - removal and installation

Note: *The cowl cover is held in place with plastic push pins. Carefully remove the push pins so they can be reused. If they are damaged, replacement push pins can be found at most auto parts stores.*

1   Remove the windshield wiper arms (see Chapter 12).
2   Remove the fasteners securing the right and left side cowl cover extensions (if applicable), then remove the covers **(see illustration)**.
3   Remove the fasteners securing the left side of the cowl cover, then remove the cover **(see illustration)**.

25.2 After removing the mat at the front of the console, remove the two console fasteners

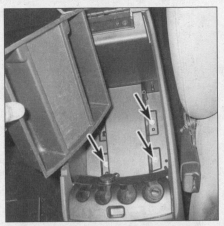

25.3 Remove the bolts securing the console to the floorpan

25.4a Using a trim removal tool, carefully pry off the ashtray trim panel

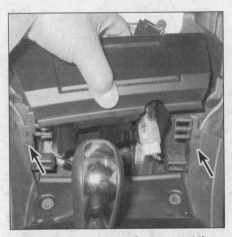

25.4b Remove the two fasteners at the front of the center console

25.8 The outside fasteners have a trim cap over the screws

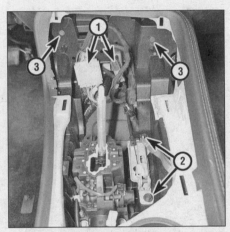

25.9 Center console forward mounting details

1   *Electrical connector(s)*
2   *Manual Park release screws (automatic transmission models)*
3   *Mounting screws*

4   Challenger models - The cowl panel has an extra fastener in each outside corner attaching the panel to the dash area (just below the windshield). Turn the fasteners a 1/4 turn to detach the fasteners.
5   Disconnect the windshield washer hose.
6   Installation is the reverse of removal.

25   Center console - removal and installation

Warning: *The models covered by this manual are equipped with a Supplemental Restraint System (SRS), more commonly known as airbags. Always disable the airbag system before working in the vicinity of any airbag system components to avoid the possibility of accidental deployment of the airbags, which could cause personal injury (see Chapter 12).*

300, Charger and Magnum models

1   Place the shift lever in the Neutral position.
2   Remove the mat at the front of the console, then remove the fasteners securing the shifter trim bezel **(see illustration)**. Remove the shifter trim bezel.
3   Open the lid to the console bin and remove the mat, then remove the bolts securing the console to the floorpan **(see illustration)**.
4   Remove the fasteners at the front of the console **(see illustrations)**.
5   Unplug any electrical connectors and remove the console from the vehicle.
6   Installation is the reverse of removal.

Challenger models

7   Remove the shift knob, shifter bezel trim/ shifter boot and center console bezel trim

(see Chapter 7A or Chapter 7B). On models with a manual transmission, also remove the shift lever.
8   Remove the fasteners from inside the storage bin and the two outside fasteners **(see illustration)**.
9   Remove the fasteners securing the console to the instrument panel on each side **(see illustration)**. On models with an automatic transmission, also remove the screws securing the manual Park release mechanism.
10   Slide the console rearward slightly, just enough to gain access to all of the electrical connections. Disconnect the electrical connections.
11   Lift the console upward and remove it from the vehicle.
12   Installation is the reverse of removal.

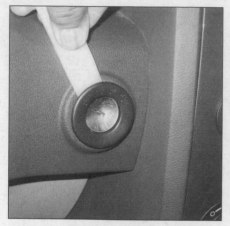

26.1 Carefully remove the ignition switch bezel

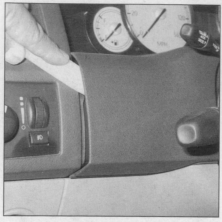

26.2a Pry around the perimeter of the instrument cluster bezel to release the clips

26.2b The instrument cluster bezel is held on with pressure clips (Challenger model shown)

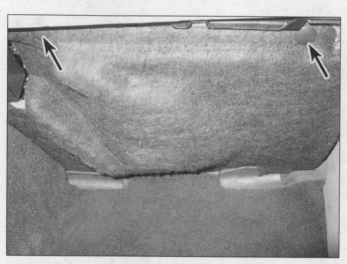

26.5a Remove the fasteners securing the panel insulator (300 shown)

26.5b Panel insulator fasteners (Challenger shown)

26.5c With the panel down, disconnect any electrical connections attached to the panel

26 Dashboard trim panels - removal and installation

Warning: *The models covered by this manual are equipped with a Supplemental Restraint System (SRS), more commonly known as airbags. Always disable the airbag system before working in the vicinity of any airbag system component to avoid the possibility of accidental deployment of the airbags, which could cause personal injury (see Chapter 12).*

Instrument cluster bezel

1   On models with a bezel-mounted ignition switch, use a trim removal tool or a flat-blade screwdriver with the tip taped to carefully remove the ignition switch bezel **(see illustration)**.
2   Use a trim removal tool or a flat-blade screwdriver with the tip taped to pry around the complete edge of the instrument cluster bezel **(see illustrations)**.

3   Pull the panel out far enough to disconnect any electrical connectors.
4   Installation is the reverse of removal.

Lower instrument panel insulator

5   Remove the fasteners securing the panel insulator **(see illustrations)**.
6   Remove the lower instrument panel insulator.
7   Installation is the reverse of removal.

Knee bolster

8   Remove the fasteners securing the driver's knee bolster, then grasp the bolster securely and pull out sharply to detach the retaining clips **(see illustrations)**.
9   Detach the parking brake release cable and disconnect any electrical connectors, then remove the bolster from the vehicle.
10   Installation is the reverse of removal.

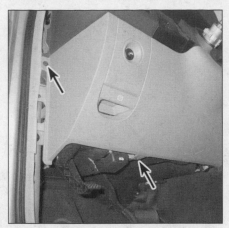

26.8a Remove the fasteners securing the driver's knee bolster

26.8b On later models you'll need to pry the dash end cap off . . .

26.8c . . . remove the screws . . .

Instrument panel center bezel

11   Use a dull, flat-bladed tool to pry around the edges of the panel to release it from the clips **(see illustration).**
12   Disconnect any electrical connectors, then remove the bezel from the vehicle.
13   Installation is the reverse of removal.

Glove box door - removal and installation

14   Open the glove box door.
15   Squeeze the two sides of the glove compartment bin together and pull the door down until the bumpers on the bin have cleared the stops **(see illustration).**
16   Detach the glove box support strap **(see illustration).**
17   Pull the glove box door away from the instrument panel.
18   Installation is the reverse of removal.

26.8d . . . then pry the knee bolster panel off with a flat trim tool

26.11 Pry around the edges of the panel to release it from the clips

26.15 Push on the sides of the glove box until the stops clear the instrument panel

26.16 Detach the glove box support strap

26.20 Use a flat trim tool to release the pressure clips on the end cap

26.21 The door sill trim is held in place with pressure clips. Pry straight out to release them

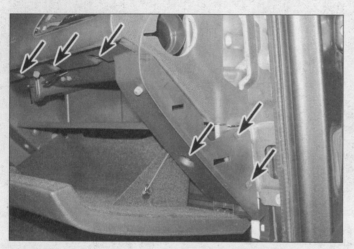

26.22 Glove box screws (not all screws are shown)

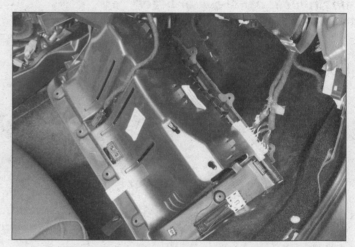

26.23 Be sure to disconnect all the connections before removing the glove box assembly

27.5a Pry the top of the A-pillar trim away from the pillar, then disconnect the roof rail airbag electrical connector . . .

Glove box assembly - removal and installation

19   Remove the glove box door.
20   Remove the instrument panel end cap (**see illustration**).
21   Remove the passenger's side door sill trim (**see illustration**).
22   Remove the eight screws securing the glove box assembly to the instrument cluster (**see illustration**).
23   Lower the glove box assembly, then disconnect the electrical connections and wire harness straps (**see illustration**).
24   Installation is the reverse of removal.

27   Instrument panel - removal and installation

Warning: *The models covered by this manual are equipped with a Supplemental Restraint System (SRS), more commonly known as airbags. Always disable the airbag system before working in the vicinity of any airbag system component to avoid the possibility of accidental deployment of the airbags, which could cause personal injury (see Chapter 12).*
Note: *This procedure depicts instrument panel removal on a Challenger model. The procedure on other models is similar.*

Removal

1   Disconnect the cable from the negative terminal of the battery (see Chapter 5, Section 3).
2   Remove the center console (see Section 25).
3   Remove the glove box (see Section 26).
4   Remove the steering column (see Chapter 10).
5   Remove the A-pillar trim panels (**see illustrations**).
6   Remove the defroster grille panel (**see illustration**).

27.5b . . . then detach the trim
from the pillar

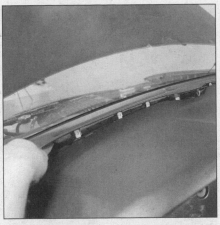

27.6 Carefully pry the defroster grille up
and detach it from the instrument panel

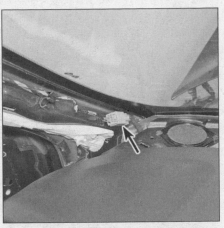

27.7 Unplug the electrical connectors
at the bottom of each A-pillar

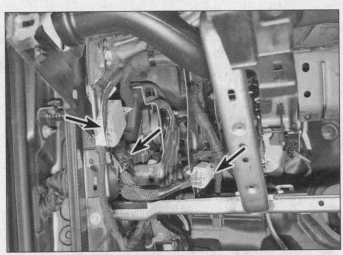

27.9a At the left end of the instrument panel, disconnect
these electrical connectors

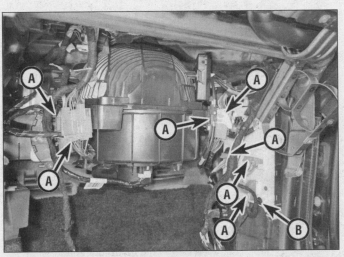

27.9b At the right end of the instrument panel, disconnect the
electrical connectors (A) and free the harness from the clip (B)

7   Disconnect the electrical connector at
the bottom of each A-pillar **(see illustration)**.
8   Remove the driver's knee bolster (see
Section 26).
9   Disconnect the electrical connectors
under each end of the instrument panel **(see
illustrations)**.
10   Remove the floor air duct and the bolts
from the instrument panel center brackets
(see illustration).
11   Remove the two bolts from the forward
edge of the instrument panel on the left side
(see illustration).

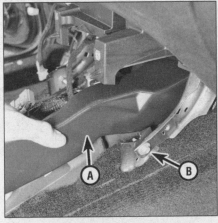

27.10 Floor air duct (A) and center bracket
bolt (B, left side identical)

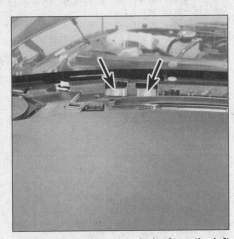

27.11 Remove these two bolts from the left
side of the instrument panel

27.12 Remove these screws from the center and right side of the instrument panel

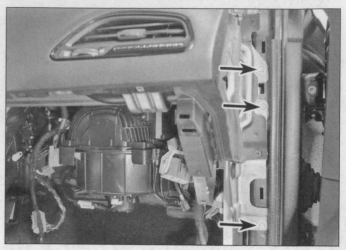

27.13a Remove the bolts from the right . . .

27.13b . . . and left end of the instrument panel

28.3a Remove the rear bolts from the front seat . . .

12   Remove the screws from the center and right forward edge of the instrument panel **(see illustration)**.

13   Remove the instrument panel mounting bolts from each end of the instrument panel **(see illustrations)**.

14   With the help of an assistant, lift the instrument panel up and off of the support hooks at each end, make sure nothing is still attached, then guide it out of the vehicle.

15   Installation is the reverse of the removal procedure.

---

28 Seats - removal and installation

Warning: *These models are equipped with seat belt pre-tensioners, which are pyrotechnic (explosive) devices that tighten the seat belts during an impact of sufficient force to deploy the airbags. Always disable the airbag system before working in the vicinity of any restraint system component to* avoid the possibility of accidental deployment of the airbag(s) and seat belt pre-tensioners, which could cause personal injury (see Chapter 12).

Warning: *Vehicles equipped with passenger airbag and passenger presence systems require some special procedures for working around the passenger seat. Any time any component or part of the ORC (the front passenger seat is one of those components) is removed or replaced, new data needs to be configured in the ORC (Occupant Restraint Controller). This requires the use of factory equivalent scanner. If after replacing or removing the front passenger seat the air bag light remains on, see your local dealer or qualified independent shop for service.*

Warning: *Vehicles with seat air bags (SAB) need to handled with care. Always disconnect the battery and let the vehicle sit idle for approximately two minutes before proceeding with any repair procedure. Keep the disconnected air bag components away from any electrical devices while they are not in the* vehicle. Be sure to reinstall any trim that was removed from the seat. Any trim that is not correctly installed can greatly affect the SAB operation.

Front seat

1   Disconnect the cable from the negative battery terminal (see Chapter 5, Section 1).

2   Remove the plastic covers to expose the anchor bolts. If applicable, remove the seat belt anchor bolt.

Note: *To gain access to the front or rear seat anchoring bolts you may want to temporarily connect the battery and move the seat in the direction needed to access the bolts.*

3   Remove the seat track-to-floor anchor bolts and remove the front seat assembly. Disconnect the electrical connections **(see illustrations)**.

Note: *After removing the anchor bolts you can tilt the seat backwards to make it easier to disconnect the electrical connections.*

Note: *Lifting the seat out of the car may require two people.*

28.3b. . . then remove the front bolts

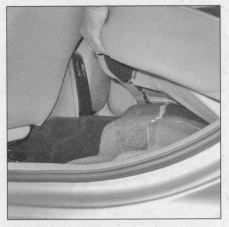

28.5 Pull up on the front of the seat cushion, to release the retainer loops

28.6 Remove the fasteners securing the seat backs

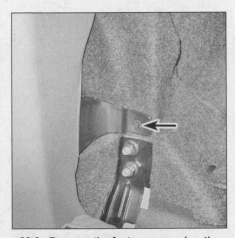

29.3a Remove the fastener securing the sides of the seat back

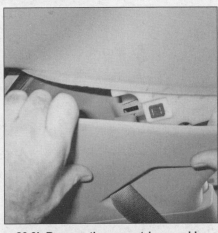

29.3b Remove the upper trim panel by pulling it outward carefully to free it from the spring clips

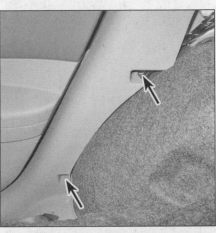

29.3c Remove the fasteners securing the lower portion of the C-pillar trim

4   Installation is the reverse of removal. Tighten the bolts to the torque listed in this Chapter's Specifications.

Rear seat

5   Pull up on the front of the seat cushion, to release the retainer loops (see illustration). Remove the cushion.

6   Fold the left and right seat backs forward (60/40 split seat back), then remove the fasteners securing each unit (see illustration).

7   Remove the seat assembly.

8   Installation is the reverse of removal. Tighten the fasteners to the torque listed in this Chapter's Specifications.

29   Rear package tray - removal and installation

1   Remove the rear seat cushion (see Section 28).

2   Unlock and flip forward the seat backs.

3   Remove the C-pillar trim panels (see illustrations).

4   Remove the fastener securing the pack-

age tray, then remove the tray from the vehicle (see illustrations).

5   Installation is the reverse of removal.

29.4a Remove the fastener at the center portion of the package tray. . .

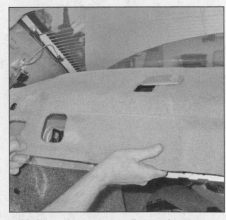

29.4b. . . then remove the package tray

30   Rear spoiler - removal and installation

1   Remove the liftgate upper trim panel (see Section 21).
2   Disconnect the rear washer fluid hose.
3   Disconnect any electrical connectors, then remove the spoiler mounting fasteners.
4   Remove the spoiler from the liftgate.
5 Installation is the reverse of removal.

Chapter 12
Chassis electrical system

Contents

Specifications

Torque specifications — Ft-lbs (unless otherwise indicated)

Wiper system	
Wiper arm mounting nuts	13
Motor mounting screws	66 in-lbs
Linkage mounting nuts	77 in-lbs
Seat Belts	
Front	
Seat belt buckle mounting bolt	32
Seat belt tensioner assembly to seat	29
Retractor mounting bolt	
2014 and earlier	23
2015 and later	33
Lower anchoring bolt	
2012 and earlier	27
2013 and later	33
Rear	
Seat belt buckle mounting bolt	
2014 and earlier	25
2015 and later	41
Retractor mounting bolt	
2014 and earlier	25
2015 and later	37
Lower anchoring bolt	
2012 and earlier	23
2013 and 2014	33
2015 and later	41

1 General Information

1 The electrical system is a 12-volt, negative ground type. Power for the lights and all electrical accessories is supplied by a lead/acid battery, which is charged by the alternator.

2 This Chapter covers repair and service procedures for the various electrical components not associated with the engine. Information on the battery, alternator and starter motor can be found in Chapter 5.

3 It should be noted that when portions of the electrical system are serviced, the negative battery cable should be disconnected from the battery to prevent electrical shorts and/or fires.

2 Electrical troubleshooting - general information

1 A typical electrical circuit consists of an electrical component, any switches, relays, motors, fuses, fusible links or circuit breakers related to that component and the wiring and connectors that link the component to both the

battery and the chassis. To help you pinpoint an electrical circuit problem, wiring diagrams are included at the end of this Chapter.

2   Before tackling any troublesome electrical circuit, first study the appropriate wiring diagrams to get a complete understanding of what makes up that individual circuit. Trouble spots, for instance, can often be narrowed down by noting if other components related to the circuit are operating properly. If several components or circuits fail at one time, chances are the problem is in a fuse or ground connection, because several circuits are often routed through the same fuse and ground connections.

3   Electrical problems usually stem from simple causes, such as loose or corroded connections, a blown fuse, a melted fusible link or a failed relay. Visually inspect the condition of all fuses, wires and connections in a problem circuit before troubleshooting the circuit.

4   If test equipment and instruments are going to be utilized, use the diagrams to plan ahead of time where you will make the necessary connections in order to accurately pinpoint the trouble spot.

5   The basic tools needed for electrical troubleshooting include a circuit tester or voltmeter (a 12-volt bulb with a set of test leads can also be used), a continuity tester, which includes a bulb, battery and set of test leads, and a jumper wire, preferably with a circuit breaker incorporated, which can be used to bypass electrical components **(see illustrations)**. Before attempting to locate a problem with test instruments, use the wiring diagram(s) to decide where to make the connections.

Voltage checks

6   Voltage checks should be performed if a circuit is not functioning properly. Connect one lead of a circuit tester to either the negative battery terminal or a known good ground. Connect the other lead to a connector in the circuit being tested, preferably nearest to the battery or fuse **(see illustration)**. If the bulb of the tester lights, voltage is present, which means that the part of the circuit between the connector and the battery is problem free. Continue checking the rest of the circuit in the same fashion. When you reach a point at which no voltage is present, the problem lies

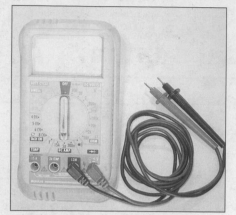

2.5a The most useful tool for electrical troubleshooting is a digital multimeter that can check volts, amps, and test continuity

2.5b A simple test light is a very handy tool for testing voltage

between that point and the last test point with voltage. Most of the time the problem can be traced to a loose connection. **Note:** *Keep in mind that some circuits receive voltage only when the ignition key is in the Accessory or Run position.*

Finding a short

7   One method of finding shorts in a circuit is to remove the fuse and connect a test light or voltmeter in place of the fuse terminals. There should be no voltage present in the circuit. Move the wiring harness from side-to-side while watching the test light. If the bulb goes on, there is a short to ground somewhere in that area, probably where the insulation has rubbed through. The same test can be performed on each component in the circuit, even a switch.

Ground check

8   Perform a ground test to check whether a component is properly grounded. Disconnect the battery and connect one lead of a continuity tester or multimeter (set to the ohms scale), to a known good ground. Connect the other lead to the wire or ground connection being tested. If the resistance is low (less than 5 ohms), the ground is good. If the bulb on a self-powered test light does not go on, the ground is not good.

Continuity check

9   A continuity check is done to determine if there are any breaks in a circuit - if it is passing electricity properly. With the circuit off (no power in the circuit), a self-powered continuity tester or multimeter can be used to check the circuit. Connect the test leads to both ends of the circuit (or to the power end and a good ground), and if the test light comes on the circuit is passing current properly **(see illustration)**. If the resistance is low (less than 5 ohms), there is continuity; if the reading is 10,000 ohms or higher, there is a break somewhere in the circuit. The same procedure can be used to test a switch, by connecting the continuity tester to the switch terminals. With the switch turned On, the test light should come on (or low resistance should be indicated on a meter).

Finding an open circuit

10   When diagnosing for possible open circuits, it is often difficult to locate them by sight because the connectors hide oxidation or terminal misalignment. Merely wiggling a connector on a sensor or in the wiring harness may correct the open circuit condition. Remember this when an open circuit is indicated when troubleshooting a circuit. Intermittent problems may also be caused by oxidized or loose connections.

11   Electrical troubleshooting is simple if you keep in mind that all electrical circuits are basically electricity running from the battery, through the wires, switches, relays, fuses and fusible links to each electrical component (light bulb, motor, etc.) and to ground, from which it is passed back to the battery. Any electrical problem is an interruption in the flow of electricity to and from the battery.

Connectors

12   Most electrical connections on these vehicles are made with multiwire plastic connectors. The mating halves of many connectors are secured with locking clips molded into the plastic connector shells. The mating halves of large connectors, such as some of

2.6 In use, a basic test light's lead is clipped to a known good ground, then the pointed probe can test connectors, wires or electrical sockets - if the bulb lights, the circuit being tested has battery voltage

those under the instrument panel, are held together by a bolt through the center of the connector.

13 To separate a connector with locking clips, use a small screwdriver to pry the clips apart carefully, then separate the connector halves. Pull only on the shell, never pull on the wiring harness as you may damage the individual wires and terminals inside the connectors. Look at the connector closely before trying to separate the halves. Often the locking clips are engaged in a way that is not immediately clear. Additionally, many connectors have more than one set of clips.

14 Each pair of connector terminals has a male half and a female half. When you look at the end view of a connector in a diagram, be sure to understand whether the view shows the harness side or the component side of

2.9 With a multimeter set to the ohm scale, resistance can be checked across two terminals - when checking for continuity, a low reading indicates continuity, a high reading or infinity indicates high resistance or lack of continuity

Electrical connectors

Most electrical connectors have a single release tab that you depress to release the connector

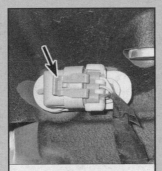

Some electrical connectors have a retaining tab which must be pried up to free the connector

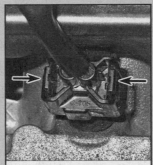

Some connectors have two release tabs that you must squeeze to release the connector

Some connectors use wire retainers that you squeeze to release the connector

Critical connectors often employ a sliding lock (1) that you must pull out before you can depress the release tab (2)

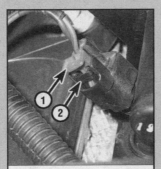

Here's another sliding-lock style connector, with the lock (1) and the release tab (2) on the side of the connector

On some connectors the lock (1) must be pulled out to the side and removed before you can lift the release tab (2)

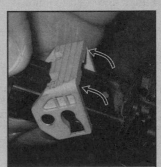

Some critical connectors, like the multi-pin connectors at the Powertrain Control Module employ pivoting locks that must be flipped open

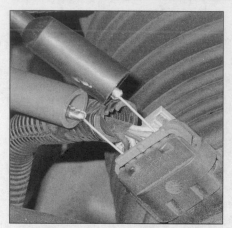

2.15 To backprobe a connector, insert a small, sharp probe (such as a straight-pin) into the back of the connector alongside the desired wire until it contacts the metal terminal inside; connect your meter leads to the probes - this allows you to test a functioning circuit

3.1a The engine compartment fuse and relay box is located at the right front corner of the engine compartment. To open the lid, pull out the latch and swing up the lid

3.1b The rear fuse and relay box is located in the trunk, beneath the spare tire access panel. To open the lid, pull back the latch on the left end of the lid and swing the lid up

the connector. Connector halves are mirror images of each other, and a terminal shown on the right side end view of one half will be on the left side end view of the other half.

15   It is often necessary to take circuit voltage measurements with a connector connected. Whenever possible, carefully insert a small straight pin (not your meter probe) into the rear of the connector shell to contact theterminal inside, then clip your meter lead to the pin. This kind of connection is called "backprobing" (see illustration). When inserting a test probe into a male terminal, be careful not to distort the terminal opening. Doing so can lead to a poor connection and corrosion at that terminal later. Using the small straight pin instead of a meter probe results in less chance of deforming the terminal connector.

3   Fuses and fusible links - general information

Fuses

1   The electrical circuits of the vehicle are protected by a combination of fuses and relays. The engine compartment fuse and relay box is located at the right front corner of the engine compartment (see illustrations). You can access either fuse and relay box by simply opening the cover.

2   At a dealer parts department you might hear the phrase "Integrated Power Module" or "Power Distribution Center." These are the Chrysler/Dodge terms for the fuse and relay boxes. In this manual, we use the simpler term "fuse and relay box." However, you should know the location of each specific fuse and relay box by its factory name so that you can communicate with a Chrysler or Dodge parts department if you have to buy fuses or relays.

The Integrated Power Module is located at the right front corner of the engine compartment; it's also referred to as the "Front Power Distribution Center." The Rear Power Distribution Center is located in the trunk, under the spare tire access panel.

3   Each fuse is designed to protect a specific circuit, and the various circuits are identified (in a highly abbreviated way) on the fuse panel itself. Different sizes of fuses are employed. There are "mini" and "maxi" sizes, with the larger located in the fuse and relay box. You'll need to use electronics needle-noise pliers or a small plastic fuse-puller tool to remove most fuses. There should be one of these fuse-puller tools in the fuse and relay box. If an electrical component fails, always check the fuse first. The best way to check the fuses is with a test light. Check for power at the exposed terminal tips of each fuse. If power is present at one side of the fuse but not the other, the fuse is blown. A blown fuse can also be identified by visually inspecting it (see illustration).

4   Be sure to replace blown fuses with the correct type. Fuses of different ratings are physically interchangeable, but only fuses of the proper rating should be used. Replacing a fuse with one of a higher or lower value than specified is not recommended. Each electrical circuit needs a specific amount of protection. The amperage rating of each fuse is molded into the fuse body.

5   If the replacement fuse immediately fails, don't replace it again until the cause of the problem is isolated and corrected. In most cases, the cause will be a short circuit in the wiring caused by a broken or deteriorated wire.

Fusible links

6   The wiring between the alternator and the starter is protected by two fusible links. These links function like a fuse, in that they melt when the circuit is overloaded, but they resemble a large-gauge wire. To replace a fusible link, first disconnect the negative cable from the battery. Disconnect the burned-out link and replace it with a new one (available from your dealer or auto parts store). Always determine the cause for the overload that melted the fusible link before installing a new one.

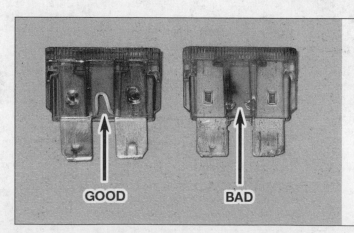

3.3 When a fuse blows, the element between the terminals melts

GOOD   BAD

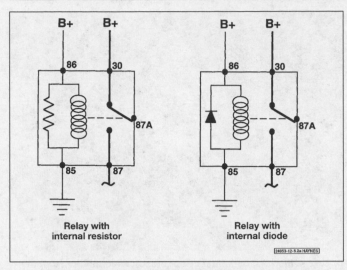

5.5a Typical ISO relay designs, terminal numbering and circuit connections

5.5b Most relays are marked on the outside to easily identify the control circuit and power circuits - this one is of the four-terminal type

4 Circuit breakers - general information

1 Circuit breakers protect certain heavy-load circuits. On some models there will be one to three circuit breakers located in the rear fuse and relay box.

2 Because the circuit breakers reset automatically, an electrical overload in a circuit-breaker-protected system will cause the circuit to fail momentarily, then come back on. If the circuit does not come back on, check it immediately.

3 For a basic check, pull the circuit breaker up out of its socket on the fuse panel, but just far enough to probe with a voltmeter. The breaker should still contact the sockets.

4 With the voltmeter negative lead on a good chassis ground, touch each end prong of the circuit breaker with the positive meter probe. There should be battery voltage at each end. If there is battery voltage only at one end, the circuit breaker must be replaced.

5 Relays - general information and testing

1 Many electrical accessories in the vehicle utilize relays to transmit current to the component. If the relay is defective, the component won't operate properly.

2 Most relays are located in the engine compartment fuse and relay box **(see illustration 3.1a or 3.1b)**.

3 Some relays are located in other parts of the vehicle or in various wiring harnesses underneath the instrument panel.

4 If a faulty relay is suspected, it can be removed and tested using the procedure below or by a dealer service department or a repair shop. Defective relays must be replaced as a unit.

Testing

5 Most of the relays used in these vehicles are of a type often called "ISO" relays, which refers to the International Standards Organization. The terminals of ISO relays are numbered to indicate their usual circuit connections and functions. There are two basic layouts of terminals on the relays used in the vehicles covered by this manual **(see illustrations)**.

6 Refer to the wiring diagram for the circuit to determine the proper connections for the relay you're testing. If you can't determine the correct connection from the wiring diagrams, however, you may be able to determine the test connections from the information that follows.

7 Two of the terminals are the relay control circuit and connect to the relay coil. The other relay terminals are the power circuit. When the relay is energized, the coil creates a magnetic field that closes the larger contacts of the power circuit to provide power to the circuit loads.

8 Terminals 85 and 86 are normally the control circuit. If the relay contains a diode, terminal 86 must be connected to battery positive (B+) voltage and terminal 85 to ground. If the relay contains a resistor, terminals 85 and 86 can be connected in either direction with respect to B+ and ground.

9 Terminal 30 is normally connected to the battery voltage (B+) source for the circuit loads. Terminal 87 is connected to the ground side of the circuit, either directly or through a load. If the relay has several alternate terminals for load or ground connections, they usually are numbered 87A, 87B, 87C, and so on.

10 Use an ohmmeter to check continuity through the relay control coil.

a) *Connect the meter according to the polarity shown in illustration 5.5a for one check; then reverse the ohmmeter leads and check continuity in the other direction.*

b) *If the relay contains a resistor, resistance will be indicated on the meter, and should be the same value with the ohmmeter in either direction.*

c) *If the relay contains a diode, resistance should be higher with the ohmmeter in the forward polarity direction than with the meter leads reversed.*

d) *If the ohmmeter shows infinite resistance in both directions, replace the relay.*

11 Remove the relay from the vehicle and use the ohmmeter to check for continuity between the relay power circuit terminals. There should be no continuity between terminal 30 and 87 with the relay de-energized.

12 Connect a fused jumper wire to terminal 86 and the positive battery terminal. Connect another jumper wire between terminal 85 and ground. When the connections are made, the relay should click.

13 With the jumper wires connected, check for continuity between the power circuit terminals. Now there should be continuity between terminals 30 and 87.

14 If the relay fails any of the above tests, replace it.

6 Front Control Module (FCM) - replacement

Note: *The Front Control Module (FCM) is located on the front of the front Power Distribution Center (the engine compartment fuse and relay box). Chrysler/Dodge refers to the FCM/Power Distribution Center assembly as the Integrated Power Module (IPM). The IPM is connected directly to the battery and provides the primary means of circuit protection and power distribution for all vehicle electrical systems, including:*

a) *Air conditioning condenser cooling fan*

b) *Daytime Running Lights (DRL) system*

6.2 Pull back the tang and swing the Power Distribution Center up . . .

6.3 . . . then, to disconnect each electrical connector from the Front Control Module (FCM), slide the red locks toward the wiring harness, then depress the release tabs (A) and pull off the connectors

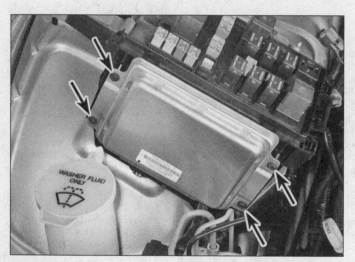

6.4 To detach the FCM from the engine compartment fuse and relay box, remove these four screws

7.3 To keep the clockspring from uncentering itself, loosen - BUT DON'T REMOVE - at least one of the three clockspring screws

c) Front and rear hazard warning light system (there is no conventional hazard flasher relay)
d) Front turn signal lights
e) Horns
f) Radiator fans
g) Rear window defroster power and timing
h) Brake lights, turn signal lights and taillights
i) Windshield, and on Magnums, liftgate wiper/washer systems

1   Disconnect the cable from the negative battery terminal (see Chapter 5, Section 1).
2   Disconnect the positive battery cable from the front Power Distribution Center, then pull back the tang and lift the Power Distribution Center up (see illustration).
3   Disconnect the electrical connectors from the FCM (see illustration).
4   Remove the FCM mounting screws (see

illustration) and remove the FCM.
5   Installation is the reverse of removal.

---

7   Steering Column Control Module and steering column switches - replacement

---

Steering Column Control Module (SCCM)

Warning: *The models covered by this manual are equipped with a Supplemental Restraint System (SRS), more commonly known as airbags. Always disarm the airbag system before working in the vicinity of any airbag system component to avoid the possibility of accidental deployment of the airbag, which could cause personal injury (see Section 26). Do not*

use a memory-saving device to preserve the PCM's memory when working on or near airbag system components.
Note: *The Steering Column Control Module (SCCM) includes the clockspring for the steering wheel airbag, the multi-function switch, the speed control switch, the Steering Angle Sensor (SAS), if equipped with Electronic Stability Program, and the Steering Column Module (SCM).*
1   Disconnect the cable from the negative battery terminal (see Chapter 5, Section 1).
2   Remove the steering wheel airbag and the steering wheel (see Chapter 10).

2014 and earlier models
3   To prevent the clockspring from uncentering itself, loosen - BUT DON'T REMOVE - at least one clockspring screw (see illustration).

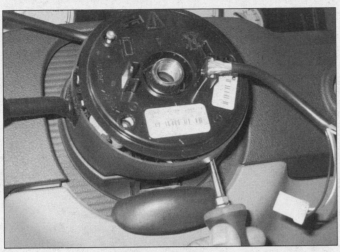

7.4 To release the SCCM from the steering column, insert a screwdriver through the access hole in the bottom of the SCCM and back out the setscrew

7.5 To remove the SCCM from the steering column, pull it to the rear

7.6 To detach the clockspring from the SCCM, loosen these three screws just enough to pull off the clockspring, but not so much that the clockspring unwinds

7.7 Separate the clockspring from the SCCM

4   Back out the setscrew through the access hole in the bottom of the SCCM **(see illustration)**.

5   Remove the SCCM from the steering column **(see illustration)** and place the SCCM on a clean workbench.

6   Loosen all three clockspring screws **(see illustration)** just enough to detach the clockspring from the SCCM.

Caution: *Do NOT completely unscrew the clockspring screws or the clockspring will unwind.*

7   Remove the clockspring from the SCCM **(see illustration)**.

8   Remove the Steering Angle Sensor (SAS) retaining screw and remove the SAS **(see illustrations)**.

7.8a To detach the Steering Angle Sensor (SAS) from the SCCM, remove this retaining screw. . .

7.8b. . . then lift the SAS out of the SCCM

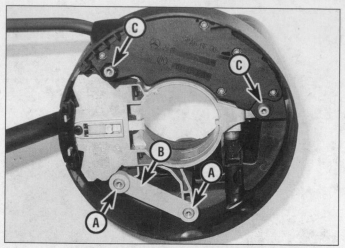

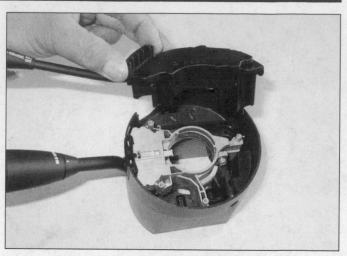

7.9 On vehicles with a manual telescoping steering column, remove these two Phillips screws (A) and remove the strut (B). On all vehicles, remove these two speed control switch screws (C). . .

7.10. . . and remove the speed control switch

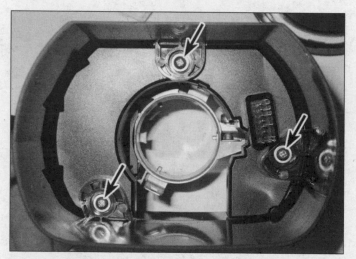

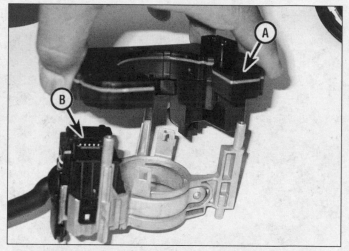

7.11 To detach the multi-function switch and Steering Control Module (SCM) from the SCCM shroud, remove these three screws, then pull the SCM and multi-function switch out of the shroud

7.12 Separate the SCM (A) from the multi-function switch (B)

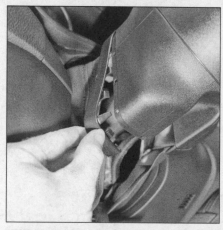

7:16 Detach the gap hider bezel from the steering column shrouds

9   If the vehicle is equipped with a manual telescoping steering column, remove the strut screws **(see illustration)** and remove the strut. If the vehicle is equipped with an electronic telescoping steering column, remove the switch mounting screws and remove the switch and stalk.

10   Remove the speed control switch retaining screws **(see illustration 7.9)** and remove the speed control switch **(see illustration)**.

11   On vehicles with an SCCM assembly on which the steering column tilt/telescoping lever is an integral part of the SCCM assembly, remove the two-lever assembly retaining screws and remove the lever assembly **(see illustration)**. (On some models, the tilt lever is located underneath the steering column and its removal is not necessary for this procedure.)

12   Remove the three screws at the bottom of the SCCM shroud **(see illustration)** for the multi-function switch and the Steering Control

Module (SCM), then remove the switch and SCM as a single assembly.

13   Separate the multi-function switch from the SCM.

14   No further disassembly is possible.

15   Installation is the reverse of removal.

2015 and later models

16   Using a flat-bladed trim tool, pry the steering column gap hider bezel from the steering column covers **(see illustration)**.

17   Sharply tug on the SCCM/shroud assembly and slide it off of the steering colum, then disconnect the wiring harness electrical connector from the SCCM **(see illustration)**.

18   Remove the three screws from the back side of the assembly, sepaarate the cover halves, then remove the SCCM **(see illustrations)**.

19   Installation is the reverse of the removal procedure.

7.17 Pull the SCCM/shroud assembly off of the steering column and disconnect the electrical connector

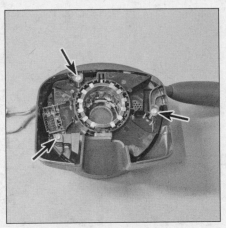

7.18a Remove the screws . . .

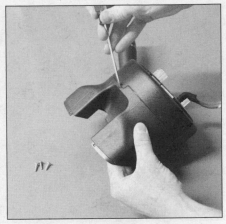

7.18b . . . carefully pry the cover halves apart . . .

7.18c . . . and remove the SCCM

7.21a Pry off the trim cover(s) . . .

7.21b . . . and remove the switch retaining screw(s)

Steering wheel switches

20   Remove the steering wheel airbag and the steering wheel (see Chapter 10).

21   Turn the steering wheel upside down on a flat surface and remove the trim cover to expose the retaining screw (**see illustrations**).

22   Remove the switch on the back side of the steering wheel, then remove the screw securing the switch to the front side of the steering wheel (**see illustration**).

23   Turn the steering wheel over and remove the remaining screw securing the switch(es) to the front side of the steering wheel (**see illustration**). Feed the harness through the opening and remove the switch(es).

24   Installation is the reverse of removal.

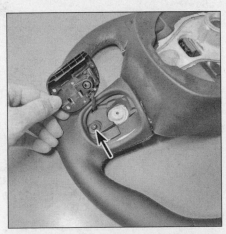

7.22 Separate the switch from the steering wheel, then remove the exposed screw to detach the switch on the opposite side

7.23 Switch retaining screws

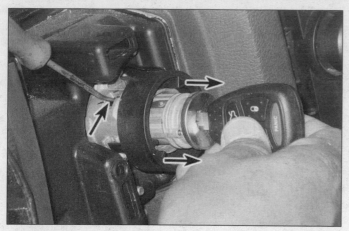

8.4 To remove the key lock cylinder, turn the ignition key to the On position, depress the spring-loaded retaining lug with a small screwdriver, awl, punch or scribe, and pull out the key lock cylinder

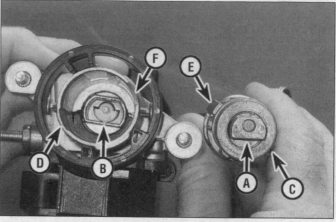

8.5 Lock cylinder installation details

A must align with B
C must align with D

E must align with F

8   Key lock cylinder and ignition switch - replacement

Warning: *The models covered by this manual are equipped with a Supplemental Restraint System (SRS), more commonly known as airbags. Always disarm the airbag system before working in the vicinity of any airbag system component to avoid the possibility of accidental deployment of the airbag, which could cause personal injury (see Section 26). Do not use a memory-saving device to preserve the PCM's memory when working on or near airbag system components.*

Note: *If the key is difficult to turn, the problem might not be in the ignition switch or the key lock cylinder. The transmission shift cable or the Brake Transmission Shift Interlock (BTSI) cable might be out of adjustment (see Chapter 7B).*

Key lock cylinder

Caution: *Make sure that you obtain the correct key lock cylinder for your ignition key BEFORE you install the lock cylinder. Otherwise,* you will find that once you have installed a new key lock cylinder the key won't fit! Take your ignition key and your vehicle's Vehicle Identification Number (VIN) with you when you purchase a new key lock cylinder. If you lose an ignition key and have to replace it, you might be in trouble. The ignition key contains a transponder with its own unique encrypted Radio Frequency (RF) signal, and a new key will have to be programmed to work with the Sentry Key Remote Entry Module (SKREEM) (see below).

1   Disconnect the cable from the negative battery terminal (see Chapter 5, Section 1).
2   Using a trim stick, carefully pry off the ignition switch bezel (the small circular trim piece around the key lock cylinder) and remove the instrument cluster bezel (see Chapter 11).
3   Insert the ignition key into the key lock cylinder and turn it to the On position.
4   Insert a small awl, punch or scribe into the hole for the key lock cylinder retaining pin, depress the retaining pin and pull the key lock cylinder out of the steering column **(see illustration)**.

5   Before installing the key lock cylinder, make sure that the ignition key is still in the On position. Align the D-shaped lug on the end of the key lock cylinder with the D-shaped socket in the bottom of the lock cylinder housing **(see illustration)** and push the lock cylinder into the ignition switch until it clicks into place.
6   Installation is otherwise the reverse of removal.

Ignition switch

7   Disconnect the cable from the negative battery terminal (see Chapter 5, Section 1).
8   Disable the airbag system (see Section 26).
9   Remove the knee bolster and the knee bolster reinforcement panel (see Chapter 11).
10   Remove the lower mounting nut and the two front mounting bolts **(see illustration)** and remove the key lock cylinder/ignition switch assembly.
11   Disconnect the electrical connectors from the ignition switch and from the Sentry Key Remote Entry Module **(see illustrations)**.
12   Disconnect the Brake Transmission Shift

8.10 To detach the key lock cylinder/ ignition switch housing from the instrument panel, remove this nut (1) and these two bolts (2)

8.11a To disconnect the electrical connector from the ignition switch, slide the red lock sideways, then depress the release tab (A) and pull off the connector

8.11b To disconnect the electrical connector from the Sentry Key Remote Entry Module (SKREEM), depress this release tab and pull off the connector

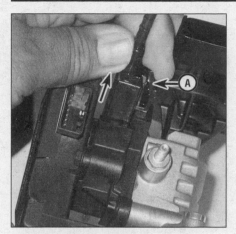

8.12 To disconnect the Brake Transmission Shift Interlock (BTSI) cable from the ignition switch/key lock cylinder assembly, depress this release tab and pull off the cable

8.13 To detach the ignition switch from the key lock cylinder housing, remove this Torx screw

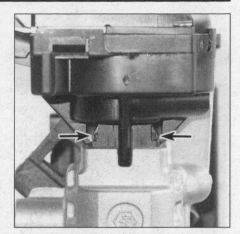

8.14a To remove the ignition switch from the key lock cylinder housing, depress these two release tabs. . .

Interlock (BTSI) cable **(see illustration)**.
13   Remove the ignition switch mounting screw **(see illustration)**.
14   Depress the release tabs **(see illustrations)** and remove the ignition switch
15   Installation is the reverse of removal.

Keyless start button

16   Insert a small screwdriver under the chrome ring at the bottom of the button. Gently pry the button from the instrument panel and disconnect the electrical connector **(see illustration)**.

Sentry Key Remote Entry Module (SKREEM)/Wireless Control Module (WCM)

17   The SKREEM, also sometimes referred to as the Wireless Control Module (WCM), is the small black plastic module mounted

on the left side of the ignition switch/key lock cylinder housing. The SKREEM/WCM is part of the Sentry Key Remote Entry System (SKREES). When you press the Unlock button on the ignition key, a transponder inside the key produces its own unique encrypted Radio Frequency (RF) signal, which the SKREEM/WCM recognizes as a valid signal. The SKREEM/WCM unlocks the door locks (one push to unlock the driver's door, two pushes to unlock the other doors), activates the vehicle's electronic modules via the Controller Area Network (CAN) and allows you to start the engine. When you press the Lock button, the transponder's RF signal tells the SKREEM to lock the doors, activate the security system and disable all electronic modules. The SKREEM/WCM and the ignition key are programmed to work together. If you replace the key or the SKREEM/WCM, you must have them reprogrammed at a dealership or other

qualified repair shop, so we don't recommend trying to replace the SKREEM at home.

9   Instrument panel switches and clock - replacement

Warning: *The models covered by this manual are equipped with a Supplemental Restraint System (SRS), more commonly known as airbags. Always disarm the airbag system before working in the vicinity of any airbag system component to avoid the possibility of accidental deployment of the airbag, which could cause personal injury (see Section 26). Do not use a memory-saving device to preserve the PCM's memory when working on or near airbag system components.*
1   Disconnect the cable from the negative battery terminal (see Chapter 5, Section 3).

8.14b. . . with a pair of needle-nose pliers

8.16 Even though the factory manual describes the removal by gently prying it out, we found it was easier to squeeze the tangs inward from the back side while you push the switch through the instrument panel

9.3 To remove the headlight switch from the instrument panel, push it out from behind

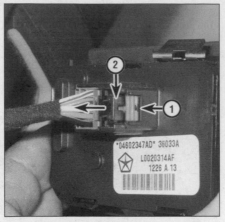

9.4 To disconnect the electrical connector from the headlight switch, slide the red lock (1) toward the wiring harness until it snaps out and overlaps the release tab (2), then depress the red lock and release tab and pull off the connector

9.10 To remove the instrument panel switch pod from the instrument panel center bezel, depress the release tab (1) and disconnect the electrical connector, then remove the two switch pod mounting screws (2) and remove the pod from the center bezel

Headlight switch

2005 through 2010 300, Charger and Magnum models/2008 through 2014 Challenger models

2   Remove the steering column opening cover.

3   Reach behind the instrument panel and push the retaining tab on the bottom of the switch up, then push the switch out of the instrument panel from behind **(see illustration)**.

4   Disconnect the electrical connector from the headlight switch **(see illustration)**.

5   Installation is the reverse of removal.

2011 and later 300 and Charger/ 2015 and later Challenger models

6   Turn the headlight switch to the Off position.

7   Push the switch knob inward and rotate the switch clockwise to the Park Lamps On position, then pull the switch out and disconnect the electrical connector.

8   Installation is the reverse of removal.

Instrument panel switch pod

9   Remove the instrument panel center bezel (see Chapter 11).

10   Disconnect the electrical connector from the instrument panel switch pod **(see illustration)**.

11   Remove the instrument panel switch pod mounting screws.

12   Remove the instrument panel switch pod from the instrument panel center bezel.

13   Installation is the reverse of removal.

Clock

14   Remove the instrument panel center bezel (see Chapter 11).

15   Disconnect the electrical connector from the clock **(see illustration)**.

16   Remove the clock mounting screws.

17   Remove the clock from the instrument panel center bezel.

18   Installation is the reverse of removal.

10   Instrument cluster - removal and installation

Warning: *The models covered by this manual are equipped with a Supplemental Restraint System (SRS), more commonly known as airbags. Always disarm the airbag system before working in the vicinity of any airbag system component to avoid the possibility of accidental deployment of the airbag, which could cause personal injury (see Section 26). Do not use a memory-saving device to preserve the PCM's memory when working on or near airbag system components.*

1   Disconnect the cable from the negative battery terminal (see Chapter 5, Section 1).

2   Remove the instrument cluster bezel (see Chapter 11).

3   Remove the two instrument cluster mounting screws **(see illustrations)**.

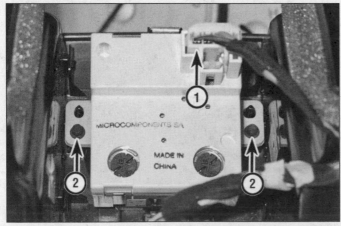

9.15 To remove the clock from the instrument panel center bezel, depress the release tab (1) and disconnect the electrical connector, then remove the clock mounting screws (2) and remove the clock from the center bezel

10.3a Instrument cluster screws - early models (300 shown)

10.3b Instrument cluster screws - later models
(Challenger shown)

10.4 Pull out the instrument cluster, depress the release tabs
and disconnect these two electrical connectors (there's another
smaller connector, not shown in this photo, at the
other end of the cluster)

4   Pull out the cluster and disconnect the electrical connectors from the backside of the cluster **(see illustration)**.
5   Installation is the reverse of removal.

---

11   Wiper and washer systems - component removal and installation

---

Windshield wiper motor

1   Disconnect the cable from the negative battery terminal (see Chapter 5, Section 1).
2   Remove the protective trim cap from each wiper arm, then remove the wiper arm retaining nuts **(see illustration)**.
3   Mark the relationship of the wiper arms to their shafts **(see illustration)**, then remove the wiper arms.
4   Remove the cowl cover (see Chapter 11).
5   Disconnect the electrical connector from the windshield wiper motor **(see illustration)**.
6   Remove the windshield wiper module mounting bolts **(see illustration)** and remove the module.

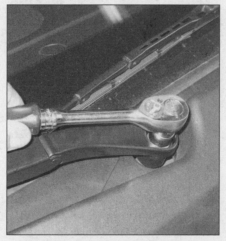

11.2 Carefully pry off the protective trim cap from each windshield wiper arm. . .

11.3. . . then remove the wiper arm retaining nuts and mark the relationship between each wiper arm and its shaft

11.5 To disconnect the electrical connector from the windshield wiper motor, depress the release tab and pull off the connector

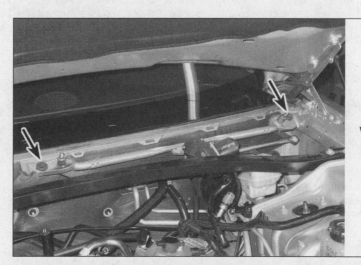

11.6 To detach the windshield wiper module from the cowl area, remove these two mounting bolts

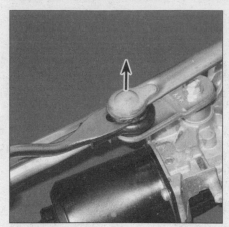

11.7 To separate the two link rods from the bellcrank pin, pry them off, one at a time, with a trim removal tool or large screwdriver

11.8a To detach the bellcrank from the windshield wiper motor shaft, remove this nut

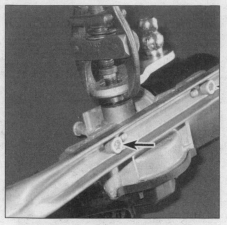

11.8b Use a small puller to separate the bellcrank from the wiper motor nut. To detach the motor from the wiper module assembly, remove these two Torx screws

11.10 Inspect the condition of the wiper motor's rubber insulator. If it's cracked, torn, deteriorated or otherwise damaged, replace it

11.14 Remove the rear wiper arm retaining nut

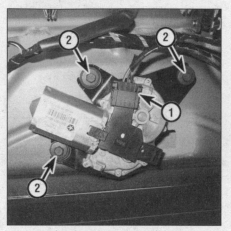

11.18 Disconnect the electrical connector (1) from the rear wiper motor, then remove the three wiper motor mounting bolts (2) and remove the wiper motor from the rear liftgate

7   Disconnect the two link rods from the bellcrank **(see illustration)**.

8   Remove the nut that secures the bellcrank to the motor shaft **(see illustration)**, then use a small puller to remove the bellcrank from the wiper motor shaft **(see illustration)**.

9   Remove the windshield wiper motor mounting screws and remove the wiper motor from the module.

10   Inspect the condition of the wiper motor's rubber insulator **(see illustration)**.

11   Installation is the reverse of removal.

Rear wiper motor (Magnum models)

12   Disconnect the cable from the negative battery terminal (see Chapter 5, Section 1).

13   Remove the trim cap from the rear wiper arm.

14   Remove the rear wiper arm retaining nut **(see illustration)**.

15   Mark the relationship of the wiper arm to the wiper motor shaft.

16   Remove the arm from the motor shaft.

17   Remove the rear liftgate trim panel (see Chapter 11).

18   Disconnect the electrical connector from the rear wiper motor **(see illustration)**.

19   Remove the rear wiper motor mounting bolts **(see illustration 11.18)**.

20   Remove the rear wiper motor.

21   Installation is the reverse of removal.

Washer fluid reservoir and pump

22   The washer bottle is located forward of the right front wheel and mounted on the chassis. The washer pump motor and the washer level indicator are located on the washer reservoir.

23   Disconnect the cable from the negative terminal of the battery (see Chapter 5, Section 3).

24   Remove the fasteners from the front por-

tion of the right inner fender splash shield (see Chapter 11).

Note: *Be sure to have a catch pan under the reservoir before removing the pump or level indicator.*

25   Disconnect the electrical connector from the washer pump (and from the fluid level sensor if the reservoir is to be removed) **(see illustration)**.

26   Disconnect the washer pump hoses.

Note: *Be sure to have a catch pan under the reservoir before removing the pump or level indicator.*

27   If you're replacing the washer pump, pull it straight out of the reservoir grommet. To install, make sure the grommet is seated properly, then push the new pump straight into the grommet until it seats.

28   To remove the reservoir, unscrew the mounting fasteners and guide it out through the fenderwell.

29   Installation is the reverse of removal.

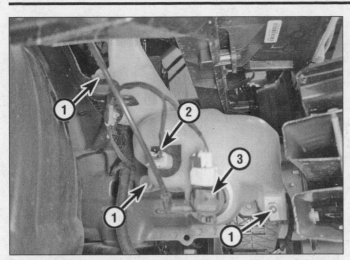

11.25 Washer fluid reservoir and fluid pump details - bumper cover removed for clarity (Challenger model shown)

1   Reservoir fasteners   3   Fluid pump
2   Fluid level sensor

12.3a To detach the radio from the dash, remove these four mounting screws (early model shown)

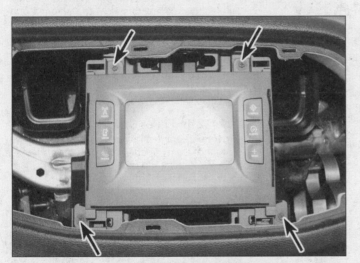

12.3b Later models just slightly different

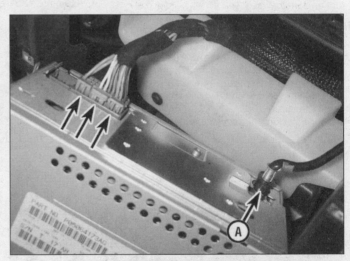

12.4 Pull the radio out of the dash far enough to disconnect the antenna cable (A) and the electrical connectors

12   Radio and speakers - removal and installation

Warning: *The models covered by this manual are equipped with a Supplemental Restraint System (SRS), more commonly known as airbags. Always disarm the airbag system before working in the vicinity of any airbag system component to avoid the possibility of accidental deployment of the airbag, which could cause personal injury (see Section 26). Do not use a memory-saving device to preserve the PCM's memory when working on or near airbag system components.*

Radio

1   Disconnect the cable from the negative battery terminal (see Chapter 5, Section 1).
2   Remove the instrument panel center bezel (see Chapter 11).

3   Remove the radio mounting screws **(see illustrations)** and pull the radio out of the dash.
4   Disconnect the antenna cable and the electrical connectors from the radio **(see illustration)**.
5   Installation is the reverse of removal.

Speakers
Door speakers

6   Disconnect the cable from the negative battery terminal (see Chapter 5, Section 1).
7   Remove the door trim panel (see Chapter 11).
8   Remove the four speaker mounting screws **(see illustration)**.
9   Pull out the speaker and disconnect the electrical connector.
10 Installation is the reverse of removal.

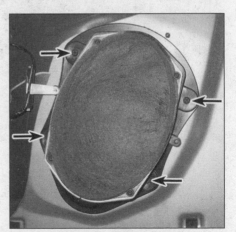

12.8 To detach the speaker from the front door, remove these four mounting screws (rear door speakers similar)

12.12 On 2011 and earlier models, carefully pry off the speaker cover with a trim panel removal tool

Instrument panel speakers

11 Disconnect the cable from the negative battery terminal (see Chapter 5, Section 1).

12 **2011 and earlier models:** Remove the speaker cover **(see illustration)**.

13 **2012 and later models:** Remove the A-pillar trim panels and the defroster grille panel (see Chapter 11, Section 27).

14 Remove the speaker mounting screws. Pull out the speaker and disconnect the electrical connector.

15 Installation is the reverse of removal.

Rear speakers

Sedans

16 Disconnect the cable from the negative battery terminal (see Chapter 5, Section 1).

17 Remove the upper C-pillar trim and the lower C-pillar trim from both sides, then remove the rear shelf trim (see Chapter 11).

18 Remove the rear speaker mounting screws **(see illustration)**.

19 Pull out the speaker and disconnect the electrical connector.

20 Installation is the reverse of removal.

Wagons

Note: *There are two rear speakers, which are located at the rear corners of the cargo area, on small panels mounted on the rear D-pillars.*

21 Disconnect the cable from the negative battery terminal (see Chapter 5, Section 1).

22 Remove the screw from the speaker bezel, carefully pry the front end of the bezel up, then release the two spring clips holding the bezel to the quarter trim panel.

23 Pass the seat belt trough the slit in the bezel, then move the speaker bezel forward to free it from the trim panel.

24 Remove the rear speaker mounting screws.

25 Pull out the speaker and disconnect the electrical connector.

26 Installation is the reverse of removal.

Subwoofers

Note: *The subwoofer, if equipped, is located behind the rear seat, under the rear shelf trim, between the two rear speakers.*

27 Follow the procedure for replacing a rear speaker (see Steps 16 through 20).

13   Antenna and antenna module - replacement

Antenna

1 The grid-type antenna is bonded to and is an integral component of the rear window glass. If the antenna is faulty the rear glass must be replaced. This is a job best left to a professional.

Antenna module

2 Disconnect the cable from the negative terminal of the battery (see Chapter 5, Section 3).

300, Charger and 2011 and earlier Challenger models

Note: *The antenna module is mounted on the inside of the right C-pillar.*

3 Remove the right upper C-pillar trim (see Chapter 11).

4 Disconnect the two antenna module electrical connectors from the antenna grid and disconnect the antenna coaxial cable and electrical connector from the module **(see illustration)**.

5 Remove the antenna module mounting bolt **(see illustration 13.3)** and remove the module.

6 Installation is the reverse of removal.

2012 and later Challenger models

Note: *The antenna module is mounted to the rear of the roof panel, above the headliner.*

7 Remove the rear seat back (see Chapter 11, Section 28).

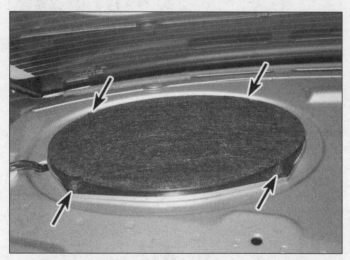

12.18 To detach a rear speaker, remove these four screws

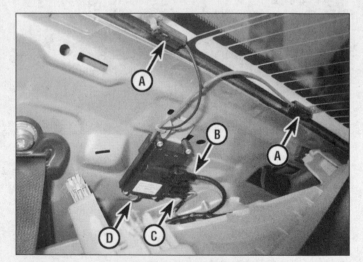

13.4 To remove the antenna module (300, Charger and 2011 and earlier Challenger models):

A   *Disconnect the module electrical connectors from the two antenna grid terminals*

B   *Disconnect the coaxial antenna cable (this cable goes to the radio)*

C   *Disconnect the electrical connector*

D   *Remove the module mounting bolt*

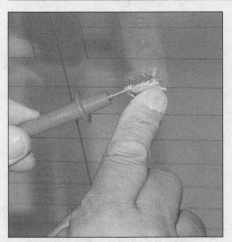

14.5 When measuring voltage at the rear window defogger grid, wrap a piece of aluminum foil around the positive probe of the voltmeter and press the foil against the wire with your finger

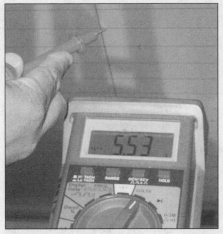

14.6 To determine if a heating element has broken, check the voltage at the center of each element - if the voltage is 6-volts, the element is unbroken

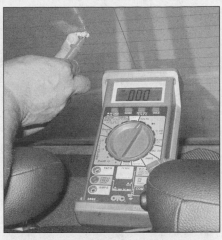

14.8 To find the break, place the voltmeter negative lead against the defogger ground terminal, place the voltmeter positive lead with the foil strip against the heat wire at the positive terminal end and slide it toward the negative terminal end. The point at which the voltmeter deflects from several volts to zero volts is the point at which the wire is broken

8   Remove the grab handles and coat hooks from the sides of the headliner.
9   Remove the quarter trim panels.
10   Using a trim panel tool, carefully pry the platic retainers from the roof panel at the rear edge of the headliner.
11   Disconnect the antenna module lead from the antenna grid terminal on the rear glass. Detach the lead from the slot in the roof panel.
12   Disconnect the wire harness electrical connector from the antenna module.
13   Remove the module mounting nut and slide the module out from the welded stud on the left end.
14   Installation is the reverse of the removal procedure.

Wagons

Note: *The antenna module is mounted on the inside of the liftgate.*
15   Remove the liftgate trim panel (see Chapter 11).
16   Disconnect the two antenna module electrical connectors from the antenna grid and disconnect the antenna coaxial cable and electrical connector from the module (see illustration 13.3).
17   Remove the antenna module mounting bolt and remove the module.
18   Installation is the reverse of removal.

14   Rear window defogger - check and repair

1   The rear window defogger consists of a number of horizontal heating elements baked onto the inside surface of the glass. Power is supplied through a relay and fuse from the interior fuse/relay box. A defogger switch on the instrument panel controls the defogger grid.
2   Small breaks in the element can be repaired without removing the rear window.

Check

3   Turn the ignition and defogger switches to the On position.
4   Using a voltmeter, place the positive probe against the defogger grid positive side and the negative probe against the ground side. If battery voltage is not indicated, check that the ignition switch is On and that the feed and ground wires are properly connected. Check the two fuses, defogger switch, defogger relay and related wiring. The dealer can scan the body control module if necessary. If voltage is indicated, but all or part of the defogger doesn't heat, proceed with the following tests.
5   When measuring voltage during the next two tests, wrap a piece of aluminum foil around the tip of the voltmeter positive probe and press the foil against the heating element with your finger (see illustration). Place the negative probe on the defogger grid ground terminal.
6   Check the voltage at the center of each heating element (see illustration). If the voltage is 5 to 6 volts, the element is okay (there is no break). If the voltage is 0 volts, the element is broken between the center of the element and the positive end. If the voltage is 10 to 12 volts, the element is broken between the center of the element and the ground side. Check each heating element.
7   If none of the elements are broken, connect the negative probe to a good chassis ground. The voltage reading should stay the same; if it doesn't, the ground connection is bad.
8   To find the break, place the voltmeter negative probe against the defogger ground terminal. Place the voltmeter positive probe with the foil strip against the heating element at the positive side and slide it toward the negative side. The point at which the voltmeter deflects from several volts to zero is the point where the heating element is broken (see illustration).

Repair

9   Repair the break in the element using a repair kit specifically for this purpose, available at most auto parts stores. The kit includes conductive plastic epoxy.
10   Before repairing a break, turn off the system and allow it to cool for a few minutes.
11   Lightly buff the element area with fine steel wool; then clean it thoroughly with rubbing alcohol.
12   Use masking tape to mask off the area being repaired.
13   Thoroughly mix the epoxy, following the kit instructions.
14   Apply the epoxy material to the slit in the masking tape, overlapping the undamaged area about 3/4-inch on either end (see illustration).
15   Allow the repair to cure for 24 hours before removing the tape and using the system.

15   Headlight housing - removal and installation

Warning: *This procedure does not apply to the headlight housings on models which use a High Intensity Discharge (HID) headlight bulb. If you own one of these models, we don't recommend that you try to remove the headlight housing yourself because you could be accidentally electrocuted when disconnecting the electrical connector from one of these bulbs, which you must do to remove the headlight housing (see the Warning in*
Section 17).
1   On Charger, Magnum and Challenger models, remove the bumper cover (see Chapter 11).

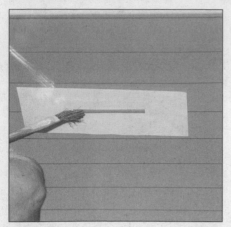

14.14 To use a defogger repair kit, apply masking to the inside of the window at the damaged area, then brush on the special conductive coating

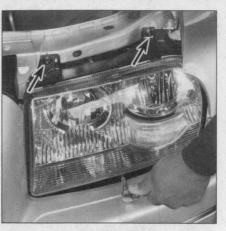

15.2a Remove the headlight housing bolts (300 model shown; lower bolt not visible)

15.2b Challenger models - remove this upper bolt . . .

15.2c . . . and these two lower bolts

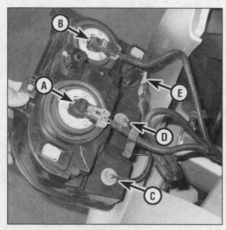
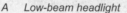

15.3 Pull out the headlight housing and disconnect the electrical connectors for all five headlight housing lights (300 model shown):

A   Low-beam headlight
B   High-beam headlight
C   Outer parking light
D   Inner parking light
E   Park/turn signal light

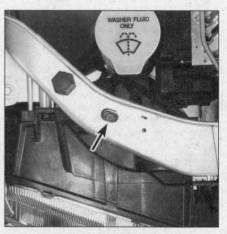

16.1a Headlight vertical adjustment screw location

16.1b Challenger headlight adjuster is in a similar location

2   Remove the the headlight housing mounting bolts **(see illustrations)**. On 300 models there are two upper and one lower bolt. On Charger and Magnum models there are two upper and two lower bolts. On Challenger models there are one upper and two lower bolts.
Note: *Most models will have the upper outer corner as a ball-and-socket attachment. To remove, pull the housing away from the radiator support and the ball and socket will separate.*
3   Pull out the headlight housing far enough to disconnect the electrical connectors **(see illustration)**, then remove the headlight housing.
4   Installation is the reverse of removal.

16   Headlights - adjustment

Headlights

Warning: *The headlights must be aimed correctly. If adjusted incorrectly, they could temporarily blind the driver of an oncoming vehicle and cause an accident or seriously reduce your ability to see the road. The headlights should be checked for proper aim every 12 months and any time a new headlight is installed or front-end bodywork is performed. The following procedure is only intended to provide temporary adjustment until you can have the headlights professionally adjusted by a dealer service department.*

1   Open the hood and locate the headlight vertical adjusting screw **(see illustrations)** for each headlight in the upper radiator crossmember. Each headlight has a vertical adjustment screw; there are no horizontal adjustment screws.
2   There are several ways to adjust the headlights. The simplest method requires an

open area with a blank wall and a level floor (**see illustration**).

3   Position masking tape vertically on the wall in reference to the vehicle centerline and the centerlines of both headlights.

4   Position a horizontal tape line in reference to the centerline of the headlights.

Note: *It might be easier to position the tape on the wall with the vehicle parked only a few inches away.*

5   Adjustment should be made with the vehicle parked 25 feet from the wall, sitting level, the gas tank full and no unusually heavy load in the vehicle.

6   The high intensity zone should be vertically centered with the exact center, about three inches below the horizontal line.

7   Have the headlights adjusted by a qualified technician at the earliest opportunity.

Fog lights

Note: *This procedure applies only to vehicles equipped with fog lights.*

8   Park the vehicle 25 feet from the wall.

9   Tape a horizontal line on the wall that represents the height of the fog lights and tape another line four inches below that line.

10   Using the adjusting screw (**see illustration**) on each fog light, adjust the pattern on the wall so that the top of the fog light beam meets the lower line on the wall. Repeat this procedure for the other fog light.

---

17   Headlight bulb - replacement

All headlight bulbs except HID low-beam bulb models

Warning: *Halogen bulbs are gas-filled and under pressure and they can shatter if the surface is scratched or the bulb is dropped. Wear eye protection and handle the bulbs carefully, grasping only the base whenever possible. Don't touch the surface of the bulb with your fingers because the oil from your skin could cause it to overheat and fail prematurely. If you do touch the bulb surface, clean it with rubbing alcohol.*

Note: *This procedure applies to both the high- and low-beam headlight bulbs on all models except models with High Intensity Discharge (HID) units (see Warning below with "High Intensity Discharge bulbs).*

Note: *This procedure simplifies headlight bulb replacement by showing how to replace a headlight bulb without removing the headlight housing. However, if there's not enough space to work on your vehicle or if your hands are too big, remove the headlight housing first (see Section 15), then replace the bulb.*

1   If you're replacing one of the left headlight bulbs, remove the air filter housing (see Chapter 4).

2   On 2011 and later models, remove the dust cap (**see illustration**).

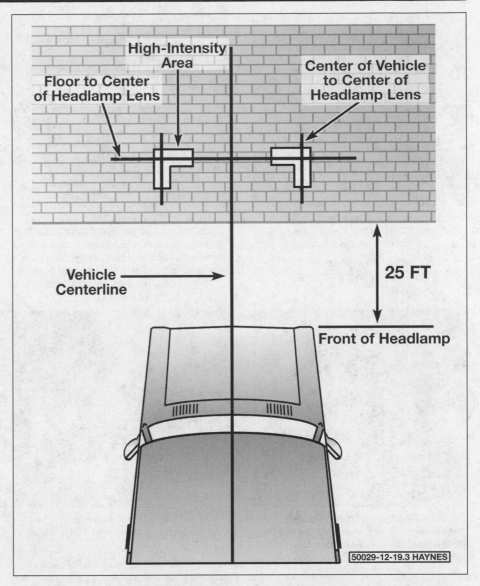

16.2 Headlight adjustment screen details

16.10 Fog light adjustment screw location

17.2 Turn the dust cap counterclockwise to remove it

17.3a To replace either headlight bulb, disconnect the electrical connector (2010 and earlier models shown)

A   *Low-beam headlight connector*
B   *High-beam headlight connector*

17.3b To disconnect either headlight electrical connector, grasp the connector firmly, then pull it out of the bulb socket terminal. You must pull firmly enough to spread the two halves of the split locking tab apart so that they release from the locking lug (A)

17.4 To remove a headlight bulb socket from the headlight housing, rotate it counterclockwise about 30 degrees, then pull it out of the housing

18.2 To disconnect the electrical connector from the front park/turn signal bulb holder, slide the red lock toward the harness, then depress the release tab (A) and pull off the connector

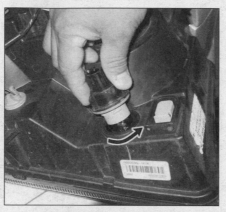

18.3 Turn the bulb holder counterclockwise and pull it out

18.4 To remove a front park/turn signal bulb from the holder, simply pull it straight out

3   Disconnect the electrical connector from the headlight bulb that you want to replace (**see illustrations**).
4   To remove either headlight bulb from the headlight housing, turn it counterclockwise about 30-degrees (**see illustration**). The headlight bulb and the bulb socket are a single assembly. No further disassembly is possible. The new bulb will have its own socket.
5   Installation is the reverse of removal.

High Intensity Discharge (HID) low-beam bulb models

Warning: *Some models use a High Intensity Discharge (HID) low-beam bulb instead of a conventional halogen low-beam bulb. According to the manufacturer, high voltage can remain in the circuit even after the headlight switch has been turned to Off and the ignition key has been removed. Therefore, for your safety, we don't recommend that you try to replace one of these bulbs yourself. Instead, have this service performed by a dealer service department.*
6   According to the manufacturer, these bulbs

put out three times as much light of a standard halogen bulb, use less energy doing so, and last 10 times longer. HID bulbs use an alternating current (AC) electrical charge to ignite xenon gas inside the sealed bulb. They're similar in operation to vapor-filled streetlights. Instead of a filament, the gas inside the bulb is ignited by creating an arc between two electrodes. A ballast module converts battery voltage to alternating current, stepping it up from 12 volts DC to 800 volts AC. The HID bulb takes about 10-15 seconds to warm up before it operates normally. The igniter (part of the HID bulb) controls the voltage applied to the electrodes. The igniter steps up the 800 volt AC input to about 25,000 volts to start up the light; once ignited, it reduces voltage to about 85 volts.

18   Bulb replacement

Exterior lights

Front park/turn signal light bulbs

Warning: *If you're replacing a front park/turn signal bulb on a model with HID headlights, you will have to do it with the headlight hous-*

ing installed because we don't recommend removing the headlight housing on these models because of the danger of electrocution. Please refer to the Warning in Section 17.
Note: *This procedure applies to 2014 and earlier models. On 2015 and later models the front Park/turn signals utilize LED bulbs which must be replaced as a unit.*
Note: *This procedure simplifies front park/turn signal bulb replacement by showing how to replace a park/turn signal bulb without removing the headlight housing. However, if there's not enough space to work on your vehicle or if your hands are too big, remove the headlight housing first (see Section 15), then replace the bulb.*
1   If you're replacing the front park/turn signal bulb in the left headlight housing, remove the air filter housing (see Chapter 4).
2   Disconnect the electrical connector from the holder for the front park/turn signal light bulb (**see illustration**).
3   Remove the front park/turn signal bulb holder from the headlight housing (**see illustration**).
4   Remove the front park/turn signal bulb from its holder by simply pulling it straight out of the socket (**see illustration**). To install a

new bulb into the holder, push it straight into the socket until it's fully seated.

5   Installation is the reverse of removal.

Inner and outer front parking light bulbs (Chrysler 300 models)

Note: *This procedure simplifies front parking light bulb replacement by showing how to replace a parking light bulb without removing the headlight housing. However, if there's not enough space to work on your vehicle or if your hands are too big, remove the headlight housing first (see Section 15), then replace the bulb.*

Note: *This procedure applies to any of the four parking light bulbs (two in each headlight housing) - they're identical.*

6   If you're replacing a front parking light bulb in the left headlight housing, remove the air filter housing (see Chapter 4).

7   Remove the front parking light bulb holder **(see illustration)**.

8   Remove the front parking light bulb from its holder by pulling it straight out. To install a new bulb into the holder, push it straight into the socket until it's fully seated.

9   Installation is the reverse of removal.

Front sidemarker lights

Note: *This procedure applies to 2010 and earlier 300 models and 2014 and earlier Charger and Challenger models. On later models the sidemarker lights use LEDs and must be replaced as a unit.*

Chrysler 300 models

Note: *The front sidemarker lights are located in the front bumper cover. They can be accessed from underneath the vehicle or by removing the headlight housing.*

10   Remove the headlight housing (see Section 15).

11   Or, raise the vehicle and place it securely on jackstands, then remove the front engine

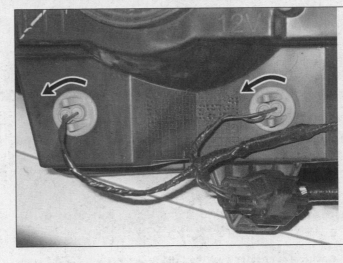

18.7 To remove the holder for either front parking light bulb, simply twist it counterclockwise and pull it out of the housing

splash shield (see Chapter 2A).

12   Disconnect the electrical connector from the sidemarker light bulb holder **(see illustration)**.

13   To remove the front sidemarker light bulb holder, turn it counterclockwise and pull it out.

14   To remove a front sidemarker light bulb from its holder, pull it straight out of the holder. To install a new bulb, push it straight into the socket until it's fully seated.

15   Installation is the reverse of removal.

Charger and Magnum models

Note: *The front sidemarker light bulbs are located in the headlight housing.*

16   Unbolt and pull out the headlight housing (see Section 15). Disconnect the electrical connector from the sidemarker light bulb holder.

17   To remove the front sidemarker light bulb holder, turn it counterclockwise and pull it out. To remove a front sidemarker light bulb from its holder, pull it straight out of the holder.

To install a new bulb, push it straight into the socket until it's fully seated.

18   Installation is the reverse of removal.

Challenger models

19   Using a trim stick, pry the rear of the lens and housing from the bumper cover, then turn the bulb holder counterclockwise to remove it from the housing.

20   Pull the bulb straight out of the holder. To install a new bulb, push it straight into the socket until it's fully seated.

Fog light bulbs

21   Raise the vehicle and place it securely on jackstands.

22   Remove the front engine splash shield (see Chapter 2A).

23   Disconnect the electrical connector from the fog light bulb holder **(see illustration)**.

24   Remove the bulb holder from the fog light housing **(see illustration)**.

25   Installation is the reverse of removal.

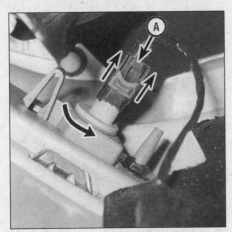

18.12 To disconnect the electrical connector from the front sidemarker light bulb holder on a Chrysler 300 model, slide the red lock toward the harness, then depress the release tab (A) and pull off the connector. To remove the holder, turn it counterclockwise and pull it out

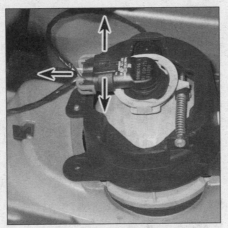

18.23 To disconnect the electrical connector from the fog light bulb holder, grasp the connector firmly and pull it off. You must pull firmly enough to spread the two halves of the split locking tab apart so that they release from the locking lug

18.24 To remove the fog light bulb holder, turn it counterclockwise and pull it out of the fog light housing

18.27 To disconnect the electrical connector from the CHMSL housing, depress the release tab and pull off the connector

18.28 To detach the CHMSL housing from the vehicle, remove these two screws

Center high-mounted stop lights - CHMSL

Note: *The center high-mounted brake light assembly uses LEDs (light emitting diodes). You cannot replace individual bulbs.*

Note: *The third brake light is commonly known as the Center High Mount Stop Lamp or CHMSL. For clarity we will refer to it as the CHMSL throughout this manual.*

300 and Charger models

26 Remove the upper and lower C-pillar trim panels and remove the rear shelf trim panel (see Chapter 11).
27 Disconnect the electrical connector from the CHMSL housing (see illustration).
28 Remove the CHMSL housing mounting screws (see illustration).
29 Remove the CHMSL housing.
30 Installation is the reverse of removal.

Magnum models

31 Remove the rear spoiler (see Chapter 11).

32 Disconnect the electrical connector from the CHMSL housing.
33 Remove the fasteners that secure the CHMSL housing to the rear spoiler.
34 Remove the CHMSL housing.
35 Installation is the reverse of removal.

Challenger models

36 Wedge a flat trim tool between the light fixture cover and the cover base far enough to release the pressure clips, while twisting the trim tool (see illustration).
37 If you're needing just a bit more room, try removing the five stamped nuts that secure the CHMSL to the headliner substrate.
38 Carefully pull down the rear part of the headliner to gain access to the wire connector for the CHMSL (see illustrations).
39 Pull the rear part of the CHMSL downward to release the tabs that hold the front part of the CHMSL to the inner part of the roof.

40 Remove the CHMSL.
41 Installation is the reverse of removal. Be sure to fully align the CHMSL to the base before snapping it into place.

Taillight bulbs

300, Charger and Challenger models

42 Open the trunk, unscrew the fastener that secures the carpeting (see illustration), then pull back the carpeting to expose the taillight electrical connector and retainers.
43 Disconnect the electrical connector from the taillight assembly (see illustration).
44 Remove the remaining taillight assembly retainers (see illustration), then pull out the taillight housing to the rear.
45 There are three bulbs in the taillight housing (see illustration).

Note: *This illustration depicts the taillight housing on a Chrysler 300 model. If you're replacing a bulb on a Charger or Challenger model, refer to the bulb replacement section of*

18.36 Wedge the flat trim tool around the outside edges. As you twist the trim tool the pressure clips will release

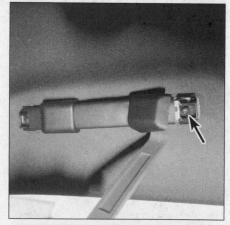

18.38a To gain more access to the CHMSL and to avoid damaging the headliner, remove the grab handles on either side (trim cover moved out of the way to expose the hidden screw) . . .

18.38b . . . and remove the clothing hanger hooks

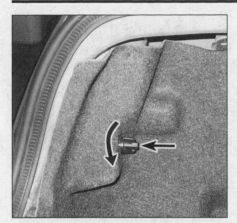

18.42 To access the taillight electrical connector and retainers, unscrew this fastener and peel back the carpeting

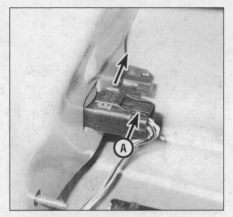

18.43 To disconnect the electrical connector from the taillight assembly, slide the red lock sideways, then depress the release tab (A) and pull off the connector

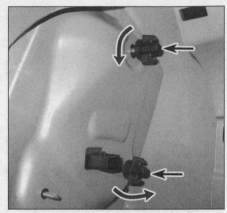

18.44 To detach the taillight assembly from the body, unscrew these two retainers, then pull out the taillight housing to the rear

your owner's manual for the location of each bulb.

Magnum models

46   Open the liftgate, then pry out the two pushpin fasteners from the taillight housing with a flat-blade screwdriver or some other suitable tool with a flat blade.

47   Pull off the taillight housing to the rear.

All models

48   Remove the holder for the bulb that you want to replace **(see illustration)**.

49   Remove the bulb from its holder by pulling it straight out of the holder. To install a new bulb, insert it straight into the socket and push it in until it's fully seated.

50   Installation is the reverse of removal.

License plate light bulb

51   Remove the license plate light housing retaining screws **(see illustrations)** and pull the housing out of the trunk lid.

52   Remove the license plate light bulb socket from the housing **(see illustration)**.

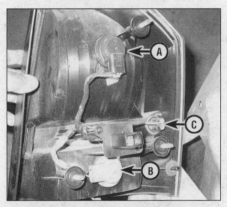

18.45 There are three bulbs in the taillight housing (Chrysler 300 shown, others slightly different):

A   *Brake light/turn signal/taillight bulb*
B   *Backup light bulb*
C   *Rear sidemarker light bulb*

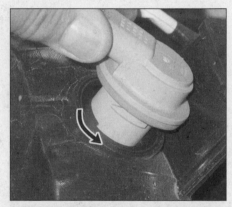

18.48 To remove a bulb holder from the taillight housing, turn it counterclockwise and pull it out

18.51a To detach the license plate light assembly from the rear bumper cover, remove these two screws

18.51b Challenger license plate light fixture is held in with a pressure clip. Press in on the edge and pull downward to release the fixture. The bulb is pulled straight out (don't twist the bulb)

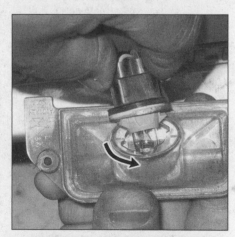

18.52 To remove the bulb holder from the license plate housing, rotate it counterclockwise 30-degrees and pull it out

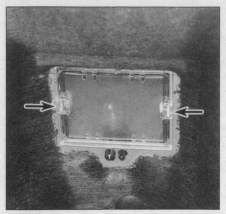

18.56a To remove the trunk light lens on models with retaining tangs, push one of the retaining tangs inward (300 shown)

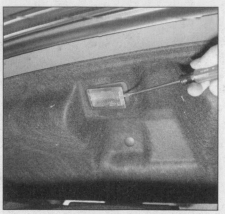

18.56b To remove the trunk light lens (light housing) on models without retaining tangs, pry out the right side of the lens and remove the housing (Challenger shown)

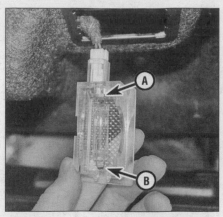

18.58 To replace a festoon bulb, detach the terminal closest to the connector from its contact (A), then unhook the other terminal from its contact (B)

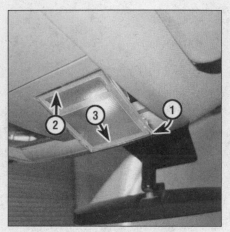

18.60a To remove a reading light lens:

1   *Carefully pry down the leading edge of the lens with a small screwdriver or a small trim removal stick*
2   *Push up on the trailing edge of the lens to disengage the rear tab*
3   *Slide the lens forward until the rear tab clears the light switch*

18.60b On the Challenger models, simply pry the lens cover off with a flat trim tool

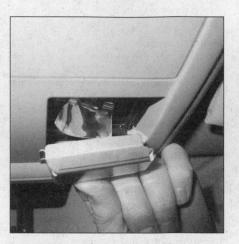

18.61 Before removing the small metal shroud that surrounds the reading light bulb, note its exact location and orientation; it's held in by friction, so you must place it in the exact same spot when installing it (or you won't be able to install the lens)

53   Remove the license plate light bulb from its socket, by pulling it straight out.
54   To install a new bulb, insert it into the socket and push it in until it's fully seated.
55   Installation is the reverse of removal.

Interior lights

Warning: *The models covered by this manual are equipped with a Supplemental Restraint System (SRS), more commonly known as airbags. Always disarm the airbag system before working in the vicinity of any airbag system component to avoid the possibility of accidental deployment of the airbag, which could cause personal injury (see Section 26). Do not use a memory-saving device to preserve the PCM's memory when working on or near airbag system components.*

Trunk light

56   Open the trunk lid and locate the trunk light on the ceiling of the trunk space, ahead of the trunk lid **(see illustrations)**.
57   On models with standard bulbs, pull the bulb straight out of its socket, then push the new one into place.
58   On models with festoon-type bulbs, detach the end terminals from the contacts in the housing, then clip the new bulb into place **(see illustration)**.
59   Installation is the reverse of removal.

Reading light bulbs

Caution: *Removing the lens from one of these reading lights can be tricky. But if you get frustrated and try to force the lens out, you will break it. So proceed carefully.*

60   Using a small, thin-blade screwdriver or a trim stick, carefully pry out the front edge (the edge nearest the windshield) of the reading light lens **(see illustrations)**, disconnect the switch electrical connectors and remove the switch housing.
61   Pull off the small metal housing that surrounds the bulb **(see illustration)**.
62   Remove the reading light bulb from its terminals **(see illustration)**.
63   To install a reading light bulb, push it up into its terminals until its snaps into place.
64   Installation is the reverse of removal.

Vanity light bulbs

65   Using a small, thin-blade screw-driver, carefully pry off the vanity light lens **(see illustrations)**.

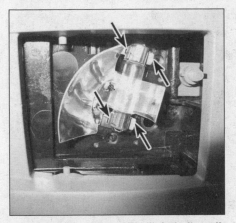

18.62 To remove a reading light bulb, pull it straight down out of its terminals (if it's necessary to pry the bulb out, pry only on the metal ends, not the glass)

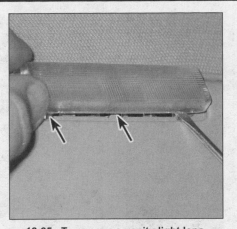

18.65a To remove a vanity light lens, carefully pry it loose along its lower edge at these three spots only (where there are small recesses to insert a small screwdriver). Pry it off just far enough to disengage the two small retaining tabs at each end . . .

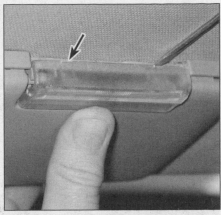

18.65b . . . then flip the visor farther down, carefully insert the screwdriver into these two slots on the upper edge of the lens and pry the lens down until it comes loose

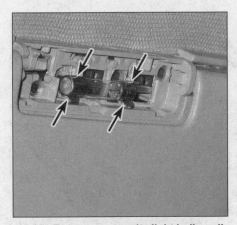

18.66a To remove a vanity light bulb, pull it out of these spring-type conductors (if it's necessary to pry the bulb out, pry only on the metal ends, not the glass)

18.66b On some models (Challenger for one) you'll find a dual vanity light in the visor. Simply pry the lens cover off to expose the bulb. The bulb pulls straight out of the housing

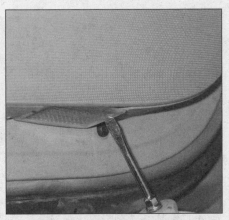

18.69 Pry the front end of the door courtesy light out of the door panel. . .

66   Remove the vanity light bulb from its contacts **(see illustrations)**.
67   Push the new bulb into its terminals.
68   Installation is the reverse of removal.

Door courtesy lights

69   Pry the forward end of the lens out of the door trim panel and pull the light/lens assembly down **(see illustration)**.
70   Remove the bulb from the door courtesy light **(see illustration)**.
71   Installation is the reverse of removal.

Rear reading lights

72   Insert a small flat-bladed screwdriver into the gap between the forward edge of the lens and the light housing and carefully pry out the lens **(see illustration)**. When the forward end is free, carefully disengage the tab at the rear end of the lens and pull out the lens.

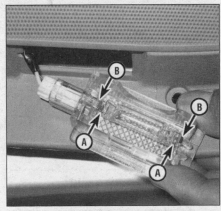

18.70. . . then disengage the wire loops (A) on the bulb from the hooked end of each conductor (B)

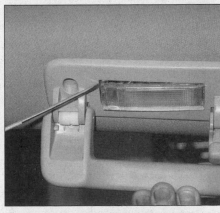

18.72 To remove the lens from a rear reading light, insert a small screwdriver into the gap between the forward end of the lens and the light housing, carefully pry the lens down, then move it forward to disengage the tab at the rear

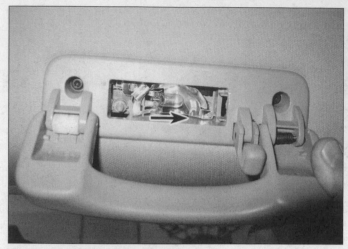

18.73 To remove a bulb from a rear reading light housing, just pull it straight out of the socket

19.2a The horns are located below two small plastic trim covers that span the gap between the front bumper cover and the upper radiator support (trim covers and front bumper cover removed for clarity)

19.2b Disconnect the electrical connector from the horn. . .

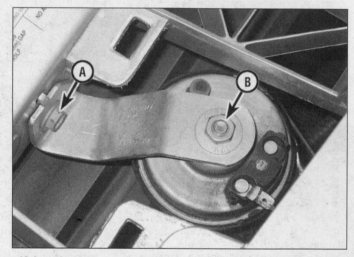

19.3. . . then remove the horn mounting bracket bolt (A), remove the horn and bracket, then remove the nut (B) to separate the horn from the bracket

73   Remove the bulb from the rear reading light housing **(see illustration)**. To install a new bulb, insert the bulb straight into its socket until it's fully seated.

74   Installation is otherwise the reverse of removal.

19 Horn - replacement

Note: *The horns are located under the plastic covers between the fascia (front bumper cover) and the upper radiator support.*

1   Remove the plastic covers that span the gap between the front bumper cover and the upper radiator support (see *Radiator - removal and installation* in Chapter 3).

2   Disconnect the electrical connector from the horn that you want to replace **(see illustrations)**.

3   Remove the horn mounting bracket bolt **(see illustration)**.

4   Remove the horns and mounting bracket as a single assembly, then disconnect the electrical connectors from the horns.

5   To replace either horn, remove the retaining nut for the horn that you want to replace, then remove that horn from the mounting bracket.

6   Installation is the reverse of removal.

20 Electric side-view mirrors - general information

1   The electric side-view mirrors can be adjusted up-and-down and left-to-right by a driver's side switch located on the left door trim panel. On models with factory-installed dual power mirrors, each mirror

is also equipped with a heater grid behind the mirror glass to clear the mirror surface of fog, ice or snow. On these models, the mirror heater grid is an integral component of each mirror. If a heater grid fails, replace the mirror (see Chapter 11). The heater grid switches and the heated mirror system indicator light are integral components of the heater/air conditioning control panel on the dash. If one of these components fails, replace the heater/air conditioning control assembly (see Chapter 3). The heated mirror relay is located in one of the fuse and relay boxes.

2   The power mirror control switch has a Left-Right selector switch that allows you to send voltage to the side-view mirror that you want to adjust. With the ignition switch in the ACC position, roll down the windows and operate the mirror control switch through all

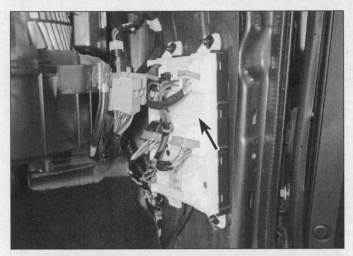

22.1a The body control module (BCM) is mounted next to the blower motor in the front corner of the passenger area. The BCM has many functions, one of which includes window operation

22.1b Each door has its own door control module. These modules work with the BCM and several other microprocessors to determine the operating status of the windows, door locks, and other door related functions

functions (left-right and up-down) for both the left and right side-view mirrors.

3   Listen carefully for the sound of the electric motors running in the mirrors.

4   If you can hear the motors but the mirror glass doesn't move, the problem is probably a defective drive mechanism inside the mirror, which will necessitate replacement of the mirror.

5   If the mirrors don't operate and no sound comes from the mirrors, check the fuse in one of the fuse and relay boxes (see Section 3).

6   If the fuse is OK, refer to Chapter 11 and remove the door panel for access to the back of the mirror control switch, without disconnecting the wires attached to it. Turn the ignition On and check for voltage at the switch. There should be voltage at one terminal. If there's no voltage at the switch, check for an open in the wiring between the fuse panel and the switch.

7   If there's voltage at the switch, disconnect it. Check the switch for continuity in all its operating positions. If the switch does not have continuity, replace it.

8   Reconnect the switch. Locate the wire going from the switch to ground. Leaving the switch connected, connect a jumper wire between this wire and ground. If the mirror works normally with this wire in place, repair the faulty ground connection.

9   If the mirror still doesn't work, remove the mirror and check the wires at the mirror for voltage. Check with the ignition key turned to On and the mirror selector switch on the appropriate side. Operate the mirror switch in all its positions. There should be voltage at one of the switch-to-mirror wires in each switch position, except the neutral (off) position.

10   If voltage is not present in each switch position, check the wiring between the mirror and control switch for opens and shorts.

11   If there's voltage, remove the mirror and test it off the vehicle with jumper wires. Replace the mirror if it fails this test.

21   Cruise control system - general information

1   The Powertrain Control Module (PCM) controls the cruise control system electronically via the Electronic Throttle Control (ETC) system. If you have problems with the cruise control system, have it checked by a dealer service department or other qualified repair shop.

22   Power window system - general information

1   The power window system controls the electric motors, mounted inside the doors, that lower and raise the windows. The power window system consists of the control switches, the fuse, the circuit breaker, the motors, the window "regulators" (the scissor-like mechanisms that raise and lower the window glass) and the wiring connecting the switches to the motors. When the ignition switch is turned to On, current flows through the power window fuse in the engine compartment fuse and relay box to a circuit breaker located in the instrument panel wiring harness (located near the parking brake pedal). From there, current flows to the power window switches. In most cases the simplicity of the window system has been complicated with the introduction of the Body Control Module (BCM) and individual door modules. These modules take the input signal from the switches and create a digital signal that is sent to the the various control modules on a high speed data line to deter-

mine the window state, condition, position, and operating function. All this is balanced by the operator instructions and the actual window requirements and configuration (**see illustrations**).

2   The power windows are wired so that they can be lowered and raised from the master control switch by the driver or by passengers using remote switches located at each passenger's window. Each window has a separate motor that is reversible. The position of the control switch determines the polarity and therefore the direction of operation.

3   The power window system will only operate when the ignition switch is turned to On. However, there is a residual operating voltage for most window systems that allow the window to be relocated for a short time after the key has been turned off and a door has not been opened. In addition, a window lockout switch at the master control switch can, when activated, disable the power window switches on the other doors. Always check these items before troubleshooting a window problem.

4   These procedures are general in nature, so if you can't find the problem using them, take the vehicle to a dealer service department.

5   If the power windows don't work at all, check the fuse or circuit breaker.

6   If only the rear windows are inoperative, or if the windows only operate from the master control switch, check the window lockout switch for continuity in the unlocked position. If it doesn't have continuity, replace it.

7   Check the wiring between the switches and the fuse for continuity. Repair the wiring, if necessary.

8   If only one window is inoperative from the master control switch, try the control switch at the window that doesn't work.

Note: *This doesn't apply to the driver's door window.*

9   If the same window works from one switch, but not the other, check the switch for continuity.

10   If the switch tests OK, check for a short or open in the wiring between the affected switch and the window motor.

11   If one window is inoperative from both switches, remove the trim panel from the affected door (see Chapter 11), then check for voltage at the switch and at the motor while operating the switch. First check for voltage at the electrical connectors for the circuit. With the ignition key turned to On and the connectors all connected, backprobe at the designated wire (see the wiring diagrams at the end of this Chapter) with a grounded test light. Pushing the driver's window switch to the Down position, there should be voltage at one terminal. Pushing the same switch to the Up position, there should be voltage at another terminal. If these voltage checks are OK, disconnect the electrical connector at the driver's motor, and check it for voltage when the switch is operated.

Note: *A good "first check" is to have the key on and operate a window switch while watching the dome light. If the window is stuck (not moving) the dome light will dim as the current is applied to the stuck window. This is a good indication that the wiring is somewhat intact (it's not a fool proof method, but it may get you to the next step). Most likely the motor or window track is the problem. Although, you will still need to verify that none of the wiring is shorted or causing a problem.*

12   If voltage is reaching the motor and the switch is OK, disconnect the door glass from its regulator (see Chapter 11). Move the window up and down by hand while checking for binding and damage. Also check for binding and damage to the regulator. If the regulator is not damaged and the window moves up and down smoothly, replace the motor. If there's binding or damage, lubricate, repair or replace parts, as necessary.

13   If voltage isn't reaching the motor, check the wiring in the circuit for continuity between the switches and motors (see the wiring diagram at the end of this Chapter).

Note: *Keep in mind, an electric motor (which is all that the window regulator is) needs two things. A ground signal, and a positive voltage signal. If you only have one of them, the window will not work. Be sure to check for both a ground potential and a positive voltage potential at the window motor connection.*

14   If you have to replace the main power window switch, pry it out of the door trim panel, then disconnect the electrical connector(s) from the switch.

15   When you're done, test the windows to confirm that the window system is functioning correctly.

Note: *In most instances, a scanner can be the most valuable tool in diagnosing window operation problems. A good scanner can simulate the switch or the BCM control signals and make quick work of determining the exact cause of the failure. Using the proper tools to diagnose a window problem will eliminate the unnecessary replacement of assumed bad parts.*

23   Power door lock system - general information

1   The power door lock system operates the power door motors, which are integral components of the door latch units in each door. The system consists of a fuse (in the engine compartment fuse and relay box), the instrument cluster, the control switches (in each of the front doors), the power door motors and the electrical wiring harnesses connecting all of these components.

2   The lock mechanisms in the door latch units are actuated by a reversible electric motor in each door. When you push the door lock switch to Lock, the motor operates one way and locks the latch mechanism. When you push the door lock switch the other way, to the Unlock position, the motor operates in the other direction, unlocking the latch mechanism. Because the motors and lock mechanisms are an integral part of the door latch units, they cannot be repaired. If a door lock motor or lock mechanism fails, replace the door latch unit (see Chapter 11).

3   Even if you don't manually lock the doors or press the door lock switch to the Lock position before driving, the instrument cluster automatically locks the doors when the vehicle speed exceeds 15 mph, as long as all the doors are closed and the accelerator pedal is depressed. (You can turn off this feature if you don't want the doors to lock automatically. Refer to your owner's manual.)

4   Some vehicles have an optional Remote Keyless Entry (RKE) system that allows you to lock and unlock the doors from outside the vehicle. The RKE system consists of the transmitter (the electronic push-button "key") and a receiver located on the instrument cluster. The RKE receiver, which operates all the time, is protected by a fuse in the engine compartment fuse and relay box. Vehicles are shipped from the factory with two RKE transmitters, but if you want to purchase extra units, the RKE receiver can actually handle up to four vehicle access codes.

5   Some features of the door lock system on these vehicles rely on resources that they share with other electronic modules through the Programmable Communications Interface (PCI) data bus network. Professional diagnosis of these modules and the PCI data bus network requires the use of a wiTech scan tool (latest factory scan tool) and factory diagnostic information. At-home repairs are therefore limited to inspecting the wiring for bad connections and for minor faults that can be easily repaired. If you are unable to locate the trouble using the following general steps, consult your dealer service department. There are several aftermarket scanners available that have similar capabilities as the factory tool does. However, with most aftermarket scanners, they generally will not have all the factory capabilities built into one tool. It generally takes several aftermarket scanners to equal the factory tool. A J2534 interface is another

great tool, but like most scanners, aftermarket or factory, they are designed for the independent repair facility and not for home use.

Note: *Code readers are not scanners. Don't confuse reading a code with what a scanner is capable of. Keep in mind, codes don't fix cars. Codes are a direction or lead to where a problem has been detected by the software.*

6   Always check the circuit fuses (in the engine compartment fuse and relay box) first.

7   When depressed, each power door lock switch locks or unlocks all of the doors. The easiest way to verify that each door lock switch is operating correctly is to watch the door lock button in each door as you operate the switch. The door lock buttons should all go down when you push the door lock switch to the Lock position, and go up when you push the door lock switch to the Unlock position. Also, with the engine turned off so that you can hear better, operate the door lock switches in both directions and listen for the faint click of the motors locking and unlocking the latch mechanisms.

8   If there's no click, check for voltage at the switches. If no voltage is present, check the wiring between the fuse and the switches for shorts and opens (see the wiring diagrams at the end of this Chapter).

9   If voltage is present, but no clicking sound is apparent, remove the switch from the door trim panel (see Chapter 11) and test it for continuity. If there is no continuity in either direction, replace the switch.

10   If the switch has continuity but the latch mechanism doesn't click, check the wiring between the switch and the motor in the latch mechanism for continuity. If the circuit is open between the switch and the motor, repair the wiring.

11   If all but one motor is operating, remove the trim panel from the affected door (see Chapter 11) and check for voltage at the motor while operating the lock switch. One of the wires should have voltage in the Lock position; the other should have voltage in the Unlock position.

12   If the inoperative motor is receiving voltage, replace the latch mechanism.

13   If the inoperative motor isn't receiving voltage, check for an open or short in the circuit between the switch and the motor.

Note: *It's common for wires to break in the harness between the body and the door because repeatedly opening and closing the door fatigues and eventually breaks the wires.*

Key fob battery replacement

14   Remove the emergency key from the key fob **(see illustration)**.

15   Put the key fob face down and insert the tip of the emergency key into the slot and twist the key to unsnap the key fob **(see illustration)**.

16   Remove the battery and replace it with the exact type and size **(see illustration)**.

17   Be sure to have the polarity correct before closing the key fob halves.

18   Align the halves and squeeze the two section together.

23.14 Slide the emergency key out of the key fob

23.15 Use the key to separate the key fob halves

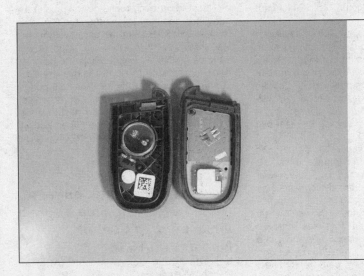

23.16 Always be sure to have the polarity correct before closing the case halves

19 Reinsert the emergency key and test your key fob operating range.

20 Programming the key fob requires a factory level scanner. For any reason the key fob fails to operate have it checked by your local dealer or independent repair facility with the appropriate equipment.

24 Power seats - general information

Warning: *The models covered by this manual are equipped with a Supplemental Restraint System (SRS), more commonly known as airbags. Additionally, some models are equipped with seat belt pre-tensioners, which are explosive devices. Always disarm the airbag/ restraint system before working in the vicinity of any airbag/restraint system component to avoid the possibility of accidental deployment of the airbag/seat belt pre-tensioners, which could cause personal injury (see Section 26). Do not use a memory-saving device to preserve the PCM's memory when working on or near airbag system components.*

1 Some models feature an optional eight-way power seat system that allows the driver and passenger to adjust the front seats up, down, front up, front down, rear up, rear down, forward and rearward. The system consists of the driver's power seat switch, the passenger power seat switch, the driver's power seat track, the passenger power seat track and, on some models, the optional power lumbar adjusters.

2 The power seat switches are located on the outboard side of the seat cushions, on the seat cushion side panels. If the vehicle is equipped with the optional power lumbar adjusters, the lumbar switches are located on the power seat switch assemblies. Each switch assembly is attached to the seat side panel by two Torx screws. Refer to your owner's manual for instructions regarding switch functions. Individual switches in the power seat switch assemblies cannot be repaired or replaced separately. If one of the switches in a power seat switch assembly fails, replace the entire switch assembly.

3 The seats are powered by three reversible motors that are attached to the upper half of the power seat track assembly. These motors are controlled by the power seat switches on the sides of the seats. Each switch changes the direction of seat travel by reversing polarity to the drive motor. The motors are an integral part of the power seat track assembly and cannot be repaired or replaced separately. If a motor fails, replace the power seat track assembly.

4 The optional power lumbar adjuster and motor are located on the back of the seat, under the seat trim cover and padding, where they're attached to a molded plastic back panel and to the seat back frame. The power lumbar adjuster and motor cannot be repaired or replaced separately from the seat back frame. If either the adjuster or the motor fails, replace the entire seat back frame unit.

5 Diagnosis is usually a simple matter, using the following procedures.

6 Look under the seat for any object which may be preventing the seat from moving.

7 If the seat won't work at all, check the fuse, which is located in the engine compartment fuse and relay box.

8 With the engine off to reduce the noise level, operate the seat controls in all directions and listen for sound coming from the seat motors.

9 If the motor doesn't work or make noise, check for voltage at the motor while an assistant operates the switch.

10 If the motor is getting voltage but doesn't run, test it off the vehicle with jumper wires. If it still doesn't work, replace it. The individual components are not available separately. The whole power-seat track must be purchased as an assembly.

11 If the motor isn't getting voltage, remove the seat side panel to access the switch and check for voltage. If there's no voltage at the switch, check the wiring between the fuse and the switch. If there's voltage at the switch, check for a short or open in the wiring between

the switch and the motor. If that circuit is okay, replace the switch. No further testing is recommended. If the power seat system is still malfunctioning at this point, have the system checked out by a dealer service department.

Seat heater

12   The seat heating system consists of a lower cushion element and a back rest element, which are all controlled by the seat heater module, operated by the seat heater two position switches.

13   When the seat heater button is pushed a signal is sent to the seat heater module which in turn directs the TIPM to close the heater relay and send battery voltage to the seat heating elements.

14   The system is designed to send a high current for approximately four minutes to bring the element up to temperature regardless of the seat heater switch position. The current flow then lowers to the preset level based on the given switch position. If the high level of heating has been selected it will remain in the high mode for approximately 30 minutes, but even then it will automatically reduce the current flow to the lower (second switch position) to avoid over heating the seat elements. If the lower setting has been selected, only the initial four-minute warm up cycle is completed and then it reverts to the lower nominal setting without any of the higher switch settings or functions.

15   If for any reason the system develops a short or an open in the circuit the light in the switch will either not come on or will flash slowly. At that point do not use the seat heater anymore and have it serviced at the next possible trip to your local repair facility.

Note: *The seat heater elements can be replaced by removing the seat and disassembling the seat. Probably not the typical job for the DIY'r, but it is possible to be done with some basic tools and a well thought out plan of attack.*

25   Daytime Running Lights (DRL) - general information

1   Canadian models are equipped with Daytime Running Lights (DRL). The DRL system illuminates the headlights whenever the engine is running and the parking brake is disengaged. The DRL system provides reduced power to the headlights so that they won't be too bright for daytime use and it prolongs the headlight bulbs' service life. It does this by modulating the pulse-width of the power to the headlights. The duration and interval of these power pulses is programmed into the Front Control Module (FCM), which is located on the instrument cluster. If you want to alter the pulse-width, you must have it done by a dealer service department.

26   Airbag system - general information

Note: *The airbag general information is for the purpose of informing you of the complexities of the modern SRS system. This is not for diagnostics or repair of any part of the SRS system.*

1   These models are equipped with a Supplemental Restraint System (SRS), more commonly called an airbag system. There are at least two airbags, one for the driver and one for the front seat passenger, on all models. The SRS system is designed to protect the driver and passenger from serious injury in the event of a head-on or frontal collision. The airbag control module is located on the transmission tunnel, right below the center of the instrument panel. Some models are also equipped with optional side curtain airbags. Vehicles with this option can be identified by the "SRS - AIRBAG" logo printed on the headliner above the B-pillar.

Warning: *Models equipped with seat belt pretensioners (pyrotechnic-type explosive) tighten the seat belts during an impact with sufficient force to hold (pull) you into the seat a fraction of a second ahead of the main airbags deployment. Under no circumstances apply any voltage to the leads, measure resistance with an ohmmeter or attempt to service the seat belt mechanism in any way.*

Warning: *Always disconnect the battery and let the vehicle set idle for approximately two minutes before proceeding with any repair procedure. Keep all of the disconnected air bag components away from any electrical devices while they are not in the vehicle. Be sure to reinstall any trim that was removed from the seats, pillars, doors, headliner, and knee bolster areas. Any trim that is not correctly installed or missing, can greatly affect the operation of the restraint system.*

Warning: *Vehicles equipped with passenger airbag and passenger presence systems require special procedures for working around the passenger's seat. Any time any component or part of the ORC (the front passenger's seat is one of those components) is removed or replaced new data needs to be configured in the ORC (Occupant Restraint Controller). This requires the use of a factory equivalent scanner. If after replacing or removing the front passenger's seat the airbag light remains on, see your local dealer or qualified independent shop for service.*

Airbag modules

2   The airbag module houses the airbag and the inflater unit. The inflater unit is mounted on the back of the housing over a hole through which gas is expelled, inflating the bag almost instantaneously when an electrical signal is received from the airbag control module. On the driver's airbag, the specially

wound wire that carries this signal to the module is called a "clockspring." The clockspring is a flat, ribbon-like electrically conductive tape that winds and unwinds as the steering wheel is turned so it can transmit an electrical signal regardless of wheel position. The procedure for removing the driver's airbag is part of *Steering wheel - removal and installation* in Chapter 10.

3   The passenger's airbag is located in the top of the dashboard, above the glove box. There's also a passenger airbag On/Off switch located at the lower right corner of the center instrument panel bezel. This switch allows you to deactivate the passenger's airbag if you're transporting an infant or a young child in a child safety seat. We don't recommend removing the passenger's airbag because there is no reason to do so unless it has been activated during an accident and needs to be replaced afterward.

4   Optional side-curtain airbags, if equipped, are located on each roof side rail, above the headliner, and they extend from the A-pillar to the C-pillar. SAB (Seat air bag) systems are also incorporated into these vehicles as well as knee bolster air bag protection. Again, we don't recommend trying to remove the side-curtain airbags or SAB for repair purposes, because there is no reason to do so unless they've been deployed in an accident and must be replaced. Keep in mind, they also will need to be configured to the ORC if they are removed or replaced.

Airbag Control Module - Occupant Restraint Controller (ACM - ORC) and Side Impact Airbag Control Modules (SIACMs)

5   The Airbag Control Module (ACM) or Occupant Restraint Controller (ORC) is the microprocessor that monitors and operates the airbag system. The ACM checks the system every time the vehicle is started. When you start the car, an Airbag indicator light comes on for about six seconds, then goes off, if the system is operating properly. If there is a fault in the system, the ACM stores a Diagnostic Trouble Code (DTC) and illuminates the Airbag indicator light, which remains on until the problem is repaired and the ACM memory is cleared of any DTCs. If the Airbag indicator light comes on at any time other than the bulb test and remains on, or doesn't come on at all, there's a problem in the system. A DRBIII scan tool is the only means by which the system can be diagnosed. Take the vehicle to your dealer immediately and have the system professionally diagnosed and repaired.

6   The ACM controls the operation of the standard driver's and passenger's airbags. Vehicles with optional side-curtain airbags are also equipped with Side Impact Airbag Control

Modules (SIACMs). There are two SIACMs, one for each side-curtain airbag. The SIACMs are located behind the B-pillar trim, above the outboard front seat belt retractor inside each B-pillar.

Impact seat belt retractors

7   All models are equipped with pyrotechnic (explosive) units in the front seat belt retracting mechanisms for both the lap and shoulder belts. During an impact that would trigger the airbag system, the airbag control unit also triggers the seat belt retractors. When the pyrotechnic charges go off, they accelerate the retractors to instantly take up any slack in the seat belt system to more fully prepare the driver and front seat passenger for impact.

8   The airbag system should be disabled any time work is done to or around the seats. **Warning:** *Never strike the pillars or floorpan with a hammer or use an impact-driver tool in these areas unless the system is disabled.*

Servicing components near the SRS system

9   There are times when you need to remove the steering wheel, the instrument cluster, the radio, the heater/air conditioning control assembly or other components on or near the dashboard. At these times you'll be working around components and wire harnesses for the SRS system. Do not use electrical test equipment on airbag system wires; it could cause the airbag(s) to deploy. ALWAYS DISABLE THE SRS SYSTEM BEFORE WORKING NEAR THE SRS SYSTEM COMPONENTS OR RELATED WIRING.

SRS component locations

10   Here is a list of the possible places to find an SRS component in a vehicle.

> *Steering wheel - driver's airbag*
> *Driver's knee bolster*
> *Passenger's dash area - passenger's airbag*
> *Passenger's knee bolster*
> *Headliner - roof rail airbags*
> *Front seats*
> *Doors*
> *Front bumper area - collision sensors*
> *SRS (ORC or ACM) module - under center console area*

Disabling the system

11   Whenever working in the vicinity of the steering wheel, steering column, floor console or other airbag system components, the system should be disarmed. To do this perform the following steps:

a) *Turn the ignition switch to the Off position.*

b) *Disconnect the cable from the negative battery terminal (see Chapter 5, Section 3).*

c) **WAIT FOR AT LEAST TWO MINUTES** *before beginning work (during this two-minute interval the capacitor that provides emergency back-up power to the system loses its reserve voltage charge).*

Enabling the system

12   To enable the airbag system, perform the following steps:

a) *Turn the ignition switch to the Off position.*

b) *Connect the cable to the negative battery terminal.*

c) *Without putting your body in front of either airbag, turn the ignition switch to the On position. Note whether the airbag indicator light glows for six seconds, then goes out. If it does, this indicates that the system is functioning properly.*

27 Wiring diagrams - general information

1   Since it isn't possible to include all wiring diagrams for every year covered by this manual, the following diagrams are those that are typical and most commonly needed.

2   Prior to troubleshooting any circuits, check the fuse and circuit breakers (if equipped) to make sure they are in good condition. Make sure the battery is properly charged and has clean, tight cable connections (see Chapter 1).

3   When checking the wiring system, make sure that all electrical connectors are clean, with no broken or loose pins. When disconnecting an electrical connector, do not pull on the wires, only on the connector housings.

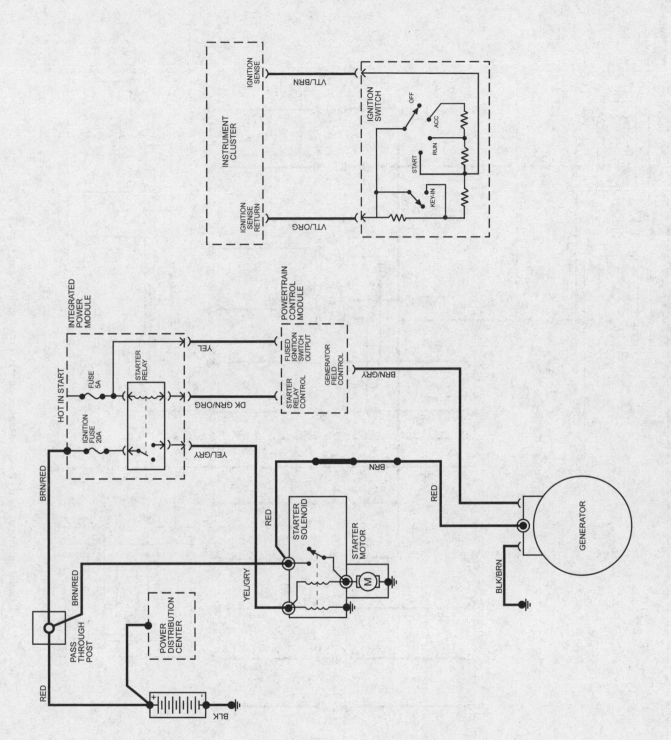

Starting and charging systems - 2010 and earlier models (except Challenger)

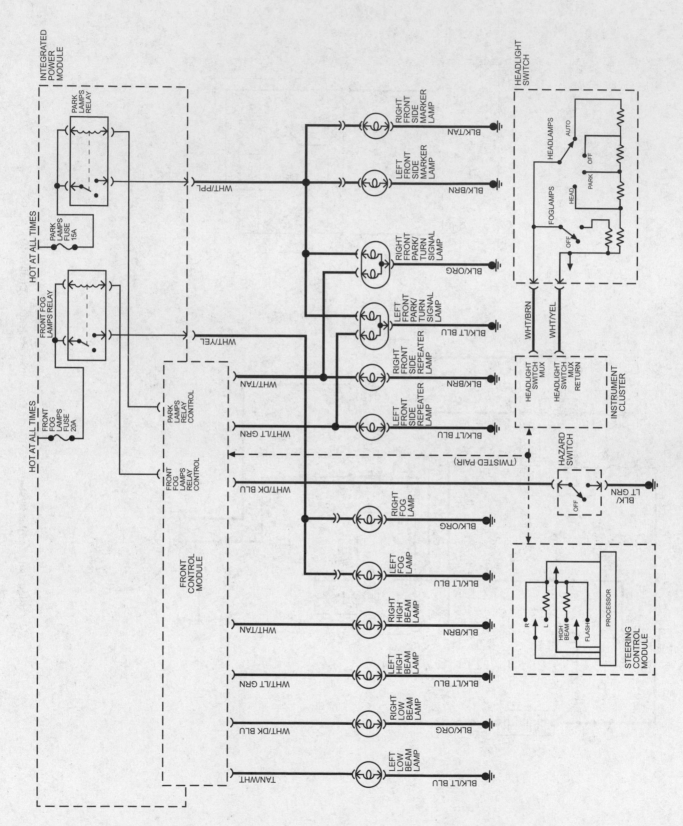

Exterior lighting system - 2010 and earlier models (except Challenger) (1 of 2)

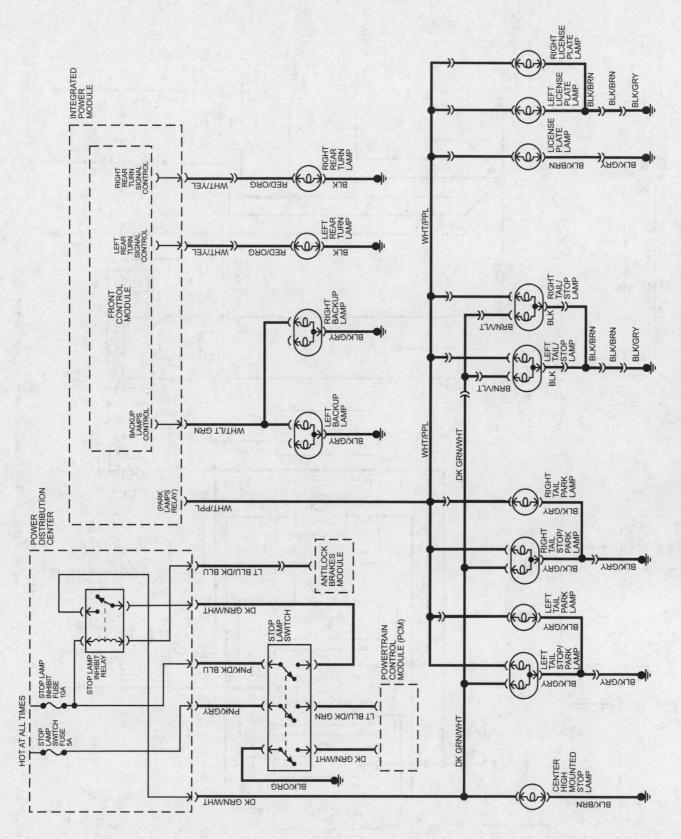

Exterior lighting system - 2010 and earlier models (except Challenger) (2 of 2)

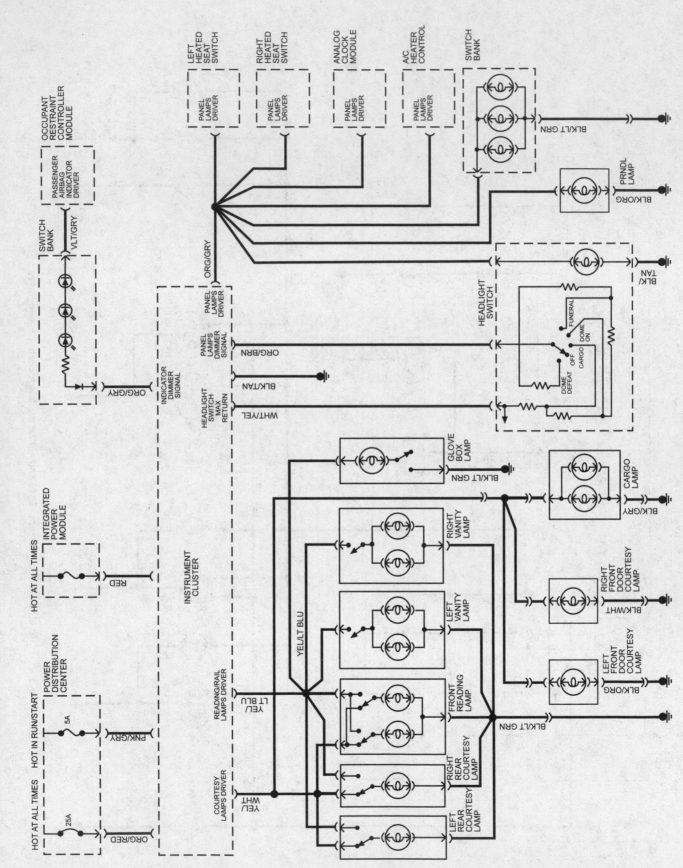

Interior lighting system - 2010 and earlier models (except Challenger)

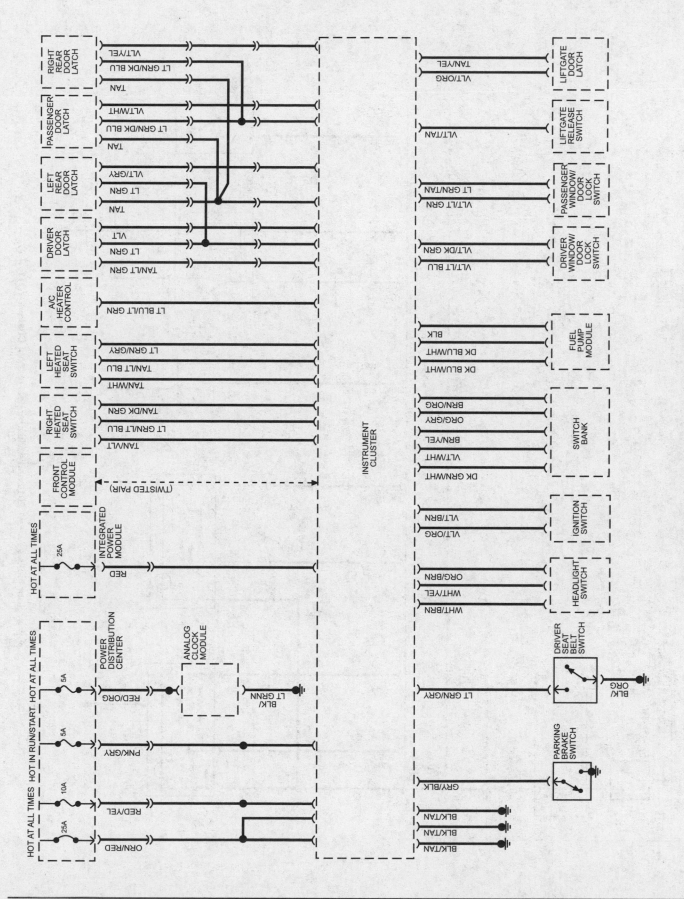

Instrument panel - 2010 and earlier models (except Challenger)

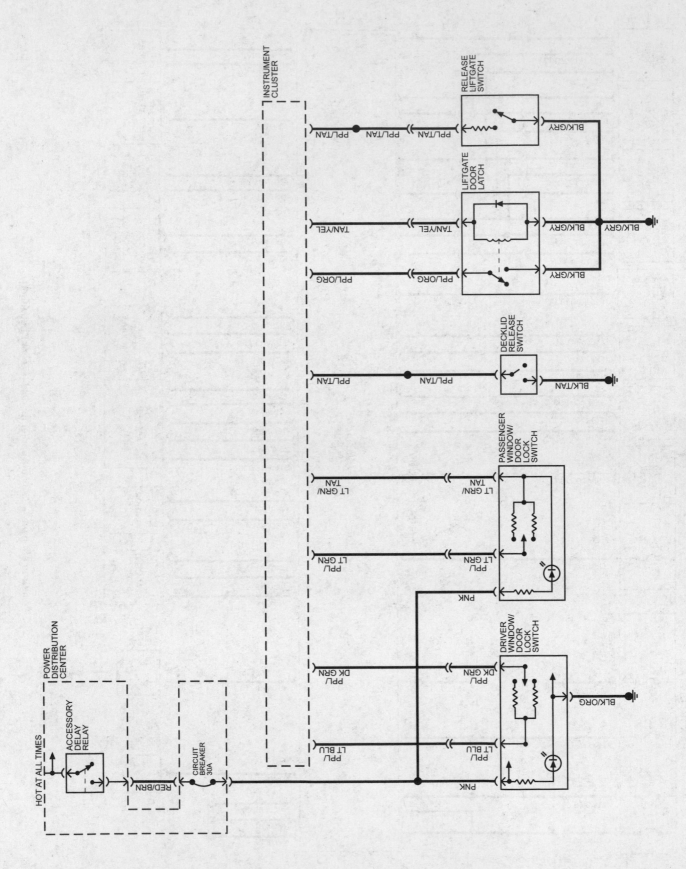

Power door locks system - 2010 and earlier models (except Challenger) (1 of 2)

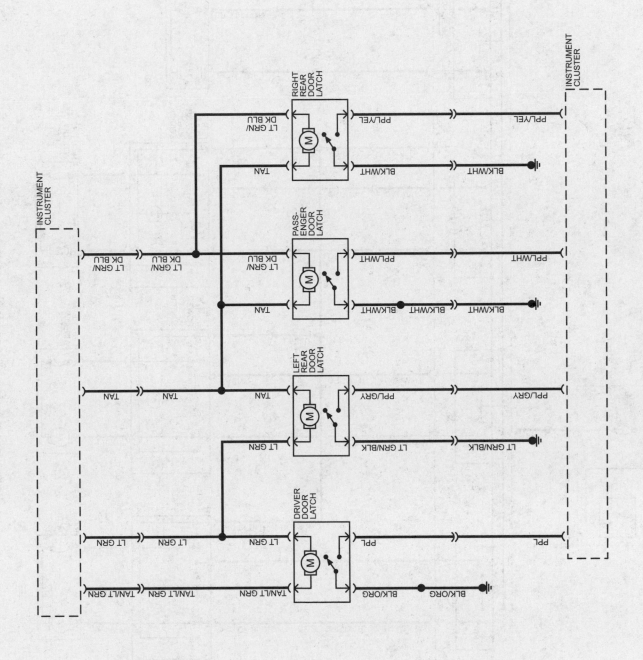

Power door locks system - 2010 and earlier models (except Challenger) (2 of 2)

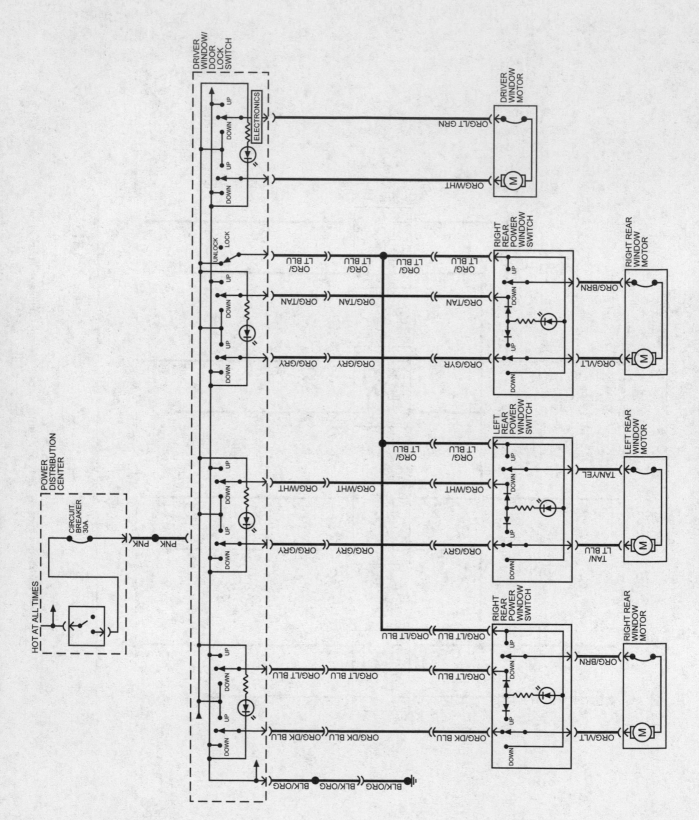

Power windows system - 2010 and earlier models (except Challenger)

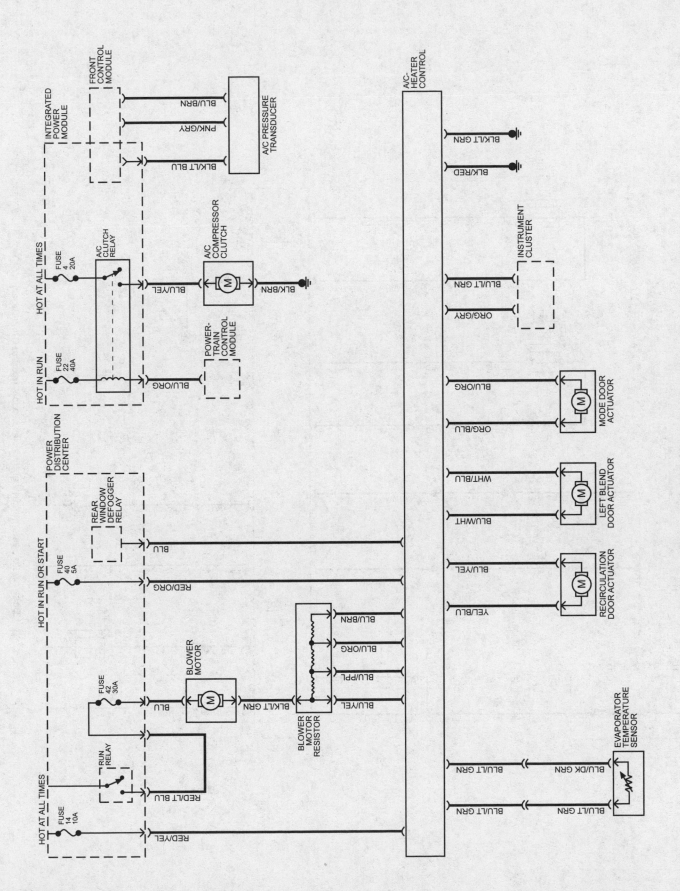

Heating and air conditioning systems - 2010 and earlier models (except Challenger)

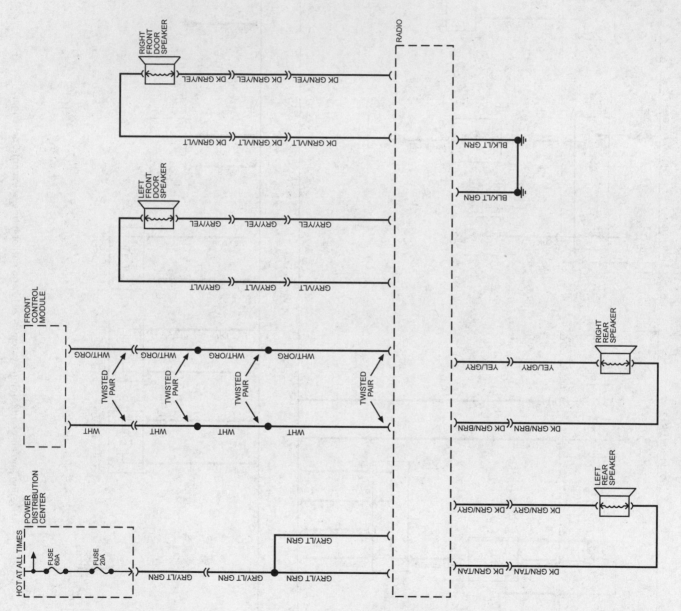

Base audio system - 2010 and earlier models (except Challenger)

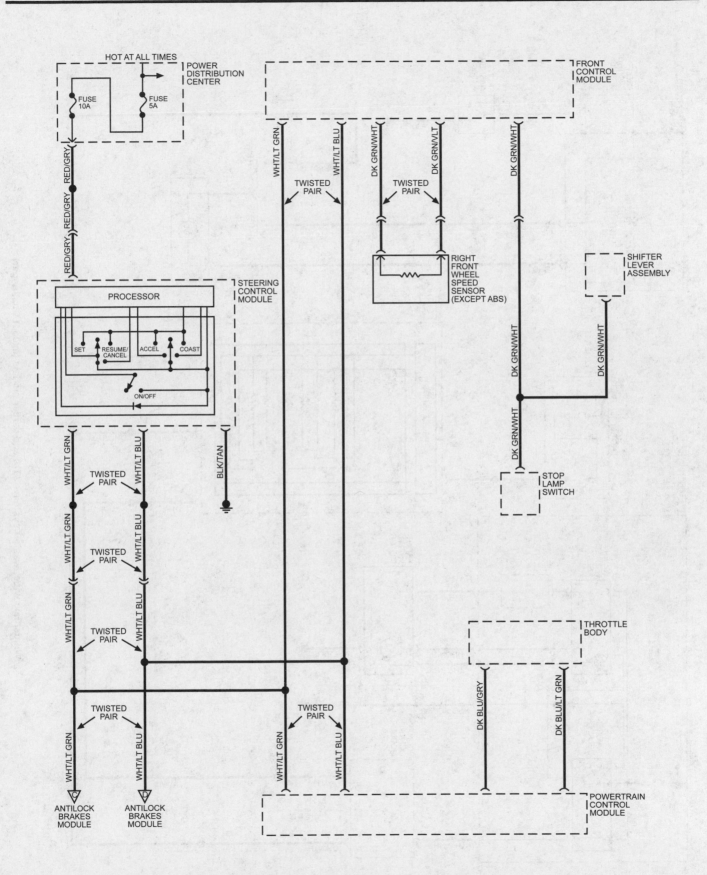

Cruise control system - 2010 and earlier models (except Challenger)

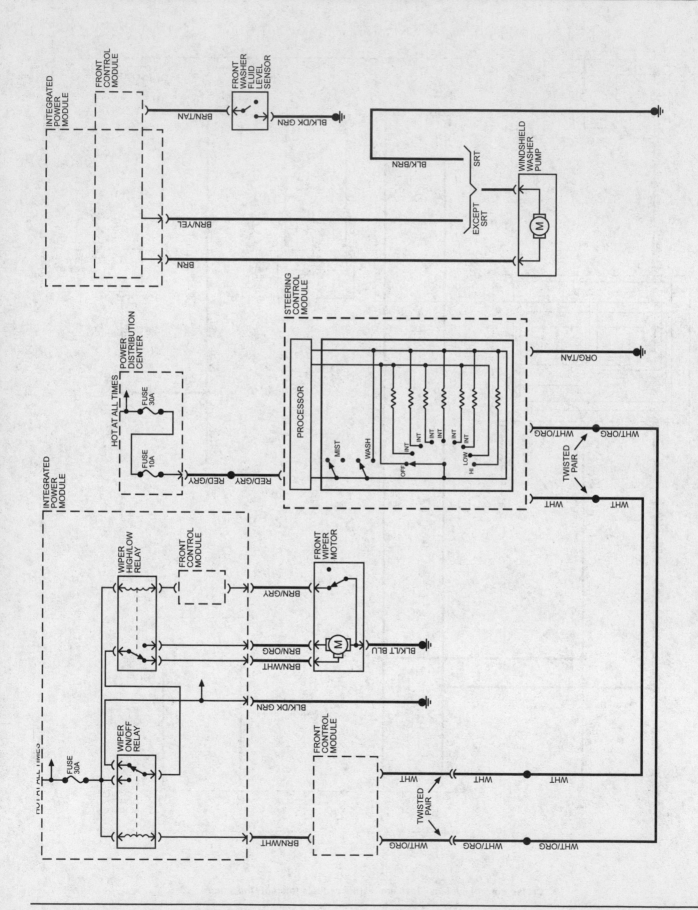

Windshield wiper/washer system - 2010 and earlier models (except Challenger)

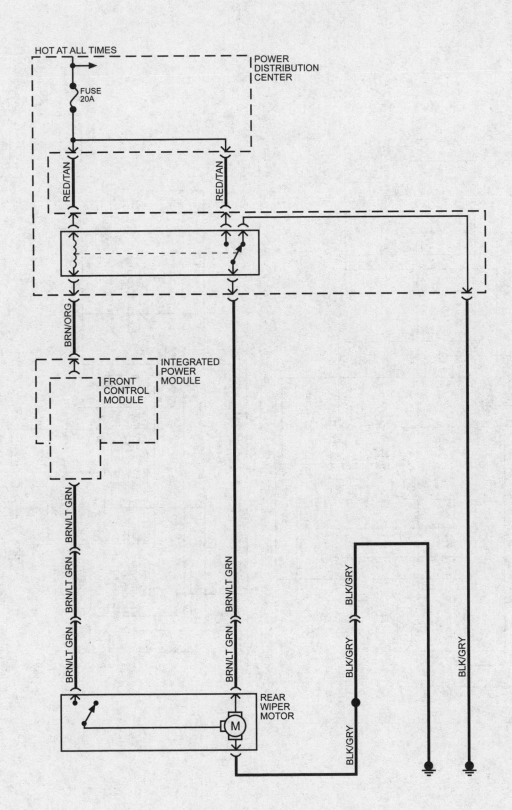

Rear wiper system (Magnum models)

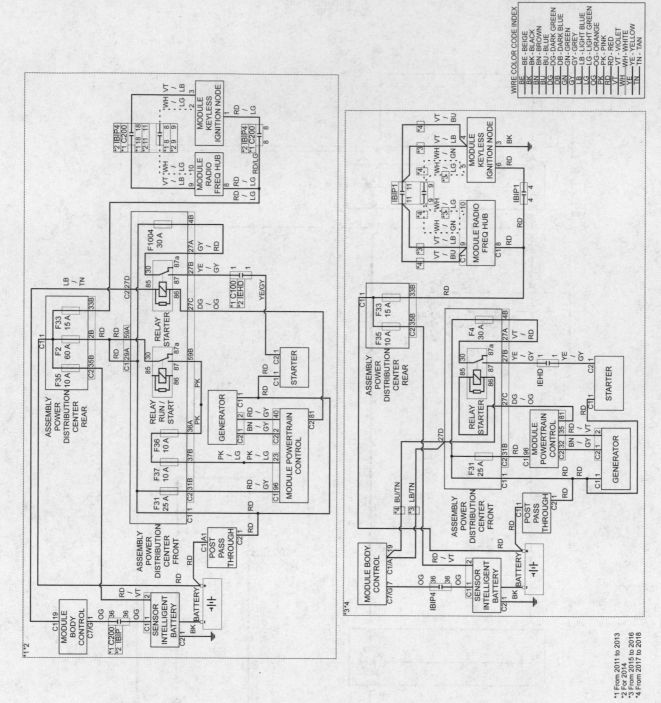

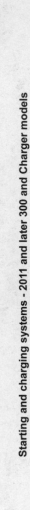

Starting and charging systems - 2011 and later 300 and Charger models

*1 From 2011 to 2013
*2 For 2014
*3 From 2015 to 2016
*4 From 2017 to 2018

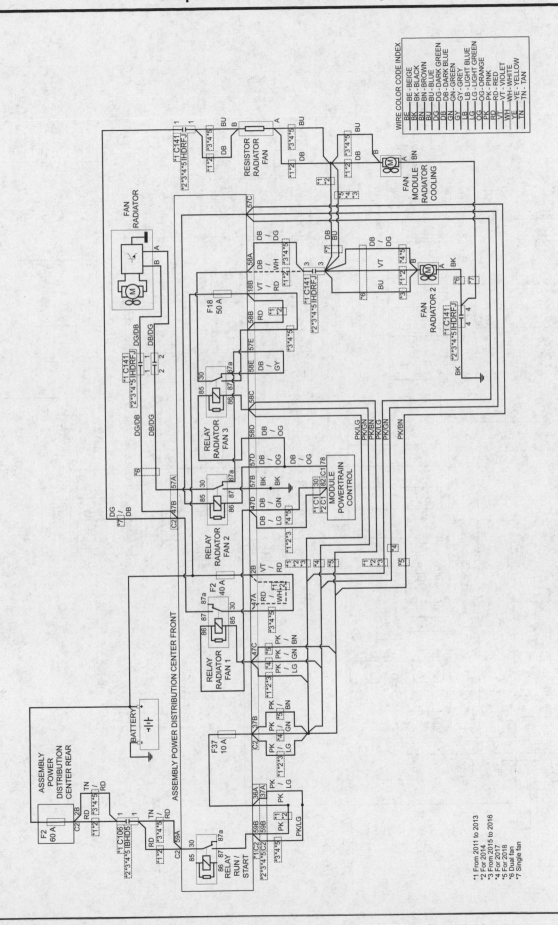

Radiator fan system - 2011 and later 300 and Charger models

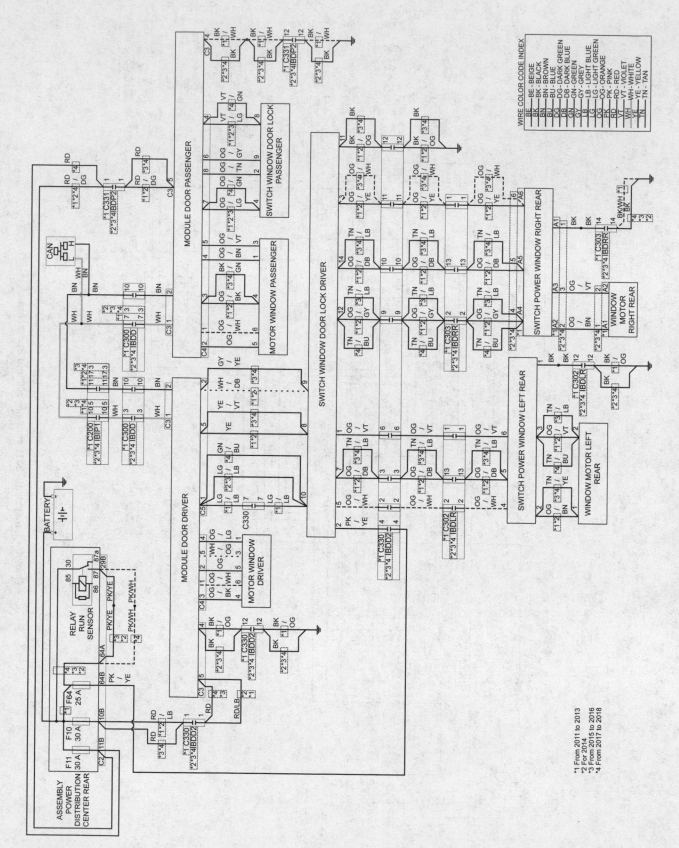

Power windows system - 2011 and later 300 and Charger models

*1 From 2011 to 2013
*2 For 2014
*3 From 2015 to 2016
*4 From 2017 to 2018

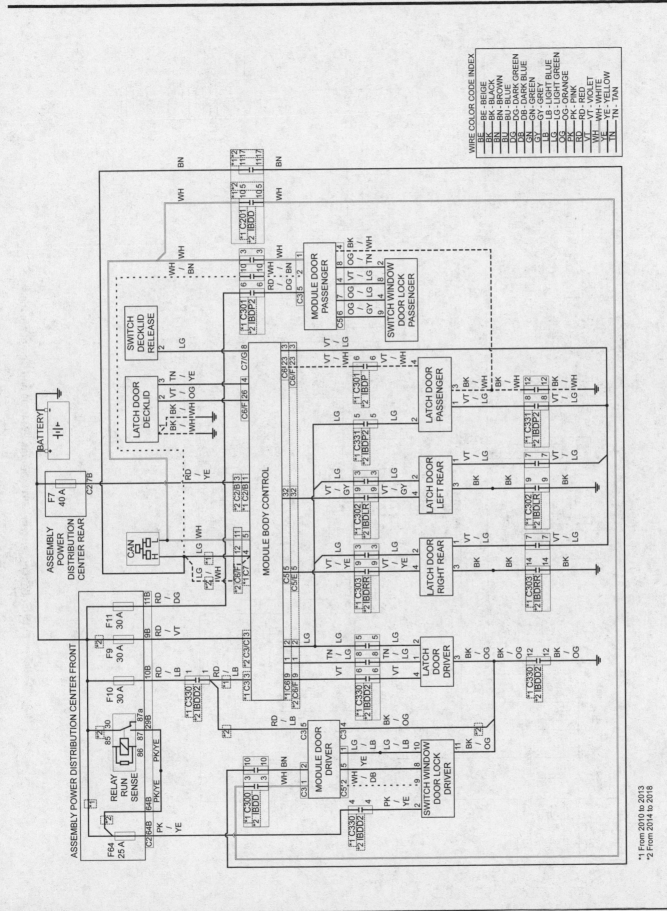

Power door locks system - 2011 and later 300 and Charger models

*1 From 2010 to 2013
*2 From 2014 to 2018

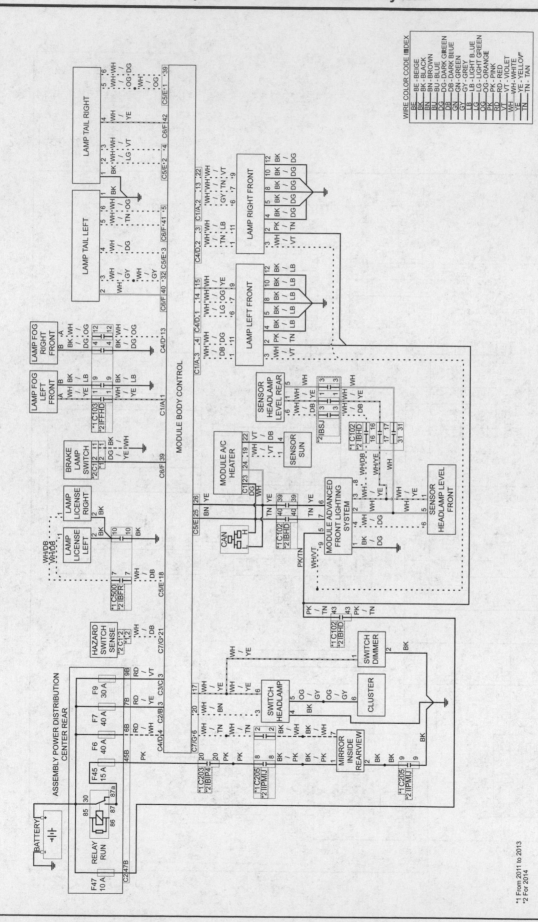

Exterior lighting system - 2011 through 2014 300 and Charger models

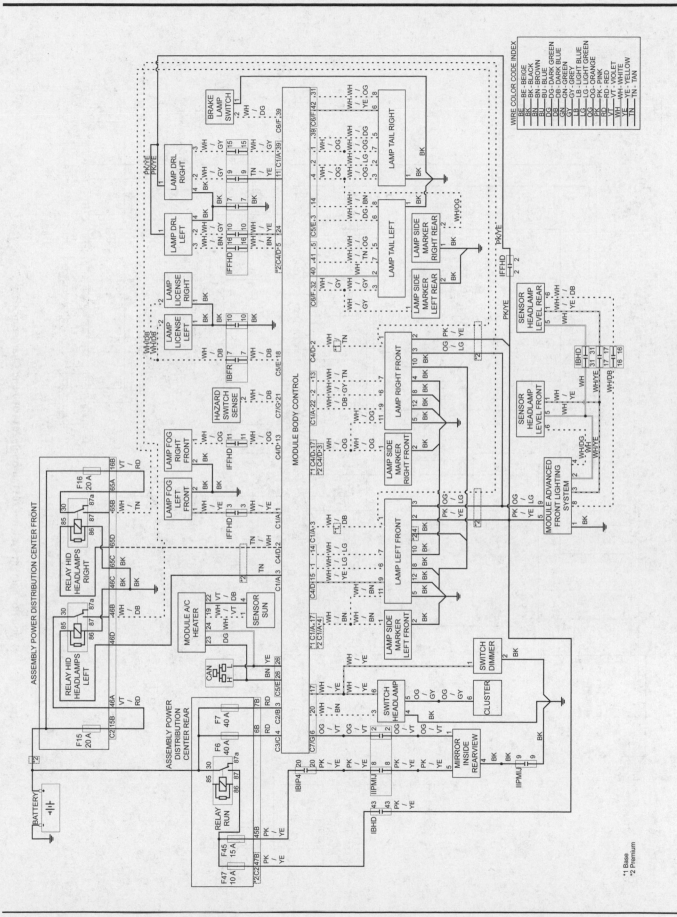

Exterior lighting system - 2015 and later 300 and Charger models

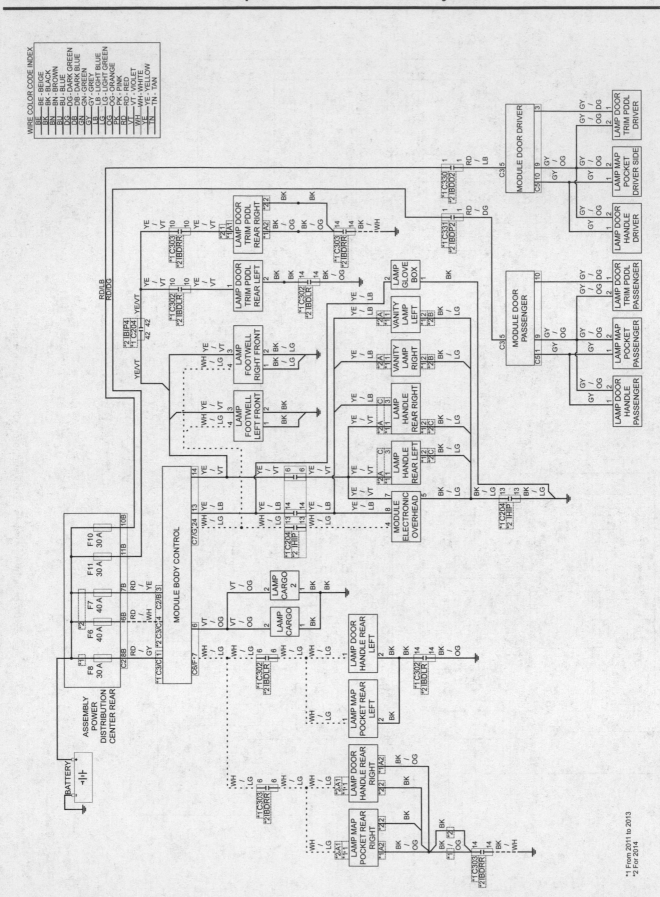

Interior lighting system – 2011 through 2014 300 and Charger models

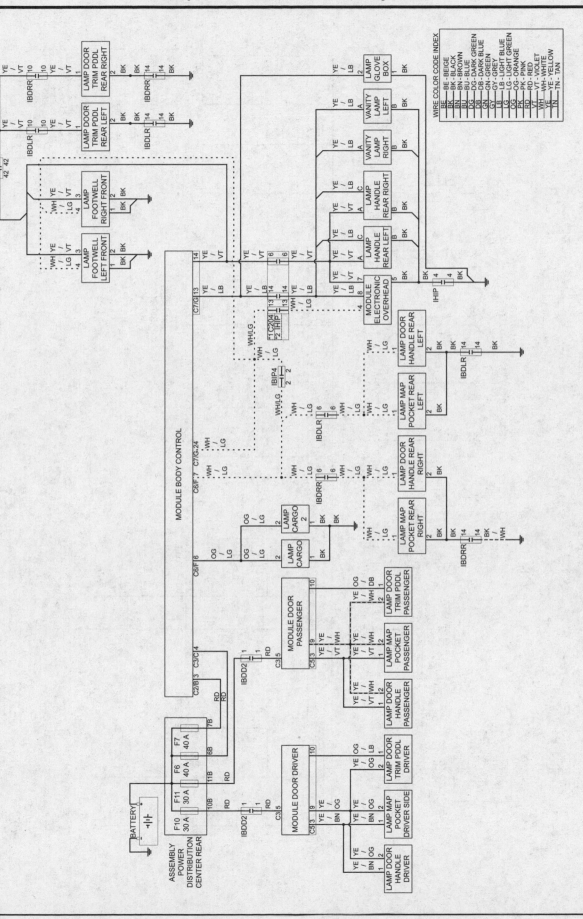

Interior lighting system - 2015 and later 300 and Charger models

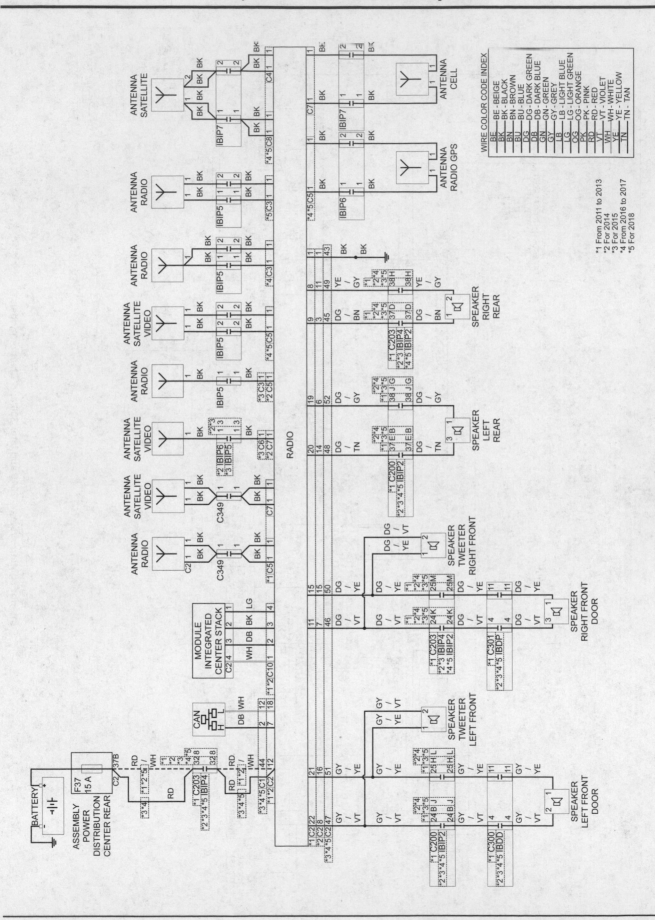

Audio system - 2011 and later 300 and Charger models

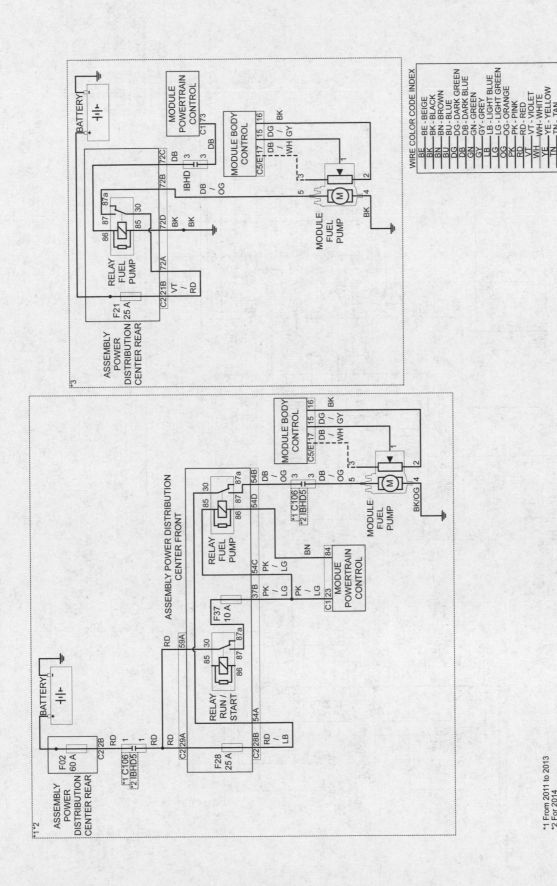

Fuel pump system - 2011 and later 300 and Charger models

*1 From 2011 to 2013
*2 For 2014
*3 From 2015 to 2018

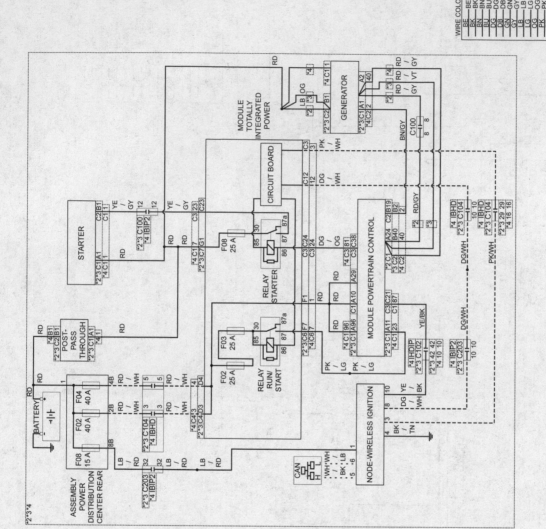

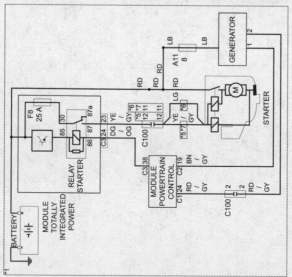

Starting and charging systems - 2014 and earlier Challenger models

*1 From 2008 to 2009
*2 For 2010
*3 From 2011 to 2013
*4 For 2014
*5 Engine: 3.5 L
*6 Engine: 3.6 L
*7 Engine: 5.7 L

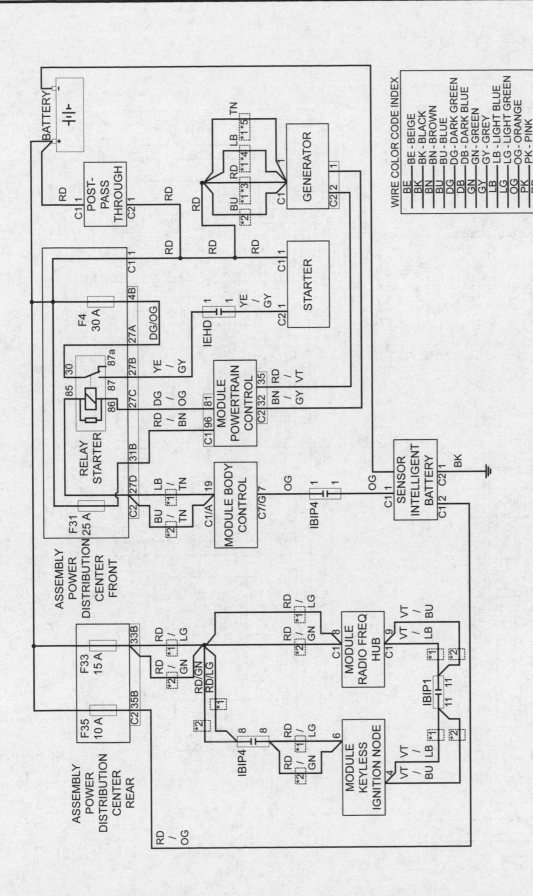

Starting and charging systems - 2015 and later Challenger models

*1 From 2015 to 2016
*2 From 2017 to 2018
*3 Engine: 3.6 L
*4 Engine: 5.6 L
*5 Engine: 6.4 L

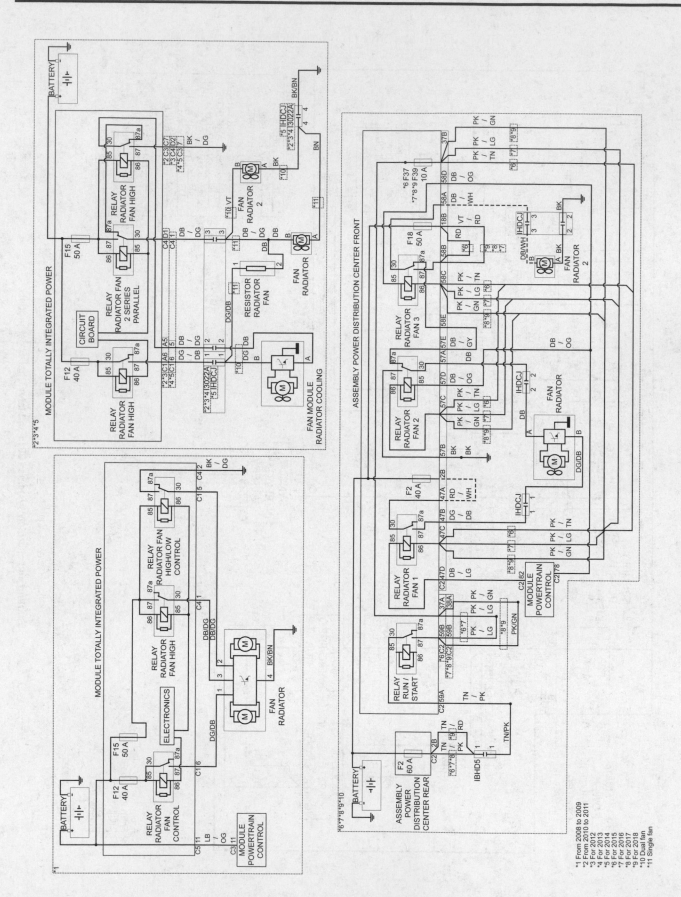

Radiator fan system - Challenger models (1 of 2)

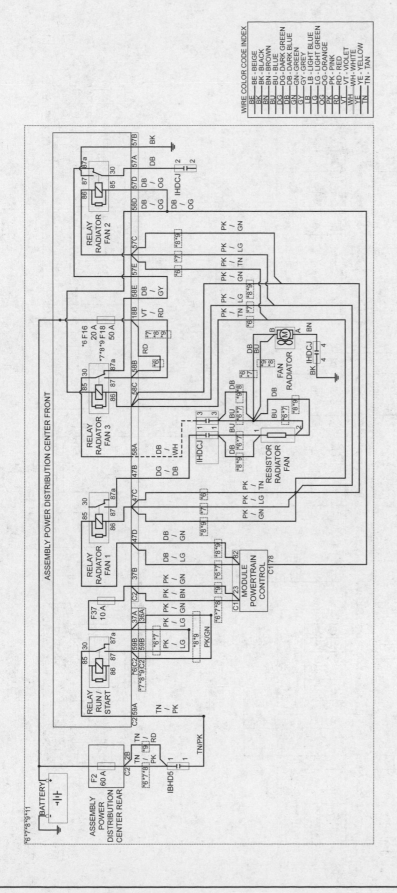

Radiator fan system - Challenger models (2 of 2)

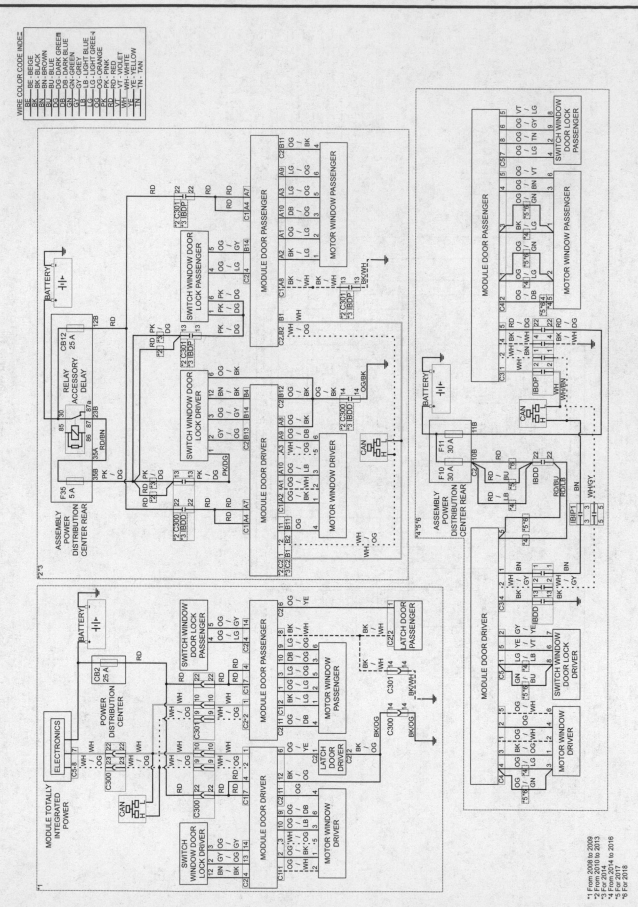

Power windows system - Challenger models

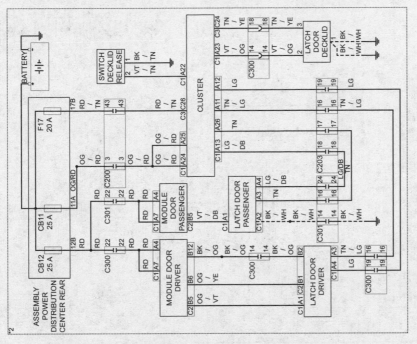

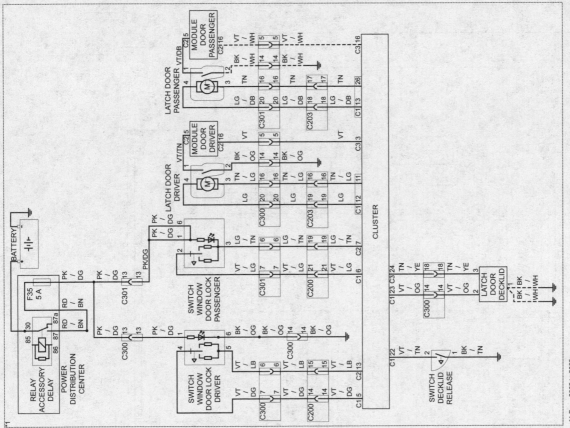

Power door locks system - Challenger models (1 of 2)

*1 From 2008 to 2009
*2 For 2010 to 2013
*3 For 2014
*4 From 2015 to 2016
*5 From 2017 to 2018

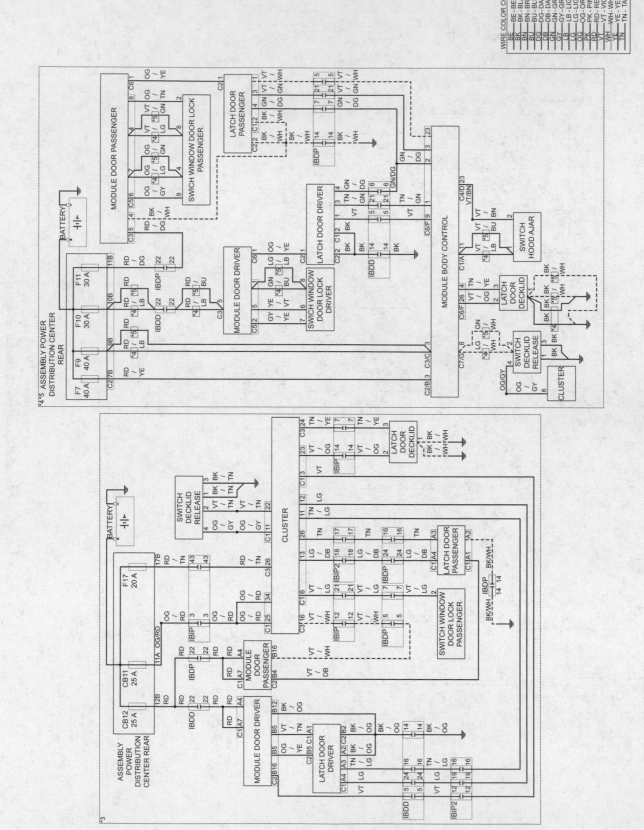

Power door locks system - Challenger models (2 of 2)

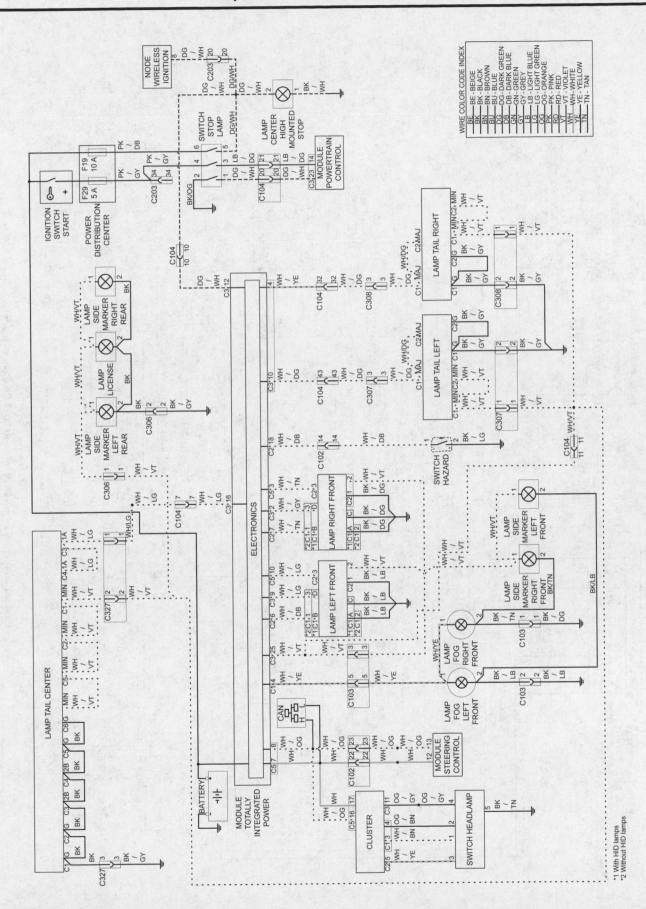

Exterior lighting system - 2008 and 2009 Challenger models

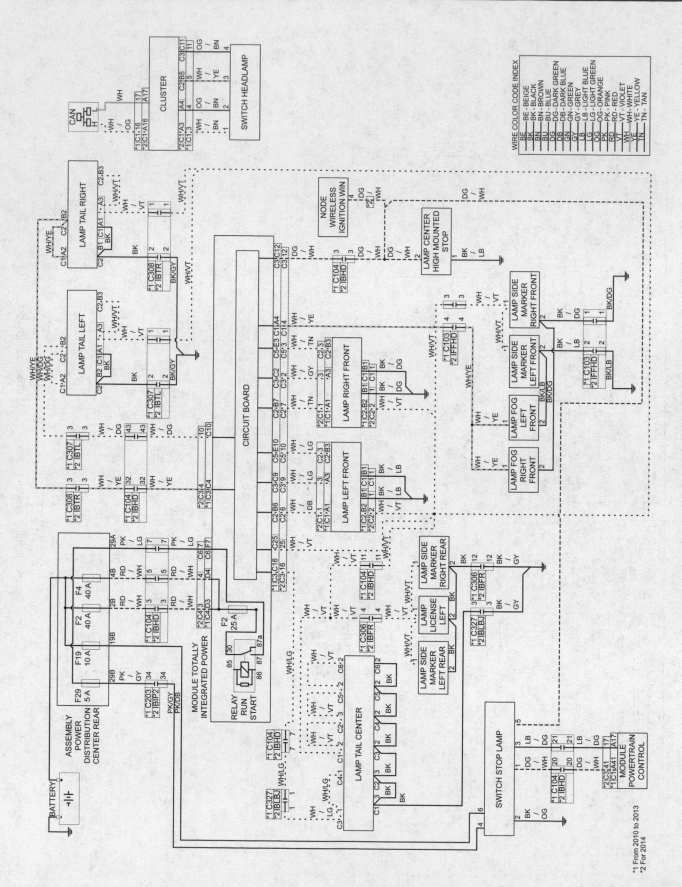

Exterior lighting system - 2010 through 2014 Challenger models

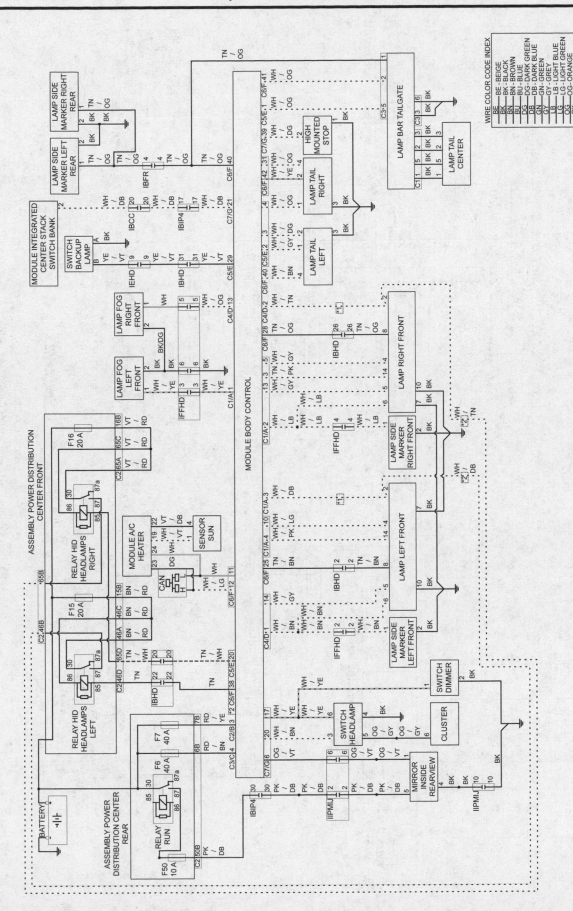

Exterior lighting system - 2015 and later Challenger models

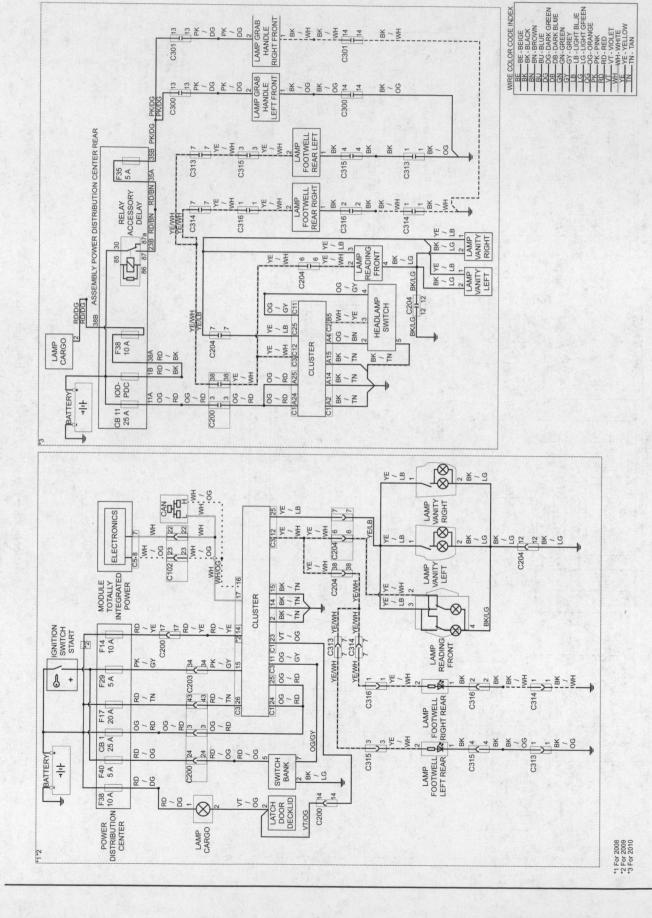

Interior lighting system - 2010 and earlier Challenger models

*1 For 2008
*2 For 2009
*3 For 2010

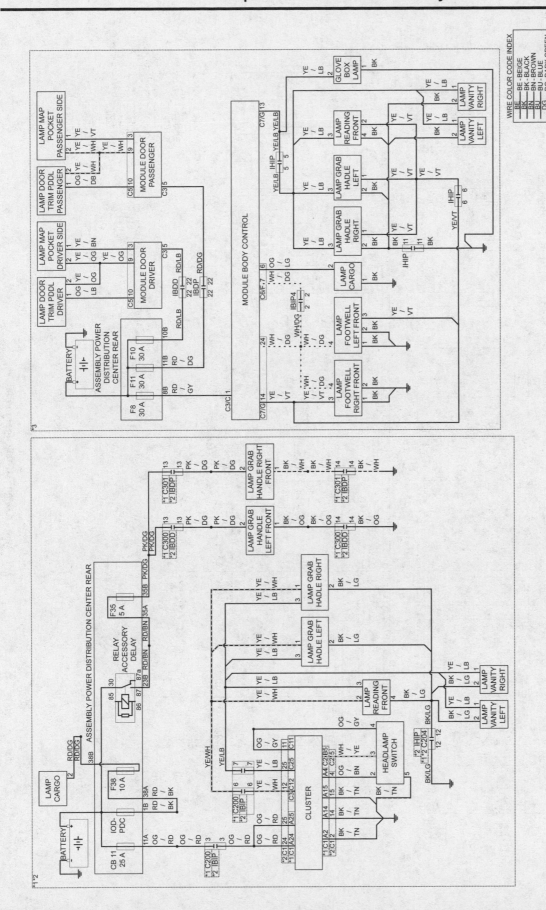

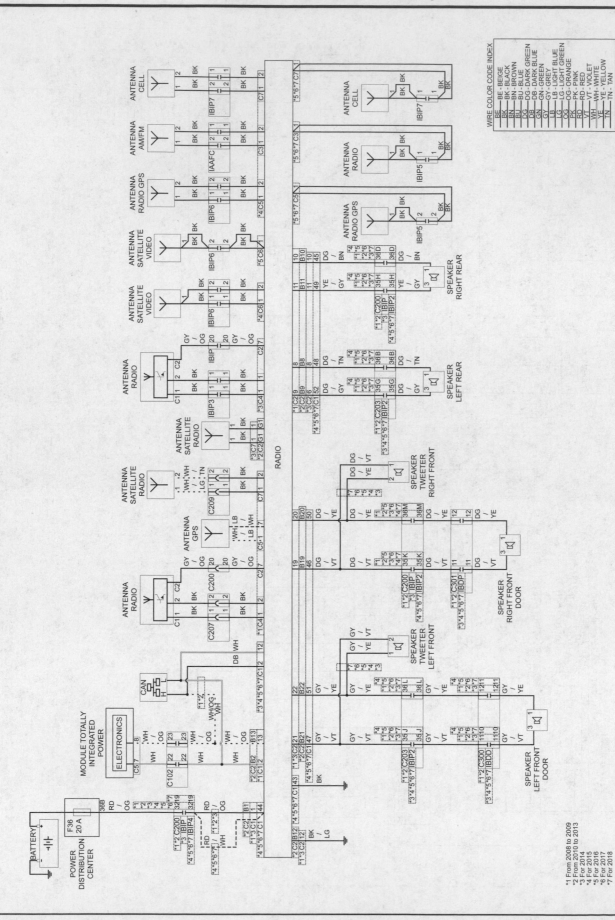

Audio system - Challenger models

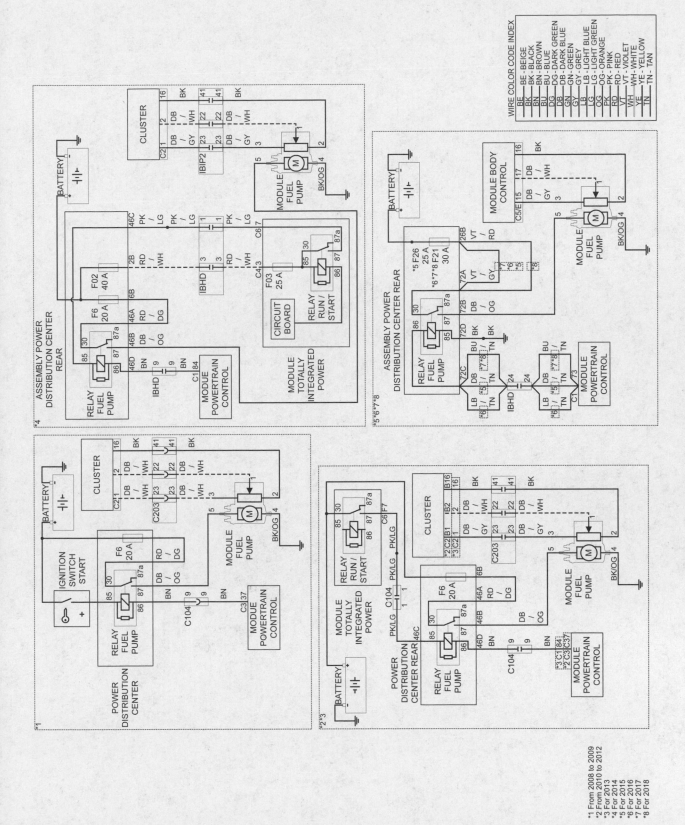

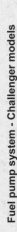

Fuel pump system - Challenger models

*1 From 2008 to 2009
*2 From 2010 to 2012
*3 For 2013
*4 For 2014
*5 For 2015
*6 For 2016
*7 For 2017
*8 For 2018

Notes

Index

Haynes Automotive Manuals

ACURA
12020 **Integra** '86 thru '89 **& Legend** '86 thru '90
12021 **Integra** '90 thru '93 **& Legend** '91 thru '95
 Integra '94 thru '00 - *see HONDA Civic (42025)*
 MDX '01 thru '07 - *see HONDA Pilot (42037)*
12050 **Acura TL** all models '99 thru '08

AMC
14020 **Mid-size models** '70 thru '83
14025 **(Renault) Alliance & Encore** '83 thru '87

AUDI
15020 **4000** all models '80 thru '87
15025 **5000** all models '77 thru '83
15026 **5000** all models '84 thru '88
 Audi A4 '96 thru '01 - *see VW Passat (96023)*
15030 **Audi A4** '02 thru '08

AUSTIN-HEALEY
 Sprite - *see MG Midget (66015)*

BMW
18020 **3/5 Series** '82 thru '92
18021 **3-Series** incl. Z3 models '92 thru '98
18022 **3-Series** incl. Z4 models '99 thru '05
18023 **3-Series** '06 thru '14
18025 **320i** all 4-cylinder models '75 thru '83
18050 **1500 thru 2002** except Turbo '59 thru '77

BUICK
19010 **Buick Century** '97 thru '05
 Century (front-wheel drive) - *see GM (38005)*
19020 **Buick, Oldsmobile & Pontiac Full-size**
 (Front-wheel drive) '85 thru '05
 Buick Electra, LeSabre and Park Avenue;
 Oldsmobile Delta 88 Royale, Ninety Eight
 and Regency; **Pontiac** Bonneville
19025 **Buick, Oldsmobile & Pontiac Full-size**
 (Rear wheel drive) '70 thru '90
 Buick Estate, Electra, LeSabre, Limited,
 Oldsmobile Custom Cruiser, Delta 88,
 Ninety-eight, **Pontiac** Bonneville,
 Catalina, Grandville, Parisienne
19027 **Buick LaCrosse** '05 thru '13
 Enclave - *see GENERAL MOTORS (38001)*
 Rainier - *see CHEVROLET (24072)*
 Regal - *see GENERAL MOTORS (38010)*
 Riviera - *see GENERAL MOTORS (38030, 38031)*
 Roadmaster - *see CHEVROLET (24046)*
 Skyhawk - *see GENERAL MOTORS (38015)*
 Skylark - *see GENERAL MOTORS (38020, 38025)*
 Somerset - *see GENERAL MOTORS (38025)*

CADILLAC
21015 **CTS & CTS-V** '03 thru '14
21030 **Cadillac Rear Wheel Drive** '70 thru '93
 Cimarron - *see GENERAL MOTORS (38015)*
 DeVille - *see GENERAL MOTORS (38031 & 38032)*
 Eldorado - *see GENERAL MOTORS (38030)*
 Fleetwood - *see GENERAL MOTORS (38031)*
 Seville - *see GM (38030, 38031 & 38032)*

CHEVROLET
10305 **Chevrolet Engine Overhaul Manual**
24010 **Astro & GMC Safari Mini-vans** '85 thru '05
24013 **Aveo** '04 thru '11
24015 **Camaro V8** all models '70 thru '81
24016 **Camaro** all models '82 thru '92
24017 **Camaro & Firebird** '93 thru '02
 Cavalier - *see GENERAL MOTORS (38016)*
 Celebrity - *see GENERAL MOTORS (38005)*
24018 **Camaro** '10 thru '15
24020 **Chevelle, Malibu & El Camino** '69 thru '87
 Cobalt - *see GENERAL MOTORS (38017)*
24024 **Chevette & Pontiac T1000** '76 thru '87
 Citation - *see GENERAL MOTORS (38020)*
24027 **Colorado & GMC Canyon** '04 thru '12
24032 **Corsica & Beretta** all models '87 thru '96
24040 **Corvette** all V8 models '68 thru '82
24041 **Corvette** all models '84 thru '96
24042 **Corvette** all models '97 thru '13
24044 **Cruze** '11 thru '19
24045 **Full-size Sedans** Caprice, Impala, Biscayne,
 Bel Air & Wagons '69 thru '90
24046 **Impala SS & Caprice and Buick Roadmaster**
 '91 thru '96
 Impala '00 thru '05 - *see LUMINA (24048)*
24047 **Impala & Monte Carlo** all models '06 thru '11
 Lumina '90 thru '94 - *see GM (38010)*
24048 **Lumina & Monte Carlo** '95 thru '05
 Lumina APV - *see GM (38035)*
24050 **Luv Pick-up** all 2WD & 4WD '72 thru '82
24051 **Malibu** '13 thru '19
24055 **Monte Carlo** all models '70 thru '88
 Monte Carlo '95 thru '01 - *see LUMINA (24048)*
24059 **Nova** all V8 models '69 thru '79

24060 **Nova and Geo Prizm** '85 thru '92
24064 **Pick-ups** '67 thru '87 - Chevrolet & GMC
24065 **Pick-ups** '88 thru '98 - Chevrolet & GMC
24066 **Pick-ups** '99 thru '06 - Chevrolet & GMC
24067 **Chevrolet Silverado & GMC Sierra** '07 thru '14
24068 **Chevrolet Silverado & GMC Sierra** '14 thru '19
24070 **S-10 & S-15 Pick-ups** '82 thru '93,
 Blazer & Jimmy '83 thru '94,
24071 **S-10 & Sonoma Pick-ups** '94 thru '04,
 including Blazer, Jimmy & Hombre
24072 **Chevrolet TrailBlazer, GMC Envoy &**
 Oldsmobile Bravada '02 thru '09
24075 **Sprint** '85 thru '88 **& Geo Metro** '89 thru '01
24080 **Vans - Chevrolet & GMC** '68 thru '96
24081 **Chevrolet Express & GMC Savana**
 Full-size Vans '96 thru '19

CHRYSLER
10310 **Chrysler Engine Overhaul Manual**
25015 **Chrysler Cirrus, Dodge Stratus,**
 Plymouth Breeze '95 thru '00
25020 **Full-size Front-Wheel Drive** '88 thru '93
 K-Cars - *see DODGE Aries (30008)*
 Laser - *see DODGE Daytona (30030)*
25025 **Chrysler LHS, Concorde, New Yorker,**
 Dodge Intrepid, **Eagle** Vision, '93 thru '97
25026 **Chrysler LHS, Concorde, 300M,**
 Dodge Intrepid, '98 thru '04
25027 **Chrysler 300** '05 thru '18, **Dodge Charger**
 '06 thru '18, **Magnum** '05 thru '08 **&**
 Challenger '08 thru '18
25030 **Chrysler & Plymouth Mid-size**
 front wheel drive '82 thru '95
 Rear-wheel Drive - *see Dodge (30050)*
25035 **PT Cruiser** all models '01 thru '10
25040 **Chrysler Sebring** '95 thru '06, **Dodge Stratus**
 '01 thru '06 **& Dodge Avenger** '95 thru '00
25041 **Chrysler Sebring** '07 thru '10, **200** '11 thru '17
 Dodge Avenger '08 thru '14

DATSUN
28005 **200SX** all models '80 thru '83
28012 **240Z, 260Z & 280Z** Coupe '70 thru '78
28014 **280ZX** Coupe & 2+2 '79 thru '83
 300ZX - *see NISSAN (72010)*
28018 **510 & PL521 Pick-up** '68 thru '73
28020 **510** all models '78 thru '81
28022 **620 Series Pick-up** all models '73 thru '79
 720 Series Pick-up - *see NISSAN (72030)*

DODGE
 400 & 600 - *see CHRYSLER (25030)*
30008 **Aries & Plymouth Reliant** '81 thru '89
30010 **Caravan & Plymouth Voyager** '84 thru '95
30011 **Caravan & Plymouth Voyager** '96 thru '02
30012 **Challenger & Plymouth Sapporo** '78 thru '83
30013 **Caravan, Chrysler Voyager &**
 Town & Country '03 thru '07
30014 **Grand Caravan &**
 Chrysler Town & Country '08 thru '18
30016 **Colt & Plymouth Champ** '78 thru '87
30020 **Dakota Pick-ups** all models '87 thru '96
30021 **Durango** '98 & '99 **& Dakota** '97 thru '99
30022 **Durango** '00 thru '03 **& Dakota** '00 thru '04
30023 **Durango** '04 thru '09 **& Dakota** '05 thru '11
30025 **Dart, Demon, Plymouth Barracuda,**
 Duster & Valiant 6-cylinder models '67 thru '76
30030 **Daytona & Chrysler Laser** '84 thru '89
 Intrepid - *see CHRYSLER (25025, 25026)*
30034 **Neon** all models '95 thru '99
30035 **Omni & Plymouth Horizon** '78 thru '90
30036 **Dodge & Plymouth Neon** '00 thru '05
30040 **Pick-ups** full-size models '74 thru '93
30042 **Pick-ups** full-size models '94 thru '08
30043 **Pick-ups** full-size models '09 thru '18
30045 **Ram 50/D50 Pick-ups & Raider and**
 Plymouth Arrow Pick-ups '79 thru '93
30050 **Dodge/Plymouth/Chrysler** RWD '71 thru '89
30055 **Shadow & Plymouth Sundance** '87 thru '94
30060 **Spirit & Plymouth Acclaim** '89 thru '95
30065 **Vans - Dodge & Plymouth** '71 thru '03

EAGLE
 Talon - *see MITSUBISHI (68030, 68031)*
 Vision - *see CHRYSLER (25025)*

FIAT
34010 **124 Sport Coupe & Spider** '68 thru '78
34025 **X1/9** all models '74 thru '80

FORD
10320 **Ford Engine Overhaul Manual**
10355 **Ford Automatic Transmission Overhaul**
11500 **Mustang** '64-1/2 thru '70 Restoration Guide
36004 **Aerostar Mini-vans** all models '86 thru '97
36006 **Contour & Mercury Mystique** '95 thru '00
36008 **Courier Pick-up** all models '72 thru '82

36012 **Crown Victoria &**
 Mercury Grand Marquis '88 thru '11
36014 **Edge** '07 thru '19 **& Lincoln MKX** '07 thru '18
36016 **Escort & Mercury Lynx** all models '81 thru '90
36020 **Escort & Mercury Tracer** '91 thru '02
36022 **Escape** '01 thru '17, **Mazda Tribute** '01 thru '11,
 & Mercury Mariner '05 thru '11
36024 **Explorer & Mazda Navajo** '91 thru '01
36025 **Explorer & Mercury Mountaineer** '02 thru '10
36026 **Explorer** '11 thru '17
36028 **Fairmont & Mercury Zephyr** '78 thru '83
36030 **Festiva & Aspire** '88 thru '97
36032 **Fiesta** all models '77 thru '80
36034 **Focus** all models '00 thru '11
36035 **Focus** '12 thru '14
36045 **Fusion** '06 thru '14 **& Mercury Milan** '06 thru '11
36048 **Mustang V8** all models '64-1/2 thru '73
36049 **Mustang II** 4-cylinder, V6 & V8 models '74 thru '78
36050 **Mustang & Mercury Capri** '79 thru '93
36051 **Mustang** all models '94 thru '04
36052 **Mustang** '05 thru '14
36054 **Pick-ups & Bronco** '73 thru '79
36058 **Pick-ups & Bronco** '80 thru '96
36059 **F-150** '97 thru '03, **Expedition** '97 thru '17,
 F-250 '97 thru '99, **F-150 Heritage** '04
 & Lincoln Navigator '98 thru '17
36060 **Super Duty Pick-ups & Excursion** '99 thru '10
36061 **F-150** full-size '04 thru '14
36062 **Pinto & Mercury Bobcat** '75 thru '80
36063 **F-150** full-size '15 thru '17
36064 **Super Duty Pick-ups** '11 thru '16
36066 **Probe** all models '89 thru '92
 Probe '93 thru '97 - *see MAZDA 626 (61042)*
36070 **Ranger & Bronco II** gas models '83 thru '92
36071 **Ranger** '93 thru '11 **& Mazda Pick-ups** '94 thru '09
36074 **Taurus & Mercury Sable** '86 thru '95
36075 **Taurus & Mercury Sable** '96 thru '07
36076 **Taurus** '08 thru '14, **Five Hundred** '05 thru '07,
 Mercury Montego '05 thru '07 **& Sable** '08 thru '09
36078 **Tempo & Mercury Topaz** '84 thru '94
36082 **Thunderbird & Mercury Cougar** '83 thru '88
36086 **Thunderbird & Mercury Cougar** '89 thru '97
36090 **Vans** all V8 Econoline models '69 thru '91
36094 **Vans** full-size '92 thru '14
36097 **Windstar** '95 thru '03, **Freestar & Mercury**
 Monterey Mini-van '04 thru '07

GENERAL MOTORS
10360 **GM Automatic Transmission Overhaul**
38001 **GMC Acadia** '07 thru '16, **Buick Enclave**
 '08 thru '17, **Saturn Outlook** '07 thru '10
 & Chevrolet Traverse '09 thru '17
38005 **Buick Century, Chevrolet Celebrity,**
 Oldsmobile Cutlass Ciera & Pontiac 6000
 all models '82 thru '96
38010 **Buick Regal** '88 thru '04, **Chevrolet Lumina**
 '88 thru '04, **Oldsmobile Cutlass Supreme**
 '88 thru '97 **& Pontiac Grand Prix** '88 thru '07
38015 **Buick Skyhawk, Cadillac Cimarron,**
 Chevrolet Cavalier, Oldsmobile Firenza,
 Pontiac J-2000 & Sunbird '82 thru '94
38016 **Chevrolet Cavalier & Pontiac Sunfire** '95 thru '05
38017 **Chevrolet Cobalt** '05 thru '10, **HHR** '06 thru '11,
 Pontiac G5 '07 thru '09, **Pursuit** '05 thru '06
 & Saturn ION '03 thru '07
38020 **Buick Skylark, Chevrolet Citation,**
 Oldsmobile Omega, Pontiac Phoenix '80 thru '85
38025 **Buick Skylark** '86 thru '98, **Somerset** '85 thru '87,
 Oldsmobile Achieva '92 thru '98, **Calais** '85 thru '91,
 & Pontiac Grand Am all models '85 thru '98
38026 **Chevrolet Malibu** '97 thru '03, **Classic** '04 thru '05,
 Oldsmobile Alero '99 thru '03, **Cutlass** '97 thru '00,
 & Pontiac Grand Am '99 thru '03
38027 **Chevrolet Malibu** '04 thru '12, **Pontiac G6**
 '05 thru '10 **& Saturn Aura** '07 thru '10
38030 **Cadillac Eldorado, Seville, Oldsmobile**
 Toronado & Buick Riviera '71 thru '85
38031 **Cadillac Eldorado, Seville, DeVille, Fleetwood,**
 Oldsmobile Toronado & Buick Riviera '86 thru '93
38032 **Cadillac DeVille** '94 thru '05, **Seville** '92 thru '04
 & Cadillac DTS '06 thru '10
38035 **Chevrolet Lumina APV, Oldsmobile Silhouette**
 & Pontiac Trans Sport all models '90 thru '96
38036 **Chevrolet Venture** '97 thru '05, **Oldsmobile**
 Silhouette '97 thru '04, **Pontiac Trans Sport**
 '97 thru '98 **& Montana** '99 thru '05
38040 **Chevrolet Equinox** '05 thru '17, **GMC Terrain**
 '10 thru '17 **& Pontiac Torrent** '06 thru '09

GEO
 Metro - *see CHEVROLET Sprint (24075)*
 Prizm - '85 thru '92 see CHEVY (24060),
 '93 thru '02 see TOYOTA Corolla (92036)
40030 **Storm** all models '90 thru '93
 Tracker - *see SUZUKI Samurai (90010)*

(Continued on other side)

Haynes Automotive Manuals (continued)

NOTE: If you do not see a listing for your vehicle, please visit **haynes.com** for the latest product information and check out our **Online Manuals!**

GMC
Acadia - see GENERAL MOTORS (38001)
Pick-ups - see CHEVROLET (24027, 24068)
Vans - see CHEVROLET (24081)

HONDA
42010 **Accord CVCC** all models '76 thru '83
42011 **Accord** all models '84 thru '89
42012 **Accord** all models '90 thru '93
42013 **Accord** all models '94 thru '97
42014 **Accord** all models '98 thru '02
42015 **Accord** '03 thru '12 & **Crosstour** '10 thru '14
42016 **Accord** '13 thru '17
42020 **Civic 1200** all models '73 thru '79
42021 **Civic 1300 & 1500 CVCC** '80 thru '83
42022 **Civic 1500 CVCC** all models '75 thru '79
42023 **Civic** all models '84 thru '91
42024 **Civic & del Sol** '92 thru '95
42025 **Civic** '96 thru '00, **CR-V** '97 thru '01 & **Acura Integra** '94 thru '00
42026 **Civic** '01 thru '11 & **CR-V** '02 thru '11
42027 **Civic** '12 thru '15 & **CR-V** '12 thru '16
42030 **Fit** '07 thru '13
42035 **Odyssey** all models '99 thru '10
Passport - see ISUZU Rodeo (47017)
42037 **Honda Pilot** '03 thru '08, **Ridgeline** '06 thru '14 & **Acura MDX** '01 thru '07
42040 **Prelude CVCC** all models '79 thru '89

HYUNDAI
43010 **Elantra** all models '96 thru '19
43015 **Excel & Accent** all models '86 thru '13
43050 **Santa Fe** all models '01 thru '12
43055 **Sonata** all models '99 thru '14

INFINITI
G35 '03 thru '08 - see NISSAN 350Z (72011)

ISUZU
Hombre - see CHEVROLET S-10 (24071)
47017 **Rodeo** '91 thru '02, **Amigo** '89 thru '94 & '98 thru '02 & **Honda Passport** '95 thru '02
47020 **Trooper** '84 thru '91 & **Pick-up** '81 thru '93

JAGUAR
49010 **XJ6** all 6-cylinder models '68 thru '86
49011 **XJ6** all models '88 thru '94
49015 **XJ12 & XJS** all 12-cylinder models '72 thru '85

JEEP
50010 **Cherokee, Comanche & Wagoneer Limited** all models '84 thru '01
50011 **Cherokee** '14 thru '19
50020 **CJ** all models '49 thru '86
50025 **Grand Cherokee** all models '93 thru '04
50026 **Grand Cherokee** '05 thru '19 & **Dodge Durango** '11 thru '19
50029 **Grand Wagoneer & Pick-up** '72 thru '91 Grand Wagoneer '84 thru '91, Cherokee & Wagoneer '72 thru '83, Pick-up '72 thru '88
50030 **Wrangler** all models '87 thru '17
50035 **Liberty** '02 thru '12 & **Dodge Nitro** '07 thru '11
50050 **Patriot & Compass** '07 thru '17

KIA
54050 **Optima** '01 thru '10
54060 **Sedona** '02 thru '14
54070 **Sephia** '94 thru '01, **Spectra** '00 thru '09, **Sportage** '05 thru '20
54077 **Sorento** '03 thru '13

LEXUS
ES 300/330 - see TOYOTA Camry (92007, 92008)
ES 350 - see TOYOTA Camry (92009)
RX 300/330/350 - see TOYOTA Highlander (92095)

LINCOLN
MKX - see FORD (36014)
Navigator - see FORD Pick-up (36059)
59010 **Rear-Wheel Drive Continental** '70 thru '87, **Mark Series** '70 thru '92 & **Town Car** '81 thru '10

MAZDA
61010 **GLC (rear-wheel drive)** '77 thru '83
61011 **GLC (front-wheel drive)** '81 thru '85
61012 **Mazda3** '04 thru '11
61015 **323 & Protegé** '90 thru '03
61016 **MX-5 Miata** '90 thru '14
61020 **MPV** all models '89 thru '98
Navajo - see Ford Explorer (36024)
61030 **Pick-ups** '72 thru '93
Pick-ups '94 thru '09 - see Ford Ranger (36071)
61035 **RX-7** all models '79 thru '85
61036 **RX-7** all models '86 thru '91
61040 **626 (rear-wheel drive)** all models '79 thru '82
61041 **626 & MX-6 (front-wheel drive)** '83 thru '92
61042 **626** '93 thru '01 & **MX-6/Ford Probe** '93 thru '02
61043 **Mazda6** '03 thru '13

MERCEDES-BENZ
63012 **123 Series Diesel** '76 thru '85
63015 **190 Series** 4-cylinder gas models '84 thru '88
63020 **230/250/280** 6-cylinder SOHC models '68 thru '72
63025 **280 123 Series** gas models '77 thru '81
63030 **350 & 450** all models '71 thru '80
63040 **C-Class:** C230/C240/C280/C320/C350 '01 thru '07

MERCURY
64200 **Villager & Nissan Quest** '93 thru '01
All other titles, see FORD Listing.

MG
66010 **MGB** Roadster & GT Coupe '62 thru '80
66015 **MG Midget, Austin Healey Sprite** '58 thru '80

MINI
67020 **Mini** '02 thru '13

MITSUBISHI
68020 **Cordia, Tredia, Galant, Precis & Mirage** '83 thru '93
68030 **Eclipse, Eagle Talon & Plymouth Laser** '90 thru '94
68031 **Eclipse** '95 thru '05 & **Eagle Talon** '95 thru '98
68035 **Galant** '94 thru '12
68040 **Pick-up** '83 thru '96 & **Montero** '83 thru '93

NISSAN
72010 **300ZX** all models including Turbo '84 thru '89
72011 **350Z & Infiniti G35** all models '03 thru '08
72015 **Altima** all models '93 thru '06
72016 **Altima** '07 thru '12
72020 **Maxima** all models '85 thru '92
72021 **Maxima** all models '93 thru '08
72025 **Murano** '03 thru '14
72030 **Pick-ups** '80 thru '97 & **Pathfinder** '87 thru '95
72031 **Frontier** '98 thru '04, **Xterra** '00 thru '04, & **Pathfinder** '96 thru '04
72032 **Frontier & Xterra** '05 thru '14
72037 **Pathfinder** '05 thru '14
72040 **Pulsar** all models '83 thru '86
72042 **Roque** all models '08 thru '20
72050 **Sentra** all models '82 thru '94
72051 **Sentra & 200SX** all models '95 thru '06
72060 **Stanza** all models '82 thru '90
72070 **Titan pick-ups** '04 thru '10, **Armada** '05 thru '10 & **Pathfinder Armada** '04
72080 **Versa** all models '07 thru '19

OLDSMOBILE
73015 **Cutlass** V6 & V8 gas models '74 thru '88
For other OLDSMOBILE titles, see BUICK, CHEVROLET or GENERAL MOTORS listings.

PLYMOUTH
For PLYMOUTH titles, see DODGE listing.

PONTIAC
79008 **Fiero** all models '84 thru '88
79018 **Firebird** V8 models except Turbo '70 thru '81
79019 **Firebird** all models '82 thru '92
79025 **G6** all models '05 thru '09
79040 **Mid-size Rear-wheel Drive** '70 thru '87
Vibe '03 thru '10 - see TOYOTA Corolla (92037)
For other PONTIAC titles, see BUICK, CHEVROLET or GENERAL MOTORS listings.

PORSCHE
80020 **911** Coupe & Targa models '65 thru '89
80025 **914** all 4-cylinder models '69 thru '76
80030 **924** all models including Turbo '76 thru '82
80035 **944** all models including Turbo '83 thru '89

RENAULT
Alliance & Encore - see AMC (14025)

SAAB
84010 **900** all models including Turbo '79 thru '88

SATURN
87010 **Saturn** all S-series models '91 thru '02
Saturn Ion '03 thru '07- see GM (38017)
Saturn Outlook - see GM (38001)
87020 **Saturn L-series** all models '00 thru '04
87040 **Saturn VUE** '02 thru '09

SUBARU
89002 **1100, 1300, 1400 & 1600** '71 thru '79
89003 **1600 & 1800** 2WD & 4WD '80 thru '94
89080 **Impreza** '02 thru '11, **WRX** '02 thru '14, & **WRX STI** '04 thru '14
89100 **Legacy** all models '90 thru '99
89101 **Legacy & Forester** '00 thru '09
89102 **Legacy** '10 thru '16 & **Forester** '12 thru '16

SUZUKI
90010 **Samurai/Sidekick & Geo Tracker** '86 thru '01

TOYOTA
92005 **Camry** all models '83 thru '91
92006 **Camry** '92 thru '96 & **Avalon** '95 thru '96
92007 **Camry, Avalon, Solara, Lexus ES 300** '97 thru '01

92008 **Camry, Avalon, Lexus ES 300/330** '02 thru '06 & **Solara** '02 thru '08
92009 **Camry, Avalon & Lexus ES 350** '07 thru '17
92015 **Celica Rear-wheel Drive** '71 thru '85
92020 **Celica Front-wheel Drive** '86 thru '99
92025 **Celica Supra** all models '79 thru '92
92030 **Corolla** all models '75 thru '79
92032 **Corolla** all rear-wheel drive models '80 thru '87
92035 **Corolla** all front-wheel drive models '84 thru '92
92036 **Corolla & Geo/Chevrolet Prizm** '93 thru '02
92037 **Corolla** '03 thru '19, **Matrix** '03 thru '14, & **Pontiac Vibe** '03 thru '10
92040 **Corolla Tercel** all models '80 thru '82
92045 **Corona** all models '74 thru '82
92050 **Cressida** all models '78 thru '82
92055 **Land Cruiser** FJ40, 43, 45, 55 '68 thru '82
92056 **Land Cruiser** FJ60, 62, 80, FZJ80 '80 thru '96
92060 **Matrix** '03 thru '11 & **Pontiac Vibe** '03 thru '10
92065 **MR2** all models '85 thru '87
92070 **Pick-up** all models '69 thru '78
92075 **Pick-up** all models '79 thru '95
92076 **Tacoma** '95 thru '04, **4Runner** '96 thru '02 & **T100** '93 thru '08
92077 **Tacoma** all models '05 thru '18
92078 **Tundra** '00 thru '06 & **Sequoia** '01 thru '07
92079 **4Runner** all models '03 thru '09
92080 **Previa** all models '91 thru '95
92081 **Prius** all models '01 thru '12
92082 **RAV4** all models '96 thru '12
92085 **Tercel** all models '87 thru '94
92090 **Sienna** all models '98 thru '10
92095 **Highlander** '01 thru '19 & **Lexus RX330/330/350** '99 thru '19
92179 **Tundra** '07 thru '19 & **Sequoia** '08 thru '19

TRIUMPH
94007 **Spitfire** all models '62 thru '81
94010 **TR7** all models '75 thru '81

VW
96008 **Beetle & Karmann Ghia** '54 thru '79
96009 **New Beetle** '98 thru '10
96016 **Rabbit, Jetta, Scirocco & Pick-up** gas models '75 thru '92 & Convertible '80 thru '92
96017 **Golf, GTI & Jetta** '93 thru '98, **Cabrio** '95 thru '02
96018 **Golf, GTI, Jetta** '99 thru '05
96019 **Jetta, Rabbit, GLI, GTI & Golf** '05 thru '11
96020 **Rabbit, Jetta & Pick-up** diesel '77 thru '84
96021 **Jetta** '11 thru '18 & **Golf** '15 thru '19
96023 **Passat** '98 thru '05 & **Audi A4** '96 thru '01
96030 **Transporter 1600** all models '68 thru '79
96035 **Transporter 1700, 1800 & 2000** '72 thru '79
96040 **Type 3 1500 & 1600** all models '63 thru '73
96045 **Vanagon Air-Cooled** all models '80 thru '83

VOLVO
97010 **120, 130 Series & 1800 Sports** '61 thru '73
97015 **140 Series** all models '66 thru '74
97020 **240 Series** all models '76 thru '93
97040 **740 & 760 Series** all models '82 thru '88
97050 **850 Series** all models '93 thru '97

TECHBOOK MANUALS
10205 **Automotive Computer Codes**
10206 **OBD-II & Electronic Engine Management**
10210 **Automotive Emissions Control Manual**
10215 **Fuel Injection Manual** '78 thru '85
10225 **Holley Carburetor Manual**
10230 **Rochester Carburetor Manual**
10305 **Chevrolet Engine Overhaul Manual**
10320 **Ford Engine Overhaul Manual**
10330 **GM and Ford Diesel Engine Repair Manual**
10331 **Duramax Diesel Engines** '01 thru '19
10332 **Cummins Diesel Engine Performance Manual**
10333 **GM, Ford & Chrysler Engine Performance Manual**
10334 **GM Engine Performance Manual**
10340 **Small Engine Repair Manual**, 5 HP & Less
10341 **Small Engine Repair Manual**, 5.5 thru 20 HP
10345 **Suspension, Steering & Driveline Manual**
10355 **Ford Automatic Transmission Overhaul**
10360 **GM Automatic Transmission Overhaul**
10405 **Automotive Body Repair & Painting**
10410 **Automotive Brake Manual**
10411 **Automotive Anti-lock Brake (ABS) Systems**
10420 **Automotive Electrical Manual**
10425 **Automotive Heating & Air Conditioning**
10435 **Automotive Tools Manual**
10445 **Welding Manual**
10450 **ATV Basics**

Over a 100 Haynes motorcycle manuals also available

10/22